Sustainable Aquafeeds

Sustainable Aquafeeds

Technological Innovation and Novel Ingredients

Edited by
Jose M. Lorenzo and Jesus Simal-Gandara

CRC Press
Taylor & Francis Group
Boca Raton London New York

CRC Press is an imprint of the
Taylor & Francis Group, an **informa** business

First edition published 2022
by CRC Press
2 Park Square, Milton Park, Abingdon, Oxon, OX14 4RN

and by CRC Press
6000 Broken Sound Parkway NW, Suite 300, Boca Raton, FL 33487-2742

CRC Press is an imprint of Informa UK Limited

British Library Cataloguing-in-Publication Data
A catalogue record for this book is available from the British Library

ISBN: 9780367354442 (hbk)
ISBN: 9781032120720 (pbk)
ISBN: 9780429331664 (ebk)

DOI: 10.1201/9780429331664

Typeset in Times
by Deanta Global Publishing Services, Chennai, India

Contents

Preface

Aquaculture is poised to meet current demands. Fishmeal and fish oil, the drivers of fed aquaculture, are efficient and nutritionally complete sources of nutrients, and the industry has been learning to stretch a relatively consistent supply. But fishmeal and fish oil are ultimately a limited resource and a potential future bottleneck (physical or economic) for aquaculture. Without alternatives, the future growth of aquaculture will be kept in check. To date, aquaculture has met these challenges. Between 2008 and 2018, global aquaculture increased 62 percent, while the global supply of fishmeal decreased 12 percent. Aquaculture must continue to innovate and not be satisfied with the *status quo*.

The future of aquafeeds will be based on the use of alternative dietary ingredients that will allow the global aquaculture industry to increase production in a sustainable manner. It is key to evolve feeds with advances in technologies and ingredients to produce *aquafoods* for the increasing world population. This book deals with the important aspects related to aquaculture, the youngest, fastest-growing, and most dynamic protein-producing industry. It was grown by a composite annual growth rate of more than 6% in the last decade. This growth is driven by technological advances and demand factors, such as the trend of consumption of healthy proteins both low in fat and rich in omega-3. The efficient use of feed is a key advantage of aquaculture compared to other proteins, which allows farmed fish to have a competitive price against terrestrial proteins, despite the fact that this industry has a much smaller scale of production and maturity. Long-term aquaculture could be one of the key solutions to feed our growing world population with a healthier diet, while using less of the planet's resources.

In Chapter 1 the current situation of the use of antibiotics, its consequences, and available measures to avoid them are summarized. Antibiotics are commonly used in aquaculture systems to prevent diseases due to management under intensive conditions. In this regard, some agents as bacteriophages, vaccines, probiotics, and others can fight directly with bacteria or help to prevent diseases that cause problems in aquaculture. Chapter 2 shows the use of alternative ingredients and probiotics in aquafeeds formulation. In this sense, it is important to develop aquafeed formulations with shipper animal byproducts, highly digestible microorganisms and insects biomass, or abundant plant ingredients, to substitute fish products partially or completely.

Chapter 3 discusses a number of invasive alien fish species, such as *Chitala ornata, Pterygoplichthys* spp., *Pangasius* spp., and Asian carp, among others, that are introduced to non-native regions, how they affect the native ecosystem, and mitigation strategies used to control their populations. This chapter also focuses on their nutritive profile when processed as fishmeal or fish oil replacement as well as their potential and actual use, benefits, and disadvantages when used as an animal feed, specifically as an aquafeed. In Chapter 4, the new strategies, focused on antioxidant supplementation from vegetable origin to improve oxidative status of fishes, are presented. In this regard, a large number of macro- and micro-algae compounds with antioxidant properties, such as sulfated polysaccharides, carotenoids, or phenolic compounds, are being used.

Chapter 5 shows the impacts on animal performance and digestibility of both common plant as well as novel ingredient (such as meals from insects, yeasts, or algae) use on fish performance and waste production. In this regard, reduction of waste outputs can be potentially achieved through the improvement of feed formulation, palatability, digestibility, and nutrient retention. On the other hand, Chapter 6 gives a perspective of ornamental fish feeding, by adopting a comparative approach, thus comparing ornamental aquaculture with conventional aquaculture and the companion feeds industry.

Chapter 7 is focused on the identification and evaluation of the environmental, economic, and social impacts of the aquafeed production system using the life cycle assessment (LCA) methodology to introduce improvements and mitigate these impacts which would allow accomplishing more sustainable production and consumption of seafood. Chapter 8 discusses the need for high-quality sources that are available all year, at low cost, and with minimal handling, transport, and processing.

Chapter 9 focuses on the several limitations with regard to the use of FO (fish oil) in aquafeeds and introduces the principle of alternative lipid sources (ALS). In addition, the effects of partial or total replacement of dietary FO with ALS on growth performance, feed utilization, nutrient digestibility, fatty acids profile and organoleptic properties of the fillet, general health and welfare of cultured aquatic species, as well as risk assessment in the application of ALS in aquafeeds, are systematically reviewed and discussed. On the other hand, a better understanding of what parameters can be used to monitor nutrient use, challenges associated with diet formulation and selecting fish, and a glimpse of the study of fish metabolism to identify pathways and markers associated with better feed conversion and nutrient retention are discussed in Chapter 10. Finally, in Chapter 11, several studies that demonstrate the positive effects of prebiotics, probiotics, and synbiotics on several relevant species for aquaculture are presented. In order to improve and standardize probiotic administration procedures, the durability of probiotic effects in the host and the best method of administration in each particular case are important aspects to be determined.

Editors

José Manuel Lorenzo is Head of Research at the Meat Technology Centre of Galicia (CTC), Ourense, Spain. He received his M.S. in Food Science and Technology (University of Vigo) and earned his Ph.D. in Food Science and Technology (University of Vigo) in 2006. He started his scientific career in the Department of Food Science and Technology at the University of Vigo, first as researcher scholarship, then, since April 2006, as academic researcher. In 2005–2006 from October to March, he completed a stage period for his research project at the *Stazione Sperimentale per L'Industria delle Conserve Alimentaria*, (Parma, Italy). He has published more than 524 research papers in well-recognized peer-reviewed international journals (SCI), with 60% of them in the first quartile (number of publications in Q1 is higher than 310), and 245 communications to congresses, mostly international. His h-index is 49 with a number of cites of 9088 in Scopus. He was principal researcher in three European projects, three national projects, and 38 regional projects (Galicia, NW Spain), and he was lead investigator in 25 project developments with meat enterprises and industry. In addition, he has participated in more than 65 projects as research collaborator. He has written 20 international books as editor and another national book as editor, and he has written chapters in 64 international and 8 national books, respectively. In addition, he has one national patent as inventor and 295 communications at congress. Finally, he has supervised five doctoral theses. He is associate editor of *Food Research International, Journal of the Science and Food and Agriculture, Food Analytical Methods, Canadian Journal of Animal Science*, and *Animal Science Journal.*

Jesus Simal-Gandara is Full Professor in Nutrition and Food Science at the Faculty of Food Science and Technology, University of Vigo (Spain). He was the first recipient of the Spanish Award of Completion of Pharmacy and PhD Prize at the Faculty of Pharmacy, University of Santiago de Compostela (Spain). He was Associate Professor in 1991 at the University of Vigo, and full Professor since 1999. He is a corresponding member of the Royal Academy of Medicine and Surgery of Galicia (1991), member of the Scientific Committee of the Spanish Agency for Consumption, Food Safety and Nutrition (2013–2016). Additional honors include the Research Medal of the Royal Galician Academy of Sciences 2020 Antonio Casares Rodriguez (Chemistry and Geology), President of the International Association of Dietary Nutrition and Safety (2020), Member of the Royal Academy of Pharmacy of Galicia (2021), and Associate Editor in *Food Science and Nutrition* (Wiley) and *Food Chemistry* (Elsevier). He leads a research group of excellence, and was leading CIA^3 - Environmental, Agricultural and Food Research Center (2008–2018). He was the Head of the Department of Analytical Chemistry and Food Science (2013–2018), and Vice-Chancellor for Internationalization at the University of Vigo (2018). He was nominated by Clarivate Analytics for highly-cited research (2018 & 2020) and was invited to research stays at the Université de Paris-Sud (France), University of Delaware (USA), Fraunhofer-Institut für Lebensmitteltechnologie und Verpackung (Germany), Central Science Laboratory (UK), TNO-Voeding (The Netherlands), Packaging Industries Research Association (UK) and The Swedish Institute for Food and Biotechnology (Sweden). He has 16000 citations in 500 papers= 32 per paper; h-index= 67 (http://scholar.google.es/citations?user=rmeHFXIAAAAJ&hl=es&oi=ao).

Contributors

Rubén Agregán
Centro Tecnológico de la Carne de Galicia
Parque Tecnológico de Galicia
Ourense, Spain

Gonzalo Astray
Department of Physical Chemistry, Faculty of Science
University of Vigo
Ourense, Spain

Fernando G. Barroso
Dept. Biology and Geology
University of Almeria
Spain

Roberto Bermúdez
Centro Tecnológico de la Carne de Galicia
Parque Tecnológico de Galicia
Ourense, Spain

Ruben Dominguez
Centro Tecnológico de la Carne de Galicia
Parque Tecnológico de Galicia
Ourense, Spain

Dmitri Fabrikov
Dept. Biology and Geology
University of Almeria
Spain

Eleni Fountoulaki
Hellenic Centre for Marine Research
Institute of Marine Biology, Biotechnology and Aquaculture
Anavissos, Attikis, Greece

Sara García-González
Department of Chemical Engineering, School of Engineering
University of Santiago de Compostela
Santiago de Compostela, Spain

Enric Gisbert
IRTA
Centre de Sant Carles de la Rápita (IRTA-SCR)
Sant Carles de la Rápita, Spain

Patricia Gullón
Centro Tecnológico de la Carne de Galicia
Ourense, Spain

Fatemeh Hekmatpour
South Iran Aquaculture Research Centre
Iranian Fisheries Science Institute (IFSRI)
Agricultural Research Education and Extension organization (AREEO)
Ahwaz, Iran

Morgane Henry
Hellenic Centre for Marine Research
Institute of Marine Biology, Biotechnology and Aquaculture
Anavissos, Attikis, Greece

Md. Sakhawat Hossain
Hagerman Fish Culture Experiment Station
University of Idaho
Hagerman, Idaho

Fotini Kokou
Animal Sciences Department, Aquaculture and Fisheries group
Wageningen University
Netherlands

Vikas Kumar
Hagerman Fish Culture Experiment Station
University of Idaho
Hagerman, ID
and
Aquaculture Research Institute, Department of Animal and Veterinary Science
University of Idaho
Moscow, Idaho

José Manuel Lorenzo
Centro Tecnológico de la Carne de Galicia
Ourense, Spain
and
Área de Tecnología de los Alimentos, Facultad de Ciencias de Ourense
Universidad de Vigo
Ourense, Spain

Victor Mercado
Department of Marine Biotechnology
Centro de Investigación Científica y de Educación Superior de Ensenada
Ensenada, B.C., México

Mansour Torfi Mozanzadeh
South Iran Aquaculture Research Centre, Iranian Fisheries Science Institute (IFSRI)
Agricultural Research Education and Extension organization (AREEO)
Ahwaz, Iran

Paulo E.S. Munekata
Centro Tecnológico de la Carne de Galicia
Parque Tecnológico de Galicia
Ourense, Spain

Jorge Olmos
Department of Marine Biotechnology
Centro de Investigación Científica y de Educación Superior de Ensenada
Ensenada, B.C., México

Mirian Pateiro
Centro Tecnológico de la Carne de Galicia
Parque Tecnológico de Galicia
Ourense, Spain

Amalia Pérez-Jiménez
Dept. Zoology
University of Granada
Spain

Rodrigo Pulgar
Laboratorio de Genómica y Genética de Interacciones Biológicas (LG2IB), Instituto de Nutrición y Tecnología de los Alimento
Universidad de Chile
Santiago, Chile
and
CRIA, Center for Research and Innovation in Aquaculture
Universidad de Chile
Santiago, Chile

Janice Alano Ragaza
Department of Biology, School of Science and Engineering
Ateneo de Manila University
Quezon City, Metro Manila, Philippines

Marta F. Riesco
Cell Biology Area, Department of Molecular Biology
Universidad de León
León, Spain

Vanesa Robles
Cell Biology Area, Department of Molecular Biology
Universidad de León
León, Spain

Ruth Rodríguez-Bermúdez
Centro Tecnológico de la Carne de Galicia
Parque Tecnológico de Galicia
Ourense, Spain

Eva E. Rufino-Palomares
Dept. Biochemistry and Molecular Biology
University of Granada
Spain

Maria José Sánchez-Muros
Dept. Biology and Geology
University of Almeria
Spain

Benedetto Sicuro
Department of Veterinary Sciences
University of Torino
Italy

Cristina E. Trenzado
Dept. Cellular Biology
University of Granada
Spain

David G. Valcarce
IEO, Spanish Institute of Oceanography
Santander, Spain

Jurij Wacyk
Laboratorio de Nutrición Animal (LABNA), Facultad de Ciencias Agronómicas, Producción Animal
Universidad de Chile
La Pintana, Chile
and
CRIA, Center for Research and Innovation in Aquaculture
Universidad de Chile
Santiago, Chile

Jose Manuel Yañez M.V.
Laboratorio de Genómica en Acuicultura, Facultad de Ciencias Veterinarias y Pecuarias
Universidad de Chile
La Pintana, Chile
and
CRIA, Center for Research and Innovation in Aquaculture
Universidad de Chile
Santiago, Chile

1 Antibiotics in Aquaculture Systems

Effects on Environment and Human Health

Ruth Rodríguez-Bermúdez, Paulo E.S. Munekata, Mirian Pateiro, Ruben Dominguez, and José Manuel Lorenzo

CONTENTS

INTRODUCTION

Aquaculture comprises all forms of culture of aquatic animals and plants in fresh, brackish, and marine environments (Pillay and Kutty, 2005). This industry is a rapidly growing sector in many developed and developing countries (Cabello, 2006). It is expected that this growth will increase at an even faster rate in the future, stimulated by the depletion of fisheries and the market forces that globalize the sources of food supply (Goldburg and Naylor, 2005). According to the FAO (Food and Agriculture Organization), in 2014, global aquaculture production has doubled over the past decade, now accounting for approximately 50% of fishery products for human consumption. This rate is expected to grow due to the fact that aquatic resources captured from the environment have leveled off in the past 20 years (FAO, 2014). Globally, fish currently represent approximately 16.6% of animal protein supply and 6.5% of all protein for human consumption (FAO, 2014), providing an important protein and nutrient source, especially for coastal populations; hence, production

DOI: 10.1201/9780429331664-1

from the aquaculture industry will play a more important role in food security (Mo et al., 2015; Liu et al., 2017).

Antibiotics are commonly used in aquaculture, both prophylactically and to treat disease (Liu et al., 2017). Fishes, especially salmon and trout, are exposed to stressors during breeding, e.g. high population densities, crowding of farming sites in coastal waters, lack of sanitary barriers, and failure to isolate fish farming units with infected animals (Naylor and Burke, 2005). These conditions affect fishes' immune systems, depressing them and leading to inhibition of immune response, so it has become common to use prophylactic antibiotics (Cabello, 2006; Defoirdt et al., 2011). This has resulted in large amounts of veterinary drugs getting into the environment (Boxall et al., 2004) and causing increased antibiotic resistance of bacteria in the environment (Zou et al., 2011; Liu et al., 2017), which then causes increased antibiotic resistance in fish pathogens (Sørum, 2006) and the bacteria of terrestrial animals and human beings (Cabello, 2006). Even though the emergence of bacteria resistance to antibiotics is the main consequence of indiscriminate use of antibiotics, there are other consequences as the presence of residues can induce human diseases, e.g., allergy or toxicity and environmental impact (Zou et al., 2011).

The massive use of antibiotics in the aquaculture industry can lead to practices that are damaging to human and animal health (Goldburg and Naylor, 2005) and have an impact on the environment (Seoane et al., 2014; Pereira et al., 2015). The objective of this chapter is to analyze the current situation of antibiotic use, its consequences, and measures available to avoid them.

ANTIBIOTICS USED IN AQUACULTURE

An antibiotic is a chemotherapeutic agent that inhibits or abolishes the growth of microorganisms, such as bacteria, fungi, and protozoa. The first antibiotics were of natural origin, e.g., penicillin, produced by fungi in the genus Penicillium, and streptomycin, from bacteria of the genus Streptomyces. Currently, antibiotics are obtained through chemical synthesis or chemical modifications of compounds of natural origin (Kümmerer, 2009). In aquaculture as in other breeding systems, like cattle raising or poultry farming, antibiotics are used mainly as prophylactics (Cabello, 2003). In fact, in aquaculture, they are often used for disease prophylaxis, treatment, and/or growth promotion (Zou et al., 2011). Antibiotics are mainly used as prophylactics in all breeding systems for the prevention of illnesses, as farmed animals are more susceptible to diseases caused by the alteration of the immune system due to increased fish stocking density, lack of sanitary barriers between farming sites, failure to isolate infected fish, inadequate handling, and malnourishment because of the industrial breeding system (Sørum and L'Abée-Lund, 2002; Naylor and Burke, 2005; Cabello, 2006). Pollution and crowding stress are well-known immunosuppressive factors resulting in higher disease susceptibility in fish (Köllner et al., 2002), with a global estimate of disease loss of 3 billion US dollars per year (Defoirdt et al., 2007a). Control of bacterial infections with antibiotics has become one of the solutions for fish farmers against pathogens (Xu et al., 2006) because of their low cost, convenience of use, and remarkable curative effect (Yuan and Chen, 2012). It has been estimated that bacterial infections account for a 15–20% loss of annual total production, so feeding antibiotics to animals has become necessary in intensive aquaculture activities (Xu et al., 2006).

The common form of administration of antibiotics to fish is the oral one, as a component of their food and occasionally in baths or injections (Sarmah et al., 2006; Sørum, 2006; Liu et al., 2017). The majority of antibiotics used in aquaculture are broad-spectrum bacteriostatic agents, active against both gram-positive and gram-negative bacteria (Zhang, 2004). The purpose of its use is mainly treating intestinal parasites and systemic diseases, the dosages of antibiotics used in oral administration varied greatly (Table 1.1), generally less than 30 mg were used per kg of fish (body weight; BW). Bath treatment (immersion in short or long baths) and pond sprinkle are two other ways to use antibiotics with the dosages of 1–25 mg/L and 0.05/0.1 mg/L, respectively. Injection, expressed as i.u./fish (i.u. indicates the international unit), is an occasional antibiotic application method in aquaculture (Liu et al., 2017).

TABLE 1.1
Antibiotics: Diseases, Doses, and Susceptible Species of Application

Antibiotic	Disease	Dose	Period (Days)	Species Treated	Reference
Amoxicillin	Furunculosis	40-80 mg/kg of f.w./day	5-10	Trout	Lalumera et al., 2004
	Gill disease	60-80 mg/kg BW/day	10	NA	Minh et al., 2009; Austin and Austin, 2016
	Infectious diseases	0.2 mg/kg BW	5		Yuan and Chen, 2012
	Pasteurella	40-80 mg/kg of f.w./day	5-10	Sea-bass	Lalumera et al., 2004
Ciprofloxacin	NA	50 mg/kg BW	NA	Eel	Wu et al., 2003
Enrofloxacin	Bacterial disease	2500 mg/kg feed	NA	NA	Yuan and Chen, 2012
	Streptococcicosis	5-10 mg/kg orally	10	Hybrid stiped bass	Xu et al., 2013
	Vibriosis	10 mg/kg	5	Rainbow trout	
	NA	4 mg/kg BW	5-7	Shrimp	Rico et al., 2013
Erythromycin	Bacterial kidney disease	25-100 mg/kg of fish/day	4-21	NA	Gulkowska et al., 2007; Minh et al., 2009; Xu et al., 2013; Zhang et al., 2013; Austin and Austin, 2016
	Gill rot disease	1-5 mg/kg WB	NA		Yuan and Chen, 2012
	Streptococcicosis	25-100 mg/kg of fish/day	4-21		Gulkowska et al., 2007; Minh et al., 2009; Zhang et al., 2009; Xu et al., 2013; Austin and Austin, 2016
Florfenicol	Bacterial Infection	10-15 mg/kg BW	3-5	NA	Yuan and Chen, 2012
		0.5 mg/kg BW	3		
	Furunculosis Vibriosis	10 mg/kg BW of fish/day	10	NA	Jiang et al., 2011; Yan et al., 2013; Austin and Austin, 2016
	NA	6 mg/kg BW	5-7	Shrimp	Rico et al., 2013
		3.8-3.9 mg/kg BW	6	Tilapia	
Flumequine	Furunculosis	12 mg/kg of f.w./day	5	Trout	Lalumera et al., 2004
	Pasteurella			Sea-bass	
	Red mouth	75 mg/kg of f.w./day	5	Trout	
	Vibriosis	12 mg/kg of f.w./day			

(*Continued*)

TABLE 1.1 (CONTINUED)
Antibiotics: Diseases, Doses, and Susceptible Species of Application

Antibiotic	Disease	Dose	Period (Days)	Species Treated	Reference
Furazolidone	NA	100-200 mg/kg BW	NA	Fish	Yan and Cai, 2004
		0.1-0.15% feed		Prawn	
Gentamycin S	NA	4 mg/kg BW	3	Tilapia	Rico et al., 2013
Flumequine	Furunculosis	12 mg/kg of f.w./day	5	Trout	Lalumera et al., 2004
Neomycin	Septicaemia	2.5-5 mg/kg BW	NA	NA	Yan and Cai, 2004
Norfloxacin	NA	1670-2000 mg/kg feed, 2 times per day	5-7	Eel	Yuan and Chen, 2012
		1000-1250 mg/kg feed	3-5		
Oxolinic acid	Red fin/skin	10-20 mg/kg BW	4-7	NA	Yuan and Chen, 2012
Oxytetracycline	Enteritis	2-10 mg/kg BW	3-7	NA	Yuan and Chen, 2012
		4% in fsh feed	3-5		
	Furunculosis	75 mg/kg of f.w./day	7	Trout	Lalumera et al., 2004
	Pasteurella			Sea-bass	
	Red mouth (Xu et al., 2013; Zhang et al., 2013; Austin and Austin, 2016)			Rainbow trout	
	Vibriosis			Sea-bass	
				Trout	
	Many diseases**	50-75 mg/kg of fish/day	10	NA	Minh et al., 2009; Zou et al., 2011; Austin and Austin, 2016
	NA	50-100 mg/kg BW	NA	Shrimp	Jiang, 1996
		2-3 mg/L WS		Prawn	
		25 mg/L WS		Fish	
Penicillin G	Bacterial Infection	2.4-18 mg/L WS	NA	NA	Jiang, 1996
	Postpartum infection	105 i.u./fish			Yan and Cai, 2004
Streptomycin	NA	105 i.u./fish	NA	Fish	Jiang, 1996
Sulfadiazine*	Red skin disease	100 mg/kg BW	3-5	NA	Chao, 2002
	Enteric Red Mouth	30 mg/kg of fish/day	10		Xu et al., 2013; Zhang et al., 2013; Austin and Austin, 2016
	Furunculosis haemorrhagic septicaemia				
	Vibriosis				

(Continued)

TABLE 1.1 (CONTINUED)
Antibiotics: Diseases, Doses, and Susceptible Species of Application

Antibiotic	Disease	Dose	Period (Days)	Species Treated	Reference
Sulfamerazine	Furunculosis	120-220 mg/kg of f.w./day	10-21	Sea-bass Trout	Lalumera et al., 2004
	Red mouth	75 mg/kg of f.w./day	7	Trout	
	Vibriosis	120-220 mg/kg of f.w./day	10-21		
Sulfamerazine S	Trichuriasis	100-200 mg/kg BW	5-10	NA	Chao, 2002
Sulfamethoxazole	Bacterial Enteritis	100-200 mg/kg BW	5	NA	Chao, 2002
	Enteric Red Mouth Furunculosis haemorrhagic septicaemia Vibriosis	30 mg/kg of fish/day	10		Zhang et al., 2009, 2013; Zou et al., 2011; Xu et al., 2013; Austin and Austin, 2016
Sulfamidine	Enteric Red Mouth Furunculosis haemorrhagic septicaemia Vibriosis	30 mg/kg of fish/day	10	NA	Minh et al., 2009; Xu et al., 2013; Zhang et al., 2013; Austin and Austin, 2016
Sulfisoxazole	Gill rot disease	200-500 mg/kg BW	4-6	NA	Chao, 2002
Thiamphenicol	Furunculosis	120-220 mg/kg of f.w./day	10-21	Sea-bass	Lalumera et al., 2004
	Pasteurella	80 mg/kg of f.w./day	10		
Trimethoprim	Enteric Red Mouth Furunculosis haemorrhagic septicaemia Vibriosis	30 mg/kg of fish/ day	10	NA	Gulkowska et al., 2007; Minh et al., 2009; Zou et al., 2011; Austin and Austin, 2016

BW: body weight; NA: not available; i.u.: international unit; fw: fish weight; WS; water solution; *Sulfonamides are usually used together with trimethoprim as potentiated sulphonamide; **acinetobacter disease, carp erythrodermatitis, coldwater disease, columnaris, edwardsiellosis, emphysematous putrefactive disease, enteric red mouth, enteric septicaemia, fin rot, furunculosis, gill disease, haemorrhagic septicaemia, red pest, salmonid blood spot, saltwater columnaris, Stretococcicosis, ulcer disease

HOW ANTIBIOTICS GET INTO THE ENVIRONMENT

Antibiotics get into the environment as a result of fish feces and unconsumed antibiotic in feed. They diffuse into the sediment so they can be washed by currents to distant sites (Boxall et al., 2004; Sørum, 2006; Kümmerer, 2009; Zou et al., 2011). It has been estimated that a minimum of

75% of most of the antibiotics in feed used in aquaculture systems are transmitted to the surrounding environment and can accumulate in the sediment (Halling-Sørensen et al., 1998). Once in the environment, these antibiotics can be ingested by wild fish and other organisms, including shellfish (Boxall et al., 2004; Sørum, 2006).

Antibiotics could also enter aquaculture farms via the application of fertilizers. The use of organic pond fertilizer is common in Asian countries, particularly in China (Xiong et al., 2015). Swine and chicken dung are often used to maintain the fertility of fish ponds because they can enhance fish growth (Dhawan and Kaur, 2002) as well as provide more food for filter feeders. The use of manure in fish ponds is believed to be an ecologically safe and economical farming practice (Su et al., 2011). However, a large proportion of antibiotics are also used in agriculture activities, and high concentrations of antibiotics have been detected in animal manure (Zhao et al., 2010). Researchers pointed out that about 30% to 90% of antibiotics are excreted unchanged in feces and urine due to the poor absorption of antibiotics in the animal gut (Sarmah et al., 2006). This practice leads to the spread of antibiotic resistance genes (ARGs) and antibiotic resistance. They may also change the bacterial community composition in the aquaculture environment since the manure often contains antimicrobials used as growth promoters for terrestrial animals including swine and chicken (Hoa et al., 2011; Xiong et al., 2015).

It is a common practice to lower the water level of a fish pond by discharging the pond water directly to a nearby surface. Fish ponds are drained periodically, and pond sediment is removed. As pond sediment is considered a good fertilizer, it is applied on vegetation. Furthermore, the waste from fish farmed in cages is continuously discharged into the environment throughout the year. In this way, the area of land affected by antibiotics derived from aquaculture could be extensive (Mo et al., 2015).

Lalumera et al. (2004) have found that different farm technologies, pond or tank-based, affect concentration of antibiotics modifying the residence times of the antibacterial agents in sediments and hence, affect the impact on the surrounding environment. In pond-based tanks oxytetracycline was abundant in the pond sediments but was not detected in the sediments surrounding the fish farm. These findings suggest that, in the pond-based system, the possible side effects of oxytetracycline are mainly directed toward the microorganisms present within the farm. In the case of flumequine, the authors have found that the antibiotic may have side effects on the external bacterial population. With cement tank farms, the presence of flumequine and oxytetracycline in the tanks and the surrounding environment suggests that the washing out via water flow is the main process responsible for any occurrence and possible impact of these antibiotics in the aquatic ecosystems near the fish farms.

CONSEQUENCES OF ANTIBIOTICS USE

Antibiotic Resistance

The development of resistance to antibiotics seems to be the most critical problem associated with antibacterial agents (Andersson and Levin, 1999; Zou et al., 2011). The emergence of antibiotic resistance in bacterial populations is not just a biological problem, but also a medical, social, economic, and ethical problem because infections produced by resistant bacteria cause higher mobility and mortality. Economic expenses caused by those types of infections affect all society, as patients generally need prolonged hospitalizations and special treatments which are more expensive (Cosgrove et al., 2002). It is well known that the continuous application of antibiotics in the aquatic environment can result in the appearance of resistance among human pathogens forming part of its microbiota (Angulo, 2000; Rico et al., 2012).

The genes that code for resistance in bacteria, ARGs, are mainly located on mobile genetic elements such as plasmids, transposons, integrons, gene cassettes, and bacteriophages, indirectly transporting the ability for resistance from non-pathogens to pathogenic microorganisms

(Kemper, 2008). Numerous studies have indicated that fish ponds are reservoirs of ARGs, since monitored bacteria in the surrounding intensive aquaculture sites contain more ARGs as compared to those in sites without aquaculture activity (Gao et al., 2012). The results presented by Xiong et al. (2015) indicate that fish ponds are reservoirs of ARGs, and the presence of potential resistant pathogen-associated taxonomic groups in fish ponds might imply a potential risk to human health. Bacteria exposed to antibiotics in the environment may accelerate the emergence of ARGs that pose potential harm to both the ecosystem and human health (Kemper, 2008; Zhang et al., 2009; Mo et al., 2015). In fact, resistance to antibiotics is well documented since some decades ago, showing high fractions of aeromonads resistant to oxytetracycline (44%), sulphadiazine-trimethoprim (72%), and multiresistencies (50%) (Nygaard et al., 1992; Kerry et al., 1996). There is epidemiologic and molecular evidence that genes able to produce resistance can be transmitted by aquatic bacteria to other bacteria capable of producing infections in human and terrestrial animals (Sørum and L'Abée-Lund, 2002).

Many cases of antibiotic resistance have been documented since then. Angulo and Griffin (2000) have shown that the DNA sequence of the transmissible element harboring the antibiotic resistance determinants has an important similarity to a plasmid of *Pasteurella piscicida*, which is also a fish pathogen. This evidence suggests that there was a horizontal transmission of antibiotic resistance determinants from bacteria in aquaculture systems to a human and terrestrial veterinary pathogen. Later, Angulo et al. (2004) demonstrated molecular and epidemiological evidence that antibiotic resistance determinants of resistance *Salmonella enterica* serotype Typhimurium DT104, an emergent pathogen and the cause of several outbreaks in Europe and the United States of salmonellosis in humans and animals, probably originated in aquaculture settings of the Far East. The epidemiology of the dissemination of *S. Typhimurium* DT104 also suggests this pathogen could have been spread by fish meal as with the Salmonella *agona* outbreak in Peru several years ago (Angulo and Griffin, 2000; Boyd et al., 2001). In a review, Defoirdt et al. (2007a) summarized the antibiotics to which *Vibrio harveyi* is resistant to e.g. amikacin, ampicillin, β-lactams, carbenicillin, cotrimoxazole, chloramphenicol, erythromycin, furazolidone, gentamycin kanamycin, nitrofurantoin, novobiocin, oxolinic acid, oxytetracycline, sulfonamide, streptomycin, and tetracycline.

Antibiotic-resistant bacteria have been detected in aquatic products purchased from Chinese seafood markets (Broughton and Walker, 2009; He et al., 2015). They can enter the human body via dietary intake and, as has been demonstrated in long-term exposure at low level experiments, can lead to resistance in intestinal bacteria. ARGs can be conjugated to human pathogens through horizontal gene transfer, resulting in infection of humans (Zheng and Su, 2010).

The determinants of antibiotic resistance have emerged, and, being selected in an aquatic environment, they have the potential of being transmitted by horizontal gene transfer to bacteria of the terrestrial environment, including human and animal pathogens (Sørum, 2006). The exchange of resistance determinants between the aquatic and terrestrial environment can also stem from the movement of antibiotic-resistant bacteria between two environments, a result of transporting fish between bodies of freshwater and the ocean, a necessary step, for example, in salmonid breeding (Goldburg and Naylor, 2005; Naylor and Burke, 2005).

In many aquaculture settings in developing countries, the possibility of these exchanges has been amplified by the high level of contamination of seawater and freshwater with untreated sewage and agricultural and industrial wastewater containing normal intestinal flora and pathogens of animals and humans usually resistant to antibiotics (Sørum, 2006).

Residues and Effects on Human Health

Another problem drawn from the excessive use of antibiotics in industrial aquaculture is the presence of residual antibiotics in commercialized fish and shellfish products (Angulo et al., 2004; Sørum, 2006). These residual antibiotics can be consumed in drinking water (Kümmerer, 2009) or eaten by consumers of fish with the added potential alteration of their normal flora, increasing their susceptibility to bacterial infections and also selects for antibiotic-resistant bacteria (Cabello, 2004; Salyers et al., 2004). Furthermore, undetected consumption of antibiotics in food can generate other health problems, such as allergy and toxicity, which can be difficult to diagnose because of a lack of previous information (Cabello, 2004). Consumers are not the only ones affected by allergies or toxicity caused by antibiotic residues; workers in the aquaculture industry are exposed to large amounts of antibiotics that come in contact with the skin and intestinal and bronchial tracts when administering the medicated food to fish (Lillehaug et al., 2003).

There exist great differences in the types and concentrations of antibiotics detected in different areas and countries – even in different regions of the same country. This can be explained by the prevalence of diseases, treatment habits, or simply market and economic reasons (Zou et al., 2011). Residues of antibiotics have been detected in many countries; for example, China, where 234 cases of antibiotic residues have been recorded in Chinese aquatic products, including 24 fish species, 8 crustacean species, and 4 mollusc species. Thirty-two antibiotics have been detected in aquatic products: quinolones and sulphonamides were the dominant residual chemical; the highest concentrations were found for ciprofloxacin, norfloxacin, and sulfisoxazole (Rico et al., 2013; Liu et al., 2017). Recent studies show that antibiotic residues are still detected in farmed or wild fish (He et al., 2012; Li et al., 2012). Other recent studies reflect the presence of antibiotic residues (sulfonamides, fluoroquinolones (e.g. ciprofloxacin), tetracyclines (e.g. oxytetracycline), and macrolides (e.g. erythromycin)) in receiving waters or sediments near different types of aquaculture (Burridge et al., 2010; Cháfer-Pericás et al., 2010; Zou et al., 2011; Zheng et al., 2012; Xu et al., 2013; Xue et al., 2013; Chen et al., 2014).

The risk of antibiotics to human health generally appears in two ways: adverse drug reaction (ADR) and potential prevalence of antibiotic resistance. Allergic reaction is one type of ADR, and the common symptoms are urticaria, angioneurotic edema, gastrointestinal reactions, aplastic anemia, as well as shock and death, if serious (Bousquet, 2009; Solensky, 2012). A large proportion of antibiotics (e.g. penicillin G, methicillin, tetracycline, and sulfonamides) have antigenicity, and consumption through aquatic products may cause allergic symptoms (Li, 2008; Kümmerer, 2009). Antibiotics can accumulate in the human body, causing another type of ADR, called chronic toxicity, that destroys organs, causing lesions due to prolonged consumption at low doses (Zheng and Su, 2010). Many examples of antibiotics used in aquaculture that have toxic effects in humans if consumed have been recorded through the years. For instance, quinolones and tetracyclines negatively influence teeth development in children, and chronic exposure to tetracycline could lead to steatosis by altering genes related to lipid metabolism and transportation (Anthérieu et al., 2011). Gentamicin is nephrotoxic (Kümmerer, 2009), and chloramphenicol has immunotoxicity and may cause aplastic anemia, leukemia, and agranulocytosis (Issaragrisil et al., 2006). Erythromycin may cause deafness and peripheral neuritis, and furazolidone may lead to hemolytic anemia and polyneuritis (Wang and Xiong, 2007; Tian, 2010). Other antibiotics can be metabolized in the human body and their metabolites can be more toxic than the original active principle (Sapkota et al., 2008; Heuer et al., 2009; Kümmerer, 2009). In summary, Liu and Wong (2013) reviewed human health risk studies and concluded that antibiotics could pose a low possibility of acute toxicity to humans; however, in extreme cases, the consumption of aquaculture products that are heavily contaminated with antibiotics could pose a human health risk (Mo et al., 2015).

Environmental Impact

A large amount of antibiotics ends up in the environment, water, and sediments and have the potential to affect the presence of normal flora and plankton in those niches, resulting in shifts in the diversity of microbiota (Hunter-Cervera et al., 2005; Sørum, 2006). In fact, bacteria, fungi, and microalgae are the organisms primarily affected by antibiotics because these principles are designed to affect microorganisms (Kümmerer, 2009). Within the potential environmental impacts of aquaculture are the increase of algal production, the dissolved oxygen depletion at the water-sediment interface and the organic enrichment of the sediments, the potential toxic effect of the chemicals used for controlling fish diseases must be considered (Lalumera et al., 2004).

The presence of antibiotics in the environment may also have deleterious effects on non-target aquatic organisms, such as microalgae (Seoane et al., 2014); in fact, algae is the trophic level most sensitive to antibiotics (Pereira et al., 2015). Studies that have used aquatic organisms of different trophic levels have found that the toxicity of antibiotics is higher to cyanobacteria, probably due to their prokaryotic nature, than to unicellular eukaryotic primary producers like microalgae. Among eukaryotic organisms, multicellular species are in general less sensitive than unicellular microorganisms (González-Pleiter et al., 2013). Seoane et al. (2014) have observed that chloramphenicol, florfenicol, and oxytetracycline have inhibitory effects on the growth of *Tetraselmis suecica* leading to adverse effects on this marine microalgae. Florfenicol showed much higher toxicity than the other two antibiotics assayed. The use of this antibiotic should be more carefully monitored to reduce the potential contamination risk of the receiving waters.

The changes observed in the environment can be amplified by the eutrophication produced in the aquaculture environment by the increased input of N, C, and P generated by the non-ingested food and fish feces. This can alter ecological equilibria, creating situations that may impact fish and human health, promoting, for example, algal blooms and anoxic environments (Hunter-Cervera et al., 2005). It is important to take into consideration that these compounds have the potential to bioaccumulate, a process in which chemical compounds are accumulated in organisms due to exposure in the environment. Ecological risk from exposure to low levels of antibiotics increases the emergence of bacterial resistance even at subinhibitory levels (Cabello, 2006; Zou et al., 2011; Chen et al., 2014). Chen et al. (2014) have found that trimethoprim is bioaccumulative in fish muscles, and antibiotics are weakly bioaccumulated in molluscs.

MEASURES FOR AVOIDING ANTIBIOTICS USE

Regulation

The use of antibiotics depends on local regulations, which vary widely between different countries (Defoirdt et al., 2011) (Table 1.2). Many countries (e.g. European Union, United States, and Japan) have implemented drastic restrictions on the use of antibiotics in aquaculture attending to the evidences demonstrated by researchers on the impacts of its use (Cabello, 2004; Wang and Xiong, 2007; Smith, 2008; Hvistendahl, 2012). Since 2002, the United States and European Union have issued stringent regulations for antibiotic residues in aquatic products (Wang and Xiong, 2007; Hvistendahl, 2012; Rico et al., 2013). In Europe, the use of non-therapeutic antibiotics, including growth promoters, in livestock production has been banned since 2006 (Cogliani et al., 2011). The sales and usage of veterinary antimicrobial across the European Union are monitored under the European Surveillance of Veterinary Antimicrobial Consumption (ESVAC) project (European Medicines Agency, 2015). The restriction has applied strict control on the use of antibiotic prophylaxis (Sørum, 2006) and proscription of the use of antibiotics that are useful in the therapy of human infections (Lillehaug et al., 2003). For example, the use of quinolones has been totally restricted in aquaculture in industrialized countries, e.g. Norway and the United States (Cabello, 2004), because it is an antibiotic able to generate cross-resistance, remaining active in sediments for prolonged periods of time as they are not readily biodegradable. Moreover, quinolones are highly effective in the treatment of human infections (Cabello, 2004; Jacoby, 2005; Sørum, 2006).

However, a large proportion of the global aquaculture production (90%) takes place in countries that have no or few effective regulations of antimicrobial agents used in animals (Defoirdt et al., 2011; Pereira et al., 2015). However, not all states have restricted the use of quinolones; many underdeveloped countries, for example Chile and China, continue allowing them (Cabello, 2004; Jacoby, 2005). As mentioned in the review published by Liu et al. (2017), China allows the use of more than twice as many antibiotics (active principles) as the United Kingdom, and in Chinese aquaculture,

TABLE 1.2
Antibiotics Authorized and Non-Authorized According to the Country or Organization (Cabello, 2004; Mo et al., 2015; Liu et al., 2017)

Antibiotics	China	United Kingdom	FAO	Norway	Chile	United States
Amoxicilin		+	+	–	+	–
Ampiciline	NA	NA	NA	–	+	–
Cefotaxime	NA	NA	NA	–	+	–
Chloramphenicol	–	NA	NA	–	+	–
Doxycycline	+	–	–	NA	NA	NA
Enrofloxacin	+	–	–	NA	NA	NA
Erythromycin	NA	NA	NA	–	+	–
Florfenicol	+	–	–	NA	NA	+
Flumequine	+	–	–	NA	NA	NA
Furazolidone	NA	NA	NA	–	+	–
Gentamicin	NA	NA	NA	–	+	–
Kanamycin	NA	NA	NA	–	+	–
Metronidazol	–	NA	NA	NA	NA	NA
Nalidixic acid	NA	NA	NA	–	+	–
Neomycin	+	–	–	–	+	–
Nitrofuran	–	NA	NA	NA	NA	NA
Norfloxacin	+	–	–	+	+	–
Oxolinic acid	+	+	+	NA	NA	NA
Oxytetracycline	–	+	+	NA	NA	+
Quinolones	NA	NA	NA	–	+	–
Sarafloxacin	–	+	+	NA	NA	NA
Sulfadiazine	+	+	+	+	+	+
Sulfamethazine	+	–	–	+	+	+
Sulfamethoxazole	+	–	–	+	+	+
Sulfamonomethoxine	+	–	–	+	+	+
Tetracycline	NA	NA	NA	+	+	+
Thiamphenicol	+	–	–	NA	NA	NA
Trimethoprim	+	+	+	+	+	+

farmers use antibiotics that are not authorized (Yuan and Chen, 2012; Liu et al., 2017). In fact, Jacoby (2005) has discussed that quinolone resistance has emerged as an important public health problem in China as a result of the unrestricted use of this group of antibiotics in aquaculture and in industrial animal husbandry. Other antibiotics besides quinolones are equally underregulated in developing countries; it has been demonstrated that the Chilean aquaculture industry uses approximately 75 times more antibiotics for salmon production than the Norwegian industry (Cabello, 2003).

Probiotics

Probiotic microorganisms refer to a live microbial adjunct which confers beneficial effects on the host by modifying the host-associated or ambient microbial community, by ensuring improved use of the feed or enhancing its nutritional value and the host response toward disease, or by improving the quality of its environment (Verschuere et al., 2000). Probiotics could be used in aquaculture for promoting growth and/or preventing diseases. An example of probiotic use is the manipulation of gut bacteria in fish via the administration of antagonistic bacteria for reducing the incidences of opportunistic pathogens (Balcázar et al., 2006). The use of lactic acid bacteria like *Lactobacillus plantarum* could enhance the growth of fish as well as fish immunity, Son et al. (2009) have observed a better response against *Streptococcus* after administering *Lactobacillus*.

It was found that the addition of *Bacillus* strains against luminescent vibrios resulted in healthier prawns and lower numbers of luminescent vibrios in the pond water (Vine et al., 2006), increasing the survival of aquaculture animals. Rengpipat et al. (2003) have demonstrated that adding *Bacillus* to the culture water of black tiger shrimp larvae resulted in a 90% decrease in accumulated mortality. The use of a commercial mixture of *Bacillus* strains selected because of their ability to inhibit pathogenic vibrios improved the survival of fish larvae and performed equally well as antibiotics (Decamp et al., 2008).

Traditional Chinese medicines could also serve as an alternative to antibiotic usage in the aquaculture industry. Ardó et al. (2008) reported that the inclusion of 0.1% of *Astragalus membranaceus* significantly enhanced phagocytic and respiratory burst activities of blood phagocytic cells. Other therapies applied in Chinese medicine that have been demonstrated to increase immunity are astragalus root (1.0–1.5%, *Astragalus propinquus*), Chinese angelica root (*Angelica sinensis*), *Radix scutellariae*, *Rhizoma coptidis*, *Herba andrographis*, and *Radix sophorae flavescentis* (Jian and Wu, 2003; Choi et al., 2013). Certain products of traditional Chinese medicine have antimicrobial properties, for example, *Angelica dahurica*, *Lycium barbarum*, *Scutellaria barbata*, and *Zingiber officinale* (Yu et al., 2004). In general, herbal medicines pose effects on growth, antimicrobials, disease resistance, stimulate appetite, and result in anti-stress (Choi et al., 2013).

The most important limitation to the use of probiotics is that, in many cases, they are not able to maintain themselves, and so need to be added regularly and at high concentrations, which makes this technique less cost-effective. Furthermore, probiotics that were selected *in vitro* based on the production of inhibitory compounds might fail to produce these compounds *in vivo* (Vine et al., 2006). Moreover, vibrios might develop resistance if the production of growth-inhibitory compounds is the only action method as has occurred with numerous antibiotics. Therefore, it may be useful to select probiotics with more than one antagonistic characteristic, or to apply a mixture of probiotics with different modes of action, to maximize the chance of success (Defoirdt et al., 2007a, 2011).

Vaccines

Vaccination refers to the administration of weakened or dead pathogenic bacteria or parts of them, with the aim of conferring long-lasting protection through immunological memory (Gudding et al., 1999). Traditionally, it was believed that only vertebrates had the mechanisms to develop immune response (Smith et al., 2003). However, some authors think that invertebrates also possess the capacity to be immunized (Arala-Chaves and Sequeira, 2000).

The results of some studies prove that vaccination against luminescent vibriosis in several fish species has increased survival significantly when experimentally infected. Unfortunately, vaccinations are not possible in fish larvae, which are more susceptible than adults to the disease, because they are nearly impossible to handle, and it is believed they are not able to develop specific immunity (Crosbie and Nowak, 2004).

Bacteriophages

Bacteriophages were discovered in the early 1920s, as viral infections of bacteria, their value for antibacterial therapy and prophylaxis was almost immediately recognized (Defoirdt et al., 2007a). However, they were proposed as candidates for therapies in aquaculture only in recent years (Nakai and Park, 2002). Many phages are strain-specific rather than species-specific (Defoirdt et al., 2007a); the advantage of using a specific method is that it will not affect beneficial bacteria (Defoirdt et al., 2011). Furthermore, knowing that the pool of bacteria in which resistance can originate will be smaller, the risk of resistance development will be lower. Currently, there are bacteriophages specific for aquaculture pathogens as *Edwardsiella ictuari*, *Flavobacterium psychrophilum*, and *V. harveyi* (Vinod et al., 2006; Stenholm et al., 2008). Many of the isolated phages are strain-specific rather than species-specific (Defoirdt et al., 2011). The selection of bacteriophages for aquaculture should be based on their capability to infect a wide range of strains of the target pathogen species (Vinod et al., 2006). Vinod et al. (2006) found that the use of a bacteriophage against vibriosis increased shrimp survival from 17% to 86%, yielding a better performance than antibiotics. It is interesting to note that the phage treatment performed much better than the daily addition of antibiotics, in which survival was only 40%.

However, using bacteriophages has also disadvantages: bacteriophages might transfer virulence factors (Austin et al., 2003), so, before using them, it is important to test if they carry virulence genes and if it is safe to use them (Defoirdt et al., 2011); another disadvantage is the rapid development of resistance of bacteria to the phages (Fischetti et al., 2006), so it would be necessary to isolate novel phages with some frequency.

Growth Inhibition

Another strategy against bacteria is to inhibit their growth rather than killing them. Short-chain fatty acids are used in diets to control pathogens (*Salmonella*, vibrios, etc.) (Immerseel et al., 2002; Ayude-Vázquez et al., 2005). The administration of short-chain fatty acids is beneficial to the microbial community in the digestive tract because of the depletion in intestinal pH. The main problem of using short-chain fatty acids is that they would leach into the culture water, and, consequently, high doses would be needed to retain sufficient activity (Defoirdt et al., 2011). Ayude-Vázquez et al. (2005) have reported that short-chain fatty acids are responsible for the inhibitory effect of lactic acid bacteria toward pathogenic vibrios.

Another method of growth inhibition is the use of polyhydroxyalkanoates. Different studies demonstrate that adding these polymers to fish food would result in biocontrol effects similar to those provided by short-chain fatty acids. There are evidences they are effective against *Vibrio campbellii* (Defoirdt et al., 2007b, 2011). It is possible to produce poly-β-hydroxybutyrate (PHB) *in situ* in the culture by adding carbonaceous compounds or by increasing the C:N ratio of the feed. Such practice, called biofloc technology, is currently getting more attention as a means to remove inorganic nitrogen from culture water through assimilation into microbial biomass, which can be used as a feed source by animals (Avnimelech, 1999).

Immune Stimulation

Immunostimulants are substances that stimulate the immune system by inducing activation or increasing activity of any of its components. They could increase the resistance of fish to infectious diseases by enhancing non-specific defence mechanisms (Vaseeharan and Thaya, 2013). Imnunostimulants can be administered by injection, bathing, or orally, with the latter appearing to be the most practicable (Yin et al., 2006). In this sense, the use of immunostimulants as a dietary supplement to larval fish could be of considerable benefit in boosting the animal's innate defenses with little detriment to the developing animal (Bricknell and Dalmo, 2005). The addition of various food additives such as vitamins, carotenoids, and herbal remedies to the fish feed has been tested in aquaculture. The reduction of the stress response, increase on the activity of innate parameters, and improved disease resistance are the overall beneficial effects (Cerezuela et al., 2009; Yin et al., 2009). There are many studies reporting a variety of substances, including synthetic, bacterial, animal, and plant products that can be used as immunostimulants to enhance non-specific immune system of cultured fish species (Vaseeharan and Thaya, 2013).

Several reports have mentioned that the use of immunostimulants to control luminescent vibriosis has increased survival after experimental infections (Marques et al., 2006). One limitation of immunostimulants is that they can hurt, or even kill, the host (Smith et al., 2003). Moreover, they have a short duration, hence would need to be administered repeatedly, and long-term administration seems to decrease the immunostimulant effect (Defoirdt et al., 2007a).

Recently, leaves, parts of plants, essential oils, and herbal drugs have been used as stimulants (Vaseeharan and Thaya, 2013), showing good results in the control of bacterial and viral diseases. Examples of leaves showed in the review by Vaseeharan and Thaya (2013) are: *Psidium guajava*, active against *Aeromonas hydrophila*; *Astragalus membranaceus* extract, significantly enhances the phagocytic activity of leucocytes; and *Ocimum sanctum*, increases the resistance against experimental infection with *A. hydrophila*. However, they are just examples because plant extracts used for aquaculture therapy are nearly unlimited. These authors have concluded that plant materials have potential applications as immunostimulants in fish culture because they are not expensive, act against a broad spectrum of pathogens, and preparation of plant extract is much easier and inexpensive. Moreover, plant extracts do not endanger the environment, as they are biodegradable. Hence, due to their beneficial effects, it is proved that herbal immunostimulants increase the immune response, survival, and growth rate of the fish and can be used in fish farming as alternatives to vaccines, antibiotics, and chemical drugs.

Antivirulence Therapy

An alternative to killing the pathogens is the specific inhibition of functions required to infect the host, which are usually referred as virulence factors. Quorum sensing, bacterial cell-to-cell communication, is a gene regulation mechanism by which bacteria coordinate the expression of certain genes in response to the presence of small signal molecules. Quorum sensing has been found to regulate the virulence gene expression, and, consequently, research efforts nowadays are directed toward finding quorum sensing-disrupting techniques (Defoirdt et al., 2011). Compounds that function as quorum sensing-disrupters are halogenated furanones, which were the first ones discovered. They were isolated from red marine alga *Delisea pulchra* and have effect against vibrios; however, their utilization is not safe because they have a toxic effect (Defoirdt et al., 2006). Other substances that have the ability to block quorum sensing are two metabolites of *Halobacillus salinus*, cinnamaldehyde, substances isolated from *Colpomenia sinuosa* (Kanagasabhapathy et al., 2009; Teasdale et al., 2009).

Another strategy to attack microorganisms is to inactivate the signal molecules. Researchers have found that N-acyl-homoserine lactones degrading enrichment cultures isolated from the intestinal tract of healthy shrimp and fish increase survival of different aquaculture species (Tinh et al., 2008). The development of resistance to quorum-sensing disruption is probably smaller than for conventional antibiotics because it only poses selective pressure under those environmental conditions where quorum sensing is essential. Conventional antibiotics, in contrast, pose strong selective pressure in any environment (Defoirdt et al., 2011).

Another regulatory mechanism that controls the expression of virulence is ToxR regulon in vibrios, which controls the expression of the cholera toxin. A recent study has demonstrated that the virulence of pathogenic vibrios is correlated with the expression of ToxR (Ruwandeepika et al., 2011). Virstatin is a small molecule that inhibits the transcriptional regulator ToxT and controls the expression of virulence genes of the ToxR regulon in vibrios (Tendencia and De La Peña; Hung et al., 2005).

Green Water

The technique called green water is the culture of fish in water in which microalgae grow abundantly. It was demonstrated that *V. harveyi* disappeared from seawater containing Chlorella after two days of incubation (Tendencia and De la Peña 2003). Lio-Po et al. (2005) have identified eight bacterial and 12 fungal isolates that were associated with green water and had a growth-inhibitory effect toward *V. harveyi*. The algae *Chaetoceros calcitrans* and *Nitzchia sp.* have an even higher growth-inhibitory effect against luminous vibrios as they completely disappeared from cultures after one or two days (Defoirdt et al., 2007a).

Hygiene Measures and Disease Prevention

According to the FAO, disease prevention is the preferred health management option in aquaculture production because preventive measures are most cost-effective (Defoirdt et al., 2007a). Pollution is one of the factors that could suppress the immune system of fish, so improving water quality could remove the immunosuppressive effect (Wang and Xiong, 2007). Disease prevention could be achieved by improving management, as e.g. prevention of the transmission of pathogens between farms (e.g. quarantine) and good hygiene (e.g. disinfection of culture tanks, water, and eggs) (Brock and Bullis, 2001).

As some bacteria are opportunistic microorganisms, it is also important to control all factors that contribute to immune system depression. It is important to eliminate stress, avoiding stress factors like high stoking rate, handling, temperature, and salinity changes (Brock and Bullis, 2001). It is also important to improve water quality, as some bacteria are capable of reducing nitrate and nitrite to nitrogen, such as *Microbacterium* and *Bacillus sp.* (Wang and Xiong, 2007). However, it is not always economically profitable to produce animals in the most optimal conditions and give them optimal feed, so there will always be a risk of infection and a need for effective biocontrol techniques (Defoirdt et al., 2007a).

CONCLUSIONS

Antibiotics are commonly used in aquaculture, mainly prophylactically, to avoid diseases produced due to management in intensive aquaculture systems that leads to an increase of stressors and higher densities. The large amounts of veterinary drugs reach the environment as a result of fish feces and unconsumed antibiotic feed. The effects of transmitting huge quantities of antibiotics to the environment are increased antibiotic resistance in fish pathogens that also affect animals and human beings, residues in food, and environmental impacts. It has been demonstrated that if preventive measures are implemented, such as the use of bacteriophages, vaccines, and probiotics, among others, the prophylactic use of antibiotics is not necessary. In fact, researchers encourage the use of alternative treatments, even though it is necessary to search for new alternatives and to develop the ones mentioned in this chapter. Although preventive medicine plays an important role, other measures have to be implemented to reduce resistance to antibiotics. In this sense, states should make an effort to legislate for reduction and control of the use of antibiotics in aquaculture, especially those countries that have few regulations related to this topic.

REFERENCES

Andersson, D.I., and B.R. Levin. 1999. The biological cost of antibiotic resistance. *Curr. Opin. Microbiol.* 2:489–493.

Angulo, F. 2000. Antimicrobial agents in aquaculture: potential impact on health. *APUA Newsl.* 18:1–6.

Angulo, F., and P. Griffin. 2000. Changes in antimicrobial resistance in Salmonella enterica serovar typhimurium. *Emerg. Infect. Dis.* 6:436–438.

Angulo, F., V. Nargund, and T. Chiller. 2004. Evidence of an association between use of anti-microbial agents in food animals and anti-microbial resistance among bacteria isolated from humans and the human health consequences of such resistance. *J. Vet. Med. B, Infect. Dis. Vet. Public Heal.* 51:374–379.

Anthérieu, S., A. Rogue, B. Fromenty, A. Guillouzo, and M.A. Robin. 2011. Induction of vesicular steatosis by amiodarone and tetracycline is associated with up-regulation of lipogenic genes in heparg cells. *Hepatology* 53:1895–1905.

Arala-Chaves, M., and T. Sequeira. 2000. Is there any kind of adaptive immunity in invertebrates? *Aquaculture* 191:247–258.

Ardó, L., G. Yin, P. Xu, L. Váradi, G. Szigeti, Z. Jeney, and G. Jeney. 2008. Chinese herbs (Astragalus membranaceus and Lonicera japonica) and boron enhance the non-specific immune response of Nile tilapia (Oreochromis niloticus) and resistance against Aeromonas hydrophila. *Aquaculture* 275:26–33.

Austin, B., and D.A. Austin. 2016. *Bacterial Fish Pathogens. Disease of Farmed and Wild Fish.* 6th edition. Springer International Publishing: London.

Austin, B., A. Pride, and G. Rhodie. 2003. Association of a bacteriophage with virulence in Vibrio harveyi. *J. Fish Dis.* 26:55–58.

Avnimelech, Y. 1999. Carbon/nitrogen ratio as a control element in aquaculture systems. *Aquaculture* 176:227–235.

Ayude-Vázquez, J., M.P. González, and M. Anxo-Murado. 2005. Effects of lactic acid bacteria cultures on pathogenic microbiota from fish. *Aquaculture* 245:149–161.

Balcázar, J.L., I. de Blas, I. Ruiz-Zarzuela, D. Cunningham, D. Vendrell, and J.L. Múzquiz. 2006. The role of probiotics in aquaculture. *Vet. Microbiol.* 114:173–186.

Bousquet, P. 2009. Drug allergy and hypersensitivity: still a hot topic. *Allergy* 64:179–182.

Boxall, A., L. Fogg, P. Blackwell, P. Kay, E. Pemberton, and A. Croxford. 2004. Veterinary medicines in the environment. *Rev. Environ. Contam. Toxicol.* 180:1–91.

Boyd, D., G.A. Peters, A. Cloeckaert, K.S. Boumedine, E. Chaslus-Dancla, H. Imberechts, and M.R. Mulvey. 2001. Complete Nucleotide sequence of a 43-kilobase Genomic Island associated with the multidrug resistance region of. *J. Bacteriol.* 183:5725–5732.

Bricknell, I., and R.A. Dalmo. 2005. The use of immunostimulants in fish larval aquaculture. *Fish Shellfish Immunol.* 19:457–472.

Brock, J., and R. Bullis. 2001. Disease prevention and control for gametes and embryos of fish and marine shrimp. *Aquaculture* 197:137–159.

Broughton, E., and D. Walker. 2009. Prevalence of antibiotic-resistant Salmonella in fish in Guangdong, China. *Foodborne Pathog. Dis.* 6:519–521.

Burridge, L., J.S. Weis, F. Cabello, J. Pizarro, and K. Bostick. 2010. Chemical use in salmon aquaculture: A review of current practices and possible environmental effects. *Aquaculture* 306:7–23.

Cabello, F.C. 2003. Antibióticos y acuicultura. Un análisis de sus potenciales impactos para el medio ambiente y la salud humana y animal Chile. *Análisis Políticas Públicas. Organ. Terram* 17:13–18.

Cabello, F.C. 2004. Antibióticos y acuicultura en Chile: consecuencias para la salud humana y animal. *Rev. Med. Chil.* 132:1001–1006.

Cabello, F.C. 2006. Heavy use of prophylactic antibiotics in aquaculture: A growing problem for human and animal health and for the environment. *Environ. Microbiol.* 8:1137–1144.

Cerezuela, R., A. Cuesta, J. Meseguer, and M. Ángeles Esteban. 2009. Effects of dietary vitamin D3 administration on innate immune parameters of seabream (Sparus aurata L.). *Fish Shellfish Immunol.* 26:243–248.

Cháfer-Pericás, C., Á. Maquieira, R. Puchades, B. Company, J. Miralles, and A. Morena. 2010. Multiresidue determination of antibiotics in aquaculture fish samples by HPLC–MS/MS. *Aquaculture Research* 41:217–225.

Chao, L. 2002. The application of sulfonamides in aquaculture. *Reserv. Fish* 22:50–51.

Chen, H., S. Liu, X.R. Xu, S.S. Liu, G.J. Zhou, K.F. Sun, J.L. Zhao, and G.G. Ying. 2014. Antibiotics in typical marine aquaculture farms surrounding Hailing Island, South China: Occurrence, bioaccumulation and human dietary exposure. *Mar. Pollut. Bull.* 90:181–187.

Choi, W., W.Y. Mo, S. Wu, N. Mak, Z. Bian, X. Nie, and M. Wong. 2013. Effects of traditional Chinese medicines (TCM) on the immune response of grass carp (Ctnopharyngodon idellus). *Aquac. Int.* 22:361–377.

Cogliani, C., H. Goossens, and C. Greko. 2011. Restricting antimicrobial use in food animals: lessons from Europe. *Microbe* 6:274.

Cosgrove, S., K. Saye, G. Eliopoulos, and Y. Carmeli. 2002. Health and economic outcomes of the emergence of third generation cephalosporin resistance in Enterobacter species. *Arch. Intern. Med.* 162:185–90.

Crosbie, P., and B. Nowak. 2004. Immune responses of barramundi, Lates calcarifer (Bloch), after administration of an experimental Vibrio harveyi bacterin by intraperitoneal injection, anal intubation and immersion. *J. Fish Dis.* 27:623–632.

Decamp, O., D.J. Moriarty, and P. Lavens. 2008. Probiotics for shrimp larviculture: review of field data from Asia and Latin America. *Aquac. Res.* 39:334–338.

Defoirdt, T., N. Boon, P. Sorgeloos, W. Verstraete, and P. Bossier. 2007a. Alternatives to antibiotics to control bacterial infections: luminescent vibriosis in aquaculture as an example. *Trends Biotechnol.* 25:472–479.

Defoirdt, T., R. Crab, T.K. Wood, P. Sorgeloos, W. Verstraete, and P. Bossier. 2006. Quorum Sensing-Disrupting Brominated Furanones Protect the Gnotobiotic Brine Shrimp Artemia franciscana from Pathogenic Vibrio harveyi , Vibrio campbellii , and Vibrio parahaemolyticus Isolates †. *Appl. Environ. Microbiol.* 72:6419–6423.

Defoirdt, T., D. Halet, H. Vervaeren, N. Boon, T. Van De Wiele, P. Sorgeloos, P. Bossier, and W. Verstraete. 2007b. The bacterial storage compound poly-β-hydroxybutyrate protects Artemia franciscana from pathogenic Vibrio campbellii. *Environ. Microbiol.* 9:445–452.

Defoirdt, T., P. Sorgeloos, and P. Bossier. 2011. Alternatives to antibiotics for the control of bacterial disease in aquaculture. *Curr. Opin. Microbiol.* 14:251–258.

Dhawan, A., and S. Kaur. 2002. Pig dung as pond manure: effect on water quality, pond productivity and growth of carps in polyculture system. *Naga, ICLARM Q.* 25:11–14.

EMA (European Medicines Agency). 2015. *European Surveillance of Veterinary Antimicrobial Consumption.* https://www.ema.europa.eu/en/veterinary-regulatory/overview/antimicrobial-resistance/european-surveillance-veterinary-antimicrobial-consumption-esvac

FAO. 2014. *The State of World Fisheries and Aquaculture.* Page in Food and Agriculture Organization of the United Nations, Rome (Italy). http://www.fao.org/3/i9540en/i9540en.pdf

Fischetti, V.A., D. Nelson, and S. Raymond. 2006. Reinventing phage therapy: are the parts greater than the sum?. *Nat. Biotechnol.* 24:1508–1511.

Gao, P., D. Mao, Y. Luo, L. Wang, B. Xu, and L. Xu. 2012. Occurrence of sulfonamide and tetracycline-resistant bacteria and resistance genes in aquaculture environment. *Water Res.* 46:2355–2364.

Goldburg, R., and R. Naylor. 2005. Future seascapes, fishing, and fish farming. *Frontiers in Ecology and the Environment.* 3(1):21–28.

González-Pleiter, M., S. Gonzalo, I. Rodea-Palomares, F. Leganés, R. Rosal, K. Boltes, E. Marco, and F. Fernández-Piñas. 2013. Toxicity of five antibiotics and their mixtures towards photosynthetic aquatic organisms: Implications for environmental risk assessment. *Water Res.* 47:2050–2064.

Gudding, R., A. Lillehaug, and Ø. Evensen. 1999. Recent developments in fish vaccinology. *Vet. Immunol. Immunopathol.* 72:203–212.

Gulkowska, A., Y. He, M.K. So, L.W.Y. Yeung, H.W. Leung, J.P. Giesy, P.K.S. Lam, M. Martin, and B.J. Richardson. 2007. The occurrence of selected antibiotics in Hong Kong coastal waters. *Mar. Pollut. Bull.* 54:1287–1293.

Halling-Sørensen, B., S. Nors Nielsen, P.F. Lanzky, F. Ingerslev, H.C. Holten Lützhøft, and S.E. Jørgensen. 1998. Occurrence, fate and effects of pharmaceutical substances in the environment- A review. *Chemosphere* 36:357–393.

He, X., Z. Wang, X. Nie, Y. Yang, D. Pan, A.O.W. Leung, Z. Cheng, Y. Yang, K. Li, and K. Chen. 2012. Residues of fluoroquinolones in marine aquaculture environment of the Pearl River Delta, south China. *Environ. Geochem. Heal.* 34:323–335.

He, Y., Y. Tang, F. Sun, and L. Chen. 2015. Detection and characterization of integrative and conjugative elements (ICEs)-positive Vibrio cholerae isolates from aquacultured shrimp and the environment in Shanghai, China. *Mar. Pollut. Bull.* 101:526–532.

Heuer, O., P. Collignon, I. Karunasagar, H. Kruse, K. Grave, and F. Angulo. 2009. Human health consequences of use of antimicrobial agents in aquaculture. *Clin. Infect. Dis. Off. Publ. Dis. Soc. Am* 49:1253.

Hoa, P.T.P., S. Managaki, N. Nakada, H. Takada, A. Shimizu, D.H. Anh, P.H. Viet, and S. Suzuki. 2011. Antibiotic contamination and occurrence of antibiotic-resistant bacteria in aquatic environments of northern Vietnam. *Sci. Total Environ.* 409:2894–2901.

Hung, D.T., E.A. Shakhnovich, P. Emily, and J.J. Mekalanos. 2005. Small-molecule inhibitor of vibrio cholerae virulence and intestinal colonization. *Science* 310:670–674.

Hunter-Cervera, J., D. Karl, and M. Buckley. 2005. *Marine Microbial Diversity: The Key to Earth's Habitability.* (A report from the American Academy of Microbiology). Page in Marine Microbial Diversity. American Academy of Microbiology, San Francisco (USA).

Hvistendahl, M. 2012. China takes aim at rampant antibiotic resistance. *Science* (80) 336:795.

Immerseel, F. Van, K. Cauwerts, L. Devriese, F. Haesebrouck, and F. Ducatelle. 2002. Feed additives to control Salmonella in poultry. *Worlds. Poult. Sci. J.* 58:201–513.

Issaragrisil, S., D.W. Kaufman, T. Anderson, K. Chansung, P.E. Leaverton, S. Shapiro, and N.S. Young. 2006. The epidemiology of aplastic anemia in Thailand. *Blood* 107:1299–1307.

Jacoby, G.A. 2005. Mechanisms of resistance to Quinolones. *Clin. Infect. Dis.* 41(Supplement 2):120–126.

Jian, J., and Z. Wu. 2003. Effects of traditional Chinese medicine on nonspecific immunity and disease resistance of large yellow croaker, Pseudosciaena crocea (Richardson). *Aquaculture* 218:1–9.

Jiang, L., X. Hu, D. Yin, H. Zhang, and Z. Yu. 2011. Occurrence, distribution and seasonal variation of antibiotics in the Huangpu River, Shanghai, China. *Chemosphere* 82:822–828.

Jiang, Y. 1996. The use of chemicals in aquaculture in the People's Republic of China. Pages 141–154 in *Use of Chemicals in Aquaculture in Asia*. Southeast Asian Fisheries Development Center Aquaculture Department.

Kanagasabhapathy, M., G. Yamazaki, A. Ishida, H. Sasaki, and S. Nagata. 2009. Presence of quorum-sensing inhibitor-like compounds from bacteria isolated from the brown alga Colpomenia sinuosa. *Lett. Appl. Microbiol.* 49:573–579.

Kemper, N. 2008. Veterinary antibiotics in the aquatic and terrestrial environment. *Ecol. Indic.* 8:1–13.

Kerry, J., R. Coyne, D. Gilroy, M. Hiney, and P. Smith. 1996. Spatial distribution of oxytetracycline and elevated frequencies of oxytetracycline resistance in sediments beneath a marine salmon farm following oxytetracycline therapy. *Aquaculture* 145:31–39.

Köllner, B., B. Wasserrab, G. Kotterba, and U. Fischer. 2002. Evaluation of immune functions of rainbow trout (Oncorhynchus mykiss) how can environmental influences be detected?. *Toxicol. Lett.* 131:83–95.

Kümmerer, K. 2009. Antibiotics in the aquatic environment – A review – Part I. *Chemosphere* 75:417–434.

Lalumera, G.M., D. Calamari, P. Galli, S. Castiglioni, G. Crosa, and R. Fanelli. 2004. Preliminary investigation on the environmental occurrence and effects of antibiotics used in aquaculture in Italy. *Chemosphere* 54:661–668.

Li, W., Y. Shi, L. Gao, J. Liu, and Y. Cai. 2012. Occurrence of antibiotics in water, sediments, aquatic plants, and animals from Baiyangdian Lake in North China. *Chemosphere* 89:1307–1315.

Li, Z. 2008. Advantages and disadvantages and strategies of antibiotic application in aquaculture. *Guizhou Anim. Sci. Vet. Med.* 32:23–24.

Lillehaug, A., B.T. Lunestad, and K. Grave. 2003. Epidemiology of bacterial diseases in Norwegian aquaculture – A description based on antibiotic prescription data for the ten-year period 1991 to 2000. *Dis. Aquat. Organ.* 53:115–125.

Lio-Po, G., E. Leaño, M. Peñaranda, A. Villa-Franco, C. Sombito, and J.N. Guanzon. 2005. Anti-luminous Vibrio factors associated with the "green water" grow-out culture of the tiger shrimp Penaeus monodon. *Aquaculture* 250:1–7.

Liu, J.-L., and M.-H. Wong. 2013. Pharmaceuticals and personal care products (PPCPs): A review on environmental contamination in China. *Environ. Int.* 59:208–224.

Liu, X., J.C. Steele, and X.Z. Meng. 2017. Usage, residue, and human health risk of antibiotics in Chinese aquaculture: A review. *Environ. Pollut.* 223:161–169.

Marques, A., J. Dhont, P. Sorgeloos, and P. Bossier. 2006. Immunostimulatory nature of β-glucans and baker's yeast in gnotobiotic Artemia challenge tests. *Fish Shellfish Immunol.* 20:682–692.

Minh, T.B., H.W. Leung, I.H. Loi, W.H. Chan, M.K. So, J.Q. Mao, D. Choi, J.C.W. Lam, G. Zheng, M. Martin, J.H.W. Lee, P.K.S. Lam, and B.J. Richardson. 2009. Antibiotics in the Hong Kong metropolitan area: Ubiquitous distribution and fate in Victoria Harbour. *Mar. Pollut. Bull.* 58:1052–1062.

Mo, W.Y., Z. Chen, H.M. Leung, and A.O.W. Leung. 2015. Application of veterinary antibiotics in China's aquaculture industry and their potential human health risks. *Environ. Sci. Pollut. Res.* 24:8978–8989.

Nakai, T., and S.C. Park. 2002. Bacteriophage therapy of infectious diseases in aquaculture. *Res. Microbiol.* 153:13–18.

Naylor, R., and M. Burke. 2005. Aquaculture and ocean resources: raising tigers of the sea. *Annu. Rev. Environ. Resour.* 30:185–218.

Nygaard, K., B.T. Lunestad, H. Hektoen, J.A. Berge, and V. Hormazabal. 1992. Resistance to oxytetracycline, oxolinic acid and furazolidone in bacteria from marine sediments. *Aquaculture* 104:31–36.

Pereira, A.M.P.T., L.J.G. Silva, L.M. Meisel, and A. Pena. 2015. Fluoroquinolones and Tetracycline Antibiotics in a Portuguese Aquaculture System and Aquatic Surroundings: Occurrence and Environmental Impact. *J. Toxicol. Environ. Heal. – Part A Curr. Issues* 78:959–975.

Pillay, T., and M. Kutty. 2005. *Aquaculture: Principles and Practices.* 2nd edition. Blackwell Publishing Limited, New Jersey, USA.

Rengpipat, S., A. Tunyanun, A.W. Fast, S. Piyatiratitivorakul, and P. Menasveta. 2003. Enhanced growth and resistance to Vibrio challenge in pond-reared black tiger shrimp Penaeus monodon fed a Bacillus probiotic. *Dis. Aquat. Organ.* 55:169–173.

Rico, A., T.M. Phu, K. Satapornvanit, J. Min, A.M. Shahabuddin, P.J.G. Henriksson, F.J. Murray, D.C. Little, A. Dalsgaard, and P.J. Van den Brink. 2013. Use of veterinary medicines, feed additives and probiotics in four major internationally traded aquaculture species farmed in Asia. *Aquaculture* 412–413:231–243.

Rico, A., K. Satapornvanit, M.M. Haque, J. Min, P.T. Nguyen, T.C. Telfer, and P.J. Van den Brink. 2012. Use of chemicals and biological products in Asian aquaculture and their potential environmental risks: a critical review. *Aquaculture* 4:75–93.

Ruwandeepika, D.H., T. Defoirdt, P.P. Bhowmick, I. Karunasagar, I. Karunasagar, and P. Bossier. 2011. In vitro and in vivo expression of virulence genes in Vibrio isolates belonging to the Harveyi clade in relation to their virulence towards gnotobiotic brine shrimp (Artemia franciscana). *Environ. Microbiol.* 13:506–517.

Salyers, A., A. Gupta, and Y. Wang. 2004. Human intestinal bacteria as reservoirs for antibiotic resistance genes.. *Trends Microbiol.* 12:412–416.

Sapkota, A., A. Sapkota, M. Kucharski, J. Burke, S. Mckenzie, P. Walker, and R. Lawrence. 2008. Aquaculture practices and potential human health risks: current knowledge and future priorities. *Environ. Int* 34:1215–1226.

Sarmah, A.K., M.T. Meyer, and A.B.A. Boxall. 2006. A global perspective on the use, sales, exposure pathways, occurrence, fate and effects of veterinary antibiotics (VAs) in the environment. *Chemosphere* 65:725–759.

Seoane, M., C. Rioboo, C. Herrero, and Á. Cid. 2014. Toxicity induced by three antibiotics commonly used in aquaculture on the marine microalga Tetraselmis suecica (Kylin) Butch. *Mar. Environ. Res.* 101:1–7.

Smith, P. 2008. Antimicrobial resistance: The use of Antimicrobials in the Livestock aector. *Rev. Sci. Tech.* 27:243–264.

Smith, V., J. Brown, and C. Hauton. 2003. Immunostimulation in crustaceans: does it really protect against infection. *Fish Shellfish Immunol.* 15:71–90.

Solensky, R. 2012. Allergy to β-lactam antibiotics. *J. Allergy Clin. Immunol.* 130:1–5.

Son, V.M., C.-C. Chang, M.-C. Wu, Y.-K. Guu, C.-H. Chiu, and W. Cheng. 2009. Dietary administration of the probiotic, Lactobacillus plantarum, enhanced the growth, innate immune responses, and disease resistance of the grouper Epinephelus coioides. *Fish Shellfish Immunol.* 26:691–698.

Sørum, H. 2006. *Antimicrobial Drug Resistance in Fish Pathogens.* American Society for Microbiology Press, Washington, DC.

Sørum, H., and T.M. L'Abée-Lund. 2002. Antibiotic resistance in food-related bacteria—a result of interfering with the global web of bacterial genetics. *Int. J. Food Microbiol.* 78:43–56.

Stenholm, A.R., I. Dalgaard, and M. Middelboe. 2008. Isolation and characterization of bacteriophages infecting the fish pathogen Flavobacterium psychrophilum. *Appl. Environ. Microbiol.* 74:4070–4078. doi:10.1128/AEM.00428-08

Su, H.C., G.G. Ying, R. Tao, R.Q. Zhang, L.R. Fogarty, and K.D. W. 2011. Occurrence of antibiotic resistance and characterization of resistance genes and integrons in Enterobacteriaceae isolated from integrated fish farms in South China. *J. Environ. Monit.* 13:3229–3236.

Teasdale, M.E., J. Liu, J. Wallace, F. Akhlaghi, and D.C. Rowley. 2009. Secondary metabolites produced by the Marine Bacterium. *Appl. Environ. Microbiol.* 75:567–572.

Tendencia, E.A., and M. De la Peña. 2003. Investigation of some components of the greenwater system which makes it effective in the initial control of luminous bacteria. *Aquaculture* 218:115–119.

Tian, F. 2010. Discussions on the disease prevention of aquatic animals and problems of fishery drug residues. *Jiangxi Fish. Sci. Technol.* 3:25–39.

Tinh, N., V. Yen, K. Dierckens, P. Sorgeloos, and P. Bossier. 2008. An acyl homoserine lactone-degrading microbial community improves the survival of first-feeding turbot larvae (Scophthalmus maxiums L.). *Aquaculture* 285:59–62.

Vaseeharan, B., and R. Thaya. 2013. Medicinal plant derivatives as immunostimulants: An alternative to chemotherapeutics and antibiotics in aquaculture. *Aquac. Int.* 22:1079–1091.

Verschuere, L., G. Rombaut, P. Sorgeloos, and W. Verstraete. 2000. Probiotic bacteria as biological control agents in aquaculture. *Microbiol. Mol. Biol. Rev.* 64:655–671.

Vine, N.G., W.D. Leukes, and H. Kaiser. 2006. Probiotics in marine larviculture. *FEMS Microbiol. Rev.* 30:404–427.

Vinod, M.G., M.M. Shivu, K.R. Umesha, B.C. Rajeeva, G. Krohne, I. Karunasagar, and I. Karunasagar. 2006. Isolation of Vibrio harveyi bacteriophage with a potential for biocontrol of luminous vibriosis in hatchery environments. *Aquaculture* 255:117–124.

Wang, S., and L. Xiong. 2007. Detriment and control of fishery drug residue. *China Anim. Heal.* 6:58–60.

Wu, H., W. Yin, J. Shao, and L. Xiang. 2003. Depletion role of ciprofloxacin residues in eel (Anguilla japonica) tissues. *Bull. Sci. Technol.* 19:448–451.

Xiong, W., Y. Sun, T. Zhang, X. Ding, Y. Li, M. Wang, and Z. Zeng. 2015. Antibiotics, Antibiotic Resistance Genes, and Bacterial community composition in fresh water aquaculture environment in China. *Microb. Ecol.* 70:425–432.

Xu, W., W. Yan, X. Li, Y. Zou, X. Chen, W. Huang, L. Miao, R. Zhang, G. Zhang, and S. Zou. 2013. Antibiotics in riverine runoff of the Pearl River Delta and Pearl River Estuary, China: Concentrations, mass loading and ecological risks. *Environ. Pollut.* 182:402–407.

Xu, W., X. Zhu, X. Wang, L. Deng, and G. Zhang. 2006. Residues of enrofloxacin, furazolidone and their metabolites in Nile tilapia (Oreochromis niloticus). *Aquaculture* 254:1–8.

Xue, B., R. Zhang, Y. Wang, X. Liu, J. Li, and G. Zhang. 2013. Antibiotic contamination in a typical developing city in south China: Occurrence and ecological risks in the Yongjiang River impacted by tributary discharge and anthropogenic activities. *Ecotoxicol. Environ. Saf.* 92:229–236.

Yan, C., Y. Yang, J. Zhou, M. Liu, M. Nie, H. Shi, and L. Gu. 2013. Antibiotics in the surface water of the Yangtze Estuary: Occurrence, distribution and risk assessment. *Environ. Pollut.* 175:22–29.

Yan, L., and X. Cai. 2004. Application and announcement of antibiotics in aquaculture. *Feed Res.* 4:35–37.

Yin, G., L. Ardó, K.D. Thompson, A. Adams, Z. Jeney, and G. Jeney. 2009. Chinese herbs (Astragalus radix and Ganoderma lucidum) enhance immune response of carp, Cyprinus carpio, and protection against Aeromonas hydrophila. *Fish Shellfish Immunol.* 26:140–145.

Yin, G., G. Jeney, T. Racz, P. Xu, X. Jun, and Z. Jeney. 2006. Effect of two Chinese herbs (Astragalus radix and Scutellaria radix) on non-specific immune response of tilapia, Oreochromis niloticus. *Aquaculture* 253:39–47.

Yu, J., J. Lei, H. Yu, X. Cai, and G. Zou. 2004. Chemical composition and antimicrobial activity of the essential oil of Scutellaria barbata. *Phytochemistry* 65:881–884.

Yuan, X., and W. Chen. 2012. Use of veterinary medicines in Chinese aquaculture: current status. In *Improving Biosecurity through Prudent and Responsible Use of Veterinary Medicines in Aquatic Food Production*, 547, 51-67.

Zhang, R., J. Tang, J. Li, Q. Zheng, D. Liu, Y. Chen, Y. Zou, X. Chen, C. Luo, and G. Zhang. 2013. Antibiotics in the offshore waters of the Bohai Sea and the Yellow Sea in China: Occurrence, distribution and ecological risks. *Environ. Pollut.* 174:71–77.

Zhang, X.X., T. Zhang, and H.H. Fang. 2009. Antibiotic resistance genesin water environment. *Appl. Microbiol. Biotechnol.* 82:397–414.

Zhang, Z. 2004. Antibiotics and antimicrobials phylogeny. *Cap. Med.* 2:14–16.

Zhao, L., Y.H. Dong, and H. Wang. 2010. Residues of veterinary antibiotics in manures from feedlot livestock in eight provinces of China. *Sci. Total Environ.* 408:1069–1075.

Zheng, B., and S. Su. 2010. Status, effects and strategies of antibiotic residues *Chinese. J. Anal. Lab.* 29:287.

Zheng, Q., R. Zhang, Y. Wang, X. Pan, J. Tang, and G. Zhang. 2012. Occurrence and distribution of antibiotics in the Beibu Gulf, China: Impacts of river discharge and aquaculture activities. *Mar. Environ. Res.* 78:26–33.

Zou, S., W. Xu, R. Zhang, J. Tang, Y. Chen, and G. Zhang. 2011. Occurrence and distribution of antibiotics in coastal water of the Bohai Bay, China: Impacts of river discharge and aquaculture activities. *Environ. Pollut.* 159:2913–2920.

2 Use of Alternative Ingredients and Probiotics in Aquafeeds Formulation

Jorge Olmos and Victor Mercado

CONTENTS

INTRODUCTION

During the past sixty years, the increase in seafood consumption had been as much as 3.2% annually; this percentage is greater than the average annual population growth rate of 1.6%, and it has exceeded the meat consumption rate of terrestrial animals (World Bank 2013). Currently, the number of species for human consumption produced by aquaculture is similar to that produced from the seas; in this sense, aquaculture has been growing 8.8% annually during the last years (FAO 2018). Therefore, species with great commercial value have attracted attention to the development of a more profitable activity (Tacon and Metian 2008). However, successful production and commercialization of these species will depend on highly digestible, economically viable, ecologically sustainable, and environmentally friendly aquafeed formulations (Olmos and Paniagua 2014, 2020).

DOI: 10.1201/9780429331664-2

Fishmeal and fish oil have represented an excellent protein and energy source for aquafeeds formulation for more than 30 years, especially for the aquaculture of carnivorous species (Naylor et al. 2009). However, as the demand for these ingredients for aquafeeds and other livestock activities has increased, so has their cost, making its utilization in aquafeeds unprofitable (Tacon and Metian 2008). Additionally, alterations in the food chain have occurred due to the overfishing of pelagic species; moreover, fishmeal and fish oil could be a considerable source of pollution due to their high nitrogen and phosphate content. Therefore, the aquafeeds production industry is searching for alternative ingredients like (1) shipper animal byproducts, (2) highly digestible microorganism and insect biomass, and (3) abundant vegetable ingredients, to avoid these inconveniences. In this sense, the aquaculture industry's main objective is to produce aquafeeds with smaller amounts of fish ingredients; hence, it is important to develop new formulations using alternative and inexpensive ingredients. However, the alternative ingredients mentioned above (1) do not contain enough nutrient concentration and could be the source of pathogens, (2) are not cost-effective for aquafeeds formulation, and (3) all contain complex macromolecules difficult to digest by the enzymes produced for aquacultured animals, respectively. Therefore, to improve the production of carnivorous species, it is necessary to include probiotic bacteria in aquafeed formulations (Olmos 2017). Innocuous *Bacillus*, a spore-forming, multifunctional probiotic bacterium, can be added to aquafeeds in order to improve aquaculture profitability and sustainability (Olmos and Paniagua 2014).

ANIMAL BYPRODUCTS, MICROORGANISM BIOMASS, AND PLANT INGREDIENTS AS SOURCES FOR AQUAFEED DEVELOPMENT

Fish meal (FM) and fish oil (FO) are indispensable ingredients in aquafeeds formulation, as they contribute essential amino acids and lipids for the growth of marine organisms. However, the increasing demand for FM and FO has produced the overexploitation of small pelagic species, inducing an ecological disorder and increasing FM and FO prices (Indexmundi.com 2020b). The increasing costs of fish ingredients have induced the search for more economical alternatives, which could satisfy the nutritional demand of aquacultured animals. Therefore, aquafeed producers are evaluating byproducts from the livestock, poultry, and fishing industries, as well as carbohydrates, protein, and lipids from plants, insects, and phytoplankton sources. But byproducts and alternative ingredients have their own advantages and disadvantages in formulating digestible, economic, sustainable, and healthy aquafeeds. This chapter will describe several sources of carbohydrates, protein, and lipids that are being used as alternative ingredients to FM and FO, as well as their performance in aquacultured animals.

Animal Byproducts as Protein and Lipid Sources for Aquafeed Formulation

The Association of American Feed Control Officials (AAFCO) defines animal byproducts as secondary products obtained from a finished product that can be used to feed other animals (AAFCO 2012). Byproducts are not for human consumption; instead they are processed to produce feed formulations for livestock and aquaculture.

Poultry Byproducts (PBP)

The poultry industry produces around 75 million tons of product for human consumption each year (Indexmundi.com 2020a). In addition, this industry produces a considerable amount of PBP, which are then used to produce animal feeds. Feathers, guts, heads, legs, and egg byproducts are mostly used to obtain protein and oil; however, these byproducts must be free of pathogens to produce safe feeds and avoid animal diseases. PBP can be divided into gut and feather meals (Table 2.1); however, their nutritional quality depends on the raw materials and the methodology used to obtain them (Dozier and Dale 2005).

TABLE 2.1
Proximate Composition (%) of Guts and Feather Meal (feedtables.com)

Composition (%)	Guts	Feather Meal
Protein	57.5	85.5
Lipids	30.8	9.2
Ash	10.9	5.9
Energy (kcal/kg)	6070	5620
SFA	6.68	2.2
MUFA	10.42	2.07
PUFA	4.68	0.25
Ω-6	4.12	0.23
Ω-3	0.56	0.02
Essential Amino Acids		
Phenylalanine	2.26	4.06
Histidine	1.03	0.72
Isoleucine	2.31	4.23
Leucine	4.01	6.84
Lysine	2.64	1.83
Methionine	0.81	0.58
Threonine	2.3	3.98
Tryptophan	0.50	0.51
Valine	3.16	6.26

PBP contains a great protein profile; however, it does not contain lysine and methionine, important amino acids for the development of marine species (Castillo-Lopez et al. 2016). With respect to the lipid profile, PBP is rich in saturated fatty acids, palmitic acid (16:0) being the predominant molecule. However, monounsaturated fatty acids (MUFA), like oleic acid [18:1(Ω-9)], and polyunsaturated fatty acids (PUFA), like linoleic [18:2(Ω-6)] and α-linolenic [18:3(Ω-3)], are present in much lower concentrations. In this sense, fishes like *Centropristis striata*, *Totoaba macdonaldi*, and *Oncorhynchus mykiss* which have been fed high levels of PBP present an altered fatty acid profile (Table 2.2). In addition, PBP do not contain Ω-3 fatty acids, like eicosapentaenoic [EPA; 20:5(Ω-3)] and docosahexaenoic [DHA; 22:6(Ω-3)] acids; this lack negatively affects the development of young species (Siddik et al. 2019; Higgs et al. 2006). Furthermore, fishes cannot transform linoleic and α-linolenic acid to EPA and DHA because they lack the desaturase Δ5 enzyme (Tocher 2015).

Complete substitution of FM by PBP is not recommended for the majority of aquaculture species because aquafeeds formulated with high levels of PBP require supplementation of essential amino acids and Ω-3 fatty acids.

In addition, feather meal presents a lower protein bioavailability due to its high keratin concentration, which cannot be digested by most aquatic animals. While thermal or enzymatic procedures can produce assimilable peptides from this ingredient, these treatments can reduce its nutritional value. Likewise, food safety control is important because PBP might contain pathogens, like *Listeria monocytogenes* and *Salmonella enterica*, which can be harmful to animals and humans.

Livestock Byproducts (LBP)

The livestock industry produces around 25 million tons of byproducts (LBP) each year in the United States and around 15 million tons in Europe (Hamilton 2016). These byproducts are composed of

TABLE 2.2
Maximum Substitution of FM and PBP Inclusion (%) in Aquafeeds and the Effect Caused in Different Aquatic Species

Organism	Maximum Substitution (%)	PBP in Test Feed (%)	FM in Control Feed (%)	Crude Protein (%)	Crude Lipids (%)	Other Sources of Protein (%)	Observations	Reference
Sparus aurata L.	50	29.20	72.90	45	13	N/A	Substitution higher than 50% causes growth and feed efficiency reduction.	Nengas et al. 1999
Sunshine Bass (*Morone chrysops X M. saxatilis*)	100	30.0	30.0	42	6	SBM – 30	SBM and PBP can replace FM without causing any adverse effect in the growth, survival, and body composition of the fishes.	Webster et al. 2000
Macrobrachium nipponense	50	29.82	57.0	38	9	N/A	Higher growth ratio compared to fishes fed with FM or MBM diets. No negative effect shown on the immunological parameters.	Yang et al. 2004
Rachycentron canadum	45	22.65	50.0	45	11	SBM – 11.30	No significant difference in weight gain, but feed and protein efficiency were lower compared to the control feed.	Zhou et al. 2011
Oncorhynchus mykiss	44	44.0	66.0	43.5	12.5	CM – 5.5	High content of PBP in feeds caused a reduction in fillet EPA and DHA content.	Parés-Sierra et al. 2012
Trachinotus carolinus L.	10	4.90	N/A	40	10	SBM – 50 CG – 5	Growth parameters reduced by the decrease of FM and the lack of taurine in the diet.	Rossi and Davis 2012
Totoaba macdonaldi	67	45.0	65.2	51.6	8.1	CM – 5.5	Weight increased by 1902.3% after 86 days but decreased the Ω-3 and Ω-6 content.	Zapata et al. 2016
Centropristis striata	90	56.4	70	44	13	WG – 4	No significant difference in growth, survival, or feed utilization, but Ω-3, EPA, and DHA decreased.	Dawson et al. 2018
Dicentrarchus Labrax	60	28.41	47.35	48	14	SBM – 10.98	Substitution of FM by 60% did not affect growth parameters but the mortality ratio increased	Srour et al. 2016

CM: corn meal; FM: fish meal; MBM: meat and bone meal; SBM: soybean meal; CG: corn gluten; WG: wheat gluten; EPA: eicosatetraenoic acid; DHA: docosahexaenoic acid.

blood, bones, skin, adipose tissue, horns, and hooves that come from porcine, caprine, as well as bovine, cattle. However, the use of LBP in aquafeeds is limited, due to normative practice and laws imposed by several countries. Bovine spongiform encephalopathy (BSE) is a disease induced by feeding ruminants with raw livestock byproducts. In Europe, the European Commission banned the use of LBP (EC No. 999/2001) to avoid BSE propagation on the continent. Likewise, the Food and Drug Administration (FDA; 21 CFR Part 589.2001) restricted the use of LBP to feed ruminants. However, the EC recently modified restrictions and allowed LBP utilization in aquafeeds (EU Commission Regulation, EC No. 56/2013).

Livestock byproducts used in feed formulations can be divided into two types: meat-bone meal (MBM) and blood meal (BM) (Table 2.3). These byproducts present high protein content, and they are rich in fatty acids, like linoleic [18:2(Ω-6)], myristic (14:0), and palmitic acid (16:0). However, their methionine, tryptophan, and leucine content are lower than in FM. In addition, they lack essential PUFAs and have a considerable ash content that compromises marine animals' development.

Aquafeeds formulated with LBP have been used to feed several marine animals (Table 2.4); however, the lack of essential amino acids and Ω-3 fatty acids, like EPA and DHA, limits LBP addition. Furthermore, its high amount of ash is difficult for the intestine to absorb, decreasing the apparent digestibility coefficient (Xavier et al. 2014). Thus, the addition of high levels of LBP induces negative effects in aquatic animals. In addition to high ash and low Ω-3 fatty acid content, high SFA (saturated fatty acid) concentration inhibits feed uptake by affecting the feed's palatability (Tang et al. 2016; Millamena 2002). Nevertheless, some negative effects can be solved by adding Ω-3 fatty acids to the formulation.

In addition, LBP requires a pretreatment to avoid the propagation of diseases like BSE. Likewise, they could contain pathogen microorganisms like *Escherichia coli*, *Salmonella* spp., *Shigella*, spp., *Vibrio* spp., *Pseudomonas aeruginosa*, *Staphylococcus aureus*, *and Yersinia enterocolitica*. Thermal treatments are required to eliminate them; however, these treatments can produce amino

TABLE 2.3
Proximate Composition (%) of MBM and BM (feedtables.com)

Composition (%)	**Meat and Bone Meal (Pork)**	**Blood Meal**
Protein	56.93	93.9
Lipids	13.6	2
Ash	26.8	3.3
Energy (kcal/kg)	4400	5750
SFA	4.37	0.753
MUFA	4.71	0.6
PUFA	1.03	0.053
PUFA Ω-6	0.95	0.043
PUFA Ω-3	0.07	0.01
Essential Amino Acids		
Phenylalanine	1.93	6.48
Histidine	1.14	5.92
Isoleucine	1.75	1.17
Leucine	3.51	1.15
Lysine	2.98	8.17
Methionine	0.82	1.08
Threonine	1.93	4.35
Tryptophan	0.38	1.43
Valine	2.46	0.80

TABLE 2.4
Maximum Substitution of FM and LBP Inclusion (%) in Aquafeeds and the Effect Caused in Different Aquatic Species

Organism	Maximum Substitution (%)	LBP Content in Test Feeds (%)	FM Content in Control Feed (%)	Crude Protein (%)	Crude Lipids (%)	Other Sources of Protein (%)	Observations	Reference
Epinephelus coioides	80	40	40	45	12	SBM – 6 SM – 10 SqM – 1	The high content of SFA negatively affects the palatability of the diet, resulting in reduced feed uptake and growth	Millamena, 2002
Pseudobagrus ussuriensis	40	27.6	45	43	7.4	SBM – 6 Yst – 3 CG – 8	The increase of LBP in the diet caused a decrease in total protein and lipids and an increase of phosphate in the fillet	Tang et al. 2016
Sparus aurata	50	40.9	57.4	53	15	WM – 17 SBM – 8	Similar growth compared to the FM diet without affecting growth parameters or aminoacidic profile	Moutinho et al. 2017
Litopenaeus vannamei	60	29.0	40	41	8	SBM – 25 SM – 4 Yst 3	The feed conversion ratio and carcass composition were not significantly affected by dietary treatments	Tan et al. 2005
Oreochromis niloticus	50	12.5	36.6	89.47	0.62	SBM – 11.47 CSM – 27.07 WM – 14.49	FM substitution with BM caused the reduction of methionine, lysine, isoleucine, leucine, proline, and valine in the fishes	Kirimi and Musalia 2016

BM: blood meal; CM: Corn meal; FM: Fish meal; SBM: Soybean meal; CG: Corn gluten; SFA: saturated fatty acids; SqM: Squid meal; Yst: Yeast; CSM: Cottonseed meal.

acid and lipid degradation in LBP. Therefore, LBP requires the supplementation of EPA, DHA, and essential amino acids to be included in aquafeeds.

Fishery Byproducts (FBP)

The fishery industry is one of the fastest-growing industries in the world. In the last 30 years, fishing has increased by 14% while aquaculture activity has increased by 541% (FAO 2020). Fisheries produce over 199,921,652.16 tons of marine products each year around the world (Indexmundi.com 2020d). Thus, this industry produces high amounts of fish byproducts (FBP) like heads, guts, gills, and dark muscle; in addition, shrimp and squid wastes, which are not used for human consumption, also are used for aquafeed production.

Fish byproducts are obtained after the production of products for human consumption (Cashion et al. 2017). Unlike PBPs and LBPs, FBPs do not affect aquafeeds palatability; on the contrary, they stimulate its intake. Additionally, FBP contains essential amino acids and Ω-3 PUFAs required for animal growth (Table 2.5); however, its protein concentration is lower than FM due to the absence of muscular tissue.

In recent years, the substitution of FM with FBP has been analyzed (Table 2.6); however, in most cases, FBP must be supplemented with protein to satisfy marine animal requirements. In addition, FBP needs to be carefully analyzed in order to eliminate pathogens and avoid animal and human diseases. Another important factor that must be analyzed in FBP, before its use in aquafeeds production, is the content of heavy metals, like mercury and cadmium. High levels of these metals could limit the FBP addition due to legal restrictions in most countries.

TABLE 2.5
Proximate Composition (%) of Tuna Byproducts, Shrimp Head, Squid Byproducts, Scallop Byproducts

Composition (%)	Tuna Byproducts	Shrimp Heads	Squid Byproducts	Scallop Byproducts
Protein	68.3	37.1	40.6	50.2
Lipids	12.06	1.7	15.9	14.5
Ashes	17.4	45.3	6.6	22.2
Energy (kcal/kg)	N/A	2915.87	4804.01	4400
SFA	5.00	1.32	4.9	N/A
MUFA	2.45	0.99	4.45	N/A
PUFA	3.60	1.48	6.55	N/A
Ω-6	0.332	0.4	0.81	N/A
Ω-3	3.27	1.08	5.28	N/A
Essential Amino Acids				
Phenylalanine	2.88	0.5	0.87	1.19
Histidine	2.84	0.81	1.02	0.42
Isoleucine	3.19	0.93	0.97	1.68
Leucine	5.15	1.69	2.87	2.49
Lysine	5.36	2.95	2.49	2.57
Methionine	0.37	1.80	0.30	1.04
Threonine	3.15	1.75	1.73	0.99
Tryptophan	N/A	N/A	N/A	N/A
Valine	3.61	17.9	1.55	1.56
References	Kim et al. 2019	Liu et al. 2013; Feedtables.com	Liu et al. 2013; Adlercreutz and Lyberg 2008	Terrazas-Fierro et al. 2010

Microorganisms as Carbohydrate, Protein, and Lipid Sources for Aquafeeds Formulation

With the increasing demand for alternative protein sources for feeds formulation, the unicellular protein (also known as single-cell protein [SCP]) obtained from yeast, filamentous fungi, as well as microalgae, has emerged as a great possibility. SCPs contain high levels of amino acids, which can supply the nutritional demand of fishes and crustaceans. However, to use SCP in aquafeeds, first they must satisfy the GRAS classification suggested by the FDA, which assures that SCP can be consumed safely by animals and humans (Bajpai 2017).

Yeast Products

Yeasts are eukaryotic unicellular organisms that can be used as a protein source, as well as an immunostimulant in animals (Øverland and Skrede 2016). In this sense, β-glucans, mannoproteins, chitin, and nucleic acids, among others, induce an increased response of the innate immune system (Ritala et al. 2017; Øverland and Skrede 2016). In addition to the immunostimulatory properties, yeasts like *Candida utilis*, *Kluyveromyces marxianus*, and *Saccharomyces cerevisiae* are rich in essential amino acids, like methionine and lysine, as well as palmitic (16:0), stearic (18:0), oleic [18:1(Ω-9)], linoleic [18:2(Ω-6)], and α-linolenic acids [18:3(Ω-3)] (Table 2.7). Therefore, yeast could be considered a promising alternative in the production of aquafeeds.

Although yeast presents an excellent amino acid profile, its high saturated fatty acids (SFA), monounsaturated fatty acids (MFA), and nucleic acid content can produce negative effects in feed palatability, reducing animal growth. Besides, yeast produces neither EPA nor DHA; thus, it is necessary to add Ω-3 to aquafeeds formulation when yeast is the main protein source. The use of yeast-based aquafeeds has been analyzed in different aquatic animals (Table 2.8); in this sense, results demonstrated that 45% of FM can be substituted with yeast SCP.

Although the use of yeast SCP in aquafeeds seems to be a good alternative, this can produce negative effects by increasing ash and nitrogen content in the meat; also, yeast can increase the production of uric acid in animals, due to its high nucleic acid concentration. Likewise, it can reduce feed digestibility due to the complex composition of the yeast cell wall and fish enzymes' inability to digest them. Another negative factor is the high level of SCP needed to produce aquafeeds; in this sense, it is important to formulate inexpensive culture media to produce high biomass levels and compete with less expensive protein sources.

Filamentous Fungi Products

Filamentous fungi have been previously considered as a protein alternative for livestock; however, their application in aquaculture is quite new. These fungi can produce protein using vegetal residues obtained from palm oil, starch extraction byproducts, vinasse, and lignocellulose. Therefore, filamentous fungi represent an economical alternative for obtaining unicellular protein to produce aquafeeds (Karimi et al. 2018). Fungi like *Paecilomyces variotii*, *Fusarium venenatum*, *Aspergillus oryzae*, *Neurospora intermedia*, and *Rhizopus oryzae* have been used in livestock feed production due to their high protein content (Table 2.9). Likewise, these fungi contain a high palmitic acid (16:0), oleic acid (18:1), and linolelaidic acid [*trans 18*:2(Ω-6)] concentration. However, they lack Ω-3 fatty acids and do not synthesize EPA or DHA; in addition, their cell wall is composed of complex polysaccharides, like α (1–3)-glucans and chitin. Therefore, they cannot be degraded by aquatic organisms due to a lack of specialized enzymes, making them anti-nutritional factors that could negatively affect animal health.

For these reasons, the use of filamentous fungi in aquafeeds production is scarce, despite their high protein content. However, a few known examples reported a growth increment of 12% compared to the control diet formulated with FM, when *Paecilomyces variotii* and *Fusarium venenatum* biomass were used in *O. niloticus* (Alriksson et al. 2014). Nevertheless, more studies are needed about the inclusion of filamentous fungi biomass in aquafeeds to produce new knowledge about the utilization of these kinds of microorganisms in aquatic species.

TABLE 2.6
Maximum Substitution of FM and FBP Inclusion (%) in Aquafeeds and the Effect Caused in Different Aquatic Species

Organism	Byproduct Source	Maximum Substitution (%)	FBP Content in Test Feeds (%)	FM Content in Control Feed (%)	Crude Protein (%)	Crude Lipids (%)	Other Sources of Protein (%)	Observations	Reference
Litopenaeus vannamei	Whitefish	100	16.3	14.7	69.3	7.6	N/A	Final weight increased by 10 times compared to the initial weight	Foster et al. 2004
Sebastes schlegelii	Tuna	75	77.5	64.8	52	10	N/A	Content of FBP higher than 75% caused a reduction in feed uptake and growth	Kim et al. 2018
Oreochromis niloticus	Tuna	100	70	N/A	50.64	9.23	N/A	The substitution did not cause any negative effect on growth; mercury and cadmium levels were low	Kim et al. 2019
Paralichthys olivaceus	Tuna	12.5	9.46	63.81	55.51	10.35	N/A	Reduced growth and feed uptake when fed with higher amounts of fermented FBP	Oncul et al. 2018
Litopenaeus vannamei	Squid Shrimp Scallop	30	30	33.6	37	9	SBM – 20	No negative effect on growth or the aminoacidic profile	Terrazas-Fierro et al. 2010

SBM: soybean meal

TABLE 2.7
Proximate Composition (%) of Yeast Single Cell Protein

Composition (%)	*C. utilis*	*K. marxianus*	*S. cerevisiae*
Protein	56.0	51.1	46.0
Nucleic acid	9.3	10.2	5.8
Lipids	0.3	0.8	0.2
Ash	0.54	0.8	1.1
Energy (kcal/kg)	5,111.30	4,896.34	4,681.37
SFA	0.16	0.56	2.13
MUFA	0.09	0.11	0.79
PUFA	0.10	0.10	0.09
Ω-6	0.09	0.095	0
Ω-3	0.012	0.00392	0
Essential Amino Acids			
Phenylalanine	4.16	3.71	3.29
Histidine	1.96	1.71	1.82
Isoleucine	4.22	3.90	3.41
Leucine	6.64	6.05	5.44
Lysine	6.95	6.51	6.19
Methionine	1.36	1.54	1.23
Threonine	3.94	4.47	4.23
Tryptophan	1.08	0.92	1.08
Valine	4.51	4.04	3.96
References	Øverland et al. 2013; Rodríguez et al. 2012	Øverland et al. 2013; Mejía-Barajas et al. 2018	Øverland et al. 2013; feedtable.com

Microalgae Products

The great interest of using microalgae in the biofuel and pharmaceutical industries is due to their high lipid content as well as their high levels of bioactive compounds, like carotenoids and vitamins. Additionally, the use of microalgae biomass in aquaculture has risen due to their high protein content, which can reach 55% of their total weight. Microalgae also present a high PUFA concentration, containing more Ω-3 fatty acids than fish or vegetable oils (Olmos 2017). Furthermore, they contain high levels of EPA and DHA, as well as large amounts of minerals like Na, K, Ca, Mg, Fe, P, Zn, Mn, Se, Cr, and Cu. In this sense, microalgae are an essential nutrient source for humans and aquacultured animals; however, protein and lipid levels depend on species, strain, and growth conditions, as well as growth media composition, pH, aeration, luminous intensity, temperature, and culture age (Tonon et al. 2002). Species like *Pavlova* spp., *Nannochloropsis gaditana*, *Crypthecodinium cohnii*, *Spirulina platensis*, *Chlorella vulgaris*, and *Isochrysis galbana* are the most used in aquafeeds due to their nutritional value (Table 2.10). Also, these species are classified as GRAS by the FDA, and their intake by animals and humans is relatively safe.

Species from the genus *Pavlova* are of great interest to producers of aquafeeds because of their high EPA and DHA content (Tibbetts et al. 2019). Other microalgae produce significant levels of palmitic (16:0) and oleic acid [18:1(Ω-9)], but produce low levels of EPA and DHA (Tokuşoglu and Ünal 2003). Microalgae biomass has been used to feed aquatic animals like shrimp larvae, bivalves, and krill. However, for larger organisms, such as fish and shrimp, microalgae are mostly used as oil substitutes; nevertheless, there are some studies where microalgae substituted FM in aquafeeds (Table 2.11). Results have shown an increase in animal growth when microalgae biomass or its Ω-3 fatty acids were used in aquafeeds production.

TABLE 2.8
Maximum Substitution of FM and Yeast SCP Inclusion (%) in Aquafeeds and the Effect Caused in Different Aquatic Species

Organism	SCP Source	Maximum SCP Substitution (%)	SCP Content in Test Feeds (%)	FM Content in Control Feed (%)	Crude Protein (%)	Crude Lipids (%)	Other Sources of Protein (%)	Observations	Reference
Oreochromis niloticus	*S. cerevisiae*	15	16	25	27	10	SBM – 15 to 28	High concentration of yeast increases the ash content in fishes	Ozorio et al. 2012
Salmo salar	*C. utilis*	40	28.3	57.9	47	10.5	N/A	No significant difference in the digestibility or nitrogen retention compared to the control feed	Øverland et al. 2013
Salmo salar	*K. marxianus*	40	30.2	57.9	46.6	12.4	N/A	No significant difference in the digestibility or nitrogen retention compared to the control feed	Øverland et al. 2013
Salmo salar	*S. cerevisiae*	40	34.5	57.9	47.5	20.2	N/A	Decreased digestibility of protein and reduced growth compared to the control feed	Øverland et al. 2013
Dicentrarchus labrax	*S. cerevisiae*	30	32.9	59.6	49	15.6	N/A	Increased feed efficiency compared to the control feed	Oliva-Teles and Gonçalves 2001
Oreochromis niloticus, Linnaeus	*K. marxianus*	N/A	15	10	28	7	SBM – 38.7 to 43.7	Lower final weight but increased protein content compared to the control feed	Ribeiro et al. 2012
Thai Panga	*S. cerevisiae*	45	13.5	30	32	7.3	SBM – 36.5	Increased lysozyme and immunoglobulin production compared to the control feed	Pongpet et al. 2015

SBM: Soybean meal

TABLE 2.9
Proximate Composition (%) of Filamentous Fungi Single Cell Protein

Composition (%)	*A. oryzae*	*N. intermedia*	*R. oryzae*	*P. variotii*	*F. venenatum*
Protein	44.7	57.6	50.9	48	53
Lipids	7.0	3.5	5.5	5	7
Ashes	N/A	N/A	N/A	5	5
Energy (kcal/kg)	N/A	N/A	N/A	N/A	N/A
SFA	1.97	0.80	1.64	N/A	2.32
MUFA	2.62	1.59	1.93	N/A	1.30
PUFA	2.324	1.07	1.91	N/A	3.36
Ω-6	2.30	1.041	1.89	N/A	2.058
Ω-3	0.020	0.03	0.03	N/A	1.31
Essential Amino Acids					
Phenylalanine	1.42	1.58	1.46	N/A	N/A
Histidine	7.51	0.86	0.76	N/A	N/A
Isoleucine	1.38	1.61	1.57	2.0	2.1
Leucine	2.6	2.71	2.45	3.1	3.1
Lysine	2.14	2.44	2.29	2.9	3.0
Methionine	5.66	0.60	0.59	0.7	0.8
Threonine	1.69	1.58	1.62	1.8	2.0
Tryptophan	N/A	N/A	N/A	0.6	0.5
Valine	1.72	2.02	1.85	2.4	2.5
Reference	Karimi et al. 2019	Karimi et al. 2019	Karimi et al. 2019	Alriksson et al. 2014	Alriksson et al. 2014; Hashempour-Baltork et al. 2020.

However, the greatest problem in using microalgae in commercial activities is their high scaling cost and their low biomass production. In this sense, the development of profitable processes to produce microalgae is the main objective in the aquaculture, biofuel, and pharmaceutical industries. Therefore, it is necessary to find the appropriate conditions for their growth: temperature, agitation, luminosity, and O_2 and CO_2 concentrations. In addition, contamination by bacteria, fungi, or other invasive algae must be avoided (da Silva and Reis 2014; Zmora et al. 2013). In this sense, optimizing microalgae yields using specific media and culture conditions is crucial for the future of aquaculture.

New Alternative Ingredients: Carbohydrates, Proteins, and Lipids

New alternatives to fish-based ingredients that are being implemented in aquafeeds are the zooplankton *Calanus finmarchicus*, the black soldier fly *(Hermetia illucens)*, the domestic fly *(Musca domestica)*, and mealworms *(Tenebrio molitor)*. These organisms are of great commercial interest due to their protein and lipid content, as well as ease of collection and use in feed formulations.

Calanus finmarchicus Products

Calanus finmarchicus is a marine copepod with a size around 2–3 mm that contributes between 40% and 90% of the zooplankton biomass in the Norwegian, Labrador, and Barents Seas (Figure 2.1) (Båmstedt and Ervik 1984). This zooplankton is important for the trophic chain because it provides larger animals with high levels of Ω-3, especially EPA and DHA fatty acids.

The life cycle of *Calanus finmarchicus* is divided into 12 phases; those being N1–N6, the larvae phases, and C1–C6, the adult phases. Once the adult stage is reached, the copepod moves to the sea surface where it consumes phytoplankton and accumulates more than 100 μg of PUFAs per

TABLE 2.10
Proximate Composition (%) of Microalgae Single Cell Protein

Composition (%)	*Pavlova* sp. (Pav459)	*Pavlova viridis*	*Nannochloropsis gaditana*	*Crypthecodinium cohnii*	*Spirulina platensis* 1	*Chlorella vulgaris*	*Isochrysis galbana*
Protein	66.5	39.8	45.8	11.5	63.26	47.82	26.99
Lipids	16.6	19.6	28.55	49	7.09	13.32	17.16
Ashes	11.3	10.0	7.78	7	7.51	6.30	16.08
Carbohydrates	6.7	N/A	14.91	29	15.17	8.08	16.98
Energy (kcal/kg)	5,636.76	5,555.55	6,360.46	N/A	3733.74	3411.32	357.62
SFA	2.93	N/A	2.4–12.24	16.2	2.42	2.95	7.43
MUFA	0.86	N/A	2.25–9.72	N/A	2.58	4.72	5.14
PUFA	7.70	N/A	4.44–6.12	32	1.58	5.18	4.18
Ω-6	1.25	N/A	0.28–1.45	9.2	1.08	1.29	0.47
Ω-3	6.21	N/A	4.15–4.68	22.8	0.49	3.89	3.71
EPA	3.37	2.5	4.15–4.93	0.5	0.161	0.43	0.33
DHA	1.64	1.0	N/A	21.8	0.23	2.78	3.25
Essential Amino Acids							
Phenylalanine	1.76	N/A	1.91	9.79	2.77	6.0	6.3
Histidine	0.51	N/A	0.75	1.71	1.01	1.6	2.1
Isoleucine	1.54	N/A	1.74	3.94	3.49	3.4	4.8
Leucine	3.26	N/A	3.24	7.1	5.4	8.2	8.7
Lysine	1.40	N/A	2.41	3.8	3.04	5.4	6.2
Methionine	0.98	N/A	0.87	2.49	1.36	2.6	2.1
Threonine	1.65	N/A	1.65	4.86	2.99	5.5	5.2
Tryptophan	0.02	N/A	0.04	1.33	0.92	0.2	1.3
Valine	2.36	N/A	2.15	5.5	3.32	10.7	6.2
References	Tibbetts et al. 2019	Haas et al. 2015.	Tibbetts et al. 2015; Hulatt et al. 2017	DSM 2013.	*Tokuşoglu and* Ünal, 2003; Belay 2013	*Tokuşoglu and* Ünal, 2003; Ursu et al. 2014	*Tokuşoglu and* Ünal, 2003; Brown 1991

TABLE 2.11
Maximum Substitution of FM/ FO and Microalgae Inclusion (%) in Aquafeeds and the Effect Caused in Different Aquatic Species

Organism	Microalgae	Addition or FM/FO Substitution (%)	Observations	Reference
Clarias gariepinus	*Nannochloropsis gaditana*	30% lyophilized biomass	Similar growth and digestibility parameters compared to the control feed.	Agboola et al. 2018
Dicentrarchus labrax L.	*Pavlova viridis*	Complete FO substitution	No negative effect on growth or fatty acid profile in fishes.	Haas et al. 2015
Oncorhynchus mykiss	*Crypthecodinium cohnii*	Addition of 9% in diets without FO	Improved growth, feed conversion, and DHA content compared with diets containing FO or vegetable oil.	Betiku et al. 2015
Litopenaeus vannamei	*Spirulina platensis*	Substitution of FM by 75%	No significative difference compared to the control feed, but there was an improvement in immunological parameters.	Macias-Sancho et al. 2014
Clarias gariepinus	*Chlorella vulgaris*	Addition of 25% in feeds without FM	Increased specific growth ratio compared to FM feeds but reduced feed ratio conversion.	Enyidi 2017

DHA: Docosahexaenoic acid; FM: fish meal; FO: fish oil

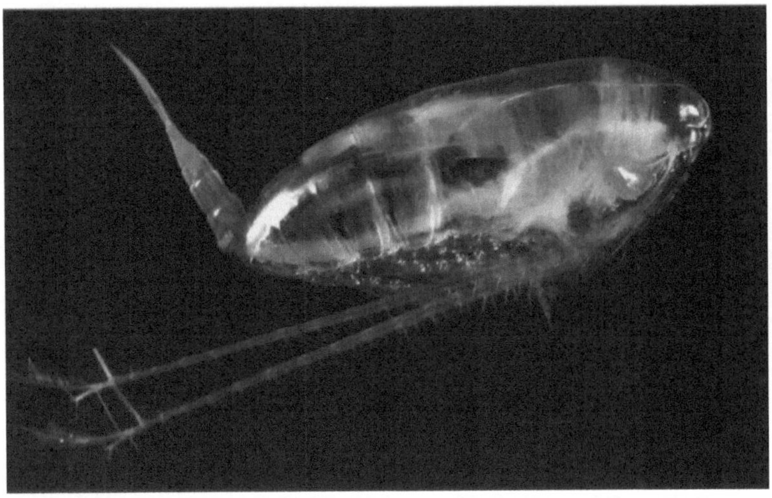

FIGURE 2.1 *Calanus finmarchicus*, taken from AQUAFEED vol. 11 Issue 4, 2019.

individual. However, these fatty acids are stored as wax esters that are difficult to digest by animals and that produce a particular taste in the fillet. In addition, this copepod also stores other lipids in lower amounts; in this sense, this could be an important organism for aquafeeds production (Table 2.12) (Parrish et al. 2012; Scott et al. 2002; Bergvik et al. 2012).

In addition to high lipid content, *C. finmarchicus* nutritional values are comparable to fish meal and soybean meal, containing essential amino acids for fish and crustacean development (Table 2.13).

The use of *C. finmarchicus* has been analyzed in different aquacultured animals (Table 2.14); it has been observed that increasing copepod oil content in aquafeeds alters the fatty acid profiles

TABLE 2.12
Proximate Composition (%) of *C. finmarchicus* Lipid and Fatty Acid Profile (Parrish et al. 2012)

Type of Lipid (%)	*Calanus finmarchicus*
Wax esters	66.4 – 66.6
Methyl esters	7.7
Triglycerides	8.9
Free fatty acids	1.5
Free ketone	3.2 – 6.6
Phospholipids	9.8
Fatty Acids	
SFA	18.2 – 35.5
MUFA	26.8 – 58.9
PUFA	22.9 – 39.7
Ω-6	0.6 – 2.3
Ω-3	19.5 – 32.8
20:5 Ω-3 (EPA)	2.3 – 8.6
22:6 Ω-3 (DHA)	9.3 – 15.5

TABLE 2.13
Proximate Composition (%) of *Calanus finmarchicus* Compared to Fish Meal and Soybean Meal (Colombo-Hixon et al. 2013)

Composition (%)	*C. finmarchicus*	Fish Meal	Soybean Meal
Moisture	9.3	6.0	9.4
Ashes	13.8	14.0	6.1
Crude Protein	60	74.7	48.0
Energy (kcal/ kg)	49713.2	48996.2	42543
Lipids	12.8	10.0	0.1
Essential Amino Acids			
Arginine	4.90	4.54	3.39
Histidine	0.93	1.65	1.19
Isoleucine	2.06	3.13	2.03
Leucine	3.90	5.19	3.49
Lysine	3.24	5.57	2.85
Methionine	1.55	2.08	0.58
Phenylalanine	1.72	2.71	2.22
Threonine	2.05	2.90	1.78
Tryptophan	N/A	0.77	0.64
Valine	2.51	4.30	2.02

of the animals, which increase the wax esters content in fillets. No other negative effect in animal growth has been observed; nevertheless, it is important to carry out more studies in other animals to rule out any undesirable effects. In addition, it is also important to know if other animals can digest and assimilate wax esters into their tissue without producing undesirable effects.

In this sense, increasing wax esters in fillet composition might induce digestive problems, like lipidic diarrhea (steatorrhea), in the consumer. Thus, use of *C. finmarchicus* oil in aquafeeds must be balanced to provide essential fatty acids to the animals and avoid possible toxic effects in consumers.

TABLE 2.14
Inclusion of *C. finmarchicus* (%) in Diets of Different Aquatic Cultures

Organism	Inclusion of *C. finmarchicus* in Diets (%)	Observations	Reference
Hippoglossus hippoglossus	30	Better protein digestibility and lipid content compared to FO.	Colombo-Hixon et al. 2013
Salmo salar	21.6	Increased PUFA wax esters in the muscle and hepatic tissue.	Olsen et al. 2004
Salmo salar	13.1 – 28.9	Incorporation of wax esters in the fillet.	Bogevik et al. 2011
Oncorhynchus mykiss	21	No significative difference in the fatty acids metabolism but stimulated the elongation and denaturalization of palmitic acid (16:0).	Oxley et al. 2005

FO: fish oil; PUFA: polyunsaturated fatty acids.

On the other hand, a responsible harvesting of *C. finmarchicus* is very important for the ecological equilibrium and sustainability of the seas; however, more than 290 tons per year are being utilized to produce different food ingredients (Bøgwald and Abrahamsen 2019). In this sense, fishes like Raitt's sand eel (*Ammodytes marinus*), Atlantic mackerel (*Scomber scombrus*), and blue whiting (*Micromesistius poutassou*) depend on this copepod. Therefore, designing the captive breeding of *C. finmarchicus* could facilitate its use in aquafeeds without causing ecological damage.

Insect Products Used in Aquafeeds Formulation

Insects have gained importance in the production of animal feeds due to their easy breeding, low maintenance costs, as well as the reduced environmental impact of this activity. In addition, insects present a similar nutritional value of animal ingredients but at a lower production cost.

Their utilization as feed ingredients in animals is limited and even prohibited by law in several countries; however, in 2017 the European Union stipulated that insect protein, as well as other ingredients listed in the EC 2017/893 regulation, can be used in aquafeeds. In this sense, insects approved to be used in aquafeeds are limited to the next species: *Hermetia illucens*, *Musca domestica*, *Tenebrio molitor*, *Alphitobius diaperinus*, *Acheta domesticus*, *Gryllodes sigillatus*, and *Gryllus assimilis*. Among these species, *H. illucens*, *M. domestica*, and *T. molitor* present the highest potential for large-scale production, due to their great protein concentration. Therefore, insect larvae are being used to transform fruits, vegetables, and animal waste into high quality protein and lipids (Table 2.15). At the moment, these larvae are being used to feed insectivorous pets, but in the near future, they may be used to feed aquacultured animals.

Insects present protein values around 30–58%, as well as an optimal amino acid profile for aquafeeds formulation, which are comparable to FM and SBM values (Table 2.16). In addition, their lipid profile is mainly composed of lauric (12:0), palmitic (16:0), vaccenic [18:1(Ω-7)], elaidic [18:1(Ω-9)], and linoleic acids [18:2(Ω-6)]; however, insects do not contain Ω-3 fatty acids (Barroso et al. 2014).

Despite their high nutritional value, insect meal must be processed to reduce the high SFA and MUFA content, since high levels of these compounds decrease feeds' half-life. In addition, they also affect the organoleptic properties of the fillet and negatively affect the palatability of aquafeeds. Another problem with using insect meal is their high chitin content because most organisms have problems digesting it; thus, thermal, or enzymatic treatments must be applied in order to increase digestibility. Insect meal performance has been studied in different aquatic animals (Table 2.17), where high levels of SFA and MUFA affected the growth parameters in fishes. Nevertheless, species like *Salmo salar* and *Pagrus major* can substitute 50% of the fish meal with insect meal in their diets without any apparent problem. However, the addition of high concentrations of insect meal

TABLE 2.15
Proximate Composition (%) of *H. illucens, M. domestica,* and *T. molitor* in Different Life Stages Compared to Fish Meal and Soybean Meal (Barroso et al. 2014)

Insect	Protein (%)	Lipids (%)	Ash (%)	SFA (%)	MUFA (%)	PUFA (%)	Ω-6 (%)	Ω-3 (%)
Hermetia illucens (Larva)	36.2	18.0	9.3	12.07	3.04	2.86	2.376	0.126
Hermetia illucens (Pupa)	40.7	15.6	19.7	10.26	5.08	0.17	0.17	0
Musca domestica (Larva)	46.9	31.3	6.5	10.2	16.5	4.75	2.38	0.16
Musca domestica (Pupa)	40.1	33.7	8.4	10.11	18.13	2.52	2.36	0.20
Tenebrio molitor (Larva)	58.4	30.1	3.5	6.68	13.8	9.5	9.5	0
Fish meal	73.0	8.2	18.0	2.9	1.64	3.06	0.22	2.84
Soybean Meal	50.4	3.0	7.8	1.02	0.45	1.66	1.66	0

TABLE 2.16
Proximate Aminoacidic Composition (%) of *H. illucens, M. domestica,* and *T. molitor* in Different Life Stages Compared to Fish Meal and Soybean Meal (Barroso et al. 2014)

Insect	Arg	His	Ile	Leu	Lys	Met	Phe	Pro	Thr	Tyr	Val
Hermetia illucens (Larva)	8.21	5.29	5.76	6.87	7.60	1.50	6.88	6.16	5.39	6.35	6.31
Hermetia illucens (Pupa)	8.05	5.16	5.34	6.83	7.31	3.26	6.22	5.56	4.95	7.14	6.34
Musca domestica (Larva)	6.83	4.68	4.89	6.75	8.36	3.00	7.01	5.33	4.87	5.79	6.08
Musca domestica (Pupa)	8.76	5.17	5.20	6.57	7.57	3.44	6.86	5.37	5.28	5.91	6.08
Tenebrio molitor (Larva)	6.14	3.64	5.87	8.56	6.03	0.64	4.29	7.17	4.49	4.18	7.61
Fish meal	7.42	7.86	5.04	7.81	8.78	2.93	5.38	4.76	6.26	3.91	5.56
Soybean Meal	8.03	3.28	5.47	8.01	6.34	1.01	5.79	4.99	4.17	2.94	5.45

has negative effects on fish organoleptic properties, which are principally induced by high content of SFA levels.

An important characteristic of insect development is the capacity to change the lipids profile when feeding ingredients are modified between breeding. In this sense, it is possible to enrich them with EPA and DHA fatty acids and use them in aquafeed production (Liland et al. 2017; Fasel et al. 2017; Fitches et al. 2018; Cullere et al. 2019). Therefore, designing large-scale breeding using economic substrates that favor the Ω-3 content will allow better performance of insect meal in aquacultured animals.

Vegetable Products as Carbohydrate, Protein, and Lipid Sources

Ingredients of vegetable origin have been the main substitute for fish meal and oil, due to their abundance and low cost. The most used plant ingredients in aquafeeds formulation are soybean, corn, wheat, sorghum, and peas, which contain different protein and lipid contents (Table 2.18). In addition, these vegetable ingredients present high carbohydrate levels (30–72%) that can be converted into a great energy source; however, carnivorous species, like fish and crustaceans, do not produce the necessary enzymes to degrade these complex carbohydrates. Therefore, a high percentage of carbohydrates in vegetable meals limit their utilization in aquafeeds production (Olmos 2017). Nowadays, it is known that *L. vannamei* can tolerate around 30% dietary carbohydrates, while most carnivorous fish can tolerate no more than 10% (Ochoa et al. 2014; Lopez et al. 2016; Olmos et al. 2011).

TABLE 2.17
Maximum Substitution of FM and IM Inclusion (%) in Aquafeeds and the Effect Caused in Different Aquatic Species

Organism	Insect	Maximum Substitution (%)	IM Content in Test Feeds (%)	FM Content in Control Feed (%)	Crude Protein (%)	Crude Lipids (%)	Other Sources of Protein (%)	Observations	Reference
Oreochromis niloticus	*Tenebrio molitor*	50	21	31	49.2	10.7	SBM – 50	Lower growth parameters compared to control feed.	Sánchez-Muros et al. 2015
Pagrus major	*Tenebrio molitor*	65	65	65	52 – 54	10 – 11	CG – 3-8	Increased feed uptake and growth.	Ido et al. 2019
Salmo salar	*Hermetia illucens*	100	39	29	39	29	SBM – 25	Growth, protein, and lipids were not affected by the substitution, but the fillet organoleptic properties were affected.	Belghit et al. 2019
Pagrus major	*Musca domestica*	100	40 – 70	40 – 70	48.90	12.84	N/A	Reduction of lipids allows higher substitution rate without affecting the growth parameters.	Hashizume *et al. 2019*

IM: insect meal; SBM: soybean meal, CG: corn gluten.

TABLE 2.18
Proximate Composition (%) of Different Vegetable Products Used in Animal Feeds (USDA 2020)

Composition (%)	Soybean (Raw)	Soybean (meal)	Soybean (Protein Isolate)	Wheat	Corn	Sorghum	Peas	Potato
Carbohydrates	30.16	33.92	0	76.31	74.26	72.09	61.63	17.49
Protein	36.49	51.46	88.32	10.33	9.42	10.62	23.12	2.05
Lipids	19.94	1.22	3.39	0.98	4.74	3.46	3.89	0.09
Ashes	4.87	N/A	3.58	0.47	1.2	1.43	2.67	1.11
SFA	2.884	0.136	0.422	0.155	0.667	0.61	0.408	0.025
MUFA	4.404	0.208	0.645	0.087	1.251	1.131	0.615	0.002
PUFA	11.55	0.533	1.648	0.413	2.163	1.558	1.022	0.042
Ω-6	9.925	0.47	1.453	0.391	2.163	1.493	0.854	0.031
Ω-3	1.33	0.063	0.195	0.022	0.065	0.065	0.163	0.01
20:5 Ω-3 (EPA)	0	0	0	0	0	0	0	0
22:6 Ω-3 (DHA)	0	0	0	0	0	0	0	0
Essential Amino Acids (%)								
Arginine	3.153	3.647	6.67	0.415	0.47	0.355	1.9	0.101
Histidine	1.097	1.268	2.303	0.23	0.287	0.246	1.15	0.035
Isoleucine	1.971	2.281	4.253	0.357	0.337	0.433	0.983	0.066
Leucine	3.309	3.828	6.783	0.71	1.155	1.491	1.68	0.098
Lysine	2.706	3.129	5.327	0.228	0.265	0.229	1.771	0.107
Methionine	0.547	0.634	1.13	0.183	0.197	0.169	0.195	0.032
Phenylalanine	2.122	2.453	4.593	0.52	0.463	0.546	1.151	0.081
Threonine	1.766	2.042	3.137	0.281	0.354	0.346	0.813	0.067
Tryptophan	0.591	0.683	1.116	0.127	0.067	0.124	0.159	0.021
Valine	2.029	2.346	4.098	0.415	0.477	0.561	1.035	0.103

Carbohydrates in Vegetables

Carbohydrates in vegetables can be divided into three main groups: sugars, composed of 1 or 2 molecules covalently linked; oligosaccharides, composed of 3–9 molecules; and polysaccharides, composed of 10 or more molecules (Figure 2.2). Therefore, carbohydrates need to be broken down by specialized enzymes to be assimilated; however, most marine animals do not produce them. Thus, aquafeeds formulations with complex carbohydrates may induce health problems in fish (Wilson 1994; López et al. 2016), as well as in crustaceans (Ochoa-Solano and Olmos-Soto 2006; Olmos et al. 2011).

Lipids in Vegetables

Lipids in vegetables are mainly composed of oleic and linoleic acids, and, in lower proportions, α-linolenic acid [18:3(Ω-3)]. Plants do not produce EPA or DHA Ω-3 fatty acids, which are very important for the development of aquatic animals. Unlike humans, fish do not produce the enzymes required to transform α-linolenic acid into EPA and DHA (Tocher 2015); therefore, feeding fish with vegetable oils could affect the product's nutritional value.

Soybean

Soybean is an important legume for the human and animal feeding market; its production reaches 225,000 tons per year with a selling cost of 319.89 USD per metric ton. China, USA, Brazil, Argentina, EU, and México are the leading producers (Indexmundi.com 2020c). Soybean is widely

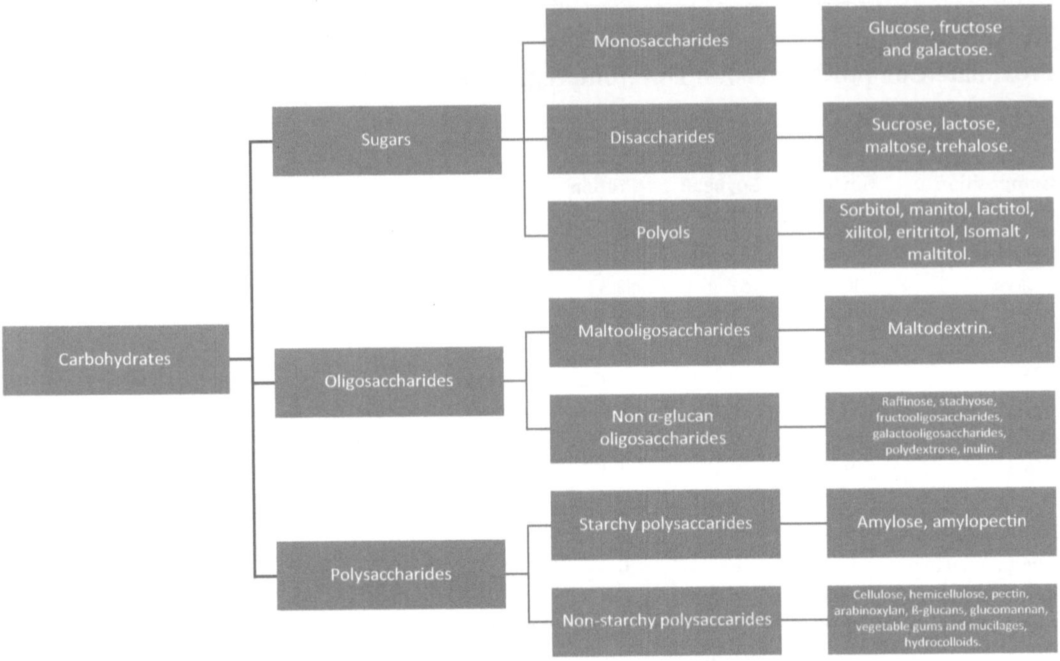

FIGURE 2.2 Carbohydrate classification, adapted from Ochoa et al 2014.

used to feed monogastric animals due to its high protein (35%) concentration; however, it is deficient in methionine and tryptophan, which could affect the development of aquatic animals (Table 2.18). On the other hand, soybean contains around 32% carbohydrates, where 12% is soluble sugars (4–5% sucrose, 3.5–4.5% stachyose, and 1–2% raffinose), while the other 20% is divided between non-starch polysaccharides (19%) and starch (1%) (Ochoa et al. 2014). However, aquatic animals do not produce the enzymes required to digest most soybean carbohydrates, limiting their use in aquafeeds. Among these carbohydrates non-starch-polysaccharides (NSP) interferes with food digestion, increasing intestinal viscosity and food's time residence, which reduces the absorption of essential nutrients (Sinha et al. 2011).

In addition to high carbohydrate and protein content, soybean contains between 18% to 20% of its weight in lipids. SFA and MUFA, like palmitic and oleic acids, represent ~3% and ~5%, respectively; however, PUFAs, like linoleic [18:2(Ω-6)] and, in much lower proportions, α-linolenic acids [18:3(Ω-3)], represent between 10% and 12%. Nevertheless, EPA and DHA Ω-3 fatty acids are not contained in soybean oil; therefore, it will be necessary to supplement aquafeeds with fish oil when soybean oil is used to grow aquatic animals.

Another limitation is that soybean contains some molecules that produce negative effects on animal health, which are classified as anti-nutritional factors. Among them, protease inhibitors can decrease the animal's enzyme capacity to digest proteins; as a result, the pancreas increases the enzyme secretion inducing a hypertrophy/hyperplasia state. Lectins are another anti-nutritional factor with harmful effects on the intestinal epithelium, reducing nutrient assimilation. Furthermore, phytic acid and allergens also limit the addition of soybean in aquafeed formulations (Mohan et al. 2016; Becker-Ritt et al. 2004).

Despite the significant amount of anti-nutritional factors, soybean has been used to feed different kinds of animals, demonstrating that soybean tolerance differs between species (Table 2.19). Therefore, the presence of complex carbohydrates and anti-nutritional factors, as well as its deficiency in EPA and DHA, precludes soybean addition to aquafeeds. Soybean could be processed to remove the carbohydrates and anti-nutritional factors to produce highly digestible aquafeeds, but

TABLE 2.19
Maximum Substitution of FM and SBM Inclusion (%) in Aquafeeds and the Effect Caused in Different Aquatic Species

Organism	Maximum Substitution (%)	SBM in Test Feeds (%)	FM in Control Feed (%)	Crude Protein (%)	Crude Lipids (%)	Other Sources of Protein (%)	Observations	Reference
Oncorhynchus mykiss	75	40	100	45.2	8.9	N/A	No significant difference observed in growth parameters, but the P and N excretion was reduced.	Cruz-Castro et al. 2011
Litopenaeus vannamei	100	48	N/A	27.41	6.46	CG – 2.0 WG – 2.0	Supplementation with *B. subtillis* allowed the complete substitution of FM, increasing the growth and hemolymphatic parameters.	Olmos et al. 2011
Litopenaeus vannamei	60	46.6	N/A	40.0	9.0	SM – 10.0	SBM supplemented with minerals improved the use of soybean as a source of energy and amino acids.	Huang et al. 2017
Diplodus puntazzo	60	60	48.2	40.0	14.0	N/A	No negative effect observed in juveniles or adults.	Hernández et al. 2007
Salmo salar	20	20	67.3	45	25	N/A	Deceased digestibility of protein, fats, starch, and amino acids compared to the control feed.	Øverland et al. 2009
Siganus rivulatus	25	19.7	52.1	40	N/A	N/A	Decreased growth: red and white blood cells count.	Monzer et al. 2017

SM: shrimp meal; SBM: soybean meal; CM: corn gluten; WG: wheat gluten.

these processes are expensive. In this sense, sustainable and more economical alternatives, like the use of probiotic bacteria, are currently being explored to increase the exogenous enzyme in animals to enhance the degradation of complex macromolecules. Thus, Olmos and coworkers (2011) replaced fish meal with soybean meal in aquafeeds formulated to grow *L. vannamei*; in addition, *Bacillus subtilis* enzymes overproducer strain was added to this diet (Olmos et al. 2011). Results show that *B. subtilis* was capable of increasing soybean meal assimilation; therefore, the growth of *L. vannamei* was comparable to that obtained with FM. In this sense, it is possible to replace FM with SBM if appropriate enzymes are added into the aquafeed formulations.

Potato Peel Byproducts

The worldwide potato industry produces around 368 million tons per year. This industry also produces between 15% and 40% of potato byproducts, especially peels (Sampaio et al. 2020; Elkahoui et al. 2018). Potato peels contain between 12% and 18% protein and 70% to 76% carbohydrates (Table 2.20) (Elkahoui et al. 2018; Feedepedia.com). However, they also have anti-nutritional factors like glycoalkaloids, which are toxic to humans and animals.

Potato peel utilization in aquafeed formulations is not so common; however, work reported by Omeregie et al. (2009) shows that *O. niloticus* can tolerate the inclusion of 15% potato peels in their diet. Likewise, Nwanna et al. (2009) used potato peel concentrate to substitute yellow corn in aquafeeds; in this work, *O. niloticus* was fed with 30% of this byproduct without experimenting adverse effects. Likewise, potato peels can be used as antioxidant compounds in aquafeeds when high lipid concentrations are added to the formulations (Sampaio et al. 2020). However, more information is required to include this byproduct in aquafeeds used to grow carnivorous marine species and to know the animal's tolerance to this byproduct.

Pea Meal Products (PMP)

Peas are legumes that contain less carbohydrates (14.4%), protein (6.3%), and lipids (0.4%) concentration than soybean; however, they are similar in protein concentration to wheat (12%) and corn (9.9%) (Table 2.18). The PMP lipid profile is mainly composed of oleic [18:1(Ω-9)], linoleic [18:2(Ω-6)], and α-linolenic [18:3(Ω-3)] acids, however, they do not produce EPA and DHA Ω-3 fatty acids. PMP have anti-nutritional factors, like protease inhibitors and lectins, which limit its use in carnivorous species. Nevertheless, they have been used in aquafeeds production as a cheaper alternative to soybean meal (Table 2.21); in this sense, obtained results show that this substitution does not affect the animal's growth. However, the substitution of fish meal by PMP produce adverse effects in most aquacultured species.

Corn and Wheat Products

Unlike soybean (30%), wheat and corn are mainly composed of carbohydrates (>70%; Table 2.18), in which amylose and amylopectin are the most abundant (Ochoa et al. 2014). Therefore, these meals

TABLE 2.20
Proximate Composition (%) of Potato Peels (Elkahoui et al. 2018; feedepedia.com)

Nutritional value	Content (%)
Moisture	3.67 – 5.66
Protein	11.98 – 17.19
Lipids	0.81 – 1.17
Carbohydrates	70 – 76
Fiber	15.97 – 22.39
Linoleic Acid (Ω-6)	0.64
α-Linolenic Acid (Ω-3)	0.20

TABLE 2.21
Maximum Substitution of FM and PMP Inclusion (%) in Aquafeeds and the Effect Caused in Different Aquatic Species

Organism	Maximum Substitution (%)	PMP Content in Test Feed (%)	FM Content in Test Feeds (%)	Crude Protein (%)	Crude Lipids (%)	Other Sources of Protein (%)	Observations	Reference
Lates calcarifer	10	10.83	40	40	10	SBM – 10 SM – 13.81 CM – 11.18	Substitutions higher than 10% cause decrease in growth.	Ganzon-Naret 2013
Salmo salar	20	20	67.38	45	26	SBM – 20	The partial substitution of FM with pea concentrate did not significantly affect growth.	Øverland et al. 2009
Cyprinus carpio L.	N/A	45/40	440	40	10	SBM – 20	Substitution of SBM with cooked PMP did not affect growth parameters.	Davies and Gouveia 2010
Dicentrarchus labrax	60	36	65	48	16	N/A	No adverse effect detected by the partial substitution of FM.	Tibaldi et al. 2005

SBM: soybean meal; SqM: squid meal; SM: shrimp meal; CM: cornmeal; PMP: pea meal product.

are used at low percentages in feed formulations due to their low digestibility in monogastric animals, especially in carnivorous marine species (Arellano-Carbajal and Olmos-Soto 2002; Lopez et al. 2016). In this sense, most aquacultured animals do not produce the α-1,6-glucosidase and α-1,4-amylase enzymes required to digest terrestrial carbohydrates (Arellano and Olmos-Soto 2002; Ochoa et al. 2014; Ochoa-Solano and Olmos-Soto 2006). Therefore, these carbohydrates could be considered anti-nutritional factors affecting fish and crustacean growth and performance; however, most commercial diets contain high concentration of these carbohydrates in their formulation (Ochoa et al. 2014). In this sense, to decrease the carbohydrate content in corn and wheat, they must be processed to obtain a protein concentrate known as gluten. These protein concentrates are composed mainly by gliadin and glutelin, which also work as binding agents through the pellet production process.

Wheat gluten contains up to 80% crude protein, 5.5% lipids, and 7.0% starch and provides around 5353.728 kCal of energy (Storebakken et al. 2000). Likewise, corn gluten contains around 65.0% protein and 7.0% lipids. However, these protein concentrates lack lysine, tryptophan, and arginine as well as essential PUFAs. Therefore, supplementation of these amino acids and fatty acids is required to balance aquafeed formulations (Apper-Bossard et al. 2013). Wheat and corn gluten have been used to substitute fish meal; however, assayed organisms present adverse effects in growth and performance, mainly due to the lack of essential amino acids and PUFAs (Table 2.22).

On the other hand, grains and other plant-based ingredients present a high concentration of carbohydrates and non-digestible fibers harmful to monogastric animals; also, they contain anti-nutritional factors that limit their use in aquafeeds. In this sense, physical treatments like cooking at high temperatures are used to inactivate these compounds; nevertheless, most of them are thermostable, and prolonged heating times may reduce nutritional value.

Therefore, utilization of exogenous enzymes could allow the inclusion of carbohydrates and proteins from vegetable origin in aquafeeds. In this sense, supplementation of *B. subtilis* in *L. vannamei* and *Atractoscion nobilis* fed with high concentrations of vegetable protein and carbohydrates, respectively, produced promising growth results (Olmos et al. 2011; Lopez et al. 2016). Thus, better performance is expected when the animals are fed with plant-based ingredients and probiotic bacteria.

Probiotic Inclusion in Aquafeed Formulations

Probiotics are bacterial species with the capacity to (a) increase assimilation of animal, microbial, and vegetable ingredients; (b) carry out bioremediation in aquaculture ponds; (c) control pathogen development inside and outside the animals; (d) improve the health status of animals by stimulating the immune system; and e) increase aquaculture profitability (Olmos and Paniagua 2014; Olmos et al. 2015). Thus, probiotic bacteria chosen to be added in the aquafeed formulations must: (1) produce carbohydrases, proteases, lipases, and other enzymes; (2) be easily grown in inexpensive culture media; (3) tolerate extreme conditions of pH and temperature; (4) produce a diverse and significant number of antimicrobial compounds; and (5) be classified as GRAS by the FDA (Olmos et al. 2020). In this sense, for almost 20 years we have been working with the isolation, characterization, and identification of *Bacillus* species to be added in aquafeed formulations to solve digestive, health, ecological, environmental, and economic problems in aquaculture (Arellano-Carbajal and Olmos-Soto 2002; Ochoa-Solano and Olmos-Soto 2006; Olmos et al. 2011; Olmos and Paniagua 2014; Olmos 2017; Olmos et al. 2020).

Species of the genus *Bacillus* are among the most prevalent bacteria worldwide due to their great enzymatic and antimicrobial capacities. These species can be found in soil, fresh and seawater, air, and fermented and non-fermented foods (Sonenshein et al. 1993). Spores of *Bacillus* species can be added to aquafeed formulations because they tolerate the high temperatures utilized throughout the production process. Also, extreme pH conditions, like those found in animal digestive tracts, are well tolerated by *Bacillus* spores (Sonenshein and Hoch 1993; Valdez et al. 2014). However, once conditions allow *Bacillus* species to germinate, they take control of the ecological niche and carry

TABLE 2.22
Maximum Substitution of FM and Corn/ Wheat Inclusion (%) in Aquafeeds and the Effect Caused in Different Aquatic Species

Organism	Grain or Grain Products	Maximum Substitution (%)	Grain Content in Test Feed (%)	FM Content in Control Feed (%)	Crude Protein (%)	Crude Lipids (%)	Other Sources of Protein (%)	Observations	Reference
Psetta maxima	Corn gluten	20	20	52	40	12	N/A	Higher substitutions cause negative effects on growth and amino acid retention.	Regost et al. 1999
Salmo salar	Wheat gluten	35	16.7	50	44	32	N/A	No adverse effects in growth or intestinal constitution.	Storebakken et al. *2000*
Dicentrarchus labrax	Wheat gluten	70	41	68	50	17	N/A	Weight gain and nitrogen retention similar to the control feed.	Tibaldi et at 2003
Lates calcarifer	Corn gluten	10	10	35	40	10	SMB – 21 SM – 10	High gluten content caused an increase in lipid concentration without affecting growth.	Nandakumar et al. 2017
Atractoscion nobilis, Ayres 1860	Corn starch	N/A	21.9	63.4	55	10	KM – 3.0	The implementation of strains of *B. subtilis* allowed higher content of carbohydrates without causing any negative effect.	López et al. 2016

SBM: soybean meal; SM: shrimp meal; KM: krill meal; FO: fish oil out their beneficial effects. Thus, food digestion, immune system activation, pathogen inhibition, and pond water bioremediation all improve when *Bacillus* probiotic strains are added to aquafeeds.

B. subtilis is the best characterized specie of the genus due to a great number of methodologies developed to manipulate this bacterium molecularly and genetically (Harwood and Cutting 1990; Olmos-Soto and Contreras-Flores 2003). Recently, its genome has been sequenced (Kunst et al. 1997), allowing the construction of recombinant strains with improved probiotic capacities. In this sense, mutations that upregulate the expression of degradative enzymes, like proteases (Olmos et al. 1997), could facilitate aquafeeds assimilation and avoid water contamination. Additionally, carbohydrases and lipases could be genetically manipulated to increase animals' digestive capacity. In addition, *Bacillus* strains with great enzymatic capacities could also be isolated from different environments, using the appropriate methodologies (Arellano and Olmos-Soto 2002; Ochoa and Olmos-Soto 2006).

Bacillus species secrete a significant amount of enzymes into the culture medium, which simplifies its purification and inclusion into the aquafeed production process (Cui et al. 2018). *Bacillus* enzymes break down carbohydrates, proteins, and complex lipids from animal, microorganism, and plant origins, facilitating its assimilation by animals and humans. Moreover, these enzymes allow *Bacillus* species to grow in a great variety of inexpensive carbon and nitrogen sources, lowering the fermentation process cost (Sonenshein and Hoch 1993). Therefore, aquafeeds degradation, as well as water bioremediation activities, will be improved when *Bacillus* or its enzymes are added to aquafeed formulations (Olmos and Paniagua 2014; Olmos et al. 2015).

B. subtilis, *B. licheniformis*, *B. thuringiensis*, and *B. pumilus* can be used successfully as probiotic bacteria in aquaculture because they are not cytotoxic. Also, these species secrete a great diversity of antimicrobial peptides (AMPs) with the capacity to inhibit bacteria, fungi, and parasite development (Cutting 2011; Olmos et al. 2020). However, species like *B. anthracis* and *B. cereus* are considered some of the most dangerous bacteria for humans and are known as significant causes of food poisoning (Helgason et al. 2000). In this sense, probiotic bacteria used in aquafeeds must be carefully analyzed, *in vitro* and *in vivo*, to avoid health problems in aquacultured animals and their consumers. Thus, improper identification and wrong utilization of pathogenic species could produce severe health and ecological problems.

In summary, a GRAS and multifunctional spore producer *Bacillus* specie could be an ideal probiotic bacterium in the effort to avoid aquaculture problems induced by aquafeeds misformulation (Olmos et al. 2011; López et al. 2016).

INGREDIENT FERMENTATION TO IMPROVE CARBOHYDRATE, PROTEIN, AND COMPLEX LIPID ASSIMILATION

Fermentation is a biochemical process carried out by microorganisms of the genus *Saccharomyces*, *Lactobacillus*, and *Bacillus*, where carbohydrates, proteins, and lipids are transformed into assimilable compounds. These microorganisms produce enzymes that break down complex macromolecules, including the anti-nutritional factors; also, they produce bioactive compounds and peptides that enrich the aquafeeds. However, the microorganisms utilized to predigest aquafeed ingredients must be innocuous to animals and consumers; therefore, they must fulfill the GRAS classification imposed by the FDA. In this sense, *Bacillus* species have a great capacity to secrete enzymes, antimicrobial peptides, and bioactive compounds. In addition, *Bacillus* species produce resistant spores that tolerate the aquafeed formulation process and other extreme physical and physiological conditions (Figure 2.3).

Therefore, solid-state fermentation can be carried out with *Bacillus* species to improve ingredient nutritional value (Table 2.23). Likewise, *Bacillus* can degrade anti-nutritional factors like protease inhibitors and lectins present in vegetables. They can also carry out the digestion of complex carbohydrates, like starch, non-starch polysaccharides, raffinose, and stachyose, which cannot be digested by monogastric animals due to the lack of specialized enzymes (Olmos 2017).

Fermented ingredients have been used in aquacultured animals to increase their tolerance to complex macronutrients, and satisfactory results have been obtained with these products (Table 2.24).

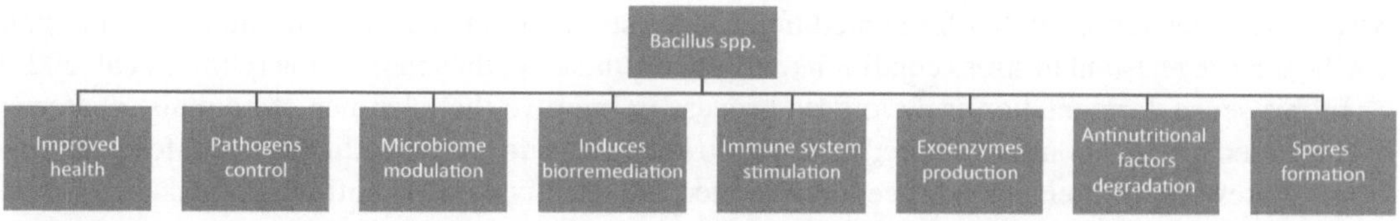

FIGURE 2.3 Beneficial characteristics of *Bacillus* spp.

TABLE 2.23
Comparison of Proximate Composition (%) of Plant Products Fermented by *Bacillus* Species

Microorganism	*B. subtilis*		*B. licheniformis*		*B. pumilus*		*B. subtilis BJ-1*	
Composition (%)	Soybean Meal		Jatropha Meal		Jatropha Meal		Cottonseed Meal	
	NF	F	NF	F	NF	F	NF	F
Protein	41.53	56.42	27.42	36.66	27.42	37.12	47.2	50.8
Lipids	N/A	N/A	5.07	5.77	5.07	5.86	0.12	0.11
Carbohydrates	32.50	25.29	40.60	43.64	40.60	43.7	N/A	N/A
Fiber	N/A	N/A	20.93	5.41	20.93	5.08	N/A	N/A
Phytic acid	N/A	N/A	0.25	0.35	0.25	2.20	0.078	0.009
Protease inhibitor (mg/g)	N/A	N/A	25.24	1.87	25.24	1.24	N/A	N/A
References	Kook et al. 2014		Hassaan et al. 2016		Hassaan et al. 2016		Sun et al. 2015	

F: fermented; NF: not fermented

TABLE 2.24
The Effect of Fermentation in the Feeds of Different Aquatic Species

Organism	Substrate	Microorganism	Observations	Reference
Epinephelus coioides, Hamilton	Soybean meal	*Bacillus subtilis-E20*	The use of fermented SBM allowed a higher substitution of FM from 18.36% to 29.32%.	Shiu et al. 2013
Oreochromis niloticus	*Jatropha curcas* meal	*Bacillus licheniformis* and *Bacillus pumilus*	Reduction of carbohydrates, protease inhibitors, and saponins allowed the substitution of FM by 50% without causing any negative effect on fishes.	Hassaan et al. 2016
Litopenaeus vannamei. Boone,1931	Cottonseed meal	*Bacillus subtilis BJ-1*	Substituting FM by 50% did not cause any negative effect on growth; it also reduced phytic acid content.	Sun et al. 2015
Litopenaeus vannamei	Soybean protein concentrate and corn gluten.	*Bacillus subtilis*	30% substitution by soybean protein concentrate and fermented corn gluten increased growth, feed uptake, immune response, and pathogen resistance.	Hamidoghli et al. 2020
Oreochromis niloticus	Sunflower seed meal.	*Bacillus subtilis* and *Saccharomyces cerevisiae*	No negative effect observed in growth and hematologic parameters when substituting FM with fermented sunflower meal by 25%.	Hassaan et al. 2018

FM: fish meal; SBM: soybean meal

Also, it has been reported that fermented ingredients stimulate the animals' immune system, making them more resistant to stress conditions and opportunistic pathogen invasion (Olmos et al. 2020).

In this sense, fermentation is a required process to improve the tolerance of animals to alternative, inexpensive, but complex, ingredients and to keep costs down by facilitating aquafeeds assimilation. Likewise, fermentation procedures induce the inactivation of anti-nutritional factors that significantly affect the quality of aquafeeds and inhibit animal growth. Therefore, the use of microorganisms to carry out fermentation of complex ingredients that are included in aquafeeds is highly recommended to increase aquaculture sustainability and profitability.

REFERENCES

Adlercreutz P., and Lyberg A.M. 2008. A polyunsaturated fatty acid (PUFA) enriched marine oil compromising eicosapentaenoic acid (EPA) and docosahexaenoic acid (DHA), and a process of production thereof. Patent WO 2008/133573A1.

Agboola J.O., Teuling E., Wierenga P.A., Gruppen H., and Schrama J.W. 2018. Cell wall disruption: an effective strategy to improve the nutritive quality of microalgae in African catfish (*Clarias gariepinus*). *Aquaculture Nutrition* 25(4):783–797. https://doi.org/10.1111/anu.12896.

Alriksson B., Hörngerg A., Gudnason A.E., Knobloch S., Arnason J., and Johannsson R. 2014. Fish feed from wood. *Cellulose Chemistry and Technology* 48(9–10):843–848.

Apper-Bossard E., Feneuil A., Wagner A., and Respondek F. 2013. Use of vital wheat gluten in aquaculture. *Aquatic Biosystems* 9(21):1–13. https://doi.org/10.1186/2046-9063-9-21.

Arellano-Carbajal F., and Olmos-Soto J. 2002. Thermostable alpha-1,4- and alpha-1,6-glucosidase enzymes from Bacillus sp. isolated from a marine environment. *World Journal of Microbiology and Biotechnology* 18:791–795. https://doi.org/10.1023/A:1020433210432.

Association of American Feed Control Officials. 2012. Byproducts. https://talkspetfood.aafco.org/byproducts (accessed September 7, 2020).

Bajpai P. 2017. Microorganisms used for single cell protein productions. In *Single cell protein production from Lignocellulosic Biomass*, 21–31. Singapore: Springer. https://doi.org/10.1007/978-981-10-5873-8.

Båmstedt U., and Ervik A. 1984. Local variations in size and activity among *Calanus finmarchicus* and *Metridia longa* (Copepoda, Calanoida) overwintering on the west coast of Norway. *Journal of Plankton Research* 6(5):843–857. https://doi.org/10.1093/plankt/6.5.843.

Barroso F.G., de Haro C., Sánchez-Muros M.J., Venegas E., Martínez-Sánchez A., and Pérez-Bañón C. 2014. The potential of various insect species for use as food for fish. *Aquaculture* 422–423:193–201. https://doi.org/10.1016/j.aquaculture.2013.12.024.

Becker-Ritt A.B., Mulinari F., Vasconcelos I.M., and Carlini C.R. 2004. Antinutritional and/or toxic factors in soybean (*Glycine max* (L) Merril) seeds: comparison of different cultivars adapted to the southern region of Brazil. *Journal of the Science of Food and Agriculture* 84(3):263–270. https://doi.org/10.1002/jsfa.1628.

Belay A. 2013. Biology and industrial production of *Arthrospira* (*Spirulina*). In *Handbook of microalgal culture: Applied phycology and biotechnology*, Second Edition, ed. Richmond A., and Hu Q., 339–358. West Sussex: Wiley-Blackwell. https://doi.org/10.1002/9781118567166.ch17.

Belghit I., Liland N.S., and Gjesdal P., et al. 2019. Black soldier fly larvae meal can replace fish meal in diets of sea-water phase Atlantic salmon (*Salmo salar*). *Aquaculture* 503:609–619. https://doi.org/10.1016/j.aquaculture.2018.12.032.

Bergvik M., Leiknes Ø., Altin D., Dalh K.R., and Olsen Y. 2012. Dynamics of the lipid content and biomass of *Calanus finmarchicus* (copepodite V) in Norwegian Fjord. *Lipids* 47:881–895. https://doi.org/10.1007/s11745-012-3700-3.

Betiku O.C., Barrows F.T., Ross C., and Sealey W.M. 2015. The effect of total replacement of fish oil with DHA -Gold® and plant oils on growth and fillet quality of rainbow trout (*Oncorhynchus mykiss*) fed a plant-based diet. *Aquaculture Nutrition* 22(1):1–11. https://doi.org/10.1111/anu.12234.

Bogevik A.S., Henderson R.J., Mundheim H., Olsen R.E., and Rocher D.R. 2011. The effect of temperature and dietary fat level on tissue lipid composition in Atlantic salmon (*Salmo salar*) fed wax ester-rich oil from *Calanus finmarchicus*. *Aquaculture Nutrition* 17(3):e781–e788. https://doi.org/10.1111/j.1365-2095.2010.00848.x.

Bøgwald I., and Abrahamsen H. 2019. A considerable resource of a tiny crustacean – *Calanus finmarchicus*. *Aquafeed: Advances in Processing & Formulation* 11(4):24–26.

Brown M.R. 1991. The amino-acid and sugar composition of 16 species of microalgae used in mariculture. *Journal of Experimental Marine Biology and Ecology* 1:79–99. https://doi.org/10.1016/0022-0981(91)90007-J.

Cashion T., Manach F.L., Zeller D., and Pauly D. 2017. Most fish destined for fishmeal production are food-grade fish. *Fish and Fisheries* 2017:1–8. https://doi.org/10.1111/faf.12209.

Castillo-Lopez E., Espinoza-Villegas E.R., and Viana M.T. 2016. *In vitro* digestion comparison from fish and poultry by-product meals from simulated digestive process at different times of the Pacific Bluefin tuna, *Thunnus orientalis*. *Aquaculture* 458:187–194. https://doi.org/10.1016/j.aquaculture.2016.03.011.

Colombo-Hixon S.M., Olsen R.E., Tibbets S.M., and Lall S.P. 2013. Evaluation of *Calanus finmarchicus* copepod meal in practical diets for juvenile Atlantic halibut (*Hippoglossus hippoglossus*). *Aquaculture Nutrition* 19(5):687–700. https://doi.org/10.1111/anu.12016.

Cruz-Castro C.A., Hernández-Hernández L.H., Fernández-Araiza M.A., Ramírez-Pérez T., and López OA. 2011. Effects of diets with soybean meal on the growth, digestibility, phosphorus and nitrogen excretion of juvenile rainbow trout *Oncorhynchus mykiss*. *Hidrobiológica* 21(2):118–125. www.scielo.org.mx/pdf/hbio/v21n2/v21n2a2.pdf.

Cui W., Han L., Suo F., Liu Z., Zhou L., and Zhou Z. 2018. Exploitation of *Bacillus subtilis* as a robust workhorse for production of heterologous proteins and beyond. *World Journal of Microbiology Biotechnology* 34(10):145. https://doi.org/10.1007/s11274-018-2531-7.

Cullere M., Woods M.J., van Emmenes L, et al. 2019. *Hermetia illucens* larvae reared on different substrates in broiler quail diets: effect on physicochemical and sensory quality of the quail meat. *Animals* 9(8):1–17. https://doi.org/10.3390/ani9080525.

Cutting S.M. 2011 *Bacillus* probiotics. *Food Microbiology* 28(2):214–220. https://doi.org/10.1016/j.fm.2010.03.007.

da Silva T.L., and Reis A. 2015. Scale-up problems for the large scale production of algae. In *Algal biorefinery: An integrated approach*, ed. Das D. Cham: Springer. https://doi.org/10.1007/978-3-319-22813-6_6.

Davies S.J., and Gouveia A. 2010. Response of common carp fry fed diets containing a pea seed meal (*Pisum sativum*) subjected to different thermal processing methods. *Aquaculture* 305(1–4): 117–123. https://doi.org/10.1016/j.aquaculture.2010.04.021.

Dawson M.R., Alam M.S., Watanabe W.O, Carroll P.M., and Seaton P.J. 2018. Evaluation of poultry by-product meal as an alternative to fish meal in the diet of juvenile black sea bass reared in a recirculating aquaculture system. *North American Journal of Aquaculture* 80:74–87. https://doi.org/10.1002/naaq.10009.

Dozier W.A.III, and Dale N.M. 2005. Metabolizable energy of feed-grade and pet food-grade poultry by-product meals. *Journal of Applied Poultry Research* 14(2):349–351. https://doi.org/10.1093/japr/14.2.349.

DSM. 2013. DHAgold for companion animals. https://www.dsm.com/content/dam/dsm/anh/en_US/documents/DHAgold_for_companion_animals.pdf. (accessed September 07,2020).

Elkahoui S., Bartley G.E., Yokoyama W. H., and Friedman M. 2018. Dietary supplementation of potato peel powders prepared form conventional and organic russet and non-organic gold and red potatoes reduces weight gain in mice on a high-fat diet. *Journal of Agricultural and Food Chemistry*. 66:6064–6072. https://doi.org/10.1021/acs.jafc.8b01987.

Enyidi U.D. 2017. *Chlorella vulgaris* as protein source in the diets of african catfish *Clarias gariepinus*. *Fishes* 2(4):1–12. https://doi.org/10.3390/fishes2040017.

Fasel N., Mene-Saffrane L., Ruczynsky I., Komar E., and Christe P. 2017. Diet induced modification of fatty-acid composition in mealworm larvae (*Tenebrio molitor*). *Journal of Food Research* 6(5):22–31. https://doi.org/10.5539/jfr.v6n5p22.

Fitches E.C., Dickinson M., De Marzo D., Wakefield M.E., Charlton A.C., and Hall H. 2018. Alternative protein production for animal feed: *Musca domestica* productivity on poultry litter and nutritional quality of processed larval meals. *Journal of Insects as Food and Feed* 5(2):77–88. https://doi.org/10.3920/JIFF2017.0061.

Food and Agricultural Organization of the United Nations. 2018. *El estado mundial de la pesca y la acuicultura: cumplir los objetivos de desarrollo sostenible*. Roma: FAO. http://www.fao.org/3/i9540es/i9540es.pdf.

Food and Agricultural Organization of the United Nations. 2020. *The state of world fisheries and aquaculture* 2020. Rome: FAO. http://www.fao.org/documents/card/en/c/ca9229en (accessed September 10, 2020).

Foster I., Babbitt J., and Smiley S. 2004. Nutritional quality of fish meals made from by-products of the Alaskan fishing industry in diets for pacific white shrimp (*Litopenaeus vannamei*). *Journal of Aquatic Food Product Technology* 13(2):115–123. https://doi.org/10.1300/J030v13n02_10.

Ganzon-Naret E.S. 2013.The use of green pea (*Pisum sativum*) as alternative protein source for fish meal in diets for Asian sea bass, *Lates calcarifer*. *AACL Bioflux* 6(4):399–406.

Haas S., Bauer J.L., Adakli A., et al. 2015. Marine microalgae *Pavlova viridis* and *Nannochloropsis sp.* as n-3 PUFA source in diets for juvenile European sea bass (*Dicentrarchus labrax* L.). *Journal of Applied Phycology* 28:1011–1021. https://doi.org/10.1007/s10811-015-0622-5.

Hamidoghli A., Won S., Nathaniel W.F., et al. 2020. Solid state fermented plant protein sources as fish meal replacers in whiteleg shrimp *Litopaeneus vannamei*. *Animal Feed Science and Technology* 264:1–10. https://doi.org/10.1016/j.anifeedsci.2020.114474.

Hamilton, C. R. 2016. Real and perceived issues involving animal proteins. In *Protein sources for the animal feed industry: expert consultation and workshop, Bangkok, 29 April–3 May 2002*, ed. FAO. Food and Agriculture Organization of the United Nations. http://www.fao.org/3/y5019e/y5019e0g.htm#bm16.

Harwood C.R., and Cutting S. M. 1990. *Molecular biological methods for Bacillus*. Chichester: Wiley.

Hashempour-Baltork F., Hosseini S.M., Assarehzadegan M.A., Khosravi-Darani K., and Hosseini H. 2020. Safety assays and nutritional values of mycoproteins produced by *Fusarium venenatum* IR372C from date waste as substrate. *Journal of Science of Food and Agriculture* 100(12): 4433–4441. https://doi.org/10.1002/jsfa.10483.

Hashizume A., Ido A., Ohta T., et al. 2019. Housefly (*Musca domestica*) larvae preparations after removing the hydrophobic fraction are effective alternatives to fish meal in aquaculture feed for red seabream (*Pagrus major*). *Fishes* 4(3):1–16. https://doi.org/10.3390/fishes4030038.

Hassaan M.S., Goda A.M.A.S., and Kumar V. 2016. Evaluation of nutritive value of fermented de-oiled physic nut, *Jatropha curca*, seed meal for Nile tilapia *Oreochromis niloticus* fingerlings. *Aquaculture Nutrition* 23(3):1–14. https://doi.org/10.1111/anu.12424.

Hassaan M.S., Soltan M.A., Mohammady E.Y., Elashry M.A., El-Haroun E.R., and Davies S.J. 2018. Growth and physiological responses of Nile tilapia, *Oreochromis niloticus* fed dietary fermented sunflower meal inoculated with *Saccharomyces cerevisiae* and *Bacillus subtilis*. *Aquaculture* 495:592–601. https://doi.org/10.1016/j.aquaculture.2018.06.018.

Helgason E., Okstad O.A., Caugant D.A., et al. 2000. *Bacillus anthracis*, *Bacillus cereus*, and *Bacillus thuringiensis*--one species on the basis of genetic evidence. *Applied and Environmental Microbiology* 66(6):2627–2630. https://aem.asm.org/content/66/6/2627.

Hernández M.D., Martínez F.J., Jover M., and García B.G. 2007. Effects of partial replacement of fish meal by soybean meat in sharpsnout seabream (*Diplodus puntazzo*) diet. *Aquaculture* 263(1–4):159–167. https://doi.org/10.1016/j.aquaculture.2006.07.040.

Higgs D.A., Balfry S.K., Oakes J.D., Rowshandeli M., Skura B. J., and Deacon G. 2006. Efficacy of an equal blend of canola oil and poultry fat as an alternate dietary lipid source for Atlantic salmon (*Salmo salar* L.) in sea water. I: effects on growth performance, and the whole body and fillet proximate and lipid composition. *Aquaculture Research* 37:180–191. https://doi.org/10.1111/j.1365-2109.2005.01420.x.

Huang F., Wang L., Zhang C., and Song K. 2017. Replacement of fishmeal with soybean meal and mineral supplements in diet of *Litopenaeus vannamei* reared in low-salinity water. *Aquaculture* 473:172–180. https://doi.org/10.1016/j.aquaculture.2017.02.011.

Hulatt C.J., Wijffels R.H., Bolla S., and Kiron V. 2017. Production of fatty acids and protein by *Nannochloropsis* in flat-plate photobioreactors. *PLOS One* 12(1):e0170440. https://doi.org/10.1371/journal.pone.0170440.

Ido A., Hashizume A., Ohta T., Takahashi T., Miura C., and Miura T. 2019. Replacement of fish meal by defatted yellow mealworm (*Tenebrio molitor*) larvae in diet improves growth performance and disease resistance in red seabream (*Pargus major*). *Animals* 9(3):1–16. https://doi.org/10.3390/ani9030100.

Indexmundi. 2020a. Broiler meat (Poultry) production by country in 1000 MT. https://www.indexmundi.com/agriculture/?commodity=broiler-meat&graph=production (accessed September 7, 2020).

Indexmundi. 2020b. Fishmeal monthly price – US Dollars per Metric Ton. https://www.indexmundi.com/commodities/?commodity=fish-meal (accessed September 10, 2020).

Indexmundi. 2020c. Soybean meal production by country in 1000 MT. https://www.indexmundi.com/agriculture/?commodity=soybean-meal (accessed September 22, 2020).

Indexmundi. 2020d. Total fisheries production (metric tons) – country ranking. https://www.indexmundi.com/facts/indicators/ER.FSH.PROD.MT/rankings (accessed September 28).

Karimi S., Soofani N.M., Lundh T., Mahboubi A., Kiessling A., and Taherzadeh M.J. 2019. Evaluation of filamentous fungal biomass cultivate on vinasse as an alternative nutrient source of fish feed: protein, lipid, and mineral composition. *Fermentation* 5(4):1–17. https://doi.org/10.3390/fermentation5040099.

Karimi S., Soofiani N.M., Mahboubi A., and Taherzadeh M.J. 2018. Use of organic wastes and industrial by-products to produce filamentous fungi with potential as aqua-feed ingredients. *Sustainability* 10(9):1–19. https://doi.org/10.3390/su10093296.

Kim K., Park Y., Je H.W., et al. 2019. Tuna byproducts as a fish-meal in tilapia aquaculture. *Ecotoxicology and Environmental Safety* 172:364–372. https://doi.org/10.1016/j.ecoenv.2019.01.107.

Kim K.D., Jang J.W., Kim K.W., Lee B.J., Hur S.W., and Han H.S. 2018. Tuna by-product meal as a dietary protein source replacing fishmeal in juvenile korean rockfish *Sebastes schlegeli. Fisheries and Aquatic Sciences* 21(29): 1–8. https://doi.org/10.1186/s41240-018-0107-y.

Kirimi J.G., and Musalia L.M. 2016. Effect of replacing fish meal with blood meal on chemical composition of supplement for nile tilapia (*Oreochormis niloticus*). *East African Agricultural and Forestry Journal* 82:1–9. https://doi.org/10.1080/00128325.2016.1158898.

Kook M.C., Cho S.C., Hong Y.H., and Park H. 2014. *Bacillus subtilis* fermentation for enhancement of feed nutritive value of soybean meal. *Journal of Applied Biological Chemistry* 57:183–188. https://doi.org/10.3839/jabc.2014.030.

Kunst F., Ogasawara N., Moszer I., et al. 1997. The complete genome sequence of the Gram-positive bacterium *Bacillus subtilis. Nature* 390:249–256. https://doi.org/10.1038/36786.

Liland N.S., Biancarosa I., Araujo P., et al. 2017. Modulation of nutrient composition of black soldier fly (*Hermetia illucens*) larvae by feeding seaweed-enriched media. *PLOS One* 12(8):e0183188. https://doi.org/10.1371/journal.pone.0183188.

Liu X.H., Ye J.D., Kong J.H., Wang AL, and Wang A.L. 2013. Apparent digestibility of 12 protein-origin ingredients for pacific white shrimp *Litopenaeus vannamei. North American Journal of Aquaculture* 75:90–98. https://doi.org/10.1080/15222055.2012.716019.

López L.M., Olmos S.J., Escamilla I.T, et al. 2016. Evaluation of carbohydrate-to-lipid ratio in diets supplemented with *Bacillus subtilis* strain on growth performance, body composition and digestibility in juvenile white seabass (*Atractoscion nobilis*, Ayres 1860). *Aquaculture Research* 47(6):1–10. https://doi.org/10.1111/are.12644.

Macias-Sancho J., Poersch L.H., Bauer W., Romano L.A., Wasielesky W., and Tesser M.B., 2014. Fishmeal substitution with Arthrospira (*Spirulina platensis*) in a practical diet for *Litopenaeus vannamei*: Effects on growth and immunological parameters. *Aquaculture* 426–427:120–125. https://doi.org/10.1016/j.aquaculture.2014.01.028.

Mejía-Barajas J., Montoya-Pérez R., Manzo-Avalos S., et al. 2018. Fatty acid addition and thermotolerance of *Kluyveromyces marxianus. FEMS Microbiology Letters* 365(7):1–5. https://doi.org/10.1093/femsle/fny043.

Millamena O.M. 2002. Replacement of fish meal by animal by-product meals in a practical diet for grow-out culture of grouper *Epinephelus coioides. Aquaculture* 204(1–2):75–84. https://doi.org/10.1016/S0044-8486(01)00629-9.

Mohan V.R., Tresina P.S., and Daffodil E.D. 2016. Antinutritional factors in legume seeds: characteristics and determination. In *Encyclopedia of food and health*, ed. Caballero B., Finglas P.M., and Toldrá F., 211–219. Oxford and Waltham, MA: Elsevier Academic Press. https://doi.org/10.1016/B978-0-12-384947-2.00036-2.

Monzer S., Nasser N., Babikian J., and Saoud P.I. 2017. Substitution of fish meal by soybean meal in diets for juvenile marbled spinefoot, *Siganus rivulatus. Journal of Applied Aquaculture* 29(2):101–116. https://doi.org/10.1080/10454438.2016.1272031.

Moutinho S., Martínez-Llorens S., Tomás-Vidal A., Jover-Cerdá M., Olivia-Teles A., and Peres H. 2017. Meat and bone meal as partial replacement for fish meal in diets for gilthead seabream (*Sparus aurata*) juveniles: growth, feed efficiency, amino acid utilization, and economic efficiency. *Aquaculture* 468(1):271–277. https://doi.org/10.1016/j.aquaculture.2016.10.024.

Nandakumar S., Ambasankar K., Raffic-Ali S.S. Syamadayal J., and Vasagam K. 2017. Replacement of fish meal with corn gluten meal in feeds for Asian seabass (*Lates calcarifer*). *Aquaculture International* 25:1495–1505. https://doi.org/10.1007/s10499-017-0133-2.

Naylor R.L., Hardy R.W., Bureau D.P., et al. 2009. Feeding aquaculture in an era of finite resources. *Proceedings of the National Academy of Sciences of the United States of America* 106(36):15103–15110. https://doi.org/10.1073/pnas.0905235106.

Nengas I., Alexis M.N., and Davies S.J. 1999. High inclusion levels of poultry meals and related byproducts in diets for gilthead seabream *Sparus aurata* L. *Aquaculture* 179(1–4):13–23. https://doi.org/10.1016/S0044-8486(99)00148-9.

Nwanna L.C., Falaye A.E., Olarewaju O.J., and Oludapo B.V. 2009. Evaluation of Nile tilapia (*Oreochromis niloticus* L.) fed dietary potato peels as replacement for yellow maize. In 24th Annual conference of the fisheries society of Nigeria (FISON) 25-28Oct 2009, 155–158. *Aquatic Commons.* http://aquaticcommons.org/id/eprint/23361.

Ochoa-Solano J.L., and Olmos-Soto J. 2006. The functional property of Bacillus for shrimp feeds. *Food Microbiology* 23(6):519–525. https://doi.org/10.1016/j.fm.2005.10.004.

Ochoa L., Paniagua-Michel J.J., and Olmos-Soto J. 2014. Chapter Twelve – Complex carbohydrates as possible source of high energy to formulate functional feeds. In *Advances in food and nutrition research vol.* 73, ed. Kim S.K., 259–288. London, UK: Elsevier Inc. https://doi.org/10.1016/B978-0-12-800268-1.00012-3.

Oliva-Teles A., and Gonçalves P. 2001. Partial replacement of fishmeal by brewer's yeast (*Saccharomyces cerevisiae*) in diets for sea bass (*Dicentrarchus labrax*) juveniles. *Aquaculture* 202(3–4): 269–278. https://doi.org/10.1016/S0044-8486(01)00777-3.

Olmos J., Acosta M., Mendoza G., and Pitones V. 2020. *Bacillus subtilis*, an ideal probiotic bacterium to shrimp and fish aquaculture that increase feed digestibility, prevent microbial diseases, and avoid water pollution. *Archive in Microbiology* 202:427–435. https://doi.org/10.1007/s00203-019-01757-2.

Olmos J., and Paniagua M.J. 2014. *Bacillus subtilis* a potential probiotic bacterium to formulate functional feeds for aquaculture. *Journal of Microbial & Biochemical Technology* 6(7):361–365. http://dx.doi.org/10.4172/1948-5948.1000169.

Olmos J., de Anda R., Ferrari E., Bolívar F., and Valle F. 1997. Effects of the sinR and degU32 (Hy) mutations on the regulation of the aprE gene in *Bacillus subtilis*. *Molecular and General Genetics* 253(5):562–567. https://doi.org/10.1007/s004380050358.

Olmos J., Ochoa L., Paniagua M.J., and Contreras R. 2011. Functional feed assessment on *Litopenaeus vannamei* using 100% fish meal replacement by soybean meal, high levels of complex carbohydrates and *Bacillus* probiotic strains. *Marine Drugs* 9(6):1119–1132. https://doi.org/10.3390/md9061119.

Olmos S.J, Paniagua-Michel J.J., Lopez L., and Ochoa L. 2015. Functional feeds in aquaculture. In *Springer handbook of marine biotechnology*, ed. Kim S.K., 1303–1319. Springer-Verlag Berlin Heidelberg: Springer. https://doi.org/10.1007/978-3-642-53971-8_59.

Olmos S.J. 2017. Chapter Two – *Bacillus* probiotic enzymes: external auxiliary apparatus to avoid digestive deficiencies, water pollution, diseases, and economic problems in marine cultivated animals. In *Advances in food and nutrition Research vol.* 80, ed. Kim S.K., and Toldá F., 15–35. London, UK: Elsevier Inc. https://doi.org/10.1016/bs.afnr.2016.11.001.

Olmos-Soto J., and Contreras-Flores R. 2003. Genetic system constructed to overproduce and secrete proinsulin in *Bacillus subtilis*. *Applied Microbiology and Biotechnology* 62:369–437. https://doi.org/10.1007/s0025 3-003-1289-4.

Olsen R.E., Henderson R.J., Sountama J., et al. 2004. Atlantic salmon, *Salmo salar*, utilizes wax ester-rich oil from *Calanus finmarchicus* effectively. *Aquaculture* 240(1–4):433–449. https://doi.org/10.1016/j.aquaculture.2004.07.017.

Omeregie E., Igoche L., Ojobe T.O., Absalom K.V., and Onusiriuka B.C. 2009. Effect of varying levels of sweet potato (*Ipomea batatas*) peels on growth, feed utilization and some biochemical responses of the cichlid (*Oreochromis niloticus*). *African Journal of Food Agriculture Nutrition and Development* 9(2):1–11.

Oncul F.O., Aya F.A., Hamidoghli A., et al. 2018. Effects of the dietary fermented tuna by-products meal on growth, blood parameters, nonspecific immune response, and disease resistance in juvenile olive flounder *Paralichthys olivaceus*. *Journal of the World Aquaculture Society* 50(1):65–77. https://doi.org/10.1111/jwas.12535.

Øverland M., Sørensen M., Storebakken T., Penn M, Krogdahl Å, and Skrede A. 2009. Pea protein concentrate substituting fish meal or soybean meal in diets for Atlantic salmon (*Salmo salar*) – effect on growth performance, nutrient digestibility, carcass composition, gut health, and physical feed quality. *Aquaculture* 288(3–4):305–311. https://doi.org/10.1016/j.aquaculture.2008.12.012.

Øverland M., and Skrede A. 2016. Yeast derived from lignocellulosic biomass as a sustainable feed resource for use in aquaculture. *Journal of the Science of Food and Agriculture* 97(3):733–742. https://doi.org/10.1002/jsfa.8007

Øverland M., Karlsson A., Mydland L.T., Romarheim O.H., and Skrede A. 2013. Evaluation of *Candida utilis*, *Kluyveromyces marxianus* and *Saccharomyces cerevisiae* yeasts as protein source in diets for Atlantic salmon (*Salmo salar*). *Aquaculture* 402–403:1–7. https://doi.org/10.1016/j.aquaculture.2013.03.016.

Oxley A., Tocher D.R., Torstensen B.E., and Olsen R.E. 2005. Fatty acid utilization and metabolism in caecal enterocytes of rainbow trout (*Oncorhynchus mykiss*) fed dietary fish or copepod oil. *Biochimica et Biophysica Acta (BBA) – Molecular and Cell Biology of Lipids* 1737(2–3):119–129. https://doi.org/10.1016/j.bbalip.2005.09.008.

Ozorio R.O.A, Portz L., Borghesi R., and Cyrino J.E.P. 2012. Effects of dietary yeast (*Saccharomyces cerevisiae*) supplementation in practical diets of tilapia (*Oreochromis niloticis*). *Animals* 2(1):16–24. https://doi.org/10.3390/ani2010016.

Parés-Sierra G., Durazo E., Ponce M.A., Badillo D., Correa-Reyes G., and Viana M.T. 2012. Partial to total replacement of fishmeal by poultry by-product meal in diets for juvenile rainbow trout (*Oncorhynchus mykiss*) and their effect on fatty acid from muscle tissue and the time required to retrieve the effect. *Aquaculture Research* 45(9):1459–1469. https://doi.org/10.1111/are.12092.

Parrish C.C., French V.M., and Whiticar M.J. 2012. Lipid class and fatty acid composition of copepods (*Calanus finmarchicus*, *C. glacialis*, *Pseudocalanus sp.*, *Tisbe furcate* and *Nitokra lacustris*) fed various combination of autotrophic and heterotrophic protists. *Journal of Plankton Research* 34(5):356–375. https://doi.org/10.1093/plankt/fbs003.

Pongpet J., Ponchunchoovong S., and Payooha K. 2015. Partial replacement of fishmeal by brewer´s yeast (*Saccharomyces cerevisiae*) in the diets of Thai panga (*Pangasianodon hypophthalmus* x *Pangasius bocourti*). *Aquaculture Nutrition* 22(3):1–11. https://onlinelibrary.wiley.com/doi/abs/10.1111/anu.12280.

Regost C., Arzel K., and Kaushik S.J. 1999. Partial or total replacement of fish meal by corn gluten meal in diet for turbot (*Psetta maxima*). *Aquaculture* 180(1–2):99–117. https://doi.org/10.1016/S0044-8486(99)00026-5.

Ribeiro C.S., Moreira R.G., Cantelmo O.A., and Esposito E. 2012. The use of *Kluyveromyces marxianus* in the diet of Red-Stirling tilapia (*Oreochromis niloticus*, Linnaeus) exposed to natural climatic variation: effects on growth performance, fatty acids, and protein deposition. *Aquaculture Research* 45(5):812–827. https://doi.org/10.1111/are.12023.

Ritala A., Häkkinen S.T., Toivari M., and Wiebe M.G. 2017. Single cell protein-state-of-the-art, industrial landscape and patents 2001–2016. *Frontiers in Microbiology* 8:1–18. https://doi.org/10.3389/fmicb.2017.02009.

Rodríguez B., Iben C., Valdivié M., and Martínez M. 2012. Profile of fatty acids from torula yeast (*Candida utilis*) grown on distiller´s vinasse. Technical note. *Cuban Journal of Agricultural Sciences* 46(2):199–201. http://cjascience.com/index.php/CJAS/article/view/83.

Rossi W.J., and Davis D. A. 2012. Replacement of fishmeal with poultry by-product meal in the diet of Florida pompano *Trachinotus carolinus* L. *Aquaculture* 338–341:160–166. https://doi.org/10.1016/j.aquaculture.2012.01.026.

Sampaio S.L., Petropoulos S.A., Alexopoulos A., et al. 2020. Potato peels as source of functional compounds for the food industry: a review. *Trends in Food Science & Technology* 103:118–129. https://doi.org/10.1016/j.tifs.2020.07.015.

Sánchez-Muros M.J., de Haro C., Sanz A., Trenzado C.E., Villareces S., and Barroso F.G. 2015. Nutritional evaluation of *Tenebrio molitor* meal as fishmeal substitute for tilapia (*Oreochromis niloticus*) diet. *Aquaculture Nutrition* 22(5):943–955. https://doi.org/10.1111/anu.12313.

Scott C.L., Kwasniewski S., Falk-Petersen S., and Sargent J.R. 2002. Species differences, origins and functions of fatty alcohols and fatty acids in the wax esters and phospholipids of *Calanus hyperboreus*, *C. glacialis* and *C. finmarchicus* from Arctic waters. *Marine Ecology Progress Series* 235:127–134. https://doi.org/10.3354/meps235127.

Shiu Y.L., Hsieh S.L., Guei W.C., Tsai Y.T., Chiu C.H., and Liu C.H. 2013. Using *Bacillus subtilis* E20-fermented soybean meal as replacement for fish meal in the diet of orange-spotted grouper (*Epinephelus coioides*, Hamilton). *Aquaculture Research* 46(6):1–14. https://doi.org/10.1111/are.12294.

Siddik M.A.B., Chungu P., Fotedar R., and Howieson J. 2019. Bioprocessed poultry by-product meals on growth, gut health and fatty acid synthesis of juvenile barramundi, *Lates calcarifer* (Bloch). *PLOSONE* 14(4):1–18. https://doi.org/10.1371/journal.pone.0215025

Sinha A.K., Kumar V., Makkar H.P.S., De Boeck G., and Becker K. 2011. Non-starch polysaccharides and their role in fish nutrition – a review. *Food Chemistry* 127(4):1409–1426. https://doi.org/10.1016/j.foodchem.2011.02.042

Sonenshein A.L., Hoch J.A., Losick R. 1993. *Bacillus subtilis and others gram-positive bacteria: biochemistry, physiology, and molecular genetics*. American Society for Microbiology. https://doi.org/10.1128/9781555818388.

Srour T.M., Essa M., Abdel-Rahim M., and Mansour M. 2016. Replacement of fish meal with poultry by-product meal (PBM) and its effects on the survival, growth, feed utilization, and microbial load of European seabass, *Dicentrarchus labrax* fry. *Global Advanced Research Journal of Agricultural Sciences* 5(7):293–301.

Storebakken T., Shearer K.D., Baeverfjord G., et al. 2000. Digestibility of macronutrients, energy and amino acids, absorption of elements and absence of intestinal enteritis in Atlantic salmon, *Salmo salar*, fed diets with wheat gluten. *Aquaculture* 184(1–2):115–132. https://doi.org/10.1016/S0044-8486(99)00316-6.

Sun H., Tang J., Yao X., Wu Y., Wang X., and Liu Y. 2015. Effects of replacement of fish meal with fermented cottonseed meal on growth performance, body composition and haemolymph indexes of Pacific white shrimp, *Litopenaeus vannamei* Boone, 1931. *Aquaculture Research* 47(8):1–10 https://doi.org/10.1111/are.12711.

Tacon A.G.J., and Metian M. 2008. Global overview on the use of fish meal and fish oil in industrially compounded aquafeeds: trends and prospects. *Aquaculture* 285(1–4):146–158. https://doi.org/10.1016/j.aquaculture.2008.08.015.

Tan B., Mai K., Zheng S., Zhou Q., Liu L., and Yu Y. 2005. Replacement of fish meal by meat and bone meal in practical diets for the white shrimp *Litopenaeus vannamai* (Boone). *Aquaculture Research* 35:439–444. https://doi.org/10.1111/j.1365-2109.2005.01223.x.

Tang B., Bu X., Lian X., et al. 2016. Effect of replacing fish meal with meat and bone meal on growth, feed utilization and nitrogen and phosphorus excretion for juvenile *Pseudobagrus ussuriensis. Aquaculture Nutrition* 24(25):1–9. https://doi.org/10.1111/anu.12625.

Terrazas-Fierro M., Civera-Cerecedo R., Ibarra-Martínez L., et al. 2010. Apparent digestibility of dry matter, protein, and essential amino acids in marine feedstuffs for juvenile whitelegs shrimps *Litopenaeus vannamei. Aquaculture* 308(1–4):166–173. https://doi.org/10.1016/j.aquaculture.2010.08.021.

Tibaldi E., Tulli F., Messina M., Franchin C., and Badini E. 2005. Pea protein concentrate as a substitute for fish meal protein in sea bass diets. *Italian Journal of Animal Sciences* 4(supp 2):597–599.

Tibaldi E., Tulli F., Piccolo G., and Guala S. 2003. Wheat gluten as a partial substitute for fish meal protein in sea bass (*D. labrax*) diets. *Italian Journal of Animal Sciences* 2(supp 1):613–615.

Tibbetts S.M., Bjornsson W.J., and McGinn P.J. 2015. Biochemical composition and amino acid profiles of *Nannochloropsis granulata* algal biomass before and after supercritical fluid CO_2 extraction at two processing temperatures. *Animal Feed Science and Technology* 204:62–71. https://doi.org/10.1016/j.anifeedsci.2015.04.006.

Tibbetts S.M., Patelakis S.J.J., Whitney-Lalonde C.G., Garrison L.L., Wall C.L., and MacQuarrie S.P. 2019. Nutrient composition and protein of microalgae meals produced from the marine prymnesiophyte *Pavlova* sp. 459 mass-cultivated in enclosed photobioreactors for potential use in salmonid aquafeeds. *Journal of Applied Phycology* 32:299–318. https://doi.org/10.1007/s10811-019-01942-2.

Tocher D.R. 2015. Omega-3 long-chain polyunsaturated fatty acids and aquaculture in perspective. *Aquaculture* 449:94–107. https://doi.org/10.1016/j.aquaculture.2015.01.010

Tokuşoglu Ö, and Ünal MK. 2003. Biomass nutrient profiles of three microalgae: *Spirulina platensis, Chlorella vulgaris*, and *Isochrysis galbana. Food and Chemistry Toxicology* 68(4):1144–1148. https://doi.org/10.1111/j.1365-2621.2003.tb09615.x.

Tonon T., Harvey D., Larson T.R., and Graham I.A. 2002. Long chain polyunsaturated fatty acid production and partitioning to triacylglycerols in four microalgae. *Phytochemistry* 61(1):15–24. https://doi.org/10.1016/S0031-9422(02)00201-7.

Ursu A.V., Marcati A., Sayd T., Sante-Lhoutellier V., Djelveh G, and Michaud P. 2014. Extraction, fractionation and functional properties of proteins from the microalgae *Chlorella vulgaris. Bioresource Technology* 157:134–139. https://doi.org/10.1016/j.biortech.2014.01.071.

Valdez A., Yepiz-Plascencia G., Ricca E., and Olmos J. 2014. First *Litopenaeus vannamei* WSSV 100% oral vaccination protection using CotC::Vp26 fusion protein displayed on *Bacillus subtilis* spores surface. *Journal of Applied Microbiology* 117(2):347–57. https://doi.org/10.1111/jam.12550.

Webster C.D., Thompson K.R., Morgan A.M., Grisby E.J., and Gannam A.L. 2000. Use of hempseed meal, poultry by-product meal, and canola meal in practical diets without fish meal for sunshine bass (*Morone chrysops×M. saxatilis*). *Aquaculture* 188(1–4):299–309. https://doi.org/10.1016/S0044-8486(00)00338-0.

Wilson R.P. 1994. Utilization of dietary carbohydrate by fish. *Aquaculture* 124(1–4): 67–80. https://doi.org/10.1016/0044-8486(94)90363-8.

World Bank. 2013. Fish to 2030: Prospects for fisheries and aquaculture. In *Agriculture and environmental services* Discussion Paper 03, ed. Kobayashi M., and Dey M.M. Washington, DC: World Bank. http://www.fao.org/3/i3640e/i3640e.pdf.

Xavier T.O., Michelato M., Vidal L.V.O., Furuya V.R.B., and Furuya W.M. 2014. Apparent protein and energy digestibility and amino acid availability of commercial meat and bone meal for nile tilapia, *Oreochromis niloticus. Journal of the World Aquaculture Society* 45(4):439–446. https://doi.org/10.1111/jwas.12127.

Yang Y., Xie S., Lei W., Zhu L., Zhu X., and Yang Y. 2004. Effect of replacement of fish meal by meat and bone meal and poultry by-product meal in diets on the growth and immune response of *Macrobrachium nipponense. Fish & Shellfish Immunology* 17(2):105–114. https://doi.org/10.1016/j.fsi.2003.11.006.

Zapata D.B., Lazo J.P., Herzka S.Z., and Viana M.T. 2016. The effect of substituting fishmeal with poultry by-product meal in diets for *Totoaba macdonaldi* juveniles. *Aquaculture Research* 47(6):1778–1789. https://doi.org/10.1111/are.12636.

Zhou Q.C., Zhao J., Li P., Wang H.L., and Wang L.G. 2011. Evaluation of poultry by-product meal in commercial diets for juvenile cobia (*Rachycentron canadum*). *Aquaculture* 322:122–127. https://doi.org/10.1016/j.aquaculture.2011.09.042.

Zmora O., Grosse D.J., Zou N., and Samocha T.M. 2013. Microalga for aquaculture: practical implications. In *Handbook of microalgal culture: Applied phycology and biotechnology*, Second edition, ed. Richmond A., and Hu Q. 628–652. West Sussex: Blackwell Publishing. https://doi.org/10.1002/9781118567166.ch34.

3 The Potential of Invasive Alien Fish Species as Novel Aquafeed Ingredients

Janice Alano Ragaza, Md. Sakhawat Hossain, and Vikas Kumar

CONTENTS

INVASIVE ALIEN FISH SPECIES

Invasive alien fish species are non-native fish species introduced into marine or freshwater systems beyond their native distribution range. The term "invasive" emphasizes the disruption of the ecological balance in a native habitat caused by non-native species (Kolar et al. 2005). Among vertebrates, freshwater fish species have topped the number of introductions (Crivelli 1995). There are two primary routes of introduction of invasive alien fish species to non-native regions – through aquaculture and the ornamental fish industry. Specifically, they are introduced for biological control, commercial fisheries, recreational fisheries, or ornamentation (Guerrero 2014). These are further classified as either intentional or accidental releases.

The escape of invasive alien species has received less attention compared to the pollution caused by aquaculture farms through the release of nutrients and organic wastes (Tuckett et al. 2016). There are several causes of the release of invasive alien fish species: (1) neglect in enclosing the invasive fish; (2) dilapidated states of facilities and enclosures on fish farms; (3) close proximity of fish farms to flood-prone areas; (4) absence of technical know-how and support about the risks of culturing invasive alien species; (5) deliberate releases due to bad management practices; (6) lack of concern for management of invasive alien species; and (7) overly simplified environmental licensing processes (Garcia et al. 2018). Hence, the escapes are largely inevitable due to lax and inefficient control systems. Once introduced in natural water bodies, invasive alien fish species can initiate significant impacts even before the establishment phase (Cunico and Vitule 2014). And those that do establish themselves in the invaded ecosystem can cause biotic homogenization (Daga et al. 2015).

Most of the highly traded ornamental fish species are usually found in places beyond their natural distribution range. Intentional release by aquarists, also known as aquarium dumping, occur commonly in small-sized aquarium fishes with lengths up to 10 cm (Garcia et al. 2018). As these

DOI: 10.1201/9780429331664-3

juveniles reach large sizes of more than 1 m, they require more space, which encourages further aquarium dumping (Magalhães 2015).

Invasive alien species are often dispersed through anthropogenic activities; most invasive alien species are successfully introduced in environments that have been degraded by people (Reid et al. 2019; Dudgeon et al. 2006). Most of these species are robust and can easily adapt. They reproduce and establish quickly in new habitats, altering the flora and fauna of the ecosystems they invade. The damage caused by these species in the ecosystem is difficult to quantify due to many factors that impact the colonized habitat (Nghiem et al. 2013). However, there are four main effects of the introduction of invasive alien species to non-native environments: (1) hybridization; (2) predation; (3) competition; and (4) ecosystem alteration.

Biotic homogenization or hybridization of invasive alien fish species with locally compatible fish species leads to loss of their biological distinctiveness or identity at functional, genetic, and taxonomic levels (Maini et al. 2016). It also reduces the fitness of native species. The "pure" native species often lose to more aggressive hybrids. Behavioral interaction, such as aggression, is also a means of competition for available resources.

Invasive alien species have direct and indirect effects on population, through competition or predation, and on changes in habitat structure, respectively (Gallardo et al. 2015). Predation causes production loss of commercial fish species. It also reduces the availability of fish protein to the local community when high-value fish are exported rather than consumed in the locality (Crivelli 1995). By occupying open niches in vulnerable ecosystems and causing competitive exclusion and eventual fish extinction, invasive alien fish species are implicated as a primary cause of displacement of native fishes. The native species are vulnerable because they have not co-evolved with the introduced species; they have not evolved the mechanisms that could have helped native species to avoid competition. If native species do not find an opportunity for refuge, then they are likely to be eliminated by predation and/or reduced survival of females due to apparent physiological stress.

The presence of invasive alien species is one of the primary factors in the loss of biodiversity across the world (Reid et al. 2019; Magbanua et al. 2017; Dudgeon et al. 2006). When invasive alien species are introduced into an environment, they cause trophic cascades that alter the ecosystem's food web (Gallardo et al. 2015). Invasive alien fish species cause habitat degradation and skew food production by changing nutrient dynamics (Hermoso et al. 2011; Capps and Flecker 2013). Introduction of invasive alien species may also lead to the transmission of infectious diseases (Reid et al. 2019). Invasive alien fish species may serve as vectors of emerging and exotic diseases.

The introduction of invasive alien fish species has drastically modified numerous ecosystems and promoted the drop and extinction of several species, including freshwater fishes. For example, three native fish species from the genus *Phoxinellus* and *Noemacheilus lendli* were driven to extinction due to the invasive pikeperch, *Stizostedion monoculus*, in the north Mediterranean zone (Crivelli 1995). In Brazil, the introduction and establishment of *Cichla monoculus*, *Astronotus ocellatus*, and *Pygocentrus nattereri* diminished the diversity and richness of the native fish community (Latini and Petrere 2004).

INVASIVE ALIEN FISH SPECIES AS AN AQUAFEED INGREDIENT

Several strategies are available for the population control of invasive alien species, such as removal by physical or mechanical means, use of chemicals, and through biological control. Invasivorism or the human consumption of invasive species, has transformed into a common approach to removing nuisance species (Nuñez et al. 2012). Nonetheless, invasivorism is still largely untested and limited by safety and consumer acceptance concerns. Recently, however, harvesting invasive alien species for utilization as an aquafeed ingredient has become a popular option. The harvesting approach is attractive because it is less costly than other mitigation strategies and revitalizes the commercial fishing industry. As the aquafeed and aquaculture industry is continuously confronted with rising costs amid the heightened pressure on the availability of feed ingredients, invasive alien species may likely fill in the shortages and demands without competing for human consumption. Moreover, since access to the free market chiefly determines whether a fish is used as food fish or rendered as feed (Tacon and Metian 2008; Tacon et al. 2006), underutilized invasive alien fish species are fit for fishmeal production.

At a growth rate of 5.8% per annum (FAO 2018), the aquafeed and aquaculture industry needs to supply sufficient feed inputs at a similar rate to meet the demand. The aquafeed and aquaculture industry could benefit from these non-traditional, underutilized, and sustainable raw material sources of fishmeal and fish oil or alternatives thereof. As both wild capture and aquaculture heavily burden global fisheries, the use of invasive alien fish species as an alternative fishmeal or fish oil source seems to be promising. The harvest-based mitigation is not only a practical approach to eradicate, or at least inhibit growth of, invasive alien species from the disturbed ecosystem, but it is also a means of income generation from its newly created market, with few environmental tradeoffs compared to other available options (Abarra et al. 2017). However, incentives including recreational and commercial opportunities, subsidies, low-interest loans, contract fisheries, and conservation harvest are necessary to support harvest-based control methods until the market is fully sustainable (Keevin and Garvey 2019; Conover et al. 2007).

The use of fishmeal in aquafeeds is expected to wane in the future; however, fishmeal will still serve as the preferred dietary protein source in aquafeeds especially for carnivorous species and emerging cultured species. Since fish demand higher protein compared to other livestock (Keembiyehetty and Gatlin 1992), fishmeal as a protein-dense ingredient meets this demand. Moreover, fishmeal has a balanced amino acid profile, feed palatability, and high digestibility (Bowzer and Trushenski 2015). However, its rising cost remains an impediment which forces feed manufacturers to search for alternative sources. Finding a replacement for fishmeal is especially daunting, as the quality of the alternative source must meet certain standards. Plant protein products are often preferred, but the chemical composition, digestibility, palatability, anti-nutrients, nutrient utilization, and functional inclusion of these products remain as hindrances to incorporating them into aquafeeds (Glencross et al. 2007).

A potential alternative to traditional fishmeal derived from marine fish is rendered fishmeal from invasive alien fish species. Fishmeal derived from these fish could be a suitable alternative, as it mirrors traditional/standard fishmeal sources and is not likely to possess the undesirable characteristics of plant- or other animal-based alternatives. Although using invasive alien fish species as aquafeed is a viable short-term approach, it is still limited due to the lack of studies on invasive alien fish species as nutritional sources for aquaculture (Maini et al. 2016). For instance, the nutrient composition of fishes rendered into fishmeal is known to vary by species, location, and season of harvest (Boran et al. 2008; Bragadóttir et al. 2004). Within species, age, sex, diet, and feed intake affect the proximate composition of the fishmeal rendered from alternative fish proteins. The quality of the fishmeal is also determined by its nutrient content, contamination levels, digestibility, and oxidative stability.

This chapter discusses a number of invasive alien fish species, such as *Chitala ornata*, *Pterygoplichthys* spp., *Pangasius* spp., and Asian carps that have been introduced to non-native regions, how they affect the native ecosystem, and mitigation strategies used to control their population. This chapter also focuses on their nutritive profile when processed as fishmeal or fish oil replacement as well as their potential and actual use, benefits, and disadvantages when used as an animal feed, specifically as an aquafeed.

INVASIVE ALIEN FISH SPECIES

CHITALA ORNATA

Chitala ornata (Gray 1831), commonly known as clown knifefish, clown featherback, or spotted knifefish is a carnivorous fish with a long and continuous anal fin extending to its caudal fin – resembling a compressed knife-shaped body (Rainboth 1996). It is a large predator belonging to the Notopteridae family (Roberts 1992) and native to the Mekong River in Vietnam (Poulsen et al. 2004). It has been introduced to the Philippines, Singapore, Sri Lanka, and other Southeast Asian countries primarily through ornamental fish aquarium trade (Guerrero 2014; Kumudinie and Wijeyaratne 2005). Floods promote its rapid dispersal (Kumudinie and Wijeyaratne 2005).

Since the introduction of *C. ornata* into various waters, there has been a decline in the abundance of native fish species followed by their eventual displacement. The abundance of native fish species, including but not limited to *Amblypharyngodon melettinus*, *Aplochielus dayi*, *Aplochielus parvus*, *Horadandiya athukorali*, *Puntuis bimaculatus*, *Puntuis vittatus*, and *Rasbora daniconius* (Gunawardena 2002), has been greatly reduced, while lacustrine species have been displaced (Corpuz 2018). In fact at one time, *C. ornata* comprised 65% of the major fish catch from a large bay in the Philippines (Cuvin-Aralar 2014). It competes for food and space, especially with carnivorous native fish species. As a carnivorous fish, it prefers large crustaceans and teleosts (Corpuz 2018), which are likely endemic to the ecosystem it invades. It has also been observed to prey on various commercial cultured fish and even on its own kind. It exhibits a territorial behavior (Kumudinie and Wijeyaratne 2005), which displaces other native fish. Moreover, hybridization with other members of the Notopteridae family may likely occur.

To mitigate its population, the fisherfolk catch both *C. ornata* embryos and adults from the invaded waters. Putting fishing pressure against the guarding *C. ornata* male makes manual retrieval or electrocution of the embryos from nests accessible (Castro et al. 2018). The adults caught are usually sold in the market. In Indonesia, Sri Lanka, and Thailand, *C. ornata* is a popular food fish (Guerrero 2014; Kumudinie and Wijeyaratne 2005). Although in other Southeast Asian countries, it has low market demand and fetches less than 0.50 USD per kg (Abarra et al. 2017). Generally, it is still considered an exotic, non-staple food fish. Despite this predicament, some national governments have launched a reward scheme for fisherfolk to continue fishing for *C. ornata* (Mayuga 2016). For example, fisherfolk receive monetary remuneration for every kg of *C. ornata* adult caught. Moreover, value-adding endeavors, such as post-harvest processing of *C. ornata* skin and carcass to leather and fishmeal, respectively, have also been proposed as an economic utilization strategy.

The use of *C. ornata* as an aquafeed ingredient is scarcely reported in literature. Abarra et al. (2017) used conventional methods of preparing fishmeal (i.e. cutting, boiling for separation of water and oil, pressing, drying, and grinding to appropriate size) to convert the fish's carcass to processed *C. ornata* fishmeal containing 8.5% moisture, 3.2% lipid, 62.3% protein, and 22.8% ash. The *C. ornata* fish were caught from the major tributaries of Laguna Bay, where most of the aquaculture farms in the Central Philippines are located. The *C. ornata* fishmeal replaced 0, 25, 50, 75 and 100% of the local fishmeal in the diets. The authors observed best fish performance when 75% *C. ornata* fishmeal replaced the local fishmeal in the diets of juvenile Nile tilapia *Oreochromis niloticus*, the second most important cultured fish in the Philippines. Feed intake remained high as well, indicating no change in feed palatability or preference of the tilapia juveniles. The processed fishmeal from *C. ornata* was safe to use as a fishmeal substitute; negative alterations in the serum and hepatic enzymes and liver morphologies were absent. Although the initial results showed potential, there is still a need to profile its other nutritive components (i.e. fatty acid and amino acid components), which are obligatory requirements in aquaculture nutrition. Its inclusion in diets of other high-value fish and shellfish species, particularly those that are carnivorous, must also be addressed.

Pterygoplichthys spp.

Pterygoplichthys spp. have been customarily sold as food fish in their native Amazon and Orinoco drainages (Hossain et al. 2018); however, these fishes are more commonly sought as aquarium fish around the world. It is highly favored by aquarists because of its unique appearance and its ability to keep tanks clean due to its algivorous feeding habit. As a natural scavenger, it prefers to stay in the murkiest part of the aquarium (Aranda et al. 2010). This behavior is also exhibited in the wild as it lives in the dark, muddy bottoms of waters.

The term Pterygoplichthys originated from the Greek words pteryg, meaning "wing" and hoplon, meaning "weapon" (Aranda et al. 2010). Like other loricariids, *Pterygoplichthys* spp. have a large bony plate and a ventral mouth (Page and Robins 2006). The genus *Pterygoplichthys* was previously divided into three subgenera: *Glyptoperichtys*, *Liposarcus*, and *Pterygoplichtys* (Weber 1991). Currently, *Liposarcus* is synonymous to *Pterygoplichthys* (Armbruster 1997). The nine or more dorsal-fin rays exhibited by the members of the genus *Pterygoplichthys*, also known as sailfin fishes, differentiate them from other loricariids (Hossain et al. 2018). These fishes are reported to grow up to 60 cm in length and 3 kg in weight (Aranda et al. 2010). Four phylogenetically related species, which invaded Southeast Asian waters, namely, *Pterygoplichthys anisitsi*, *Pterygoplichthys disjunctivis*, *Pterygoplichthys multiradiatus*, and *Pterygoplichthys pardalis* share similar external morphologies, such as a supraoccipital bone enclosed posteriorly by three scutes and an absent elevated supraoccipital process (Page and Robins 2006). Abdominal pigmentation varies among the four species.

P. anisitsi (Eigenmann and Kennedy 1903) commonly known as Southern sailfin catfish or snow pleco is native to tropical South America (i.e. Argentina, Brazil, Paraguay, and Uruguay) while *P. multiradiatus* (Hancock 1828), the radiated pleco or Orinoco sailfin catfish, is native to Venezuela and Argentina (Hoover et al. 2004). The former has light spots on a dark background while the latter has uncombined dark spots on a light background (Page and Robins 2006).

P. disjunctivus (Weber 1991) is endemic to the Amazon River basin of Brazil and Peru (Weber 2003). It is commonly called vermiculated sailfin catfish or suckermouth sailfin catfish. Likewise, another loricariid catfish species, *P. pardalis* (Castelnau 1855), which is also known as the Amazon sailfin catfish, is native to the Rio Madeira drainage of Bolivia and Brazil (Page and Robins 2006). Both species exhibit dorsally combined dark spots on a light background. *P. disjunctivus* has combined dark spots on the abdomen to form a vermiculated pattern while *P. pardalis* has an abdomen covered with separate and distinct dark spots (Page and Robins 2006). Fish with intermediary abdominal pattern shared by the two *Pterygoplichthys* spp. have been reported in the Philippines. The fish were characterized by undefined species-specific DNA barcodes, suggestive of introgression between the two *Pterygoplichthys* spp. (Quilang and Yu 2015; Jumawan et al. 2011).

Pterygoplichthys spp. have widely invaded over 20 countries spanning at least 5 continents beyond their native ranges, including Asian countries: Bangladesh, India, Indonesia, Malaysia, Philippines, Singapore, Taiwan, Thailand, and Vietnam (Hossain et al. 2018; Chaichana and Jongphadungkiet 2013; Page and Robins 2006; Chaves et al. 2006). Some are also found in Hawaii, Mexico, Puerto Rico, and the United States (Hossain et al. 2018). The main routes of introduction were through accidental or intentional aquarium release for control of aquatic weeds, or through escape from aquaculture farms, with dispersal facilitated by hurricanes, storms, floods, fishing activities, and karst geology (Nico et al. 2012).

Environmental degradation and ecological disruption have been observed after the introduction of *Pterygoplichthys* spp. Due to its tough but flexible bony armor, the fish damages aquaculture pens, nets, and other fishing implements. As a nest-building fish species, it disturbs the water's muddy bottom by its burrowing behavior, causing turbid waters and eroded riverbank soils (Jumawan and Herrera 2014), which in turn results in siltation problems in water reservoirs (Devick 1989). Its abundant fecal matter was also reported to generate a nutrient influx and promote algal bloom (Rubio et al. 2016). Large fish-eating reptiles, birds, and endangered species, such as brown pelicans *Pelecanus occidentalis* have been reported to have difficulty in swallowing this bony armored catfish (Karunarathna et al. 2008; Bunkley-Wiliiams et al. 1994). It was also reported to induce altered behaviors such as heightened activity in manatees *Trichechus manatus* through attachment and algal grazing from the marine mammal (Gibbs et al. 2010). *Pterygoplichthys* spp. also compete with native fishes and other endemic aquatic organisms for resources, resulting in a likely alteration of the food web dynamics. *Pterygoplichthys* spp. are mainly herbivorous, competing with other herbivores, such as aquatic insects and arthropods (Page and Robins 2006). A recent study, however, showed that *P. pardalis* readily ingested eggs and fry of *O. niloticus*, with greater preference for sessile eggs over mobile fry (Chaichana and Jongphadungkiet 2013).

The females of this fish are iteroparous batch spawners (Jumawan and Herrera 2014), meaning they can produce thousands of eggs multiple times. The females, particularly *P. disjunctivus* have also been observed spawning even at small body sizes (Gibbs et al. 2008). Moreover, with no known predators and a high tolerance for polluted inland waters, controlling its population has been a challenge. At one time in Laguna Bay, the Philippines, between 27% and 63% of the major fish catch was dominated by *Pterygoplichthys* spp. (Cuvin-Aralar 2014). In some parts of India, it represented 80% of the fish catch (Indrajith 2016). It is largely considered trash fish and an unmarketable by-catch (Orfinger et al. 2019).

Although the fish's meat is edible, the heavy metal contents of its flesh are still contested. Low concentrations of arsenic, cadmium, chromium, and mercury were recorded during the wet season but were absent during the dry season (Chavez et al. 2006). Moreover, lead concentrations between 0.0583 and 0.1900 mg kg^{-1} (in wet weight) were detected in the flesh (Chavez et al. 2006). Regardless, the campaign for marketing *Pterygoplichthys* spp. as food fish is still in its infancy in Asian countries. In South America, however, the fish is usually degutted and its flesh grilled or boiled to a soup stock.

Besides utilizing it as food fish, the fish's oil and bones have other unconventional uses. Oil from the fish can be converted to biodiesel (Anguebes-Franseschi et al. 2019; Amurao 2006). Approximately 1:1 (fish oil: biodiesel) conversion was achieved through transesterification of the fish oil. The product was reported to have similar properties as other fuels, such as commercial diesel, coco-diesel, denatured alcohol, and kerosene (Amurao 2006). The bones, which make up most of the fish's body mass were also converted to carbon filters for water purification (Alquitran 2006). The carbon filters were made through calcium extraction from bones (Piencenaves 2013; Aranda et al. 2010). Government subsidization and incentivization are also offered for commercial and recreational fishing of *Pterygoplichthys* spp. The fish's skin has been proposed for use as raw material for leather and décor and its carcass for fertilizer or fishmeal conversion (Hubilla et al. 2007).

Only *P. disjunctivus* and *P. pardalis* fishmeal have been evaluated for nutritive composition and as potential feed ingredient or supplement. To date, no information is available on the evaluation of *P. anisitsi* and *P. multiradiatus* fishmeal. Generally, *Pterygoplichthys* spp. fishmeal are rich in protein. The crude protein of the processed fishmeal from *P. pardalis* and *P. disjunctivus* ranged from 16.5% to 54.5% in dry matter (Panase et al 2018; Mohanty et al. 2017; Indrajith 2016; Asnawi et al. 2015; Olut and Roxas 2007). Moreover, when the fish's protein is hydrolyzed, it exhibits high antioxidant activity (Guo et al. 2019). Fishmeal from *P. disjunctivus* contained high levels of essential amino acids, such as phenylalanine and leucine at 17.55% and 15.15%, respectively and nonessential amino acids, such as alanine and serine at 7.44% and 3.67%, respectively (Mohanty et al. 2017). *P. pardalis* fishmeal, on the other hand, had lysine and methionine levels at 2.95% and 0.79%, respectively, but lacked histidine and cystine (Asnawi et al. 2015). It was also reported to contain albumin at a level of 4.29 g 100 mL^{-1} (Asnawi et al. 2015), which could impart antioxidant activity and other biological functions.

Although lipid levels from *Pterygoplichthys* spp. fishmeal are highly variable, these values are generally low. *P. pardalis* fishmeal contained 10.63–16.85% fat in dry matter (Panase et al. 2018; Ansawi et al. 2015). *P. disjunctivus* fishmeal showed low fat at 0.79% in dry matter (Mohanty et al. 2017), although another study analyzed fat levels as high as 10.2–14.6% in dry matter (Olut and Roxas 2007). Moreover, *P. disjunctivus* fishmeal was observed to have high amounts of fatty acids, such as decanoic acid, arachidonic acid, and docosahexaenoic acid at levels of 16.42, 15.32, and 13.90%, respectively (Mohanty et al. 2017).

Ash content from *Pterygoplichthys* spp. fishmeal varied from and among species to species. *P. disjunctivus* was reported to contain as low as 1.03–40.30% ash in dry matter (Mohanty et al. 2017; Olut and Roxas 2007) while *P. pardalis* had analyzed values ranging between 21.72% and 33.25%

in dry matter (Panase et al. 2018; Asnawi et al. 2015). Macro-minerals, such as Na, K, Ca, Mg, and P were all detected, with P and K at peak levels of 2220 and 1780 mg kg^{-1}, respectively; micro-minerals, such as Fe, Zn, and Mn were also present, with Fe levels reaching 22.21 mg kg^{-1} (Mohanty et al. 2017). Moisture contents of the fishmeal were at 8.04–10.63% for *P. pardalis* (Panase et al. 2018; Indrajith 2016) and at 7.17–12.0% for *P. disjunctivus* (Olut and Roxas 2007).

The differences in the nutritive components of the processed fishmeal vary due to several factors. The boiling process usually decreases the protein, lipid, and fiber contents of the fishmeal. The apparent losses in protein, lipid, and fiber from the boiling process increase the ash content.

Panase et al. (2018) first tested and used *P. pardalis* fishmeal for partial and complete replacement of commercial fishmeal in the diets of Mekong giant catfish, *Pangasianodon gigas*, juveniles. The Mekong giant catfish is one of the largest cultured freshwater fishes in the world and fetches a price point of 6–9 USD kg^{-1} (Panase et al. 2018). The *P. pardalis* fish were collected from the Phayao Lake of Thailand. The collected fish were pooled together, with visceral organs removed, and had undergone conventional fishmeal preparation to produce 300 g of fishmeal per kg of fresh *P. pardalis* (Panase et al. 2018). The *P. pardalis* fishmeal substituted 0, 25, 50, 75 and 100% of the commercial fishmeal in the diets. The growth and feed utilization efficiency of *P. gigas* fed *P. pardalis* fishmeal were on par with those fed commercial fishmeal diets; however, those fed 75% and 100% *P. pardalis* fishmeal exhibited lower performance, although insignificantly. Serum biochemical parameters were also similar between commercial fishmeal and *P. pardalis* fishmeal treatments. The results are not surprising, as dietary levels between 30–70% of animal meals are usually accepted by omnivorous and carnivorous fish (Millamena 2002). Moreover, the crude protein (54%) and lipid (6%) in the *P. pardalis* fishmeal were also comparable to those of commercial fishmeal.

To date, there is no information available on the use of *P. disjunctivus* as a fishmeal substitute in fish diets. Nonetheless, it has been tested as a broiler (chicken) feed ingredient by Olut and Roxas (2007). Two inclusion levels (i.e. 5% and 10%), which were either cooked or uncooked *P. disjunctivus* fishmeal, were tested and compared against soya- and standard fishmeal-based diets for day-old broiler chicks (Arbor Acre strain, straight run). Uncooked fishmeal lacked the process of boiling for separation of water and oil. Broilers fed *P. disjunctivus* fishmeal and standard fishmeal had similar live weight gain and feed efficiency. The growth and feed performances of the broiler chicks were independent of the cooking process. Feed intake also increased in broilers fed cooked fishmeal, suggesting high palatability and acceptability of the cooked *P. disjunctivus* fishmeal. With enhanced growth and feed efficiency, diets with the *P. disjunctivus* fishmeal showed higher returns over feed and chick costs.

In Mexico, *Pterygoplichthys* spp. have been converted to fish silage acid as animal feed for young bulls and lambs. The fish silage acid is reported to contain protein, lipid, fiber, and ash at 50.18, 21.80, 0.10, and 23.34% in dry matter, respectively (Salas et al. 2011). Its production cost was comparable to that of soybean meal at 0.14–0.35 USD kg^{-1} (Salas et al. 2011). In one study, young bulls (*Bos taurus* x *Bos indicus*) were fed diets containing 0, 12, and 18% fish silage acid (Ornelas et al. 2011). There were no differences observed in daily gain or in feed conversion. Likewise, in another study, uncastrated Criollo-Blackbelly crossed lambs fed 0, 9, 18, and 27% fish silage acid in diets showed similar average daily gain and feed intake (Tejeda-Arroyo et al. 2015). The fish silage acid in the latter study replaced plant-based protein sources in the lamb diets.

Heavy metal accumulation is a limiting factor in using *Pterygoplichthys* spp. as a fishmeal source compared to other species used for standard fishmeal. The metals enter the body through direct and indirect mechanisms. The direct mechanism occurs through metal absorption when fish absorb water and nutrients into their bodies. Indirect mechanism, on the other hand, occurs through the food chain. Along the food chain, the transfer between material and energy from the hunting organisms occurs (Puspasari 2006). The concentration of heavy metal in *Pterygoplichthys* spp. depends on water quality condition of the habitat. Therefore, *Pterygoplichthys* spp. living in heavily contaminated water contain high levels of heavy metals (Elfidasari et al. 2018). Hence, although difficult, collection of *Pterygoplichthys* spp. from non-polluted waters is highly recommended.

Pangasius spp.

Pangasius spp., often referred to as shark catfishes or Asian catfishes, are native to South and Southeast Asia (Yoğurtçuoğlu and Ekmekçi 2018). The members of the genus *Pangasius* have six branched pelvic fin rays, two maxillary barbels, and two mandibular barbels (Roberts and Vidthayanon 1991). Pangasiids are among the largest freshwater fishes in the world, with maximum length and weight reaching 3 m and 350 kg, respectively (Stone 2007; Roberts and Vidthayanon 1991).

The striped catfish *Pangasius hypophthalmus* (Sauvage 1878) or *Pangasianodon hypophthalmus* (Sauvage 1878), formerly known as *Pangasius sutchi* (Fowler 1937), is native to several river basins of Vietnam and Thailand. In its natural habitat, the striped catfish reaches up to 130 cm in length and 44 kg in weight (Garcia et al. 2018). It is a migratory fish – moving from upstream to spawn more than a million eggs during the wet, rainy season in downstream sand banks and channels (Jayaneththi 2015; Van Zalinge et al. 2002). It is omnivorous, which make eggs and larvae of native species a likely food source. It not only competes with endemic species for food (Rahim et al. 2013), it is also an active predator of native aquatic fauna.

Pangasiids migrate over long distances, which enable them to disperse more easily beyond their native ranges. They are also tolerant and resistant to extreme water quality values such as low dissolved oxygen, moderate salinity, high temperature, and a wide range of pH. They are robust air-breathing fish species that endure intense farming in highly polluted waters (Srivastava et al. 2014).

The main routes for invasion are intentional introduction for aquaculture, escapes from fish farms, and accidental introduction from the ornamental fish industry. *P. hypophthalmus* has been introduced to several regions outside of its native ranges, such as Bangladesh, China, Colombia, Guam, India, Indonesia, Israel, Malaysia, Myanmar, Philippines, Singapore, Sri Lanka, and Taiwan (Tarkan et al. 2020; Garcia et al. 2018; NACA 2005). In the Philippines, about 23% of the major fish catch was composed of *Pangasius* sp. (Cuvin-Aralar 2014). Unexpectedly, a few specimens of *P. hypophthalmus* have also been collected for the first time in a park pond in Poland (Wiecaszek et al. 2009), a canal in Iraq (Khamees et al. 2013), and a lake in Israel (Snovsky and Golani 2012) in the years 2009, 2011, and 2012, respectively. Because of its rapid growth, adaptability, and inexpensive price, it is considered an important fish protein source worldwide. Moreover, it can be cultured independently of the stocking density and the type of feeds used (Kumar et al. 2017). Its unique body shape, which is likened to a shark (hence the popular aquarium trade name "iridescent shark") makes it a favorite among ornamental fish hobbyists.

In 2016, *Pangasius* spp. comprised 3% of the aquatic species cultured in the world. More than 90% of the farmed *P. hypophthalmus* is exported to over 100 countries (Phuong and Oanh 2009). The whitefish market segment is now increasingly shared between the wild species (i.e. cod and Alaska pollock) and the lower-priced species (i.e. *Pangasius* spp. and tilapia). Vietnam remains the largest producer of *Pangasius* spp., with exports amounting to 7.3 billion USD in 2016 (FAO, 2018). The global production of *Pangasius* spp. is still rapidly growing.

Although there are several benefits from commercial aquaculture of *P. hypophthalmus*, the striped catfish is reported to be unsafe for human consumption. In India, consumption of the striped catfish has been reported to lead to high carcinogenic risk due to accumulation of heavy metals in the catfish (Srivastava et al. 2014). Elevated concentrations of mercury at an average of 0.22 mg kg^{-1} have been detected in commercially sold frozen and marinated striped catfish fillets (Rodríguez et al. 2018). Moreover, the striped catfish has served as a host to a variety of parasites, such as the trematode metacercariae (Thuy et al. 2010a), monogenean parasites (Chaudhary et al. 2014), and myxosporean species (Baska et al. 2009). Critical diseases (i.e. hemorrhagic septicemia and bacillary diseases) have been associated with intensive farming of *P. hypophthalmus* in India (Lakra and Singh 2010). Humans and other aquatic animals could risk infection when the effluent water from the intensive culture ponds is not properly treated before discharge into receiving waters or canals (Tarkan et al. 2020). The unregulated cultivation of striped catfish results in poor water quality and pollution which increases the incidence of fish disease and mortality. Fish farmers resort to the use of chemicals and antibiotics, such as malachite green, which further impedes fish exportation (Thuy et al. 2007).

Several countries have started to prohibit the import, aquarium trade, and/or culture of pangasiids into their territories. For example, the culture of *P. hypophthalmus* and *Pangasius pangasii* (Hamilton 1822) has been banned in South Africa, but their importation in aquarium trade is still permitted (Mäkinen et al. 2013). Although the importation of *P. hypophthalmus* is disallowed in Brazil, it is freely marketed throughout the country, with a strong call and support for its farming by the local aquaculture industry (Garcia et al. 2018). In Mexico, recent risk assessments for ornamental fish species have resulted in a ban on importing and culturing of *Pangasius* spp. (Golubov et al. 2017). Likewise, after risk analysis, the culturing of *P. hypophthalmus* has been discouraged and prevented in India (Singh and Lakra 2010).

P. hypophthalmus by-product meals have been tested and used as either fishmeal/fish oil substitute or an additive in livestock feeds for terrestrial animals, such as young pigs, chickens, and ducks. The studies were exclusively conducted in Vietnam, specifically in the Mekong Delta, where significant amounts of by-products from catfish cultivation and fillet processing are produced. These underutilized by-products are usually treated as wastes and release strong odors – necessitating proper disposal management.

P. hypophthalmus by-products, such as bone, head, scrap meat, skin, and visceral organs make up 62–67% of the whole catfish (Thuy et al. 2007) and are yielded in quantities of 710–900 tons d^{-1} (Thuy et al. 2007). After removal of the catfish fillet, liver, stomach, and swimming bladder, which are sold to local restaurants, the leftover by-products are ground, boiled, and the oil extracted. The oil is collected and mixed together with other marine water fish for commercial fish sauce production. Thereafter, the residue is dried to generate the catfish by-product meal (Thuy et al. 2011; Thuy et al. 2007). The final product is either classified as wet or dry/oil-extracted catfish residue meals (Tuan 2010). Wet catfish residue meals are ground, cooked, and mixed with other feed components for immediate use as animal feed (Thuy et al. 2007).

Generally, the protein and fat contents in catfish by-product meals are high, at levels of 27–71.5% and 1.7–33.8% in dry matter, respectively (Tuan and Ogle 2019; Linh et al. 2018; Thuy et al. 2010b; Thuy et al. 2010c; Thuy et al. 2007). The protein levels, however, are lower than standard fishmeal. The ash- and nitrogen-free extracts also vary among catfish by-products meals, at 3.5–45.6% and 5.7–28.9% in dry matter, respectively (Tuan and Ogle 2019; Linh et al. 2018; Thuy et al. 2010b; Thuy et al. 2010c; Thuy et al. 2007). The variation in nutritive components depends on the type of input by-product material.

Catfish by-product meals made from a combination of rejected fillet, scrap meat, skin, head, and bone provide 14.5 MJ kg^{-1} metabolizable energy (Linh et al. 2018), 1.65–1.70% acid detergent fiber (Thuy et al. 2010b; Thuy et al. 2010c), and 3.32–8.80% neutral detergent fiber (Thuy et al. 2010b; Thuy et al. 2010c). Acid and neutral detergent fibers are lower in catfish by-product meals made from scrap meat and skin or from bone and head. Calcium and phosphorus contents in meals made from bone and head vary at 7.75–13.20% and 2.1–3.1%, respectively (Thuy et al. 2007). In meals made from scrap meat and skin, calcium and phosphorus contents are lower at 3.9–7.5% and 0.7–1.6%, respectively (Thuy et al. 2007).

Regardless of the input by-product material, the meals are rich in essential amino acids, such as lysine, methionine, threonine, and phenylalanine, among others. However, meals made from scrap meat and skin have a more balanced essential amino acid profile compared to head and bone meals. The amino acid profiles of catfish by-product meals and standard fishmeal exhibit similarities. The catfish by-product meals also contain saturated and unsaturated fatty acids.

Among the different catfish by-products, meals processed from scrap meat and skin exhibit the highest values for protein and lipid but lowest for ash, calcium, crude fiber, nitrogen-free extract, and phosphorus contents. Nonetheless, this type of meal is the least produced due to the small quantities of scrap meat and skin. On the other hand, meals from the bone and head, which are the most common and abundant type of catfish by-product meal, show the highest levels of ash, calcium, and phosphorus.

The catfish by-products can be also converted to protein hydrolysates with varying molecular weights using the enzymes alcalase (Mihn 2014) or bromelain at a pH of 6.5 and a drying temperature of 55 C for 120 min (Thuy and Ha 2016; Hien et al. 2015). These protein hydrolysates contain higher protein (i.e. 60% in dry matter) and approximately double the essential amino acid levels.

In the diets of Chinese Luong Phuong chickens, catfish by-product meal replaced 0, 50, and 100% of the fishmeal ingredient (Thuy 2012). Protein digestibility in the ileal and total tract were significantly reduced when catfish by-product meal was used. The lower protein digestibility is due to the less balanced amino acid profile of the catfish by-product meal caused by heat processing. On the other hand, lipid digestibility was higher due to the polyunsaturated fatty acids in the catfish by-product meals. The lipid content in the breast meat was also higher. The major fatty acids in the breast meat were α-linolenic acid, linolenic acid, oleic acid, and palmitic acid. These four fatty acids increased slightly with increasing replacement levels. The growth and feed performance of the chickens, however, remained similar.

Likewise, fishmeal protein from diets of crossbred Muscovy ducks was replaced by catfish by-product meal at 0, 25, 50, 75, and 100% (Linh et al. 2018). Replacement levels up to 75% showed enhanced feed intake and a slight improvement in growth. The depressed growth and feed intake in ducks fed diets with complete replacement were due to the high lipid level in the diet. There was also an increase in heart weight, although the lipid contents in the breast meat remained similar.

Catfish by-product hydrolysates have replaced fishmeal at 0, 25, 50, 75 and 100% in diets of crossbred Yorkshire x Landrace weaned piglets (Thuy and Ha 2016). Piglets showed increasing growth and feed performance while decreasing diarrheal incidence and fecal scores as the levels of hydrolysates were also increased. The proteins from the catfish by-products were hydrolyzed into low molecular weight peptides and amino acids, enabling efficient digestion and absorption while maintaining the integrity of the small intestines. In theory, the piglets were given partially predigested proteins that advanced the digestion process. Complete replacement with catfish by-product hydrolysates also reduced the feed cost to 88% lesser than the cost of a fishmeal-based diet.

Ensiling of catfish by-products is also employed as a cheaper alternative to the usual drying or oil extraction. This cheaper method involves the inclusion of a fermentable carbohydrate-rich additive to the catfish by-products to rapidly decrease the pH. The resulting liquid silage after fermentation is then stored. The standard additive used in ensiling is lactic acid bacteria, but other substances such as molasses and rice bran are as effective. The nutritive profile of ensiled catfish by-product meals depends on the type and amount of additive used. Additives that contain high water-soluble carbohydrates and with rapid fermentation rates are excellent ensiling agents. Generally, the nutritive components of ensiled catfish by-product meals are lower when compared to other processed catfish by-product meals. Nonetheless, ensiled catfish by-product meals are rich in non-dissociated organic acids which inhibit spoilage microorganisms (Pahlow et al. 2003).

Crossbred Yorkshire x Landrace male pigs fed ensiled catfish by-product meals exhibited increased ileal and total tract crude protein digestibility (Thuy et al. 2010b). Bacterial fermentation improves nutrient digestibility of feed (Hong and Lindberg 2007). The low pH and high organic acid content in the fermented feed catalyze a change in the gastrointestinal microbiota. In a separate study, feed intake was also higher in pigs fed catfish by-product meals ensiled with molasses rice bran compared to those fed a basal diet containing solely plant protein sources and a fishmeal-based diet (Thuy et al. 2010c).

Processing of catfish by-product into meals generates large volumes of protein-rich wastewater which are often discarded (Thuy and Ha 2016; Thuy et al. 2011). The wastewater has also been tested and used as an ingredient in animal feeds (Thuy et al. 2011). Although the dry matter and crude protein contents of the wastewater are dependent on the process and the composition of the input by-product material, the values range between 24.2–29.2% dry matter and 13.5–38.0% crude protein (Thuy et al. 2011; Thuy et al. 2010b; Thuy et al. 2007). The dry matter content in wastewater by-product is low due to the addition of water before cooking (Thuy et al. 2010b).

Lipid and ash levels of wastewater by-product are at 6.3–19.8% and 15.4–33.8% in dry matter, respectively (Thuy et al. 2011; Thuy et al. 2010b; Thuy et al. 2007). The lipid content in wastewater by-product is high due to the inefficient oil extraction method used (Thuy et al. 2010b). The wastewater by-product provides 13.2 MJ kg^{-1} metabolizable energy, 1.5–1.6% acid detergent fiber, and 3.1–5.5% neutral detergent fiber (Thuy et al. 2011; Thuy et al. 2010b). Among the catfish by-product meals, the lowest protein and lipid while highest crude fiber and nitrogen-free extract values are observed in meals generated from wastewater and waste matter (Thuy et al. 2007).

Although essential amino acids are present in the wastewater by-product, the profile is considered less balanced. It contains 36.3–42.5, 10.6–18.9, 11.2–12.1, and 12.1–12.9 g kg^{-1} of lysine, methionine, threonine, and phenylalanine, respectively (Thuy et al. 2011; Thuy et al. 2010b). Saturated and unsaturated fatty acids are also present.

Thuy et al. (2011) tested 0, 25, 50, 75, and 100% of the crude protein from fishmeal with wastewater by-product in diets of crossbred Yorkshire x Landrace male pigs. Although feed intake and daily weight gain were reduced significantly with increasing wastewater by-product inclusion, there was a marked improvement in feed conversion ratio and feed cost. The reduced feed intake was attributed to the high fat content which is vulnerable to deterioration resulting in diminished palatability of the feed. Complete replacement with wastewater by-product brings down the feed cost to 68% less than the cost of a fishmeal-based diet.

Oil extracted from the striped catfish has also been tested and used as a lipid substitute in the diets of pigs. Crossbred Landrace x Yorkshire pigs fed diets with 5% catfish oil showed comparable weight gain and feed intake values. Backfat thickness and dry matter carcass content were lower in pigs fed 5% catfish oil compared to those fed 5% coconut oil (Men et al. 2007). The reduction is partly attributed to the difference in fatty acid compositions between catfish and coconut oils. Catfish oil is richer in polyunsaturated fatty acids, while coconut oil is known for high concentrations of saturated fatty acids. Catfish oils are reported to contain 67.7–75.0% polyunsaturated fatty acids (Men et al. 2007; Sathivel et al. 2003), which are more easily digested and absorbed than saturated fatty acids (Thuy et al. 2010b). In a separate study, pigs fed 10% catfish oil exhibited softer fat deposits (Men et al. 2003), which is linked to the polyunsaturated fatty acids in the catfish oil. Although the daily weight gain decreased in pigs fed catfish oil treatments, the highest economic benefit and lowest feed cost per kg weight gain were attained.

One study by Gustavsson (2016) reported that complete replacement of standard fishmeal with catfish bone and head by-product meal in *O. niloticus* diets resulted in positive growth and feed performance. The control diet containing conventional fishmeal as the main protein source showed similar crude protein, lipid, dry matter, and ash contents as the diet with 100% catfish bone and head by-product meal. Nonetheless, no other parameters were evaluated. There is clearly a lack of studies that use and test catfish by-product meals as an aquafeed ingredient. Moreover, this limitation is further hindered by the difficulty of identifying the types of fish used in producing commercial fishmeal in the market.

CARPS

Carps belong to a very large group of freshwater fish of the family Cyprinidae, native to Asia and Europe. Carps have invaded many environs around the world in varying degrees. For example, the common carp (*Cyprinus carpio*) has been established in US waters for more than a century. The newest carp invaders, such as the black carp, bighead carp, grass carp, and silver carp are collectively called "Asian carps".

Asian carps exhibit similar "mucking" behavior in the common carp *C. carpio*. Their foraging in the benthic environment resuspends sediments which increase turbidity, nutrient levels, and phytoplankton production (Keevin and Garvey 2019). The suspended sediments are considered the culprit for the reduction in the benthic invertebrate abundance, richness, and diversity. As the macrophyte-dominated waters become turbid, phytoplankton take over, which in turn changes the abundance and diversity of native fish species.

The silver carp, *Hypophthalmichthys molitrix* (Valenciennes in Cuvier and Valenciennes 1844), and the bighead carp, *Hypophthalmichthys nobilis* (Richardson 1845), are large, planktivorous carps widely considered food fish worldwide. These two species are native to eastern Asia, Russia, China, and northern Vietnam. They are known to be invasive in North America, where they have become abundant in major river basins and lakes, such as the Mississippi River, Illinois River, and even the Great Lakes (Bowzer et al. 2014; Tsehaye et al. 2013; McClelland et al. 2012). Both species are voracious filter feeders and feed on various plankton (Kolar et al. 2005; Cremer and Smitherman 1980). They have been introduced to many countries beyond their native ranges primarily to aid in controlling plankton population in sewage treatment lagoons, aquaculture tanks, and reservoirs (Malaypally 2013). Flooding events and escapes from confinement have led to faster spread and distribution of these two species. In the Mississippi River, the silver carp accounts for 64% of the total fish biomass (Garvey et al. 2012).

The silver carp gets the name from its very silvery-greenish colored fingerlings. It is described as having a laterally compressed body with very tiny scales except on the head and opercles. The eyes are slightly turned down and are located over the midline of the body. Likewise, the bighead carp has eyes which project downward and its head is scale-less. It is likened to the silver carp, with its dark coloration as an identifiable difference. While the silver carp can grow to 1 m and 27 kg, the bighead carp reaches the maximum length and weight at 1.4 m and 40 kg, respectively.

Hundreds, even thousands, of silver carp are known to leap out of the water to great heights when frightened and have occasionally hit and injured boaters (Vetter et al. 2017; Vetter and Mensinger 2016). When massive Asian carp fish-kills occur, their decaying carcasses have interfered with recreational activities, due to the foul smell and the unpleasant sight. Moreover, as silver carp populations increase, total catches for bigmouth buffalo and emerald shiner decrease (Hayer et al. 2014). The mesohabitat overlap between silver carp and some native fishes has increased competition for resources and reduced fitness of native species (Haupt and Phelps 2016).

Bighead carp could impact riverine ecosystems by competing for zooplankton with native species (Brugam et al. 2017; Sass et al. 2014). Bighead carp establishment has led to changes in the population of several native species in the Illinois River. Many sportfish species and catostomids became less abundant while emerald shiner *Notropis atherinoides*, grass carp *Ctenopharyngodon idella*, and shortnose gar *Lepisosteus platostomus* became more abundant after the establishment of bighead carp (Solomon et al. 2016). The population change has been linked to competition between the bighead carp and native fishes for habitat and available food sources (Tristano 2018; Solomon et al. 2016). Even in artificial ponds, paddlefish *Polyodon spatula* numbers have decreased with the establishment of bighead carp (Schrank et al. 2003). Moreover, a significant decrease in body conditions were also observed in many native fishes competing with bighead carp (Pendleton et al. 2017; Irons et al. 2007).

Different management strategies have been implemented to minimize the spreading of Asian carps, such as the use of behavioral, chemical, and physical barriers (Bowzer et al. 2015). Nonetheless, harvest enhancement seems to be the only viable short-term control strategy against established carp populations, among other integrated management strategies (Conover et al. 2007). Carp are suitable target species for new reduction fisheries due to their abundance and wide distribution (Bowzer et al. 2015).

Asian and common carp are considered low value food fishes with limited market potential in the United States. The common carp is even considered a trash fish (Keevin and Garvey 2019). Its disposal into the waste system causes pollution and emission of offensive odors. In China, on the other hand, Asian carps are the most culturally and economically important fish. These carps are also commonly consumed in European, African, and other Asian countries. Moreover, carps have contributed around 8.8 million tons in global inland aquaculture production in 2016 (FAO 2018).

Although a significant volume of Asian carps are exported from the United States to China as food fish, the transportation and logistical costs limit the sustainability of carp population control (Bowzer et al. 2013). Moreover, other industrial uses and markets are expected to generate more demand from Asian carps (Bowzer et al. 2013). Carps are currently used as fertilizers or livestock feeds (Kristinsson and Rasco 2000).

Asian carps serve as a good raw material for rendering into fishmeal. Omega-3 fatty acids and by-product proteins from Asian carps have also been used as animal feed (Malaypally 2013). These fish pose a negligible risk in fishmeal production and livestock feeding, due to their low levels of contaminants (Rogowski et al. 2009). Asian carps contain sufficient levels of long-chain polyunsaturated fatty acids independent of the harvest season; however, the highest levels observed are those harvested during autumn. Nonetheless, Asian carp meals need stabilizers to prolong shelf life. Asian carp meals are also less pricey compared to the traditional marine fish-derived fishmeal (Bowzer et al. 2014).

Asian and common carp meals are analogous to menhaden meal in terms of high digestibility and nutrient density. Although Asian carp meal is slightly less digestible than other fishmeal sources, it is still more digestible and more nutrient dense than other alternative protein sources. When carp-based meal was fed to hybrid striped bass *Morone saxatilis* x *Morone chrysops* and rainbow trout *Oncorhynchus mykiss*, its composition and digestibility was like menhaden-based fishmeal (Bowzer et al. 2014). Moreover, Asian carp meals have lower phosphorus levels than menhaden meal, while common carp meals have higher phosphorus levels. Hence, carp meals maintain or reduce the levels of discharged nitrogen and phosphorus (Bowzer et al. 2014). Asian carp meal is a more economical alternative protein source compared to menhaden fishmeal for carnivorous fishes, such as cobia *Rachycentron canadum*, hybrid striped bass, and *O. mykiss* (Bowzer and Trushenski 2015). Growth performance was similar in the three fishes, with an increased weight gain observed at higher inclusion levels, independent of its origin (i.e. carp meal or menhaden meal). However, cobia showed better performance when fed Asian carp meal-based diets than menhaden fishmeal-based diets. Only marginal differences in growth performance and organo-somatic indices were recorded.

Silver carp meals, on the other hand, could replace crystalline methionine in poultry starter and grower diets. Body weight gain, feed intake, and feed conversion ratio in poultry fed silver carp meal were on par with poultry fed a control diet. The processed meal from silver carp contained high methionine, protein, and lipid levels at 0.85, 54.9, and 4.6%, respectively (Upadhyaya et al. 2019). Moreover, of the total fatty acids in the silver carp meal, 45.6% were polyunsaturated fatty acids.

Other fish species

Sand smelt *Atherina boyeri* (Risso 1810) are small, pelagic carnivorous, schooling fish species which thrive in coastal areas and estuaries. These fish species seasonally migrate in the Atlantic region. In Turkey, sand smelt invades canals, lakes, ponds, reservoirs, and rivers. Due to its biological features, sand smelt easily dominates the invaded environment (Küçük et al. 2006). It competes with native fish for food, dissolved oxygen, and space. Generally, sand smelt is not used for human consumption and has low economic value. Sand smelt is commonly used as fresh feed in fish farms, but often leads to health-associated problems (Gumus et al. 2010).

Nutritional composition of sand smelt fish is comparable with other commercial fishmeal and soybean meal. Its crude protein, crude lipid, and fiber contents are 71.19, 9.18, and 0.40%, respectively (Gumus et al. 2010). At 75% fishmeal replacement, sand smelt meal did not negatively affect the growth and nutrient utilization of *O. niloticus* (Gumus et al. 2010).

White suckers *Catostomus commersonii* (Lacepède 1803) are distributed over 2.5 million km^2, from Canada to about 40 states in the Midwestern and Eastern United States. Moreover, white suckers are found in the Colorado River drainage basin (Page and Burr 2011). These fish species were introduced into Quebec, Canada, as baitfish for walleyes or brook trout *Salvelinus fontinalis* by sport fishermen (Mejri et al. 2019). They thrive in lakes, rivers, and streams, where they consume eggs, fish larvae, and worms. They easily tolerate murky and contaminated waters (Trippel and Harvey 1987; Beamish 1973), allowing their population to reach high numbers and broad dispersal.

The price of *C. commersonii* (600 USD per metric ton) is relatively cheaper than the price of Asian carp (World Bank Group 2013). When fed to walleye juveniles, meal from *C. commersonii* improved growth, making it an economical alternative to marine fishmeal for carnivorous fishes (Mejri et al. 2019).

CONCLUSION

Invasive alien fish species generally cause ecological damage. Nonetheless, these species and products thereof still confer evident economic returns. A formal risk assessment is imperative prior to the introduction of invasive alien fish species to evaluate the tradeoffs between their benefits and their potential impacts on the ecological integrity and biodiversity of the native waters.

Although there have been attempts to use invasive alien fish species as an alternative feed ingredient, its acceptance in the aquafeed and aquaculture industry and the global market remains a question. There are still considerable differences in the quality and composition of fishmeal from species to species, and even within species, depending on several factors, such as age, sex, environment, location, and season (Boran et al. 2008). These seasonal and geographical variations in fishmeal determine the marketing of the products derived from these sources. Moreover, the diet and feed intake of the species rendered as fishmeal affect the nutritive value of the product. For example, periods of starvation, stress, food shortages, high-energy expenditure, and decreased feed intake reduce the nutritive value of the fishmeal.

Given the obvious need for alternative feed ingredients coupled with the rapidly growing size of the aquaculture industry worldwide, a substantial amount of fishmeal or fish oil products from invasive alien fish species must be produced to support its sustainability. An industry devoted to harvesting and processing invasive alien fish species must be established to make sufficient products to supply the market. However, a commercial-scale invasive alien fish species meal production is foreseen as a gradual process. The associated initial risk of investment in a new product is high. Moreover, there is limited information about the product's nutritive value, potential volume, and contaminant content.

REFERENCES

Abarra ST, Velasquez SF, Guzman KDDC, Felipe JLF, Tayamen MM, & Ragaza JA. 2017. Replacement of fishmeal with processed meal from knife fish *Chitala ornata* in diets of juvenile Nile tilapia *Oreochromis niloticus*. *Aquac Reports*. 5:76–83.

Alquitran N. 2006. LLDA to support janitor fish oil inventor. https://www.philstar.com/nation/2006/07/16/347639/llda-support-janitor-fish-oil-inventor#gTwZQfgDUyhMVbcW.99. Accessed 14 November 2019.

Amurao RJ, inventor. 2006, June 7. Biofuel from janitor fish oil. Philippine Patent Application No. 1-2006-000295.

Anguebes-Franseschi F, Bassam A, Abatal M, May Tzuc O, Aguilar-Ucán C, Wakida-Kusunoki AT, Diaz-Mendez SE, & San Pedro LC. 2019. Physical and chemical properties of biodiesel obtained from Amazon sailfin catfish (*Pterygoplichthys pardalis*) biomass oil. *J. Chem.*, 2019: 1–12. doi:10.1155/2019/7829630.

Aranda ID, Casas EV, Peralta EK, & Elauria JC. 2010. Drying janitor fish for feeds in Laguna Lake to mitigate pollution potentials. *J Environ Sci Management*. 13(2):27–43.

Asnawi, Sjofjan O, Sudjarwo E, & Suyadi. 2015. Potency of sapu-sapu fish (*Hypostomus plecostomus*) as feed supplement for local ducks. *Int J Poult Sci*. 14(4):240–244.

Baska F, Voronin VN, Eszterbauer E, Müller L, Marton S, & Moln K. 2009. Occurrence of two myxosporean species, *Myxobolus hakyi* sp. n. and *Hoferellus pulvinatus* sp. n., in *Pangasianodon hypophthalmus* fry imported from Thailand to Europe as ornamental fish. *Parasitol Res*. 105:1391–1398.

Beamish RJ. 1973. Determination of age and growth of populations of the white sucker (*Catostomus commersoni*) exhibiting a wide range in size at maturity. *Journal of the Fisheries Research Board of Canada*. 30:607–616.

Boran G, Boran M, & Karaçam H. 2008. Seasonal changes in proximate composition of anchovy and storage stability of anchovy oil. *J Food Qual.* 31:503–513.

Bowzer J, Bergman A, & Trushenski J. 2014. Growth performance of largemouth bass fed fish meal derived from Asian carp. *N Am J Aquac.* 76(3):185–189.

Bowzer J, Trushenski J, & Glover DC. 2013. Potential of Asian carp from the Illinois River as a source of raw materials for fish meal production. *N Am J Aquac.* 75(3):404–415.

Bowzer J, Trushenski J, Rawles S, Gaylord TG, & Barrows FT. 2015. Apparent digestibility of Asian carp- and common carp-derived fish meals in feeds for hybrid striped bass *Morone saxatilis* ♀ × *M. chrysops* ♂ and rainbow trout *Oncorhynchus mykiss. Aquac Nutr.* 21(1):43–53.

Bowzer J, & Trushenski J. 2015. Growth performance of hybrid striped bass, rainbow trout, and cobia utilizing Asian carp meal-based aquafeeds. *N Am J Aquac.* 77(1):59–67.

Bragadóttir M, Pálmadóttir H, & Kristbergsson K. 2004. Composition and chemical changes during storage of fish meal from capelin (*Mallotus villosus*). *J Agr Food Chem.* 52:1572–1580.

Brugam RB, Little K, Kohn L, Brunkow O, Vogel G, & Martin T. 2017. Tracking change in the Illinois River using stable isotopes in modern and ancient fishes. *River Res Appl.* 33:341– 352.

Bunkley-Williams L, Williams EH, Lilystrom CG, Corujo-Flores I, Zerbi AJ, Aliaume C, & Churchill TN. 1994. The South American sailfin armored catfish *Liposarcus multiradiatus* (Hancock), a new exotic established in Puerto Rican fresh waters. *Caribb J Sci.* 30:90–94.

Capps KA, & Flecker AS. 2013. Invasive aquarium fish transform ecosystem nutrient dynamics. *Proc R Soc B.* 280:20131520.

Castro JMC, Camacho MVC, & Gonzales JCB. 2018. Reproductive biology of invasive knifefish (*Chitala ornata*) in Laguna de Bay, Philippines and its implication for control and management. *Asian Journal of Conservation Biology.* 7(2):113–118.

Chaichana R, & Jongphadungkiet S. 2013. Assessment of the invasive catfish *Pterygoplichthys pardalis* (Castelnau, 1855) in Thailand: ecological impacts and biological control alternatives. *Trop Zool.* 25(4):173–182.

Chaudhary A, Verma C, Varma M, & Singh HS. 2014. Identification of *Thaparocleidus caecus* (Mizelle & Kritsky, 1969) (Monogenea: Dactylogyridae) using morphological and molecular tools: a parasite invasion in Indian freshwater. *BioInvasions Rec.* 3(3):195–200.

Chavez HM, Casao EA, Villanueva EP, Paras MP, Guinto MC, & Mosqueda MB. 2006. Heavy metal and microbial analysis of janitor fish (*Pterygoplichthys* spp.) in Laguna de Bay, Philippines. *J Environ Sci Management.* 9(2):31–40.

Conover G, Simmonds R, & Whalen M. 2007. *Management and control plan for Asian carps in the United States.* Washington, DC: Aquatic Nuisance Species Task Force, Asian Carp Working Group.

Corpuz MNC. 2018. Diet variation and prey composition of exotic clown featherback, *Chitala ornata* (Gray 1831) (Osteoglossiformes: Notopteridae) in Laguna de Bay, Luzon Island, Philippines. *Asian Fish Sci.* 31(4):252–264.

Cremer MC, & Smitherman R. 1980. Food habits and growth of silver and bighead carp in cages and ponds. *Aquaculture.* 20(1):57–64.

Crivelli AJ. 1995. Are fish introductions a threat to endemic freshwater fishes in the northern Mediterranean region? *Biological Conservation.* 72:311–319.

Cunico AM, & Vitule JRS. 2014. First records of the European catfish, *Silurus glanis* Linnaeus, 1758 in the Americas (Brazil). *BioInvasions Rec.* 3:117–122.

Cuvin-Aralar MLA. 2014. Fish biodiversity and incidence of invasive fish species in an aquaculture and non-aquaculture site in Laguna de Bay, Philippines. In: Biscarini C, Pierleoni A, & Naselli-Flores L, editors. *Lakes: the mirrors of the earth balancing ecosystem integrity and human well-being.* Volume 2: Proceedings of the 15th World Lake Conference. Italy: Science4Press. p. 53–57.

Daga VD, Skóra F, Padial AA, Abilhoa V, Gubiani EA, & Vitule JRS. 2015. Homogenization dynamics of the fish assemblages in Neotropical reservoirs: comparing the roles of introduced species and their vectors. *Hydrobiologia.* 746:327–347.

Devick WS. 1989. *Disturbances and fluctuations in the Wahiawa Reservoir ecosystem. Project F-14-R-13, Job 4, Study I.* Honolulu: Hawaii Department of Land and Natural Resources, Division of Aquatic Resources.

Dudgeon D, Arthington AH, Gessner MO, Kawabata ZI, Knowler DJ, Leveque C, Naiman RJ, Prieur-Richard AH, Soto D, Stiassny MLJ, & Sullivan CA. 2006. Freshwater biodiversity: importance, threats, status and conservation challenges. *Biol Rev.* 81:163–182.

Elfidasari D, Ismi LN, Shabira AP, & Sugoro I. 2018. The correlation between heavy metal and nutrient content in plecostomus (*Pterygoplichthys pardalis*) from Ciliwung River in Jakarta. *Biosaintifika: Journal of Biology and Biology Education.*10(3):597–604.

FAO. 2018. *The State of World Fisheries and Aquaculture 2018 – Meeting the sustainable development goals.* Rome. 227 pp.

Gallardo B, Clavero M, Sanchez MI, & Villa M. 2015. Global ecological impacts of invasive species in aquatic ecosystems. *Glob Change Biol.* 22(1):151–163.

Garcia DAZ, Magalhaes ALB, Vitule JRS, Casimiro ACR, Lima-Junior DP, Cunico AM, Brito MFG, Petrere-Junior MP, Agostinho AA, & Orsi ML. 2018. The same old mistakes in aquaculture: the newly-available striped catfish *Pangasianodon hypophthalmus* is on its way to putting Brazilian freshwater ecosystems at risk. *Biodivers Conserv.* 27:3545–3558.

Garvey JE, Sass GG, Trushenski J, Glover D, Charlebois PM, Levengood J, Roth B, Whitledge B, Small BC, Tripp SJ, & Secchi S. 2012. Fishing down the bighead and silver carps: Reducing the risk of invasion to the Great Lakes. *Research summary.* March 2012. 184 pp. http://www.asiancarp.us/documents/CARP2011.pdf. Accessed 15 December 2019.

Gibbs M, Futral T, Mallinger M, Martin D, & Ross M. 2010. Disturbance of the Florida manatee by an invasive catfish. *Southeast Nat.* 9:635–648.

Gibbs MA, Shields JH, Lock DW, Talmadge KM, & Farrell TM. 2008. Reproduction in an invasive exotic catfish *Pterygoplichthys disjunctivus* in Volusia Blue Spring, Florida, U.S.A. *J Fish Biol.* 73:1562–1572.

Glencross BD, Booth M, & Allan GL. 2007. A feed is only as good as its ingredients: a review of ingredient evaluation strategies for aquaculture feeds. *Aquac Nutr.* 13:17–34.

Golubov J, Aguirre-Muñoz A, Mendoza R, & Mendez F. 2017. Mexico's invasive species plan in context. *Science.* 356(6336):386.

Guerrero RD. 2014. Impacts of introduced freshwater fishes in the Philippines (1905–2013): A review and recommendations. *Philipp J Sci.* 143(1):49–59.

Gumus E, Kayal Y, Balcil BA, Aydin B, Gulle I, & Gokoglu M. 2010. Replacement of fishmeal with sand smelt (*Atherina boyeri*) meal in practical diets for Nile tilapia fry (*Oreochromis niloticus*). *Isr J Aquacult-Bamid.* 62(3):172–180.

Gunawardena J. 2002. Occurrence of *Chitala chitala* (Syn. *Chitala ornata*) in native freshwater habitats. *Sri Lanka Naturalist.* 5(1):6–7.

Guo Y, Michael N, Fonseca Madrigal J, Sosa Aguirre C, & Jauregi P. 2019. Protein hydrolysate from *Pterygoplichthys disjunctivus*, armoured catfish, with high antioxidant activity. *Molecules.* 2019, 24(8):1628. doi:10.3390/molecules24081628.

Gustavsson H. 2016. *Locally available protein sources in diets of Nile tilapia (Oreochromis niloticus) – study of growth performance in the Mekong Delta in Vietnam.* Master Thesis. p. 47.

Haupt KJ, & Phelps QE. 2016. Mesohabitat associations in the Mississippi River Basin: a long-term study on the catch rates and physical habitat associations of juvenile silver carp and two native planktivores. *Aquat Invasions.* 11(1):93–99.

Hayer CA, Breeggemann JJ, Klumb RA, Graeb BDS, & Bertrand KN. 2014. Population characteristics of bighead and silver carp on the northwestern front of their North American invasion. *Aquat Invasions.* 9:289–303.

Hermoso V, Clavero M, Blanco-Garrido F, & Prenda J. 2011. Invasive species and habitat degradation in Iberian streams: an analysis of their role in freshwater fish diversity loss. *Ecol Appl.* 21:175–188.

Hien DM, Ha NC, Thuy NT, & Cuong TH. 2015. The hydrolysis ability of red meat by-product protein from Catfish in case of high fat content using enzyme bromelain. International Proceedings of Food Ingredients Asia Conference 2015, Bitec, Bangkok, Thailand. 10–11 September 2015. pp. 105–110.

Hong TTT, & Lindberg JE. 2007. Effect of cooking and fermentation of a pig diet on gut environment and digestibility in growing pigs. *Livest Sci.* 109(1–3):135–137.

Hoover JJ, Killgore KJ, & Cofrancesco AF. 2004. Suckermouth catfishes: Threats to aquatic ecosystems of the United States? *Aquat Nuis Species Res Bull.* 4:1–9.

Hossain M, Vadas R, Ruiz-Carus R, & Galib SM. 2018. Amazon sailfin catfish *Pterygoplichthys pardalis* (Loricariidae) in Bangladesh: a critical review of its invasive threat to native and endemic aquatic species. *Fishes.* 3(1):14. doi:10.3390/fishes3010014.

Hubilla M, Kis F, & Primavera J. 2007. Janitor fish *Pterygoplichthys disjunctivus* in the Agusan Marsh: a threat to freshwater biodiversity. *J Environ Sci Management.* 10(1):10–23.

Indrajith S 2016. *NARA to turn "threat" into food for fish. The Island Home News.* June 20, 2016. p. 2.

Irons KS, Sass GG, McClelland MA, & Stafford JD. 2007. Reduced condition factor of two native fish species coincident with invasion of non-native Asian carps in the Illinois River, U.S.A. Is this evidence for competition and reduced fitness? *J Fish Biol.* 71:258–273.

Jayaneththi HB. 2015. Record of iridescent shark catfish *Pangasianodon hypophthalmus* Sauvage, 1878 (Siluriformes: Pangasiidae) from Madampa-Lake in Southwest Sri Lanka. *Ruhuna J Sci.* 6:63–68.

Jumawan JC, & Herrera AA. 2014. Ovary morphology and reproductive features of the female suckermouth sailfin catfish, *Pterygoplichthys disjunctivus* (Weber 1991) from Marikina River, Philippines. *Asian Fish Sci.* 27(1):75–89.

Jumawan JC, Vallejo BM, Herrera AH, Buerano CC, & Fontanilla IKC. 2011. DNA barcodes of the suckermouth sailfin catfish *Pterygoplichthys* (Siluriformes: Loricariidae) in the Marikina River system, Philippines: Molecular perspective of an invasive alien fish species. *Philipp Sci Lett.* 4(2):103–113.

Karunarathna DMSS, Amarasinghe AAT, & Ekanayake EMKB. 2008. Observed predation on a suckermouth catfish (*Hypostomus plecostomus*) by water monitor (*Varanus salvator*) in Bellanwila-Attidiya Sanctuary. *Biawak.* 2(1):37–39.

Keembiyehetty CN, & Gatlin DM. 1992. Dietary lysine requirement of juvenile hybrid striped bass (*Morone chrysops* × *M. saxatilis*). *Aquaculture.* 104:271–277.

Keevin TM, & Garvey JE. 2019. Using marketing to fish-down bigheaded carp (Hypophthalmichthys spp.) in the United States: Eliminating the negative brand name, "carp." *J Appl Ichthyol.* 35(5):1141–1146.

Khamees NR, Ali AH, Abed JM, & Adday TK. 2013. First record of striped catfish *Pangasianodon hypophthalmus* (Sauvage, 1878) (Pisces: Pangasiidae) from inland waters of Iraq. *Basrah Journal of Agricultural Sciences.* 26(1, special Issue):184–197.

Kolar CS, Chapman DC, Courtenay WR., Housel CM, Williams JD, & Jennings DP. 2005. *Asian carps of the genus Hypophthalmichthys (Pisces, Cyprinidae) – a biological synopsis and environmental risk assessment.* Washington DC, USA: U.S. Fish and Wildlife Service. 183 pp.

Kristinsson HG, & Rasco BA. 2000. Fish protein hydrolysates: production, biochemical, and functional properties. *Crit Rev Food Sci Nutr.* 40:43–81.

Küçük F, Gülle İ, Güçlü SS, Gümüş E, & Demir O. 2006. Effect on fishing and lake ecosystem of sand smelt (*Atherine boyeri* Risso, 1810) as an invasive species. In National Fishing and Reservoir Management Symposium, 7–9 February, Antalya, Turkey. pp. 119–128.

Kumar A, Harikrishna V, Reddy AK, Chadha NK, & Babitha Rani AM. 2017. Salinity tolerance of *Pangasianodon hypophthalmus* in inland saline water: Effect on growth, survival and haematological parameters. *Ecology, Environment and Conservation.* 23:475–482.

Kumudinie OMC, & Wijeyaratne MJS. 2005. Feasibility of controlling the accidentally introduced invasive species *Chitala ornata* in Sri Lanka. *SIL Proceedings*, 1922-2010. 29(2):1025–1027.

Lakra WS, & Singh AK. 2010. Risk analysis and sustainability of *Pangasianodon hypophthalmus* culture in India. *Aquac Asia.* 15:34–37.

Latini AO, & Petrere M. 2004. Reduction of a native fish fauna by alien species: an example from Brazilian freshwater tropical lakes. *Fisheries Manag Ecol.* 11(2):71–79.

Linh NT, Dong NTK, & Van Thu N. 2018. A study of replacing dietary crude protein of fish meal by catfish (*Pangasius hypophthalmus*) by-products on growth performance and meat quality of muscovy ducks. *Livest Res Rural Dev.* 30(12):1–4.

Magalhães ALB. 2015. Presence of prohibited fishes in the Brazilian aquarium trade: effectiveness of laws, management options and future prospects. *J Appl Ichthyol.* 31:170–172.

Magbanua FS, Fontanilla AM, Ong PS, & Hernandez MBM. 2017. 25 years (1988–2012) of freshwater research in the Philippines: what has been done and what to do next. *Philipp J Syst Biol.* 11(1):1–16.

Maini ZA, Kumar V, & Ragaza JA. 2016. Aliens from aquariums. *World Aquaculture.* 49(2):59–63.

Mäkinen T, Weyl OLF, van der Walt KA, & Swartz ER. 2013. First record of an introduction of the giant pangasius, *Pangasius sanitwongsei* Smith 1931, into an African river. *African Zool.* 48(2):388–391.

Malaypally SP. 2013. *Invasive silver carp (Hypophthalmichthys molitrix) protein hydrolysates a potential source of natural antioxidant.* ProQuest Diss Theses. p. 207.

Mayuga, J. 2016. Fish biodiversity under siege. https://businessmirror.com.ph/fish-biodiversity-under-siege/. Accessed 11 November 2019.

McClelland MA, Sass GG, Cook TR, Irons KS, Michaels NN, O'Hara TM, & Smith CS. 2012. The long-term Illinois River fish population monitoring program. *Fisheries.* 37:340–350.

Mejri S, Tremblay R, Vandenberg G, & Audet C. 2019. Novel feed from invasive species is beneficial to walleye aquaculture. *N Am J Aquacult.* 81:3–12.

Men LT, Chi HH, Nghia NV, Khang NTK, Ogle B, & Preston TR. 2003. Utilization of catfish oil in diets based on dried cassava root waste for crossbred fattening pigs in the Mekong delta of Vietnam. *Livest Res Rural Dev.* 15(4):21–26.

Men LT, Yamasaki S, Chi HH, Loan HT, & Takada R. 2007. Effects of catfish (*Pangasius hypophthalmus*) or coconut (*Cocos nucifera*) oil, and water spinach (*Ipomoea aquatica*) in diets on growth/cost performances and carcass traits of finishing pigs. *Japan Agric Res Q.* 41(2):157–162.

Millamena OM. 2002. Replacement of fish meal by animal by-product meals in a practical diet for grow-out culture of grouper *Epinephelus coioides*. *Aquaculture*. 204:75–84.

Minh NP. 2014. Utilization of *Pangasius hypophthalmus* by-product to produce protein hydrolysate using alcalase enzyme. *Journal of Harmonized Research in Applied Sciences*. 2(3):250–256.

Mohanty BP, Ganguly S, Mahanty A, Mitra T, Mahaver L, Bhowmick S, Paul SK, & Das BK. 2017. Nutritional composition of the invasive *Pterygoplichthys disjunctivus* from East Kolkata Wetland, India. *J Inland Fish Soc India*. 49(2):48–54.

NACA. 2005. The way forward: building capacity to combat impacts of aquatic invasive alien species and associated transboundary pathogens in ASEAN countries. Final report of the regional workshop, hosted by the Department of Fisheries, Government of Malaysia, on 12th–16th July 2004. Network of Aquaculture Centres in Asia-Pacific, Bangkok, Thailand.

Nghiem LT, Soliman T, Yeo DC, Tan HT, Evans TA, Mumford JD, Keller RP, Baker RH, Corlett RT, & Carrasco LR. 2013. Economic and environmental impacts of harmful non-indigenous species in Southeast Asia. *PLoS One*. 8(8):e71255.

Nico LG, Butt PL, Johnston GR, Jelks HL, Kail M, & Walsh SJ. 2012. Discovery of South American suckermouth armored catfishes (Loricariidae, *Pterygoplichthys* spp.) in the Santa De River drainage, Suwannee River basin, USA. *Bioinvasions Rec*. 1:179–200.

Nuñez MA, Kuebbing S, Dimaro, RD, & Simberloff D. 2012. Invasive species: To eat or not eat, that is the question. *Conservation Letters*. 5:334–341.

Olut MJP, & Roxas DB. 2007. The effects of feeding janitor fish (*Pterygoplichthys disjunctivus*) meal on performance of broilers. *Philippine J Vet Anim Sci*. 33(2):109–122.

Orfinger AB, Lai QT, & Chabot RM. 2019. Effects of nonnative fishes on commercial seine fisheries: Evidence from a long-term data set. *Water (Switzerland)*. 11(6):1–11. doi:10.3390/w11061165.

Ornelas S, Gutierrez E, Juarez A, Garcidueñas R, Espinoza JL, Perea M, Flores JP, & Salas G. 2011. Use of silage acid devil fish (*Pterygoplichthys* spp.) as protein supplement in finishing beef cattle. *Journal of Agricultural Science and Technology. A*. 1:1280–1283.

Page LM, & Robins RH. 2006. Identification of sailfin catfishes (Teleostei: Loricariidae) in Southeastern Asia. *Raffles Bull Zool*. 54(2):455–457.

Page LM, Burr BM. 2011. *Peterson field guide to freshwater fishes of North America and Mexico*. Boston, MA: Houghton Mifflin Harcourt. 644 pp.

Pahlow G, Muck RE, Driehuis F, Elferink SJ, & Spolestra S. 2003. Microbiology of ensiling. In: Buxton DR, Muck RE, & Harrison JH, editors. *Silage science and technology*. Madison, WI, USA: American Society of Agronomy Inc. pp. 31–93.

Panase P, Uppapong S, Tuncharoen S, Tanitson J, Soontornprasit K, & Intawicha P. 2018. Partial replacement of commercial fish meal with Amazon sailfin catfish *Pterygoplichthys pardalis* meal in diets for juvenile Mekong giant catfish *Pangasianodon gigas*. *Aquac Reports*. 12:25–29.

Pendleton RM, Schwinghamer C, Solomon LE, & Casper AF. 2017. Competition among river planktivores: Are native planktivores still fewer and skinnier in response to the silver carp invasion? *Environ Biol of Fish*. 100:1213–1222.

Phuong NT, & Oanh DTH. 2009. Striped catfish (*Pangasianodon hypophthalmus*) aquaculture in Vietnam: an unprecedented development within a decade. In: De Silva SS, & Davy FB, editors. *Success stories in Asian aquaculture*. Dordrecht: Springer, NACA and IDRC. pp. 133–149.

Piencenaves J. 2013. Production and evaluation of carbonated calcium phosphate bioceramic powder from janitor fish (*Pterygoplichthys disjunctivus*) bones. http://www.academia.edu/7770834/ Production _and_Evaluation_of_Carbonated_Calcium_Phosphate_Bioceramic_Powder_from_Janitor_Fish_P terygoplitchys_disjunctivus_Bones. Accessed November 14, 2019.

Poulsen AF, Hortle KG, Valbo-Jorgensen J, Chan S, Chhuon CK, Viravong S, Bouakhamvongsa K, Suntornratana U, Yoorong N, Nguyen TT, & Tran BQ. 2004. *Distribution and ecology of some important riverine fish species of the Mekong River Basin*. MRC Technical Paper No. 10. ISSN: 1683–1489.

Puspasari R. 2006. Logam dalam Ekosistem Perairan. *J. BAWAL*. 1(2):43–47.

Quilang JP, & Yu SCS. 2015. DNA barcoding of commercially important catfishes in the Philippines. *Mitochondrial DNA*. 23(3): 435–444.

Rahim KA, Esa Y, & Arshad A. 2013. The influence of alien fish species on native fish community Structure in Malaysian waters. *Kuroshio Sci*. 7(1):81–93.

Rainboth WJ. 1996. *Fishes of the Cambodian Mekong. FAO species identification field guide for fishery purposes*. Rome: FAO. 265 pp.

Reid AJ, Carlson AK, Creed IF, Eliason EJ, Gell PA, Johnson PT, Kidd KA, MacCormack TJ, Olden JD, Ormerod SJ, Smol JP, Taylor WW, Tockner K, Vermaire JC, Dudgeon D, & Cooke SJ. 2019. Emerging threats and persistent conservation challenges for freshwater biodiversity. *Biol Rev.* 94:849–873.

Roberts TR, & Vidthayanon C. 1991. Systematic revision of the Asian catfish family Pangasiidae, with biological observations and description of three new species. *Proceedings of the Academy of Natural Science of Philadelphia.* 143:97–144.

Roberts TR. 1992. Systematic revision of the old-world freshwater fish family Notopteridae. Ichthyol *Explor Fres.* 2:361–383.

Rodríguez M, Gutiérrez ÁJ, Rodríguez N, Rubio C, Paz S, Martín V, Revert C, & Hardisson A. 2018. Assessment of mercury content in panga (*Pangasius hypophthalmus*). *Chemosphere.* 196:53–57.

Rogowski DL, Soucek DJ, Levengood JM, Johnson SR, Chick JH, Dettmers JM, Pegg MA, & Epifanio JM. 2009. Contaminant concentrations in Asian carps, invasive species in the Mississippi and Illinois Rivers. *Environ Monit Assess.* 157:211–222.

Rubio VY, Gibbs MA, Work KA, & Bryan CE. 2016. Abundant feces from an exotic armored catfish, *Pterygoplichthys disjunctivus* (Weber, 1991), create nutrient hotspots and promote algal growth in a Florida spring. *Aquat Invasions.* 11:337–350.

Salas G, Gutierrez E, Juarez A, Flores JP, & Perea M. 2011. Use of the devil fish in animal feed as an alternative productive diversification and mitigation of environmental damage in the South and West of Mexico. *Journal of Agricultural Science and Technology. A.* 1:1232–1234.

Sass GG, Hinz C, Erickson AC, McClelland NN, McClelland MA, & Epifanio JM. 2014. Invasive bighead and silver carp effects on zooplankton communities in the Illinois River, Illinois, USA. *J Great Lakes Res.* 40:911–921.

Sathivel S, Prinyawiwatkul W, King JM, Grimm CC, & Lloyd S. 2003. Oil production from catfish viscera. *J Am Oil Chem Soc.* 80(4):377–382.

Schrank SJ, Guy CS, & Fairchild JF. 2003. Competitive interactions between age-0 bighead carp and paddlefish. *T Am Fish Soc.* 132:1222–1228.

Singh AK, & Lakra WS. 2010. Risk analysis and sustainability of *Pangasianodon hypophthalmus* culture in India. *Gener Biodivers.* 15:34–37.

Snovsky G, & Golani D. 2012. The occurrence of an aquarium escapee, *Pangasius hypophthalmus* (Sauvage, 1878), (Osteichthys, Siluriformes, Pangasiidae) in Lake Kinneret (Sea of Galilee), Israel. *BioInvasions Rec.* 1(2):101–103.

Solomon LE, Pendleton RM, Chick JH, & Casper AF. 2016. Long- term changes in fish community structure in relation to the establishment of Asian carps in a large floodplain river. *Biol Invasions.* 18:2883–2895.

Srivastava SC, Verma P, Verma AK, & Singh AK. 2014. Assessment for possible metal contamination and human health risk of *Pangasianodon hypophthalmus* (Sauvage, 1878) farming, India. *Int J Fish Aquat Stud.* 1:176–181.

Stone R. 2007. The last of the Leviathans. *Science.* 316(5832):1684–1688.

Tacon AGJ, Hasan MR, & Subasinghe RP. 2006. *Use of fishery resources as feed inputs to aquaculture development: trends and policy implications.* FAO (Food and Agriculture Organization of the United Nations) Fisheries Circular 1018.

Tacon AGJ, & Metian M. 2008. Global overview on the use of fish meal and fish oil in industrially compounded aquafeeds: trends and future prospects. *Aquaculture.* 285:146–158.

Tarkan AS, Yoğurtçuoğlu B, Ekmekçi FG, Clarke SA, Wood LE, Vilizzi L, & Copp G. 2020. First application in Turkey of the European non-native Sspecies in aquaculture risk analysis scheme to evaluate the farmed non-native fish, striped catfish *Pangasianodon hypophthalmus. Fish Manag Ecol.* 27(2):123–131.

Tejeda-Arroyo E, Cipriano-Salazar M, Camacho-Díaz LM, Salem AZM, Kholif AE, Elghandour MMMY, DiLorenzo N, & Cruz-Lagunas B. 2015. Diet inclusion of devil fish (*Plecostomus* spp.) silage and its impacts on ruminal fermentation and growth performance of growing lambs in hot regions of Mexico. *Trop Anim Health Prod.* 47(5):861–866.

Thuy DT, Kania P, & Buchmann K. 2010a. Infection status of zoonotic trematode metacercariae in Sutchi catfish (*Pangasianodon hypophthalmus*) in Vietnam: associations with season, management and host age. *Aquaculture.* 302:19–25.

Thuy NT, & Ha NC. 2016. Effect of replacing marine fish meal with catfish (*Pangasius hypophthalmus*) by-product protein hydrolyzate on the growth performance and diarrhoea incidence in weaned piglets. *Trop Anim Health Prod.* 48(7):1435–1442.

Thuy NT, Lindberg JE, & Ogle B. 2010b. Digestibility and nitrogen balance of diets that include marine fish meal, catfish (*Pangasius hypophthalmus*) by-product meal and silage, and processing waste water in growing pigs. *Asian-Australasian J Anim Sci.* 23(7):924–930.

Thuy NT, Lindberg JE, & Ogle B. 2010c. Effect of additive on the chemical composition of tra catfish (*Pangasius hypophthalmus*) by-product silages and their nutritive value for pigs. *Asian-Australasian J Anim Sci.* 23(6):762–771.

Thuy NT, Lindberg JE, & Ogle B. 2011. Effects of replacing fish meal with catfish (*Pangasius hypophthalmus*) processing waste water on the performance of growing pigs. *Trop Anim Health Prod.* 43(2):425–430.

Thuy NT, Loc NT, Lindberg JE, & Ogle B. 2007. Survey of the production, processing and nutritive value of catfish by-product meals in the Mekong Delta of Vietnam. *Livest. Res. Rur. Dev.* 19(124):1–5. http://www.lrrd.or/lrrd19/9/thuy19124.htm. Accessed 22 November 2019.

Thuy NT. 2012. Effects of inclusion of catfish (*Pangasius hypophthalmus*) by-product meal and probiotics on performance, carcass quality and ileal and total amino acid digestibility in growing chickens. *Livest Res Rural Dev.* 24(10):1–8.

Trippel EA, & Harvey HH. 1987. Abundance, growth, and food supply of white suckers (*Catostomus commersoni*) in relation to lake morphometry and pH. *Can J Zool.* 65:558–564.

Tristano EP. 2018. *Effects of invasive species introductions on nutrient pathways in aquatic food webs.* PhD Dissertation, Southern Illinois University, Carbondale, Illinois.

Tsehaye I, Ctalano M, Sass G, Glover D, & Roth B. 2013. Prospects for fishery-induced collapse of invasive Asian carp in the Illinois River. *Fisheries.* 38:445–454.

Tuan TT, & Ogle B. 2019. Performance of growing pigs given diets in which fish meal was replaced by Tra catfish by products. *Livest Res Rural Dev.* 31(1):1–6.

Tuan TT. 2010. Tra catfish (*Pangasius hypophthalmus*) residue meals as protein sources for residue meals as protein sources for growing pigs. *Swedish Univ Agric Sci Dep Anim Nutr Manag.*:ISBN 978-86197-99-5.

Tuckett QM, Ritch JL, Lawson KM, & Hill JE. 2016. Implementation and enforcement of best management practices for Florida ornamental aquaculture with an emphasis on nonnative species. *N Am J Aquac.* 78:113–124.

Upadhyaya I, Arsi K, Fanatico A, Wagle BR, Shrestha S, Upadhyay A, Coon CN, Schlumbohm M, Trushenski J, Owens-Hanning C, Riaz MN, Farnell MB, Donoghue DJ, & Donoghue AM. 2019. Bigheaded carp-based meal as a sustainable and natural source of methionine in feed for ecological and organic poultry production. *J Appl Poult Res.* 28:1131–1142.

Van Zalinge N, Sopha L, Bun NP, Kong H, & Jorgensen JV. 2002. *Status of the Mekong Pangasianodon hypophthalmus resources, with special reference to the stock shared between Cambodia and Vietnam.* MRC Technical Paper No. 1, Mekong River Commission, Phnom Penh. 29 pp.

Vetter BJ, Casper AF, & Mensinger AF. 2017. Characterization and management implications of silver carp (*Hypophthalmichthys molitrix*) jumping behavior in response to motorized watercraft. *Manag Biol Invasions.* 8:113–124.

Vetter BJ, & Mensinger AF. 2016. Broadband sound can induce jumping behavior in invasive silver carp (*Hypophthalmichthys molitrix*). Fourth International Conference on the Effects of Noise on Aquatic Life.

World Bank Group. 2013. *Fish to 2030: Prospects for fisheries and aquaculture.* Washington, DC: World Bank, Report N-83177-GLB.

Weber C. 2003. Subfamily Hypostominae (armored catfishes). In: Reis RE, Kullander SO, & Ferraris CJ, editor. *Checklist of the freshwater fishes of South and Central America.* Porto Alegre: EDIPUCRS. 729 pp.

Wiecaszek B, Kaeszka S, Sobecka E, & Boeger W. 2009. Asian pangasiids—an emerging problem for European inland waters? Systematic and parasitological aspects. *Acta Ichthyol Piscat.* 39:131–138.

Yoğurtçuoğlu B, & Ekmekçi FG. 2018. First record of the giant pangasius, *Pangasius sanitwongsei* (Actinopterygii: Siluriformes: Pangasiidae), from Central Anatolia, Turkey. *Acta Ichthyol Piscat.* 48(3):241–244.

4 New Trends in Aquafeed Formulation and Future Perspectives

Inclusion of Antioxidants from the Marine Environment

Rubén Agregán, Rubén Domínguez, Roberto Bermúdez, Mirian Pateiro, and José M. Lorenzo

CONTENTS

INTRODUCTION

The aquaculture industry focuses its efforts on fish health maintenance and production performance (Giannenas et al. 2012, 26), and achieve their targets through feeding. Current aquafeed has excellent nutritional and physical properties, thanks to the use of high-quality, properly formulated ingredients (Turchini, Trushenski, and Glencross 2019, 14). Lipids are essential ingredients in fish diets, allowing them to grow and thrive. In addition, it is known that they have a positive influence on the immune system (Song et al. 2019, 1585). The most abundant lipids in aquafeed formulation are polyunsaturated fatty acids (PUFAs), and as it is well known, this kind of lipids are sensitive to oxidation, causing the appearance of rancid flavors and odors in diets (Song et al. 2019, 1586). Therefore, the industry is forced to use antioxidants, such as butylated hydroxylanisole (BHA), butylated hydroxyltoluene (BHT), and ethoxyquin (EQ), as additives to maintain the desired physical and chemical food properties (Dawood, Koshio, and Esteban 2018, 951). Stressor factors, such as salinity or temperature, can promote reactive oxygen species (ROS) formation with harmful effects on fish tissues (Rahman et al. 2017, 236). With the purpose of minimizing this impact on cells, limiting DNA damage, and increasing the normal physiological homeostasis toward anabolism, it is essential to supplement the feed with antioxidants, such as vitamin C and E (Aklakur 2018, 387). Therefore, antioxidants are added to fish diets in order to protect fish organism and prevent

DOI: 10.1201/9780429331664-4

feed rancidity. In addition, they increase immunity and preserve the metabolic equilibrium toward anabolism (Saleh, Wassef, and Shalaby 2018, 189).

Resources from the marine environment are still minimally exploited. Recently, studies focused on the macro and micro-algae application as food additives have been carried out, since, among other components, they contain important amounts of bioactive compounds, such as polyphenols (Dawood, Koshio, and Esteban 2018, 951), with recognized antioxidant properties. The antioxidant capacity of some brown macro-algae has been tested (Ahn et al. 2007, 71; Chew et al. 2008, 1067; Norra, Aminah, and Suri 2016, 1558; Dang et al. 2017, 3161; Agregán et al. 2018, 1) with good results. These organisms produce a kind of polyphenol called phlorotannin, derived exclusively from these algae (Agregán et al. 2018, 2). On the other hand, antioxidant activity has also been found in natural pigments, such as chlorophyll and carotenoids, produced by micro-algae (Agregán et al. 2018, 2).

The exploration of the marine environment opens the door to multiple natural compounds with antioxidant properties, offering a new perspective for the aquaculture industry in the formulation of aquafeed, extending its shelf life, eliminating the use of synthetic antioxidants, and improving fish conservation as well as production performance.

BIOACTIVE COMPOUNDS WITH ANTIOXIDANT PROPERTIES FOUND IN THE MARINE ENVIRONMENT

Marine organisms are sources of novel bioactive compounds (Demirel et al. 2009, 619). Many of these compounds are chemically and biologically different, and some of them are still under investigation as new potential drugs (Ely, Supriya, and Naik 2004, 122). In the last few years, many compounds extracted from macro-algae have shown a wide range of biological properties, including being antioxidants (Demirel et al. 2009, 619). For instance, marine algae are capable of surviving adverse environmental conditions, such as extreme light, high pressure, and high salinity or oxygen concentrations, meaning that these organisms produce potent antioxidant compounds (Agatonovic-Kustrin and Morton 2018, 113). Many authors have focused their efforts on finding these molecules and studying their antioxidant abilities. Vitamin C or ascorbic acid is an available antioxidant in terrestrial plants and also presents in macro-algae. Martínez-Hernández et al. (2018) found high ascorbic acid contents in the form of dehydroascorbic acid in seven edible macro-algae: *Ulva* spp, *Laminaria ochroleuca*, *Undaria pinnatifida*, *Himanthalia elongata*, *Palmaria palmata*, *Chondrus crispus*, and *Porphyra* spp, most of them collected in the north of Spain. The total amount determined ranged among 538.4–785 mg/kg alga on dry basis. Among all the seaweeds analyzed, *L. ochroleuca* presented the highest content with amounts between 785.1 and 758.9 mg/kg dried alga. Other interesting compounds belonging to marine algae are sulfated polysaccharides. They are composed of polymers whose molecules are arranged forming different structures. Within this chemical network are sulfated groups esterified with sugars (Flores et al. 2017; 2). These polysaccharides have a wide spectrum of physiological functions, such as anticoagulant, antihyperlipidemic, antiviral, and antitumor agents (Díaz et al. 2019, 577). In addition, they are also reported to have antioxidant activity. Flores et al. (2017; 2) found an acceptable radical scavenging capacity for these compounds in the macro-algae *Spyridia filamentosa*, *Codium isabelae*, and *Rhizoclonium riparium*, harvested in the state of Sinaloa, Mexico. Díaz et al. (2019) found the same for the macro-alga *Laminaria ochroleuca* collected on the Atlantic coast of Tarifa, Spain. Along the same line, Rupérez, Ahrazem, and Leal (2002) tested the antioxidant potential of different fractions of sulfated polysaccharides extracted from the macro-alga *Fucus vesiculosus* with satisfactory results. One of these fractions examined were the polysaccharides called "fucans", composed of the monosaccharide fucose, among other components. This saccharide showed the best antioxidant potential according to the study. The antioxidant power of these compounds was later observed by other researchers. Castro et al. (2015) reported antioxidant capacity with strong inhibition of hydroxyl and of peroxide from fucans extracted from the brown alga *Lobophora variegata*. In the same line,

Sellimi et al. (2014) found that fucans extracted from the brown macro-alga *Cystoseira barbata* harvested in Tunisia have antioxidant activity. A remarkable feature of these polysaccharides was the different mechanism of action involved, such as metal ion chelating, hydrogenor electron donation and radical scavenging (Sellimi et al. 2014, 287).

Marine organisms are an abundant source of secondary metabolites with bioactive activities (Metidji et al. 2015, 3184). Algal pigments, such as carotenoids, have shown antioxidant properties in diverse studies (Stengel and Connan 2015, 11). It has been reported that the incorporation of carotenoids in human and animal diets protects the body cells against oxidative damage (Christaki et al. 2013, 7). Within this group of pigments, fucoxanthin is highlighted for its antioxidant potential and is found in macro-algae, such as *Hijikia fusiformis*, *Undaria pinnatifida*, *Fucus serratus*, and *Padina tetrastromatic* (Pangestuti and Kim 2015, 816), and in micro-algae, such as *Isochrysis galbana*, *Cylindrotheca closterium*, Cylindrotheca *fusiformis*, and *Phaeodactylum tricornutum* (Cezare-Gomes et al. 2019, 617). Another carotenoid extracted from the marine environment that exerts significant antioxidant activity on the cells of living organisms is astaxanthin (Galasso, Corinaldesi, and Sansone 2017, 3). This red pigment is widespread and can be found in various marine organisms, such as bacteria (*Agrobacterium* and *Paracoccus* genera), yeast (*Xanthophyllomyces*, *Rhodotorula*, and *Phaffia* genera), and even in aquatic animals, such as snails (*Charonia sauliae*) and sea urchin (*Arbacia lixula*) (Galasso, Corinaldesi and Sansone 2017, 18). Marine animals do not produce carotenoids naturally, but instead accumulate them by food intake or generate them through metabolic transformations (Galasso, Corinaldesi and Sansone 2017, 15–16). Carotenoids, such as canthaxathin or zeaxanthin, also possess antioxidant capacity and are found in some marine organisms. Micro-algae *Chlorella emersonii* and *zofingiensis* are examples of canthaxanthin accumulation, whose dark red color is appreciated in the food industry as colorant and as an additive in animal feed. On the other hand, many *Chlorella* species are potential sources of the yellow-orange pigment zeaxanthin. Thus, the marine micro-alga *Chlorella ellipsoidea* is a promising producer, accumulating an amount around nine times higher (4.26 mg/g) than in red pepper (a typical vegetal source of this pigment). This pigment is also found in the *Porphyridium* species, with values greater than 95% of the total carotenoids reported for the micro-algae *Porphyridium cruentum* (Cezare-Gomes et al. 2019, 615–616).

Other secondary metabolites with recognized antioxidant properties and found in the marine environment are polyphenols. Many of these antioxidant compounds come from marine macro and micro-algae (Agatonovic-Kustrin, Morton, and Ristivojević 2016, 1469). Brown algae are considered the algae with the highest content of polyphenols. Conversely, green and red algae have comparatively lower amounts of these compounds. That is why brown algae are more extensively studied (Montero et al. 2018, 4809 – 4813). Natural polyphenols can exhibit several structures, which range from simple molecules, such as phenolic acids and polyphenolic compounds, to phlorotannins, which are comprised of phloroglucinol (1, 3, 5-trihydroxybenzene), exclusive from brown algae (Agatonovic-Kustrin, Morton and Ristivojević 2016, 1469). Many phenolic compounds have been identified and isolated from marine macro-algae in the last few years. Farvin et al. (2019) determined the phenolic profile of 26 seaweed species from the Kuwait coast of the Arabian Gulf (11 brown, 5 green, and 10 red). They identified the presence of phenolic acids, such as hydroxybenzoic (e.g., gallic, protocatechuic, vanillic, syringic, and salicylic acids) hydroxycinnamic acids (e.g., caffeic, p-coumaric, and t-ferulic acid), and flavonoids, such as flavan-3-ols (e.g., catechin and epicatechin) (Table 4.1). They noted that this last compound was the most abundant in almost all the seaweeds studied. Rengasamy et al. (2015) reported similar findings in eight South African seaweed species (4 green and 4 red). They reported the presence of phenolic acids, such as hydroxybenzoic acid derivative and flavonoids. Along the same line, Agregán et al. (2017) tentatively identified 3 brown algae collected in the Atlantic Ocean in the northwest of Spain with very similar results, reporting the presence of phenolic acids including hydroxybenzoic acid derivative as well as others, such as rosmarinic or quinic acid derivative and flavonoids, such as acacetin derivative or gallocatechin derivative 4.24.1.

TABLE 4.1
Summary of Antioxidant Compounds from Marine Origin Reported in this Dissertation

Antioxidant Compound		Marine Organism	Reference
Ascorbic acid			
	Green macro-alga	*Ulva* spp.	Martínez-Hernández et al. (2018)
	Brown macro-alga	*Laminaria ochroleuca*, *Undaria pinnatifida* and *Himanthalia elongata*	
	Red macro-alga	*Palmaria palmata*, *Chondrus crispus* and *Porphyra* spp.	
Sulfated polysaccharides			
	Red macro-alga	*Spyridia filamentosa*	Flores et al. (2017)
	Green macro-alga	*Codium isabelae* and *Rhizoclonium riparium*	
	Brown macro-alga	*Laminaria ochroleuca*	Díaz et al. (2019)
Fucans			
	Brown macro-alga	*Fucus vesiculosus*	Rupérez, Ahrazem and Leal (2002)
		Lobophora variegata	Castro et al. (2015)
		Cystoseira barbata	Sellimi et al. (2014)
Carotenoids			
Fucoxanthin			
	Brown macro-alga	*Hijikia fusiformis*, *Undaria pinnatifida*, *Fucus serratus*, *Padina tetrastromatic*	Pangestuti and Kim (2015)
	Micro-alga	*Isochrysis galbana*, *Cylindrotheca closterium*, *Cylindrotheca fusiformis* and *phaeodactylum tricornutum*	Cezare-Gomes et al. (2019)
Astaxanthin			
	Bacterium	*Agrobacterium* and *Paracoccus*	Galasso, Corinaldesi and Sansone (2017)
	Yeast	*Xanthophyllomyces*, *Rhodotorula* and *Phaffia* genera	
	Snails	*Charonia sauliae*	
	Sea urchin	*Arbacia lixula*	
Canthaxanthin			
	Micro-alga	*Chlorella emersonii* and *Chlorella zofingiensis*	Cezare-Gomes et al. (2019)

(Continued)

TABLE 4.1 (CONTINUED)
Summary of Antioxidant Compounds from Marine Origin Reported in this Dissertation

Antioxidant Compound	Marine Organism		Reference
Zeaxanthin			
	Micro-alga	*Chlorella* spp. and *Porphyridium* spp.	Cezare-Gomes et al. (2019)
Phenolic compounds			
Phenolic acids			
	Brown macro-alga	*Sargassum* spp., *Feldmannia* spp., *Padina gymnospora*, *Iyengaria stellata*, *Colpomenia sinuosa* and *Canistrocarpus cervicornis*	Farvin et al. (2019)
		Ascophyllum nodosum, *Fucus vesiculosus* and *Bifurcaria bifurcata*	Agregán et al. (2017)
	Green macro-alga	*Codium* sp., *Cladophora* sp., *Bryopsis pulmosa*, and *Ulva flexosa*	Farvin et al. (2019)
		Codium capitatum, *Codium duthieae*, *Halimeda cuneata* and *Ulva faciata*	Rengasamy et al. (2015)
	Red macro-alga	*Polysiphonia* spp., *Chondria* spp., *Hypnea* spp., *Grateloupia* sp., *Laurencia obtusa*, *Murrayella periclados* and *Spyridia filamentosa*	Farvin et al. (2019)
		Amphiroa beauvoisii, *Gelidium foliaceum*, *Laurencia complanata* and *Rhodomelopsis africana*	Rengasamy et al. (2015)
Flavonoids			
	Brown macro-alga	*Sargassum* spp., *Feldmannia* spp, *Padina gymnospora*, *Colpomenia sinuosa* and *Canistrocarpus cervicornis*	Farvin et al. (2019)
		Ascophyllum nodosum, *Fucus vesiculosus* and *Bifurcaria bifurcata*	Agregán et al. (2017)
	Green macro-alga	*Cladophora* sp., *Bryopsis pulmosa* and *Ulva flexosa*	Farvin et al. (2019)
		Codium capitatum, *Codium duthieae*, *Halimeda cuneata* and *Ulva faciata*	Rengasamy et al. (2015)
	Red macro-alga	*Polysiphonia* spp., *Chondria* spp., *Hypnea* spp., *Grateloupia* sp., *Laurencia obtusa*, *Murrayella periclados* and *Spyridia filamentosa*	Farvin et al. (2019)
		Amphiroa beauvoisii, Gelidium foliaceum, Laurencia complanata and Rhodomelopsis Africana.	Rengasamy et al. (2015)
Phlorotannins			
	Brown macro-alga	*Fucus vesiculosus*	Catarino et al. (2019)
		Durvillaea Antarctica, *Lessonia spicata*	Olate-Gallegos et al. (2019)
		Cystoseira nodicaulis, *Cystoseira tamariscifolia*, *Cystoseira usneoides* and *Fucus spiralis*	Ferreres et al. (2012)

As previously commented, brown algae are the only class of seaweeds which produce the polyphenols called phlorotannins. It has been reported that these phenolic compounds have an extraordinary antioxidant power and that they are responsible for the antioxidant capacity of this group of algae. Phlorotannins are a huge group of compounds that are classified according to the way in which units of phloroglucinol are linked. According to this, there are three primary types, fucols (only phenyl bonds), phlorethols (only arylether bonds) and fucophlorethols (arylether and phenyl bonds). Depending on the length of the floroglucinol chain, there are other classification criteria (Martínez and Castañeda 2013, 825). In recent years, there have been an increasing number of attempts to isolate these compounds and study their antioxidant capacity. The genus *Fucus* accumulates the highest amounts of phlorotannin. These compounds reach 12% dry matter, although this percentage varies according to diverse factors, such as seasonality, solar exposure, salinity, and geographical origin. In this sense, Catarino et al. (2019) contributed to the knowledge of this polyphenol family, tentatively identifying the phlorotannins from the specie belonging to the previous genus, *Fucus vesiculosus*, collected on the northern Portuguese coast. The authors tentatively identified the compounds fucols (trifucol, tetrafucol, and hexafucol), fucophlorethols, and fuhalols (hydroxytetrafuhalol and pentafuhalol isomers) along with some other phlorotannin derivatives of several degrees of polymerization by using LC-ESI-MS/MS. In addition, other possible new phlorotannins, such as fucofurodiphlorethol, fucofurotriphloretol, and fucofuropentaphlorethol, could be present in this macro-alga specie (Catarino et al. 2019, 2). Little by little, the phlorotannin profile of more brown algae is being identified. Olate-Gallegos et al. (2019) studied the seaweeds *Durvillaea Antarctica* and *Lessonia spicata*, collected from the Chilean coast, searching for the presence of phlototannins. The use of LC-MS/MS led to the identification of phlorotannins, such as fucophloroethol/triphloroethol, tetraphloroethol, pentaphloroethol, and hexaphloroethol. In the same line, Ferreres et al. (2012) reported the presence of 7 different classes of phlorotannins in several brown macro-algae harvested on the Peniche coast (west Portugal). The compounds fucophloroethol, fucodiphloroethol, fucotriphloroethol, 7-phloroeckol, phlorofucofuroeckol, and bieckol/dieckol were tentatively identified using HPLC-DAD-ESI-MS^n.

AQUAFEED FORMULATION: CLASSICAL AND NOVEL APPROACHES

Typical Aquafeed Composition and Keys for Their Preparation

From the nutritional point of view, the fish diet is not different from that of most animals. The nutrients they need to grow and develop include water, proteins, lipids, carbohydrates (sugars, starch), vitamins, and minerals (Abowei and Ekubo 2011, 183). A better understanding of nutritional requirements of different fish species, along with technological advances in the field of fodder manufacturing, has led to the use of artificial diets to complement or even replace the natural foods in aquaculture.

Classical aquafeed formulation includes a variety of products, such as fishmeal, soya bean meal, oilseed meals, animal byproducts, and grains and byproducts thereof (Teves and Ragaza 2016, 154). Oilseed meals represent a nutritive alternative to be included in aquafeeds. Nevertheless, some of them can present some drawbacks. Soya bean meal is commonly used to replace fishmeal. It shows a balanced amino acid profile but high fat content and trypsin inhibition. Another example is groundnut cake or peanut, which present high protein content but are potentially poisonous when moldy due to mycotoxin contamination. Other ingredients used to formulate aquafeed are animal byproducts. They come from animal waste tissues and are usually a cheap source of proteins. However, the prohibition of inclusion in fish diets of meat and bone meal (MBM) due to the outbreak of bovine spongiform encephalopathy in the 1990s, among other inconveniences, hinder their use. On the other hand, carbohydrates are indispensable in aquafeed formulation. This ingredient can be found in high amounts in grains or byproducts thereof, including wheat bran, an extraordinary binder widely used in salmonid aquafeed; but, as with other ingredients discussed above, its use is complicated (Teves and Ragaza 2016, 154, 155).

One of the most important sensorial attributes for consumers is appearance. An atypical color in the meat or skin of a particular fish could lead to its rejection; that is why it is very important to preserve the identity of each species at the sale point. Thus, aquafeed can include synthetic pigments or carotenoids to enhance coloration in the flesh of salmonid fish and the skin of freshwater and marine ornamental fish. The astaxanthin is the synthetic pigment most used in formulations (Abowei and Ekubo 2011, 183), being responsible for the typical reddish color of salmonid meat.

The fodder is made in the form of pellets that need to be stabilized, for which binding agents are added, reducing also the leaching of nutrients into the water. Examples of these ingredients are beef heart, carbohydrates, and extracts or derivatives of these, such as gelatin, gum arabic, agar, and carrageenan (Abowei and Ekubo 2011, 183).

As in the case of food for humans, shelf life is a capital issue in the aquaculture industry. Therefore, preservatives are commonly used in fish diets. Sodium and potassium salts of propionic, benzoic, or sorbic acids are widely used antimicrobials, while BHT, BHA, and EQ are the most in-demand antioxidants for lipid oxidation prevention in these products (Abowei and Ekubo 2011, 184). In addition to these antioxidants, others, such as vitamin E and C, are added to protect the fish organism from oxidation caused by ROS and reactive nitrogen species (RNS) (Aklakur 2018, 387).

Chemoattractants and flavorings are additives commonly employed in fish feeds. Some examples are hydrosilates and condensed fish soluble. The main purpose of their use is to improve palatability and its intake (Abowei and Ekubo 2011, 184).

In addition to all the ingredients previously mentioned, fiber and ash are employed in most aquafeeds along with organic aquatic material, such as seaweed, brine shrimp, and zooplankton (Abowei and Ekubo 2011, 184).

Formulating a food intended for aquiculture requires the study of a series of variables. According to Glencross, Booth, and Allan (2007), there are several factors to consider when developing an aquafeed: ingredient characterization, ingredient digestibility, palatability, and nutrient utilization and functionality.

There are several factors to take into account to choose the right ingredients when designing feedstuff, since there could be substantial variations in important nutrients even among cultivars of the same fish species (Glencross, Booth, and Allan 2007). The source of an ingredient is an important factor to consider, since problems related to the origin of the ingredient have been reported. According to Glencross, Booth, and Allan (2007), companies should at least worry about the country of origin of the ingredient. The processing of the ingredient is not a less important point, since it is a factor that could determine the chemical composition and nutritional value of the ingredient. For instance, heating protocols in protein sources may cause the loss of some amino acids (Glencross, Booth, and Allan 2007, 19).

Ingredient digestibility is another important factor when developing aquafeeds. The benefit obtained by an ingredient intake is not directly dependent on its amount because losses may occur in the digestive and absorption processes. Other possible problems that are related to the previous one are ingredient functionality and their interferences, which must be studied and calculated by qualified staff to avoid growth deficiencies in larvae and fishes with the consequent economic loss (Glencross, Booth, and Allan 2007, 18).

The main aim of the food industry is to manufacture a nutritious aquafeed that contributes all the essential ingredients in the right amount according to variables, such as species or growth stage. However, as in the case of humans, a factor such as palatability can seriously affect the consumption of a product, which makes any previous effort to formulate it useless. Palatability is a concept which means that the food is agreeable enough to be eaten. In the case of aquaculture, there is the problem of how to know fish preferences. Nevertheless, it is possible to establish differences in quantities consumed by them. If it is observed that some ingredient reduces the intake of food by the fish, its use will be limited. Although there may be strategies to avoid or solve the palatability issue, the best option is to try not to include that ingredient in the diet (Glencross, Booth, and Allan 2007, 25).

New Strategies in the Formulation of Aquafeed

Recently, the aquaculture industry has been attempting to search for alternative ingredients to the classical ones due to the growing demand by the fish market. The current system of aquafeed production is not able to compete against the rising demand for fish by consumers. Therefore, the aquafeed industry is confronted with high costs, production shortage, and unmet demands (Teves and Ragaza 2016, 154). This reality leads to the inevitable substitution of ingredients, such as fishmeal and oil, common sources of quality protein and omega-3 fatty acids. Studies have been carried out recently with the aim of replacing fishmeal and oil with vegetable ingredients in order to maintain production levels and meet market demand.

The addition of meals that have been extracted from vegetable oilseed to aquafeeds has been studied. Seeds from *Salicornia bigelovii* and *Jatropha curcas* plants have been used as meal sources in Nile tilapia (Belal and Al-Dosari 1999, 285) and white leg shrimps (Harter et al. 2011, 542), respectively, with promising results. The leaves from some plants have also been used to this purpose, such as those from peanut (Garduño-Lugo and Olvera-Novoa 2008, 1299) or mulberry (Mondal, Kaviraj and Mukhopadhyay 2012, 72), with similar results. In addition to the previous examples, others could be mentioned, such as alfalfa (Chatzifotis, Esteban, and Divanach 2006, 1313), jojoba (Labib, Zaki, and Mabrouk 2012, 196), and kikuyu grass (Hlophe, Moyo, and Sara 2011, 1076).

Fish oil is rich in omega-3 (n-3) fatty acids (FAs), docosahexaenoic (DHA), and eicosapentaenoic (EPA), which are beneficial for optimal growth and fish health. However, as previously commented, the system of aquafeed production is unable to satisfy the fish demand. Therefore, the industry currently uses vegetable oils in order to avoid a shortage of supply. Although these oils do not have DHA and EPA in their composition, it was recently discovered that some of them have FAs that help to compensate for this issue, increasing long chain n-3 FAs in fish muscle (Teves and Ragaza 2016, 160), such as *Coriandrum sativum* and *Perilla frutescens* plant seed oils in rainbow trout and Nile tilapia, respectively (Teves and Ragaza 2016, 160 – 161).

When it comes to searching for natural alternatives in feeding, the terrestrial environment is the first to be explored, either by proximity and accessibility or by the extensive knowledge we have of it. Recently, the number of researches in the marine environment has increased, since it offers a multitude of natural solutions to the food industry. In the same way that terrestrial plants have been used, seaweeds are also being employed in fish diets due to their richness in essential FAs, particularly n-3, a balanced amino acid profile with non-starch polysaccharides, vitamins, minerals, and bioactive compounds (Teves and Ragaza 2016, 155). However, despite these advantages, many studies report similar fish growth performances compared to those of control diets when seaweeds are added at low levels (10–15%); the exception is shrimps (<10%), which show worse performances at higher levels. This behavior could be attributed, on the one hand, to the low protein, high fiber, and ash contents often present in seaweeds and, on the other hand, to the high carbohydrate levels (≥40%), which present digestive and usability issues (Teves and Ragaza 2016, 155). Despite this fact, Marinho-Soriano et al. (2007) found that amounts of up to 50% *Gracilaria cervicornis* meal in shrimp diets provided growth performances and weight increases comparable to those obtained through a commercial diet after 30 days of study.

Antioxidant Supplementation

As previously commented, it is essential to include some antioxidant source in fish diets to prevent oxidation and premature deterioration of their cells (Figure 4.1). Recent studies have tested the effectiveness of including natural extracts in the feed of diverse fish species. Esteban et al. (2014) and Guardiola et al. (2016) found that the addition of date palm fruit extracts to Gilthead seabream and European seabass diets satisfactorily protected their organisms from oxidation. Similar results were found by other authors when including natural compounds to fish feeding. The addition of banana peel flour to the rohu diet (Giri et al. 2016, 4–5) and of phenolic compounds, such as carvacrol and thymol, in the rainbow trout diet (Giannenas et al. 2012, 30) positively affected the fish antioxidant status.

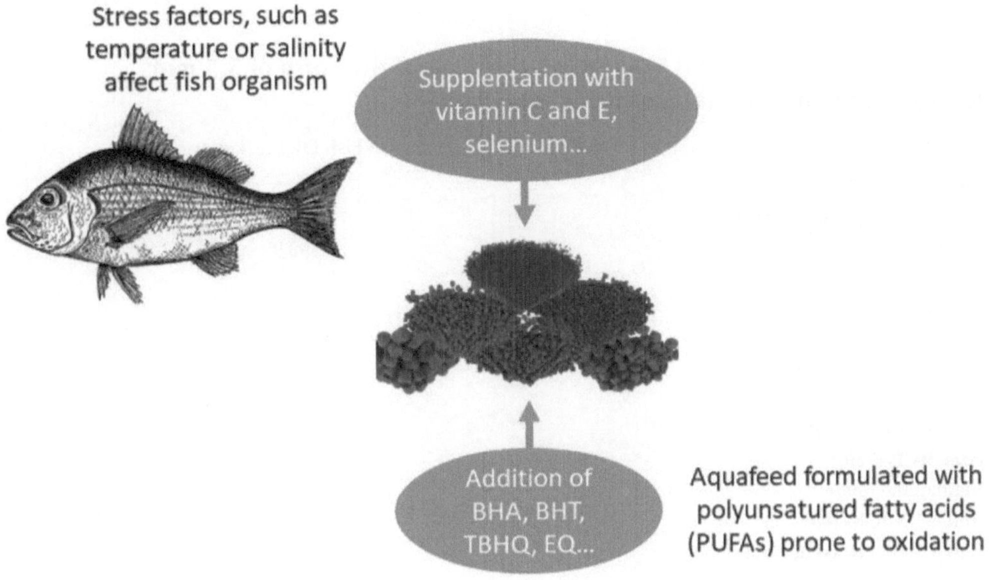

FIGURE 4.1 Antioxidant addition to aquafeed.

It is known that marine macro-algae are an important source of bioactive compounds. There are many studies in the literature that report their antioxidant potential. (Agregán et al. 2018, 1; Farvin and Jacobsen 2013; 1670; Auezova et al. 2013, 1189). *Sargassum horneri* brown seaweed addition to juvenile black sea bream improved the hepatic antioxidant status (Shi et al. 2019). In contrast, the mixture of *Gracilaria* sp. and *Ulva* sp. did not reveal significant protection against oxidation in European seabass juveniles compared to a control diet without the addition of the mentioned algae (Lobo et al. 2018, 321). The activities of various specific compounds from seaweeds to inhibit or slow down oxidation in fish tissues have also exhibited positive effects. The administration of the sulfated polysaccharide fucoidan, extracted from *Saccharina japonica* and *Sargassum horneri* brown algae, to the yellow catfish showed a positive effect from its antioxidant capacity (Yang et al. 2014, 268). According to the authors, this polysaccharide might be increasing the antioxidant enzymes causing the decrease of ROS and, consequently, the MDA content. On the other hand, carotenoids are commonly used in aquaculture during the finishing period of salmonids in order to provide the classical red-to-pink flesh color (Elia et al. 2019, 1). In addition, carotenoids, including those extracted from the marine environment, may work as antioxidants. Along this line, Elia et al. (2019) successfully stimulated the antioxidant defense in the liver and kidney of rainbow trout. adding xanthophylls, such as astaxanthin and canthaxanthin, to the fish diet (Table 4.2).

FUTURE PERSPECTIVE OF THE USE OF NATURAL ANTIOXIDANTS FROM MARINE ORIGIN IN AQUACULTURE FOODS

The necessity of, on the one hand, to preserve aquafood from lipid oxidation due to the fat content in their formulation, and, on the other hand, the maintenance of fish oxidative status to avoid premature aging, led to the inevitable use of antioxidants. There is a growing interest in the search for bioactive compounds from marine origin to replace the synthetic ones in the field of food, among others. At this point, the marine environment is an exceptional media with multiple vegetable organisms in which compounds with very good antioxidant activity such as phlorotannins can be found. As discussed in this review, the use of seaweeds and their compounds as antioxidants in fish diets can provide good

TABLE 4.2
Composition of a Typical and Complete Basal Diet in the Form of 3.5 mm Pellets for Rainbow Trout (*Oncorhynchus mykiss*)

Ingredients	Fish Diet (g/kg)
Wheat grains extruded (134)[2]	225.5
Soybean full fat extruded (355)	181.1
Fishmeal (720)	398.9
Fish oil	50.0
Gluten meal (595)	97.5
Wheat bran meal (155)	7.48
Binding (high protein) (821)	8.0
Limestone	21.0
Monocalcium phosphate	3.0
L-Lysine HCL (750)	0.2
DL-Methionine (990)	3.6
Vitamin A/D3 1000/200	0.011
Vitamin A 1000 retinyl acetate	0.01
Vitamin E 50%	0.5
DL-α-tocopheryl acetate	
Vitamin K3 50% menadione	0.04
Vitamin B1	0.05
Vitamin B2 80%	0.05
Vitamin B6	0.03
Vitamin B12 1%	0.01
Niacin	0.350
Biotin 2%	0.075
Pantothenic acid 90%	0.15
Folic acid 99%	0.008
Inositol	0.2
Choline 50%	4
STAY C35 (vitamin C)	0.716
Manganese oxide 62%	0.113
Zinc oxide 78%	0.09
Copper sulphate 25%	0.02
Cobalt sulphate 20%	0.005
Selenium salt 1%	0.01
Iodine salt	0.004
Iron sulphate 30%	0.170
Endox (antioxidant)[3]	0.1

Data from Giannenas et al., "Assessment of dietary supplementation with carvacrol or thymol containing feed additives on performance, intestinal microbiota and antioxidant status of rainbow trout (*Oncorhynchus mykiss*)", Aquaculture, 350 (2012): 26 – 32.

[2] Values in parentheses are the crude protein (N×6.25) or pure amino acid content, in g/kg.

[3] Containing calcium carbonate, ground corn cob, amorphous silicon dioxide, citric acid, soybean oil, ethoxyquin (a preservative up to 3.0%), butylated hydroxytoluene, phosphoric acid, butylated hydroxyanisole, and mono and diglycerides.

protection against oxidation in their organisms. Nevertheless, there are still too few studies about this issue. Taking into account the increasing demand for natural ingredients by the market, the good results achieved in the few studies performed until now should be the starting point of a research line that develops and implements the use of these ingredients in the aquaculture industry.

Another issue in fish feeding is the rancidity of the diet and the concern of industry to ensure its stability and shelf life. The questionable use of synthetic antioxidants revealed as the result of recent food safety problems led to the study of natural antioxidants as an alternative. The antioxidant capacities of compounds from marine organisms tested in multiple studies showed the great potential for these products in feed formulation (Aklakur 2018, 395). The aquaculture industry has a huge window of bioactive marine compounds with an extraordinary antioxidant potential, such as phorotannins, to reformulate fish diets and provide more natural foods. Nevertheless, their use presents some problems for industry, such as high cost, lack of standardized techniques of production and sustainable supply (Aklakur 2018, 395). In the future, emphasis should be placed on removing these barriers to allow the introduction of these compounds in the aquafeed industry.

CONCLUSION

The impossibility of the aquaculture industry fulfilling the demand for fish by the market with the current production system has led to a search for alternative ingredients and new formulations of aquafeed. The industry is attempting to replace ingredients from fish origin, such as fishmeal and oil, with others of vegetable origin. In addition, new strategies are starting to focus in the antioxidant supplementation from vegetable origin to improve oxidative status of fishes. The growing interest in the marine environment is leading researchers to increasingly find compounds with bioactive properties such as the antioxidant. Therefore, they are testing the addition of algae and their compounds to aquaculture foods as natural antioxidants. This trend, together with the demand for natural ingredients by the food market, might be the beginning of the development and implementation of the use of these ingredients in the aquaculture industry. On the other hand, the use of synthetic antioxidants, such as EQ, to improve the shelf life of aquafeed (one of the direst concerns of the aquaculture industry) is being rejected due to food safety problems. Antioxidants from marine organisms could be a solution. However, before commercialization can happen, the aquafeed industry must implement their use on a large scale.

REFERENCES

Abowei, J. F. N., and Ekubo, A. T. 2011. A review of conventional and unconventional feeds in fish nutrition. *British Journal of Pharmacology and Toxicology* 2(4):179–191.

Agatonovic-Kustrin, S., and Morton, D. W. 2018. Quantification of polyphenolic antioxidants and free radical scavengers in marine algae. *Journal of Applied Phycology* 30(1):113–120.

Agatonovic-Kustrin, S., Morton, D. W., and Ristivojević, P. 2016. Assessment of antioxidant activity in Victorian marine algal extracts using high performance thin-layer chromatography and multivariate analysis. *Journal of Chromatography A* 1468:228–235.

Agregán, R., Munekata, P. E., Franco, D., Dominguez, R., Carballo, J., and Lorenzo, J. M. 2017. Phenolic compounds from three brown seaweed species using LC-DAD–ESI-MS/MS. *Food Research International* 99:979–985.

Agregán, R., Munekata, P., Franco, D., Carballo, J., Barba, F., and Lorenzo, J. 2018. Antioxidant potential of extracts obtained from macro-(*Ascophyllum nodosum, Fucus vesiculosus* and *Bifurcaria bifurcata*) and micro-algae (*Chlorella vulgaris* and *Spirulina platensis*) assisted by ultrasound. *Medicines* 5(2):33.

Ahn, G. N., Kim, K. N., Cha, S. H., Song, C. B., Lee, J., Heo, M. S., Yeo, I. K., Lee, N. H., Jee, W. H., Kim, J. S., Heu, M. S., and Jeon, W. J. 2007. Antioxidant activities of phlorotannins purified from *Ecklonia cava* on free radical scavenging using ESR and H_2O_2-mediated DNA damage. *European Food Research and Technology* 226(1–2):71–79.

Aklakur, M. 2018. Natural antioxidants from sea: a potential industrial perspective in aquafeed formulation. *Reviews in Aquaculture* 10(2):385–399.

Auezova, L., Najjar, F., Selivanova, O., Moussa, E. H., and Assaf, M. D. 2013. Antioxidant activity of brown alga *Saccharina bongardiana* from Kamchatka (Pacific coast of Russia). A methodological approach. *Journal of Applied Phycology* 25(4):1189–1196.

Belal, I. E., and Al-Dosari, M. 1999. Replacement of fishmeal with Salicornia meal in feeds for Nile tilapia *Oreochromis niloticus*. *Journal of the World Aquaculture Society* 30(2):285–289.

Castro, L. S. E. P. W., de Sousa Pinheiro, T., Castro, A. J. G., Santos, M. D. S. N., Soriano, E. M., and Leite, E. L. 2015. Potential anti-angiogenic, antiproliferative, antioxidant, and anticoagulant activity of anionic polysaccharides, fucans, extracted from brown algae *Lobophora variegata*. *Journal of Applied Phycology* 27(3):1315–1325.

Catarino, M. D., Silva, A., Mateus, N., and Cardoso, S. M. 2019. Optimization of phlorotannins extraction from *Fucus vesiculosus* and evaluation of their potential to prevent metabolic disorders. *Marine Drugs* 17(3):162.

Cezare-Gomes, E. A., del Carmen Mejia-da-Silva, L., Pérez-Mora, L. S., Matsudo, M. C., Ferreira-Camargo, L. S., Singh, A. K., and de Carvalho, J. C. M. 2019. Potential of microalgae carotenoids for industrial application. *Applied Biochemistry and Biotechnology* 188(3):602–634.

Chatzifotis, S., Esteban, A. G., and Divanach, P. 2006. Fishmeal replacement by alfalfa protein concentrate in sharp snout sea bream *Diplodus puntazzo*. *Fisheries Science* 72(6):1313–1315.

Chew, Y. L., Lim, Y. Y., Omar, M., and Khoo, K. S. 2008. Antioxidant activity of three edible seaweeds from two areas in South East Asia. *LWT-Food Science and Technology* 41(6):1067–1072.

Christaki, E., Bonos, E., Giannenas, I., and Florou-Paneri, P. 2013. Functional properties of carotenoids originating from algae. *Journal of the Science of Food and Agriculture* 93(1):5–11.

Dang, T. T., Van Vuong, Q., Schreider, M. J., Bowyer, M. C., Van Altena, I. A., and Scarlett, C. J. 2017. Optimisation of ultrasound-assisted extraction conditions for phenolic content and antioxidant activities of the alga *Hormosira banksii* using response surface methodology. *Journal of Applied Phycology* 29(6):3161–3173.

Dawood, M. A., Koshio, S., and Esteban, M. Á. 2018. Beneficial roles of feed additives as immunostimulants in aquaculture: a review. *Reviews in Aquaculture* 10(4):950–974.

Demirel, Z., Yilmaz-Koz, F. F., Karabay-Yavasoglu, U. N., Ozdemir, G., and Sukatar, A. 2009. Antimicrobial and antioxidant activity of brown algae from the Aegean Sea. *Journal of the Serbian Chemical Society* 74(6):619–628.

Díaz, R. T. A., Arrojo, V. C., Agudo, M. A., Cárdenas, C., Dobretsov, S., and Figueroa, F. L. 2019. Immunomodulatory and antioxidant activities of sulfated polysaccharides from *Laminaria ochroleuca*, *Porphyra umbilicalis*, and *Gelidium corneum*. *Marine Biotechnology* 21:577–587.

Elia, A. C., Prearo, M., Dörr, A. J. M., Pacini, N., Magara, G., Brizio, P., Gasco, L., and Abete, M. C. 2019. Effects of astaxanthin and canthaxanthin on oxidative stress biomarkers in rainbow trout. *Journal of Toxicology and Environmental Health, Part A* 82(13):760–768.

Ely, R., Supriya, T., and Naik, C. G. 2004. Antimicrobial activity of marine organisms collected off the coast of South East India. *Journal of Experimental Marine Biology and Ecology* 309(1):121–127.

Esteban, M. A., Cordero, H., Martínez-Tomé, M., Jiménez-Monreal, A. M., Bakhrouf, A., and Mahdhi, A. 2014. Effect of dietary supplementation of probiotics and palm fruits extracts on the antioxidant enzyme gene expression in the mucosae of gilthead seabream (*Sparus aurata* L.). *Fish & Shellfish Immunology* 39(2):532–540.

Farvin, K. S., and Jacobsen, C. 2013. Phenolic compounds and antioxidant activities of selected species of seaweeds from Danish coast. *Food Chemistry* 138(2–3):1670–1681.

Farvin, K. S., Surendraraj, A., Al-Ghunaim, A., and Al-Yamani, F. 2019. Chemical profile and antioxidant activities of 26 selected species of seaweeds from Kuwait coast. *Journal of Applied Phycology* 31:2653–2668.

Ferreres, F., Lopes, G., Gil-Izquierdo, A., Andrade, P., Sousa, C., Mouga, T., and Valentão, P. 2012. Phlorotannin extracts from fucales characterized by HPLC-DAD-ESI-MS^n: approaches to hyaluronidase inhibitory capacity and antioxidant properties. *Marine Drugs* 10(12):2766–2781.

Flores, L., Salazar, J., Rodríguez, V., and Osuna, I. 2017. Capacidad antioxidante de polisacáridos sulfatados de seis especies de macroalgas de Sinaloa. *Revista de Ciencias Ambientales y Recursos Naturales* 3:1–8.

Galasso, C., Corinaldesi, C., and Sansone, C. 2017. Carotenoids from marine organisms: Biological functions and industrial applications. *Antioxidants* 6(4):96.

Garduño-Lugo, M., and Olvera-Novoa, M. Á. 2008. Potential of the use of peanut (*Arachis hypogaea*) leaf meal as a partial replacement for fishmeal in diets for Nile tilapia (*Oreochromis niloticus* L.). *Aquaculture Research* 39(12):1299–1306.

Giannenas, I., Triantafillou, E., Stavrakakis, S., Margaroni, M., Mavridis, S., Steiner, T., and Karagouni, E. 2012. Assessment of dietary supplementation with carvacrol or thymol containing feed additives on performance, intestinal microbiota and antioxidant status of rainbow trout (*Oncorhynchus mykiss*). *Aquaculture* 350:26–32.

Giri, S. S., Jun, J. W., Sukumaran, V., and Park, S. C. 2016. Dietary administration of banana (*Musa acuminata*) peel flour affects the growth, antioxidant status, cytokine responses, and disease susceptibility of rohu, *Labeo rohita*. *Journal of Immunology Research* 2016:4086591.

Glencross, B. D., Booth, M., and Allan, G. L. 2007. A feed is only as good as its ingredients-a review of ingredient evaluation strategies for aquaculture feeds. *Aquaculture Nutrition* 13(1):17–34.

Guardiola, F. A., Porcino, C., Cerezuela, R., Cuesta, A., Faggio, C., and Esteban, M. A. 2016. Impact of date palm fruits extracts and probiotic enriched diet on antioxidant status, innate immune response and immune-related gene expression of European seabass (*Dicentrarchus labrax*). *Fish & Shellfish Immunology* 52:298–308.

Harter, T., Buhrke, F., Kumar, V., Focken, U., Makkar, H. P. S., and Becker, K. 2011. Substitution of fishmeal by *Jatropha curcas* kernel meal: Effects on growth performance and body composition of white leg shrimp (*Litopenaeus vannamei*). *Aquaculture Nutrition* 17(5):542–548.

Hlophe, S. N., Moyo, N. A. G., and Sara, J. R. 2011. Use of kikuyu grass as a fishmeal substitute in practical diets for *Tilapia rendalli*. *Asian Journal of Animal and Veterinary Advances* 6(11):1076–1083.

Isaza Martínez, J. H., and Torres Castañeda, H. G. 2013. Preparation and chromatographic analysis of phlorotannins. *Journal of Chromatographic Science* 51(8):825–838.

Labib, E. M. H., M. A. Zaki, H. A., and Mabrouk 2012. Nutritional studies on partial and total replacement of fishmeal by jojoba meal (*Simmondsia chinensis*) in Nile tilapia (*Oreochromis niloticus*) fingerlings diets. *APCBEE Procedia* 4:196–203.

Lobo, G., Pereira, L. F., Gonçalves, J. F., Peixoto, M. J., and Ozório, R. O. 2018. Effect of dietary seaweed supplementation on growth performance, antioxidant and immune responses in European seabass (*Dicentrarchus labrax*) subjected to rearing temperature and salinity oscillations. *International Aquatic Research* 10(4):321–331.

Marinho-Soriano, E., Camara, M. R., Cabral, T. D. M., and Carneiro, M. A. D. A. 2007. Preliminary evaluation of the seaweed *Gracilaria cervicornis* (Rhodophyta) as a partial substitute for the industrial feeds used in shrimp (*Litopenaeus vannamei*) farming. *Aquaculture Research* 38(2):182–187.

Martínez-Hernández, G. B., Castillejo, N., Carrión–Monteagudo, M. D. M., Artés, F., and Artés-Hernández, F. 2018. Nutritional and bioactive compounds of commercialized algae powders used as food supplements. *Food Science and Technology International* 24(2):172–182.

Metidji, H., Dob, T., Toumi, M., Krimat, S., Ksouri, A., and Nouasri, A. 2015. In vitro screening of secondary metabolites and evaluation of antioxidant, antimicrobial and cytotoxic properties of *Gelidium sesquipedale* Thuret et Bornet red seaweed from Algeria. *Journal of Materials & Environmental Science* 6(11):3184–3196.

Mondal, K., Kaviraj, A., and Mukhopadhyay, P. K. 2012. Effects of partial replacement of fishmeal in the diet by mulberry leaf meal on growth performance and digestive enzyme activities of Indian minor carp *Labeo bata*. *International Journal of Aquatic Science* 3(1):72–83.

Montero, L., del Pilar Sánchez-Camargo, A., Ibáñez, E., and Gilbert-López, B. 2018. Phenolic compounds from edible algae: Bioactivity and health benefits. *Current Medicinal Chemistry* 25(37):4808–4826.

Norra, I., Aminah, A., and Suri, R. 2016. Effects of drying methods, solvent extraction and particle size of Malaysian brown seaweed, *Sargassum* sp. on the total phenolic and free radical scavenging activity. *International Food Research Journal* 23(4):1558–1563.

Olate-Gallegos, C., Barriga, A., Vergara, C., Fredes, C., García, P., Giménez, B., and Robert, P. 2019. Identification of polyphenols from Chilean brown seaweeds extracts by LC-DAD-ESI-MS/MS. *Journal of Aquatic Food Product Technology* 28(4):375–391.

Pangestuti, R., and Kim, S. K. 2015. Carotenoids, bioactive metabolites derived from seaweeds. In *Handbook of marine biotechnology*, ed. S. K. Kim, 816–821. Berlin: Springer.

Rahman, N. A., Khatoon, H., Yusuf, N., Banerjee, S., Haris, N. A., Lananan, F., and Tomoyo, K. 2017. *Tetraselmis chuii* biomass as a potential feed additive to improve survival and oxidative stress status of Pacific white-leg shrimp *Litopenaeus vannamei* postlarvae. *International Aquatic Research* 9(3):235–247.

Rengasamy, K. R., Amoo, S. O., Aremu, A. O., Stirk, W. A., Gruz, J., Šubrtová, M., Doležal, K., and Van Staden, J. 2015. Phenolic profiles, antioxidant capacity, and acetylcholinesterase inhibitory activity of eight South African seaweeds. *Journal of Applied Phycology* 27(4):1599–1605.

Rupérez, P., Ahrazem, O., and Leal, J. A. 2002. Potential antioxidant capacity of sulfated polysaccharides from the edible marine brown seaweed *Fucus vesiculosus. Journal of Agricultural and Food Chemistry* 50(4):840–845.

Saleh, N. E., Wassef, E. A., and Shalaby, S. M. 2018. The role of dietary astaxanthin in European sea bass (*Dicentrarchus labrax*) growth, immunity, antioxidant competence and stress tolerance. *Egyptian Journal of Aquatic Biology and Fisheries* 22(5, Special Issue):189–200.

Sellimi, S., Kadri, N., Barragan-Montero, V., Laouer, H., Hajji, M., and Nasri, M. 2014. Fucans from a Tunisian brown seaweed *Cystoseira barbata*: Structural characteristics and antioxidant activity. *International Journal of Biological Macromolecules* 66:281–288.

Shi, Q., Rong, H., Hao, M., Zhu, D., Aweya, J. J., Li, S., and Wen, X. 2019. Effects of dietary *Sargassum horneri* on growth performance, serum biochemical parameters, hepatic antioxidant status, and immune responses of juvenile black sea bream *Acanthopagrus schlegelii. Journal of Applied Phycology* 31(3):2103–2113.

Song, X., Yang, C., Tang, J., Yang, H., Xu, X., Wu, K., and Sun, B. 2019. A combination of vitamins C and E alleviates oxidized fish oil-induced hepatopancreatic injury in juvenile Chinese mitten crab *Eriocheir sinensis. Aquaculture Research* 50(6):1585–1598.

Stengel, D. B., and Connan, S. 2015. Marine algae: a source of biomass for biotechnological applications. In *Natural products from marine algae*, ed. D. B. Stengel, and S. Connan, 1–37. New York: Humana Press.

Teves, J. F. C., and Ragaza, J. A. 2016. The quest for indigenous aquafeed ingredients: a review. *Reviews in Aquaculture* 8(2):154–171.

Turchini, G. M., Trushenski, J. T., and Glencross, B. D. 2019. Thoughts for the future of aquaculture nutrition: realigning perspectives to reflect contemporary issues related to judicious use of marine resources in aquafeeds. *North American Journal of Aquaculture* 81(1):13–39.

Yang, Q., Yang, R., Li, M., Zhou, Q., Liang, X., and Elmada, Z. C. 2014. Effects of dietary fucoidan on the blood constituents, anti-oxidation and innate immunity of juvenile yellow catfish (*Pelteobagrus fulvidraco*). *Fish & Shellfish Immunology* 41(2):264–270.

5 Plant and Novel Aquafeed Ingredient Impact on Fish Performance and Waste Excretion

Eleni Fountoulaki, Morgane Henry, and Fotini Kokou

CONTENTS

INTRODUCTION

The aquaculture sector is continuously growing as the global fish consumption increases, while global fishery catches remain stagnant or have been decreased in recent years (FAO 2017). This rapid increase in aquaculture production causes issues related to aquaculture sustainability, due to both its dependence on fish meal for aquafeed production and its impact on the environment from waste output. Aquaculture waste is defined as feed nutrients that are not retained in biomass and are released into the environment either as fecal or non-fecal losses; this means that waste is related to feed acceptance, nutrient retention, and digestibility. In order to mitigate feed spillage, improvements to feeding strategies, including the use of palatable, highly digestible, nutrient-dense diets, are strongly recommended (Cho and Bureau 1997) in order to minimize aquaculture waste output. Increase in waste output consequently increases nutrient enrichment of the receiving environment causing unwanted ecological changes, and thus, effective aquaculture waste management is crucial for the long-term sustainability of the sector.

In order to deliver nutrients to the growing fish via manufactured feeds, several challenges must be overcome in the aquatic environment, since conventional feeds can rapidly deteriorate and leach nutrients and solids into the culture system. Aquafeed manufacturing has been continuously evolving over the years to meet such challenges by packaging a mixture of dietary ingredients into a form that can be delivered, consumed, and effectively utilized by the fish. Feed formulation affects both the amount of feces produced and the amount of soluble nutrients released into the water (i.e. ammonia, phosphorus). Feeding extruded, high-energy feeds can reduce waste, since these are

DOI: 10.1201/9780429331664-5

highly digestible. In a properly managed farm, 15–30% of the feed used will become fecal waste. However, not only the amount but also the consistency of the waste is important, which often causes the greatest challenges for feed manufacturers, while the desirable characteristics depend on the rearing aquaculture system.

Traditionally, fish meal is the main ingredient of aquaculture feeds, mainly for carnivorous species. Despite its highly digestibility and rich-protein content (Hardy 2010), its increasing price and decreasing availability has created the need for its substitution by alternative sources, mostly of plant origin (Bureau and Hua 2010). However, plant protein meals alter nutrient digestibility, especially when at high inclusion levels in the diets and, consequently, contribute to an increase in waste production (Bureau and Hua 2010, Glencross, Rutherford et al. 2012). The lower nutrient digestibility is attributed to the presence of antinutritional factors (ANFs), including the indigestible non-starch polysaccharides (NSP) (Cho and Bureau 2001, Francis, Makkar et al. 2001, Krogdahl, Penn et al. 2010). ANFs are naturally present, endogenous compounds in feedstuffs of plant origin, which usually have negative effects on animal feed intake, growth, nutrient digestibility, and utilization. The negative effects of ANF can be counterbalanced by their reduction or elimination from the plant materials through processing such as heating, aqueous or alcohol extraction, enzyme treatment, etc. (Drew, Borgeson et al. 2007). However, in many cases, considerable concentrations of ANF in the raw materials may be maintained even after processing, thus affecting animal performance and waste output (Venou, Alexis et al. 2003, 2006, Barrows, Stone et al. 2007, Kokou, Rigos et al. 2012, Kokou, Rigos et al. 2016). Nowadays, research has been focusing on novel ingredients from alternative sources, like single-cell proteins (i.e. algae and yeast), insects, or byproducts of industrial processes, to replace fish meal but also to reduce plant meal proteins. Such ingredients have lower ANF content and a lower ecological footprint; although they are still under evaluation, they show quite promising results (Hemaiswarya, Raja et al. 2011, Henry, Gasco et al. 2015, Roy and Pal 2015, Øverland and Skrede 2017).

In this book chapter, an overview of the contribution of different dietary ingredients on fish performance and waste production is provided. The chapter is adapted from a recent review by Kokou and Fountoulaki (2018), where more detailed information is presented on the effects of plant-origin ANF, NSP, and oligosaccharides on waste output. The technological processing to reduce the ANF levels in the feeds is also discussed in this chapter. Such knowledge is important in order to be able to design both sustainable and environmentally friendly feeds for aquaculture systems that will be efficiently utilized by the fish and contribute the least to the aquaculture output.

FACTORS AFFECTING AQUACULTURE WASTE OUTPUT

Aquaculture waste released into aquatic ecosystems can result in various environmental changes, the magnitude and type of which depend highly on the receiving environment. Different types of wastes depend on the different types of aquaculture operations, and no single strategy will work to solve all the environmental issues. Therefore, the tailoring of the waste form to the type of system (e.g. recirculating aquaculture or open cage systems) should be taken into account during feed design, as discussed further in this chapter.

There are two main types of aquaculture waste, the insoluble or solid waste and the soluble waste, which are presented in the following section. It is important to highlight that gas exchange due to respiration, mainly oxygen and carbon dioxide, can also affect waste production, mainly in recirculating systems and ponds, since it is associated with fish basal metabolism; however, this type of waste will not be discussed in this chapter.

Solid waste. This type of waste constitutes a high percentage of total waste and derives, in general, from feed spillage and fish feces (Turcios and Papenbrock 2014). It can consist of fast-sinking particles (settleable solids) or particles that remain in suspension in the water column (suspended solids).

Soluble waste. This type of waste originates from both branchial and urinary losses, due to fish metabolic activity, and the solubilization of feeds and feces released into the environment. Ammonia is the main environmentally harmful compound from branchial losses, while soluble phosphate originates from urinary losses. Although these compounds contribute to water eutrophication, ammonia is of special interest for aquaculture due to its toxicity in fish (Iwama 1991, Tomasso 1994). The nutritional balance of feed significantly affects the excretion of these compounds into the surrounding environment. Nitrogen losses can vary between 30 and 65% of the feed content (Schneider, Amirkolaie et al. 2004) and can be reduced by providing diets with an optimal amino acid profile and digestible protein/digestible energy ratio (Bureau and Hua 2010, Letelier-Gordo, Dalsgaard et al. 2015). Moreover, high levels of indigestible ANF can affect the availability, digestibility, and retention of nutrients (Francis, Makkar et al. 2001).

Feed Spillage

Feed spillage is related to feed production technologies and feeding practices. Using new technologies to control the physicochemical quality of feed pellets minimizes the generation of dust. Moreover, the feed density, floatability, and sinking rate as well as the water stability can affect the feed consumption and as a result also the spillage (i.e. a fast-sinking pellet will not be consumed from fish that eat floating pellets). Feeding practices that deliver fixed rations, taking into account fish age and environmental conditions such as water temperature, can also reduce feed losses (Amirkolaie 2011, Ballester-Moltó, Sanchez-Jerez et al. 2017). To avoid feed spillage, the maximum feeding ratio should be always adjusted to the fish species feeding behavior (Van der Meer, van Herwaarden et al. 1997), as well as to the environmental (i.e. temperature) and nutritional conditions. Another important factor that may increase feed spillage is linked to the palatability of the raw materials (Francis, Makkar et al. 2001, Krogdahl, Penn et al. 2010).

Fecal Loss

Fecal loss originating from undigested feed nutrients constitutes 10–30% of the feed (Chen, Coffin et al. 1997), while their impact on waste production depends on the amount, composition, and physical properties of the fecal mass released. Depending on the rearing system, fecal waste can have a different impact and should be managed differently; for land-based aquaculture operations and RAS, traps, filters, or settling tanks can be used. In that case, more stable fecal pellets are more efficiently removed from these systems (Summerfelt, Adler et al. 1999, Cripps and Bergheim 2000, Bureau and Hua 2010). For cage aquaculture operations, there is no mechanical means for removing waste. Thus, the environmental impact is higher and accumulation of feces should be avoided, since increased biological oxygen demand, carbon dioxide, and ammonia levels can arise from settling fecal mass below and around the cages (Chou, Haya et al. 2002).

Fecal Characteristics

The factors affecting the amount, composition, and physicochemical characteristics of feces egested to the environment are mainly feed composition and digestibility (Cho and Bureau 1997). Dry matter digestibility of the feed gives a good indication of the amount of fecal material that will be produced, and a mass balance approach can be used to estimate fecal mass and proximate composition (Bureau and Hua 2010). New methods to measure fecal chemical composition, such as the use of near infrared spectroscopy, can also be an efficient way to predict the organic carbon, nitrogen, and phosphorus levels in the feces (Galasso, Callier et al. 2017). Usually, commercial feeds have a dry matter digestibility ranging between 70% and 85% (Reid, Liutkus et al. 2009). However, the use of different feed ingredients may affect the dry matter, as well as nutrient (protein, fat, carbohydrate, minerals) digestibility. This results in different fecal compositions and increased environmental nutrient load, such as nitrogen and phosphorus. In addition, low-digestible ingredients can increase the amount of fecal waste and organic matter; thus, the selection of highly digestible ingredients is crucial for the waste load.

The physical properties of egested feces are very important to the environmental output. Different parameters of feces have been studied, such as density, sinking velocity, viscosity, etc. (Pérez, Almansa et al. 2014, Tran-Tu, Hien et al. 2018), and feces can have a different effect on the environment if they are present as suspended solids or sinking particles (Reid, Liutkus et al. 2009). Moreover, proper use of indigestible ingredients, such as binders, can sustain high-nutrient digestibility (Brinker, Koppe et al. 2005) by improving fecal pellet stability. Nonetheless, when studying feces consistency, one should bear in mind that it differs depending on the species; for example, gilthead bream, *Sparus aurata*, produces light, slow-sinking feces, while the European sea bass, *Dicentrarchus labrax*, produces heavier and faster-sinking feces (Magill, Thetmeyer et al. 2006). Furthermore, dietary composition also largely affects fecal physical properties. It is thus important to consider their effects during the production of fish feed formulations, and especially alternatives to fish meal, such as plant meals that can contain high percentages of indigestible ingredients, while less information is available for novel ingredients (Storebakken 1985, Han, Rosati et al. 1996, Amirkolaie, Leenhouwers et al. 2005, Brinker, Koppe et al. 2005).

DIETARY FACTORS PRESENT IN PLANT AND NOVEL FEEDSTUFF AFFECTING FISH PERFORMANCE

The feed nutritional composition and choice of dietary ingredients have been studied largely to optimize fish performance and waste management strategies. Alteration of fecal pellet consistency, such as density and size, due to impairment of fish digestibility of ingredients of plant origin, have been addressed for several species such as Nile tilapia, *Oreochromis niloticus* (Schneider, Amirkolaie et al. 2004, Amirkolaie, Leenhouwers et al. 2005), Atlantic salmon, *Salmo salar* (Kraugerud, Penn et al. 2007), rainbow trout, *Oncorhynchus mykiss* (Glencross 2009, Glencross, Rutherford et al. 2012) and common carp, *Cyprinus caprio* (Antony Jesu Prabhu, Fountoulaki et al. 2019). Moreover, changes in soluble waste production can also occur due to alteration in the metabolic activity as a result of interference with the digestion and nutrient retention process. Several factors, such as the presence of ANFs or indigestible components and their impacts on fish performance, should be taken into account, when designing a feed formulation. Therefore, it is important to understand how and to what extent the different feed components affect the nutrient digestibility, fish metabolism, feed acceptance but also fecal consistency in order to be able to design sustainable and high-quality feeds with the lowest environmental impact and optimal effect on fish performance. Moreover, the possibility of applying technological processing to improve the digestibility of different dietary ingredients should also be taken into account since it can significantly improve the nutritional value of the feeds.

PLANT FEEDSTUFF

Most of the alternative, plant-derived nutrient sources for fish feeds are known to contain a wide variety of ANF and indigestible components; they are presented in Table 5.1.

Protease inhibitors. Proteins that interfere with nutrient absorption by inhibiting the activity of several digestive enzymes in the gastrointestinal tract, such as trypsin and chymotrypsin, are called protease inhibitors. The effects of proteinase inhibitors have been studied thoroughly in mammals and birds where their mechanism of action is better understood (Liener 1980). In the intestine, inhibitors form a stable complex with their target-digestive enzyme (e.g. trypsin), thereby reducing its activity. This in turn stimulates the secretion of cholecystokinin (CKK) from the gut wall. This

TABLE 5.1
Antinutritional Factors Present in Most Commonly Used Fish Feed Material (adapted from Kokou and Fountoulaki, 2018)

Antinutritional Factors	Raw Material	Technological Processing Applied
Protease inhibitors	Wheat, corn, soybean, rapeseed meal, sunflower meal, wheat gluten, soybean protein concentrate	Heat treatment
Lectins	Soybean meal	Heat treatment, addition of salts
Allergens	Soybean meal, soybean protein concentrate, lupins	Alcohol extraction
Phytic acid	Wheat, corn, soybean meal, rapeseed meal, wheat gluten, soybean protein concentrate, pea protein concentrate	Milling, addition of phosphate, phytase treatment, heat treatment
Saponins	Soybean meal, pea protein concentrate	Acid or alkaline hydrolysis, microbial fermentation, alcohol extraction
Polyphenols	Wheat, corn, soybean meal, rapeseed meal, sunflower meal, corn gluten meal, soybean protein concentrate, pea protein concentrate, lupins	Alcohol extraction, microbial fermentation
Alkaloids	Lupins	Aqueous extraction, extrusion
Glucosinolates	Rapeseed meal	Enzyme treatment, heat treatment, aqueous extraction
Non-starch polysaccharides	Field peas, soybean meal, soybean protein concentrate, pea protein concentrate	Enzymatic treatment, fermentation, extrusion
Chitin	Insect meals (i.e. *Tenebrio molitor*, *Musca domestica*, *Hermetiaillucens*)	Enzymatic hydrolysis
Cell wall	Microalgae (*Chlorella* sp., *Nannochloropsis* sp., *Phaeodactylum*sp) Yeast (i.e. *Saccharomyces cerevisiae*, *Candida utilis*)	Mechanical disruption, Enzymatic hydrolysis, extrusion

hormone stimulates the secretion of trypsin from pancreatic tissue and stimulates the gall bladder to empty its contents into the intestine. Trypsin synthesis in the pancreas is stimulated, resulting in an increased requirement for amino acids, and for cysteine in particular, as trypsin is very rich in cysteine (Liener 1980). In soybean meals, which is the raw material used traditionally in fish feeds, the concentrations of protease inhibitors are around 2000 inhibitor units/g defatted meal, while in rapeseed meal they are around 10 inhibitor units/g of defatted meal. Overall, high levels of these inhibitors in the diet, usually >4–5 g/kg of feed, reduce protein digestibility, contributing mostly to soluble ammonia waste output (Krogdahl, Lea et al. 1994, Olli, Hjelmeland et al. 1994). Tolerance to trypsin inhibitor levels seems to be dependent on the fish species or the source of feed ingredients used (Krogdahl, Lea et al. 1994, Olli, Hjelmeland et al. 1994, Sveier, Kvamme et al. 2001, Santigosa, Saénz de Rodrigáñez et al. 2010).

Lectins. These molecules have the capacity to bind to membrane glycosyl-groups of the cells lining the digestive tract, potentially resulting in several effects at the digestive and metabolic level. Locally, they can damage the luminal membranes of the epithelium along with the cells, interfere with nutrient digestion and absorption, and stimulate shifts in the gut microflora (D'Mello, Duffus et al. 1991). Lectins vary in concentration in the different feed ingredients; usually the highest concentration exists in soybeans, ranging from 0.22 to 0.67 mg/g (Maenz, Irish et al. 1999). Hendriks and colleagues demonstrated that soybean agglutinin (SBA) binds *in vitro* to purified brush border membrane extracts from Atlantic salmon principally in the distal portion of the gut (Hendriks, Van den Ingh et al. 1990). The hypothesis that SBA may cause intestinal disruptions in fish was proven by Buttle, Burrells et al. (2001). These authors used fish meal or semi-purified diets supplemented with SBA and showed that SBA at 3.5% dietary levels binds only to distal intestine mucosa of rainbow trout and salmon and induces pathological changes similar to a 60% soybean-meal-based diet. Moreover, it was also shown that lower levels of SBA at 0.0075% inclusion did not induce pathological changes, but the combination of SBA with other ANFs, and specifically saponins (at 0.38% level), developed pathological symptoms in the distal intestine of rainbow trout, similar to soybean meal-induced enteritis (Yamamoto, Goto et al. 2008, Iwashita, Suzuki et al. 2009). Although later studies did not focus on the effects of dietary lectins in nutrient digestibility, due to potential intestinal disruption, these compounds can lead to wasteful protein synthesis.

Allergens. These compounds are capable of inducing intestinal damage, and they are usually present in legume seeds and cereals such as soybean meal and lupins (D'Mello, Duffus et al. 1991). Soybean contains compounds that act as allergens, such as glycinin and beta-conglycinin (around 16000 ppm). Suppression of feed intake and efficiency and changes in the intestinal morphology after dietary administration of 8% b-conglycinin (which accounts for 30% of SBM) were observed in juvenile Jian carp (*Cyprinus carpio* var. Jian) (Zhang, Guo et al. 2013) and turbot (Zheng et al., 2020). In addition, activities of trypsin and chymotrypsin were diminished, suggesting that these changes may be associated with impaired absorption, digestion, and fish growth, potentially resulting in higher soluble ammonia excretion. In Atlantic salmon, symptoms of intestinal damage were attributed to soybean allergen potentially present in solvent- or alcohol-extracted soybean meals (Rumsey, Hughes et al. 1993, Krogdahl, Bakke-McKellep et al. 2000). Nonetheless, further investigation involving more fish species and plant ingredients is needed to understand their impacts.

Phytic acid or phytate. Phytic acid is the principal storage form of phosphorus bound to inositol in the fiber of many plant tissues i.e. cereal grains, oil seeds, and nuts. In soybean meal and rapeseed meal (Makkar and Becker 2009), phytic acid is present at levels ranging from 50% to 80% of total phosphorus (Tyagi Tyagi et al. 1998). Regarding its action in the gut, phytate chelates with positively charged cations such as Ca^{2+}, Mg^{2+}, Zn^{2+}, and Fe^{2+} and thus reduces the bioavailability of these minerals (Erdman 1979, Liener 1980). Moreover, phytates form digestible phytic acid–protein complexes, thus reducing the availability of dietary protein (Richardson, Higgs et al. 1985). Studies related to commonly cultured fish species, such as carp, tilapia, trout, and salmon, report that the presence of phytic acid in the diet is negatively affecting fish performance (Spinelli, Houle et al. 1983, Richardson, Higgs et al. 1985, Hossain and Jauncey 1993, Sajjadi and Carter 2003, Riche and Garling 2004, Denstadli, Skrede et al. 2006). Regarding the effects on nutrient digestibility, phytate depresses nutrient digestibility and utilization efficiency, while the severity depends on the inclusion level and fish species (Spinelli, Houle et al. 1983, Sajjadi and Carter 2003). Impaired feed and protein conversion ratios were also reported due to the presence of phytic acid, which were partially improved with supplementation of Ca, Mg, and Zn, suggesting the low bioavailability of these minerals in the presence of this compound. Moreover, the formation of insoluble complexes (phytic acid – protein) leads to high ammonia nitrogen and phosphorus release into the surrounding environment, thus increasing eutrophication risks. Therefore, taking into consideration these results, diet formulations should avoid high levels of phytic acid, and the levels should be adapted for each species. Salmonids seem to be able to tolerate dietary levels of phytate in the range of 5–6 gr/kg, while carp appears to be sensitive to these levels.

Saponins. Saponins are amphipathic molecules, and their nature is directly related to many of their biological activities. They form micelles and can intercalate into cholesterol-containing membranes, forming holes that increase their permeability. They have been reported to increase mucosal cell permeability, inhibit active mucosal transport, and facilitate uptake of substances that are normally not absorbed (Johnson, Gee et al. 1986). In addition, by binding to cholesterol and bile salts, saponins interfere with cholesterol metabolism and entero-hepatic recirculation of bile salts. These steroid or triterpenoid glycosides are naturally found in many plant-derived feed ingredients, like legumes (Francis, Makkar et al. 2001). Negative results on growth performance from the administration of soybean saponins have been reported; however, this seems to be dependent on either the raw material administered, potential interactions between dietary ingredients, and ANFs or species tolerance (Bureau, Harris et al. 1998, Chen, Ai et al. 2011, Sørensen, Penn et al. 2011, Chikwati, Venold et al. 2012, Kokou, Samartzis et al. 2012, Couto, Kortner et al. 2014). From the waste production point of view, the main sources originate from reduced feed acceptance and impaired digestibility of nutrients, mainly lipids, due to the presence of saponin. Levels of saponins higher than 2 g/kg may decrease feed palatability and result in the accumulation of unconsumed feed pellets. In addition, decreased digestibility of protein leads to nitrogen load in the environment, as well as soluble waste from decreased fecal dry matter production. On the other hand, one must take into consideration the fish species' tolerance levels and the combination of raw materials in the feeds because they seem to play a major role in fish performance.

Polyphenols. Common sources of polyphenols in fish diets are cereals, grains, legumes, and oil seeds. The main polyphenols in cereals and grain legumes are phenolic acids, flavonoids, and tannins (Bravo 1998), while lignin is also present (Knudsen 1997). In legumes, the most characteristic group of polyphenols are the isoflavones (Bravo 1998), ranging from 0 to 0.18% (Mazur, Duke et al. 1998). Soybean meal contains isoflavone levels of 0.25 to 0.42% (Pelissero and Sumpter 1992), mainly as genistein and daidzein (Messina 1999). Regarding negative effects from isoflavones, no effects on feed acceptance and fish growth at the 0.2% level were detected in gilthead seabream (Couto, Kortner et al. 2014); however, protein digestibility was decreased (from 92.3% to 90%) after 6 months' administration (Kokou, Samartzis et al. 2012). At levels higher than 0.8%, isoflavones decreased protein digestibility (from 83% to 81%), energy (from 76% to 72%), and dry matter (from 68% to 65%) in Japanese flounder *Paralichthys olivaceus* (Mai, Zhang et al. 2012), as well as feed efficiency, but there were no negative effects on feed intake. Regarding the other polyphenol groups, tannins bind and form insoluble complexes with both proteins and carbohydrates and thus reducing their digestibility, due to the high amount of hydroxyl groups in their molecules (Bravo 1998). *In vitro* studies using extracted tannin demonstrated the inhibition of the intestinal enzymes in *Labeo rohita* fingerlings and three major Indian carps, even at very low concentrations (Maitra and Ray 2003, Mandal and Ghosh 2010). Similarly, rainbow trout fed with carob seed germ meal (containing high condensed tannin) indicated a low digestibility of protein compared to that of fish meal and soybean meal-fed fish (Filioglou and Alexis 1989, Alexis 1990). In addition, tannins can decrease feed acceptability, due to their bitter taste (Bravo 1998, Naczk, Amarowicz et al. 1998). Administration of tannic acid at 2% resulted in food rejection in common carp (Becker and Makkar 1999), but not by similar inclusion of a condensed tannin. It is therefore possible that condensed tannin content of common fish feed raw materials would not decrease fish feed consumption. Therefore, the contribution of polyphenols in waste production is dependent on fish tolerance and is mostly connected to feed acceptance and nitrogen emission in the environment due to impaired protein digestion.

Glucosinolates (GLS). Glucosinolates exist in all commercially important varieties of brassica, such as rapeseed meal (RSM) and canola meal (CM), containing levels of total GLS < 30 μmoles/gr of oil-free meal (Bell 1993). There are no experiments on fish nutrition using purified GLS or their products, and thus results are based on studies using rapeseed and canola meal or their protein concentrates. Feed consumption was found to decrease in different fish species, when rapeseed or canola meals were included in fish diets, and is partially attributed to the presence of GLS (Lim, Klesius et al. 1998, Burel, Boujard et al. 2000, Luo, Ai et al. 2012). Toxicity of GLS is caused by their derivatives, such as thiocyanates, vinyloxazolidinethiones, and isothyocyanates (Mawson, Heaney et al. 1994, Mawson, Heaney et al. 1994, Mawson, Heaney et al. 1995). These compounds affect thyroid gland function through competition in iodine transport, necessary for thyroid hormone precursor synthesis, and coupling reactions of precursors to form T4 and T3 (Mawson, Heaney et al. 1994), leading to reduction in thyroid hormone levels, and thus also fish growth as well as the retention of nitrogen, phosphorus, and energy (Leatherland 1994, Burel, Boujard et al. 2001). With regard to the environmental output, the inclusion of RSM and CM in fish feeds is low (up to 8–10%), so no significant effects are expected.

Non-starch polysaccharides (NSP). A common classification of NSP is based on their water solubility, classified as insoluble ones and soluble ones. Due to their different properties and the different effects they have on the gut, the impacts of soluble and insoluble NSP on waste output are different (Amirkolaie, Leenhouwers et al. 2005, Sinha, Kumar et al. 2011, Kokou and Fountoulaki 2018).

Insoluble NSP. Most studies that investigate the effects of insoluble NSP in fish diets have focused on cellulose, as it is a widely occurring NSP in plant ingredients. Inclusion of low levels of purified cellulose in the diet of several fish species actually can have positive effects on performance, depending on the fish species, at levels up to 20–28% (Buhler and Halver 1961, Davies 1985, AlOgaily 1996). It has been suggested that insoluble NSP stimulates bile acid release into the intestine, improving lipid emulsification, and thus increasing nutrient digestibility (Hetland, Choct et al. 2004). Indeed, studies in Nile tilapia fed graded levels of wheat bran or cellulose at 6 to 12% levels showed improved protein efficiency and feed efficiency ratio compared to the control diet (Anderson, Jackson et al. 1984, Al-Asgah and Ali 1996). Thus, the improvement in growth, feed efficiency ratio, protein utilization, and nutrient digestibility due to the inclusion of low to moderate levels of cellulose in these studies may be a result of a better utilization of the protein in diets with low protein content (Davies 1985) and, therefore, a lower nitrogen release into the environment. Most importantly, related to the aquaculture systems, insoluble NSPs have been shown to improve the removal efficiency of particles by increasing feces recovery, which is desirable in recirculating systems (Amirkolaie, Leenhouwers et al. 2005).

Soluble NSP. Purified soluble NSP, such as guar gum and carboxymethylcellulose, have been studied in fish in relation to their use as feed pellet binders, showing a reduction in growth performance in several fish species. Besides growth, soluble NSPs even at low levels in the diets (2–20%) can decrease protein and dry matter digestibility (Davies 1985, Storebakken 1985, Shiau, Yu et al. 1988, Shiau, Kwok et al. 1989, Amirkolaie, Leenhouwers et al. 2005). Apparent digestibility of dry matter, protein, and energy in rainbow trout (Davies 1985, Storebakken 1985, Glencross 2009, Glencross, Rutherford et al. 2012), Nile tilapia (Schneider, Amirkolaie et al. 2004, Amirkolaie, Leenhouwers et al. 2005, Leenhouwers, Adjei-Boateng et al. 2006), and European sea bass (Leenhouwers, Santos et al. 2004, Fountoulaki, Nikolopoulou et al. 2014) were also found to be reduced by soluble NSP, but it is not clear if this is due to an increase in digesta viscosity. Despite that, viscous conditions in the intestinal tract caused by the soluble NSP fraction was reported for common carp (Hossain, Focken et al. 2001) and African catfish *Clarias gariepinus* with a concomitant decrease in dry matter, protein, and energy digestibility (Leenhouwers, Adjei Boateng et al. 2006). However, in a study of Atlantic salmon, dietary inclusion of soluble NSP did not cause any difference in digesta viscosity in the proximal small intestine (Refstie, Svihus et al. 1999). Similarly, in common carp inclusion of soluble wheat dried distillers' grain at 30% levels did not negatively affect fecal recovery (Antony Jesu Prabhu, Fountoulaki et al. 2019). Storebakken and Austreng (1987) concluded that other factors than viscosity may be involved in the reduction of digestibility, such as the water-binding properties of soluble NSP. The ingestion of soluble NSP may cause an osmotic effect leading to fluid retention and thus increased water content in fish gut (Storebakken 1985, Storebakken and Austreng 1987, Kihara and Sakata 1997, Refstie, Svihus et al. 1999), which together with the reduced digesta passage time, may reduce nutrient digestibility (Deguara, Jauncey et al. 1999).

Novel Feedstuff

Besides plant raw material, novel ingredients may also include several compounds that can affect digestibility and nutrient utilization (Table 5.1).

Chitin. Insect meals (IM) are considered promising alternatives to substitute the plant-based ingredients in fish meal; however, results in growth performance may vary according to the insect species, its development phase, the substrate used to rear it, and the processing of the meal (Henry, Gasco et al.

2015). Some deficiencies (e.g. amino acids) may be responsible for lower growth observed with some IMs; however, adequate dietary supplementation may avoid growth reduction. A recurrent problem reported with the use of insect meals in fish feeds is their high chitin content, which is the primary component of the insect exoskeletons (Finke 2007) and is often poorly digested by fish (Rust 2002, Sánchez-Muros, Barroso et al. 2014, Henry, Gasco et al. 2015, Gasco, Henry et al. 2016). The presence of enzymes, such as chitinase and chitobiase, that can break down chitin or chitosan, is especially high in some fish preying on crustaceans or insects (Ikeda et al. 2017; Sanchez-Muroz et al. 2014) and the gene for chitinase has been sequenced in several fish species (Ikeda et al. 2013; Kakizaki et al. 2015). However, in other fish species, it can be very low or totally absent (e.g. rainbow trout, Renna et al. 2017). Moreover, whenever present, these enzymes may have limited access to their substrate due to the complex matrix in which chitin is encompassed, thus reducing the overall nutrient digestibility (Sealey, Gaylord et al. 2011, Henry, Gasco et al. 2015, Gasco, Henry et al. 2016) and fish growth (Alegbeleye, Obasa et al. 2012, Elia, Capucchio et al. 2018). As shown in an *in vitro* digestibility study, crude protein digestibility is mainly affected by the chitin content (Marono et al. 2015). It is apparent in Table 5.2, which compares the digestibility results of IM at various inclusion levels in different fish species (reviewed by Nogales-Mérida, Gobbi et al. (2019), together with more recent studies), that digestibility of protein was only very slightly affected by dietary insects except for *Zophobas morio* and the housefly in Nile tilapia, *Hermetia luminescens* in turbot and *Tenebrio molitor* in gilthead seabream, which were strongly affected. It varied between 81 and 93.4% for black soldier fly, 79.2 to 98.5% for yellow mealworm, 80.1 to 87.1% for the common housefly and 50.5 to 84.1% for the other insect species investigated compared to 77.5–98.5% for FM. Digestibility of lipids was higher than protein digestibility (83–99.97% for BSF, 82.4–98.4% for Tenebrio, 91.6–96.2% for housefly, and 69.8–82.7% for the other insects compared to 69.8–98.8% for FM). Some dietary insects greatly improved lipid digestibility despite the previous assumption that dietary chitin may decrease lipid digestibility through the decrease in bile acid levels involved in the activation of lipase and lipid absorption in the gut (Tanaka, Tanioka et al. 1997, Hansen, Penn et al. 2010). Generally, the higher the inclusion of insect the more protein and lipid digestibility were affected, but few studies showed improved digestibility in fish that were fed insect meal. However, there was a notable inter-studies variability probably due to different insect-rearing substrates and meal processing, as exemplified by BSF in the Atlantic salmon showing a significant reduction (Dumas, Raggi et al. 2018) or a significant increase (Belghit, Liland et al. 2019) in digestibility values. These effects of inclusion of IM in the fish diet may potentially affect both soluble and solid waste output. Interestingly, *Pachypterus khavalchor*, which has been shown to prefer a chitin-rich diet, has been associated with chitinase-producing endosymbionts in its gut (Gosavi, Verma et al. 2019). In fish, inclusion of *H. luminescens* in the diet of Jian carp showed gut histological changes attributed to the chitin content of insect exoskeleton (Li, Ji et al. 2017). In the seawater phase of Atlantic salmon, FM replacement with IM did not compromise gut health; it reduced steatosis in the proximal intestine and increased the relative weight of the fish distal intestine (Li, Kortner et al. 2020). A recent study has shown that dietary insect meal increased the length of the digestive tract. Another recent study has shown that dietary insect meal caused a shift of microbiota in three fish species (principally in gilt-head bream and European sea bass, less marked in rainbow trout) with increased presence of Tenericutes, which are believed to be related to the metabolism of chitin (Antonopoulou et al. 2019). Thus, it may be hypothesized that dietary IM may shift the microbiota towards commensal bacteria with chitinase activity, thus helping the fish to digest chitin-rich nutrients.

Yeast cell wall. Yeasts, such as *Saccharomyces cerevisiae* and *Candida utilis*, are another promising alternative source for fish meal replacement. Although yeast cell chemical composition depends on the strain, the media in which it is grown, and the growth conditions, it generally has low fat content and high protein levels. The protein digestibility is comparable to conventional protein sources showing to be more than 80–90% in European sea bass and redclaw crayfish (Oliva-Teles and Gonçalves 2001, Pavasovic, Anderson et al. 2007). However, the results may vary depending

TABLE 5.2
Studies Using Different Insect Meals and the Impacts on Nutrient Digestibility

Insect type used	Fish Species	Average Fish Weight	FM level in Control Diet	Dietary Level	ADC Crude Protein %	ACD Crude Lipid %	ADC Gross Energy %	Reference
HM, defatted, dried, ground	Turbot, *Psetta maxima*	96g	69%	30%	63.1 (89.1–81.1)	78.0 (98.7–92.8%)	54.5 (84.9 – 75.0)	Kroeckel et al. (2012)
HM	European sea bass, *Dicentrarchus labrax*	50g	32.4%	6.5, 13, 19.5%	(91.3–93.4,92.3,91.6)	(93.0–91.5,92.6,92.3)	(78.3–83.7,81.9,81.4)	Magalhaes et al. (2017)
HM, partially defatted	Rainbow trout, *Oncorhynchus mykiss*	182g	60%	20,40%	(89–91,87)	(97–99,97)	(60–65,60)	Renna et al. (2017)
HM, defatted	Atlantic salmon, *Salmo salar*	85g	20%	20%	85 (89–88)	43 (83–73)		Dumas et al. (2018)
HM, defatted	Atlantic salmon, *Salmo salar*	1400g	10%	5,10,15%	(84–83,83,82)	(85–84,88,86)		Belghit et al. (2019)
HM	Atlantic salmon, *Salmo salar*	247g	20%	5,10,25%		high FA digestibility (82.5–100)		Lock et al. (2016)
BSF, defatted	sturgeon	24g	70%	18.5,37.5,75%	(88.5–86.5,86.6,90.4)		(83.3–81.7,81.4,85.8)	Caimi et al. (2019)
TM, full-fat	European sea bass, *Dicentrarchus labrax*	65g	70%	25,50%	(90–92)	(98–97)		Gasco et al. (2016)
TM	Gilt-head seabream, *Sparus aurata*	87g	50%	25,50%	(90–87.3,79.2)	(91.1–89.9,82.4)		Piccolo et al. (2017)
TM	Nile tilapia, *Oreochromis niloticus*	3g		20%	85.4	90.6	82.1	Fontes et al. (2019)
TM	Rainbow trout, *Oncorhynchus mykiss*	94g	20%	5,10,20%	(98.5–98.5,98.0,97.3)	(98.8–98.4,98.4,98.3)	(96.7–97.3,96.6,96.1)	Chemello et al. (2020)
HS	Carp, *Cyprinus carpio*	110g	30%	25%	84.9(86.5–87.1)	96.8(90.7–96.2)	74.9 (82.5–79.6)	Ogunji et al. (2009)
HS	Nile tilapia, *Oreochromis niloticus*	108g	30%	30%	57.7 (89.0–80.1)	86.1 (92.4–91.6)	58.1 (85.9–76.4)	Ogunji et al. (2010)

(Continued)

TABLE 5.2 (CONTINUED)
Studies Using Different Insect Meals and the Impacts on Nutrient Digestibility

Insect type used	Fish Species	Average Fish Weight	FM level in Control Diet	Dietary Level	ADC Crude Protein %	ACD Crude Lipid %	ADC Gross Energy %	Reference
Zophobas morio	Nile tilapia, *Oreochromis niloticus*				(77.5–50.5)	(91.5–69.8)		Jabir et al. (2012)
Zophobas morio larvae	Nile tilapia, *Oreochromis niloticus*	3g		20%	70	93.5	80.1	Fontes et al. (2019)
Grylus bimaculatus	African catfish, *Clarias gariepinus*							Taufek et al. (2016)
Gryllus assimilis adult	Nile tilapia, *Oreochromis niloticus*	3g		20%	39.7	87.9	47	Fontes et al. (2019)
Nauphoeta cinerea adult, dried and grinded	Nile tilapia, *Oreochromis niloticus*	3g		20%	69.6	91.6	58.4	Fontes et al. (2019)
Gromphadorhina portentosa adult	Nile tilapia, *Oreochromis niloticus*	3g		20%	61.6	98.8	47.4	Fontes et al. (2019)
silkworm oil	carp	0.9g	25%	3,6,9% oil	(82.1-84.1,81.2,79.4)	(69.8-71.4,78.2,82.7)		Nandeesha et al. (1999)

HM, *Hermetia luminescens*
BSF, Black soldier fly
TM, *Tenebrio molitor*
HS, Housefly
ZM, *Zophobas morio*
FM, Fish meal
ADC, Apparent Digestibility Coefficient

on the level of processing of the yeast. In rainbow trout, inclusion of a whole yeast mixture of *Wickerhamomyces anomalus* and *Saccharomyces cerevisiae*, at 60% led to a decreased growth performance, while dry matter and protein digestibility were decreased by 2.6 and 6%, respectively (Vidakovic, Huyben et al. 2020). It has been suggested that the thick and rigid cell walls are a major problem for dietary utilization of yeast, as they may limit enzymatic access to cellular contents, thus reducing their digestibility by the fish. Intact brewer's yeast had lower protein and energy digestibility in rainbow trout than disrupted yeast cells, yeast extract, and yeast protein isolate (Rumsey, Hughes et al. 1991). Therefore, disruption of the cell wall can increase the utilization and nutrient digestibility of yeast, while reducing the soluble and solid waste output.

Microalgae cell wall. Microalgae, such as *Chlorella* spp. and *Nannochloropsis* spp., are major photosynthetic organisms used in the production of single-cell proteins. They use cheap and very abundant sources for their growth, such as carbon dioxide and nitrates, as source of carbon and nitrogen. Similar to yeasts, micro-algal cell walls can decrease the nutrient digestibility; therefore, their removal through technological processing is essential to increase nutrient utilization (Ugbogu and Ugbogu 2016). However, despite the likeliness of the hypothesis that cell wall composition would affect the nutrient digestibility, algal cell wall hardness does not seem to have the expected negative effect (Teuling, Schrama et al. 2017). Increasing dietary levels of *Phaeodactylum tricornutum* algae from 0 to 12% reduced linearly the digestibility of dry matter, lipid, and protein in Atlantic salmon (Sørensen, Berge et al. 2016). The addition of *Schizochytrium* at 30% inclusion decreased the total short chain fatty acid and the crude fiber digestibility in Nile tilapia; however, it increased the overall amino acid digestibility (Sarker, Gamble et al. 2016).

Technological Processing

Processing of raw materials can have a significant impact on their nutritional value, by increasing their digestibility or by deactivating antinutritional factors present in the meals (Table 5.1).

A widely used processing method is heat treatment, which can significantly decrease several antinutritional factors, such as protein inhibitors and lectins (Drew, Borgeson et al. 2007). For lectins, the negative effects may also be reverted by the inclusion of bile salts, such as cholyltaurine (Iwashita, Suzuki et al. 2009), which normalize the intestinal abnormalities.

For phytate, heat treatment may also work; however, fermentation is suggested as a more efficient deactivation process based on microbial phytase activity (Mukhopadhyay and Ray 1999) or addition of commercial phytases (Hossain and Jauncey 1993, Riche and Brown 1996, Storebakken, Shearer et al. 1998, Vielma, Lall et al. 1998, Maas et al., 2021). Optimization of the hydrolysis environment needs to be further defined. Also, milling can reduce the phytate content in the feedstuff as it removes the outer layer of the seeds, where it is mostly located. The addition of minerals in the feed such as zinc and phosphate to increase the levels of available phosphorus has been shown to be partially capable of counteracting the negative effects of dietary phytate, although more research is needed to determine stable forms.

Concerning saponins, heat treatment cannot neutralize them, but they can be degraded by acid or alkaline hydrolysis and glucosidases of bacterial origin (Gestetner, Birk et al. 1968).

Regarding insect meals, due to the high fat content of the larvae, insect producers often employ a defatting process using various methods (physical or chemical extractions). This process increases the percentage of protein and, the extracted oils may be used for different purposes, such as feed inclusion (Schiavone, Cullere et al. 2017) and biodiesel production (Henry, Gasco et al. 2015, Li, Ji et al. 2016, Surendra, Olivier et al. 2016). Regarding chitin content, the levels in the meals can be reduced through extraction process (Belluco, Losasso et al. 2013, Sánchez-Muros, Barroso et al. 2014). An alternative is the use of dietary enzymes. Unexpectedly, the digestibility of *Tenebrio molitor* meal improved in terms of dry matter and crude protein compared to the fish-meal-based diet and addition of exogenous enzymes to insect-meal-based diets reduced significantly their acid detergent fiber digestibility in European sea bass, *Dicentrarchus labrax*, suggesting that exogenous enzymes may have altered the microbiota of the fish and therefore reduced the potential chitinase activity of some commensal bacteria helping the fish to digest the chitin-encompassed nutrients (Gasco, Henry et al. 2016). However, further study is needed (Henry, Gasco et al. 2015, Gasco, Henry et al. 2016).

Various chemical, enzymatical, physical, or mechanical methods can be used to rupture the yeast cell walls, thus increasing digestibility by mechanical rupturing of cell walls or enzymatic hydrolysis (Nasseri, Rasoul-Amini et al. 2011). A higher protein content of the yeast can be obtained by removal of the cell wall material, with the water-soluble cell contents remaining. Increasing inclusion of yeast extract as a replacement for fish meal in diets for shrimp (*Litopenaeus vannamei)* and Arctic charr increased the apparent digestibility of protein, most likely due to the combined effect of removal of cell walls and increased proportion of water-soluble low molecular weight proteins (Langeland, Vidakovic et al. 2016, Zhao, Wang et al. 2017). Similar processing can also apply to the algal cell walls. However, a recent study has shown that extrusion can potentially increase the digestibility of nutrients, especially dry matter and ash in certain microalgae like defatted *Nannochloropsis* sp. (Gong, Guterres et al. 2018).

Impact of Waste Type on Different Aquaculture Systems

The rapid growth of aquaculture has raised concerns with respect to waste production and subsequent changes in the surrounding environment. The effects of waste output may vary greatly, and, since feed formulation is very important for this output, it is highly probable that the form of feces/waste may be tailored by fish farmers according to the type of system (i.e. RAS or open cages)(Reid, Liutkus et al. 2009, Verdegem 2013). Open cage or flow-through systems have a direct impact on the environment since overloading with nutrients may result in eutrophication events and high levels of biological oxygen demand (BOD), while in a RAS, the impact is mostly related to increased risks for fish welfare and water quality improvement costs. For example, suspended solids, coming from feces with low density that break apart easily, can reduce the local load below cages and increase the chance of benthic regeneration; however, this same type can create problems, like clogging, in rearing units that use filters for waste removal. On the other hand, feces of high-density, sinking particles, that settle faster, can be a problem for cage aquaculture since they remain mainly below the cages, but they are desirable for RAS, as their removal by fecal traps or filters is easier. In RAS, high soluble and suspended solid waste is not preferred due to the impact on water quality and increased costs for filters, as well as fish performance (for particles <50μm). However, in open cage systems and flow-through, soluble waste is preferable, since unlike sinking solids, it does not accumulate under the cages, and thus offers better chances for benthic regeneration. Thus, according to the aquaculture operation in focus, feed formulation should be adjusted according to the form of waste produced, and for this purpose, knowledge of the dietary impact on feed acceptance, fecal physicochemical properties, and soluble waste production is important.

FUTURE CHALLENGES

In an era of increasing use of plant proteins and novel feedstuff, antinutrient levels should be redefined according to their impact on nutrient availability of protein sources, as they may contribute to impaired growth performance and waste release in the surrounding environment. The use of pure protein concentrates and cell wall free single-cell proteins can be recommended in aquafeeds due to their low amounts of antinutrient factors, better nutrient retention by fish, and, therefore, less nutrient release into the environment.

With regard to non-starch polysaccharides, these compounds constitute the major part of plant sources, which remains highly indigestible by fish. Since these ingredients mainly affect fecal physicochemical properties, their effects should be eliminated through either *in vitro* digestion or the addition of degrading enzymes in the diets. In recent years, the use of NSP-degrading enzymes has been tested in several studies, as presented in a recent review by Castillo and Gatlin (2015). Moreover, the use of these enzymes can be cost effective, as they significantly improve the feed conversion ratio and thus reduce feed costs. The use of such enzymes can also be used to improve the insect meal digestibility. By reducing the chitin content, its impact on digestibility and growth performance is diminished. To conclude, research efforts should keep focusing on the use of degrading enzymes, while the combination with pure protein concentrates can offer better nutrient retention and less waste released into the environment.

REFERENCES

Al-Asgah, N. and A. Ali (1996). "Effect of feeding different levels of wheat bran on the growth performance and body composition of *Oreochromis niloticus*." *Agribiological Research (Germany)*.

Alabaster, J. (1982). "Survey of fish-farm effluents in some EIFAC countries." EIFAC Technical Papers (FAO).

Alegbeleye, W. O., S. O. Obasa, O. O. Olude, K. Otubu and W. Jimoh (2012). "Preliminary evaluation of the nutritive value of the variegated grasshopper (*Zonocerus variegatus* L.) for African catfish *Clarias gariepinus* (Burchell. 1822) fingerlings." *Aquaculture Research 43*(3): 412–420.

Alexis, M. N. (1990). "Comparative evaluation of soybean meal and carob seed germ meal as dietary ingredients for rainbow trout fingerlings." *Aquatic Living Resources 3*(3): 235–241.

AlOgaily, S. (1996). "Effect of feeding different levels of cellulose on the growth performance and body composition of *Oreochromis niloticus*." *Arab Gulf Journal of Scientific Research 14*(3): 731–745.

Amirkolaie, A. K. (2011). "Reduction in the environmental impact of waste discharged by fish farms through feed and feeding." *Reviews in Aquaculture 3*(1): 19–26.

Amirkolaie, A. K., J. I. Leenhouwers, J. A. J. Verreth and J. W. Schrama (2005). "Type of dietary fibre (soluble versus insoluble) influences digestion, faeces characteristics and faecal waste production in Nile tilapia (*Oreochromis niloticus* L.)." *Aquaculture Research 36*(12): 1157–1166.

Anderson, J., A. Jackson, A. Matty and B. Capper (1984). "Effects of dietary carbohydrate and fibre on the tilapia *Oreochromis niloticus* (Linn.)." *Aquaculture 37*(4): 303–314.

Antonopoulou, E., E. Nikouli, G. Piccolo, L. Gasco, F. Gai, S. Chatzifotis, E. Mente and K. A. Kormas (2019). "Reshaping gut bacterial communities after dietary *Tenebrio molitor* larvae meal supplementation in three fish species." *Aquaculture 503*: 628–635.

Antony Jesu Prabhu, P., E. Fountoulaki, R. Maas, L. T. N. Heinsbroek, E. H. Eding, S. J. Kaushik and J. W. Schrama (2019). "Dietary ingredient composition alters faecal characteristics and waste production in common carp reared in recirculation system." *Aquaculture 512*: 734357.

Ballester-Moltó, M., P. Sanchez-Jerez, J. Cerezo-Valverde and F. Aguado-Giménez (2017). "Particulate waste outflow from fish-farming cages. How much is uneaten feed?" *Marine Pollution Bulletin 119*(1): 23–30.

Barrows, F. T., D. A. Stone and R. W. Hardy (2007). "The effects of extrusion conditions on the nutritional value of soybean meal for rainbow trout (*Oncorhynchus mykiss*)." *Aquaculture 265*(1–4): 244–252.

Becker, K. and H. Makkar (1999). "Effects of dietary tannic acid and quebracho tannin on growth performance and metabolic rates of common carp (*Cyprinus carpio* L.)." *Aquaculture 175*(3): 327–335.

Belghit, I., N. S. Liland, P. Gjesdal, I. Biancarosa, E. Menchetti, Y. Li, R. Waagbø, Å. Krogdahl and E.-J. Lock (2019). "Black soldier fly larvae meal can replace fish meal in diets of sea-water phase Atlantic salmon (*Salmo salar*)." *Aquaculture 503*: 609–619.

Bell, J. (1993). "Factors affecting the nutritional value of canola meal: a review." *Canadian Journal of Animal Science 73*(4): 689–697.

Belluco, S., C. Losasso, M. Maggioletti, C. C. Alonzi, M. G. Paoletti and A. Ricci (2013). "Edible insects in a food safety and nutritional perspective: A critical review." *Comprehensive Reviews in Food Science and Food Safety 12*(3): 296–313.

Bravo, L. (1998). "Polyphenols: chemistry, dietary sources, metabolism, and nutritional significance." *Nutrition Reviews 56*(11): 317–333.

Brinker, A., W. Koppe and R. Rösch (2005). "Optimised effluent treatment by stabilised trout faeces." *Aquaculture 249*(1): 125–144.

Brinker, A., W. Koppe and R. Rösch (2005). "Optimizing trout farm effluent treatment by stabilizing trout feces: a field trial." *North American Journal of Aquaculture 67*(3): 244–258.

Buhler, D. R. and J. E. Halver (1961). "Nutrition of salmonoid fishes. 9. Carbohydrate requirements of *Chinook salmon*." *Journal of Nutrition 74*: 307–318.

Bureau, D. and K. Hua (2010). "Towards effective nutritional management of waste outputs in aquaculture, with particular reference to salmonid aquaculture operations." *Aquaculture Research 41*(5): 777–792.

Bureau, D. P., A. M. Harris and C. Y. Cho (1998). "The effects of purified alcohol extracts from soy products on feed intake and growth of chinook salmon (*Oncorhynchus tshawytscha*) and rainbow trout (*Oncorhynchus mykiss*)." *Aquaculture 161*(1): 27–43.

Burel, C., T. Boujard, S. Kaushik, G. Boeuf, K. Mol, S. Van der Geyten, V. Darras, E. Kühn, B. Pradet-Balade and B. Quérat (2001). "Effects of rapeseed meal-glucosinolates on thyroid metabolism and feed utilization in rainbow trout." *General and Comparative Endocrinology 124*(3): 343–358.

Burel, C., T. Boujard, S. J. Kaushik, G. Boeuf, S. Van Der Geyten, K. A. Mol, E. R. Kühn, A. Quinsac, M. Krouti and D. Ribaillier (2000). "Potential of plant-protein sources as fish meal substitutes in diets for turbot (*Psetta maxima*): growth, nutrient utilisation and thyroid status." *Aquaculture 188*(3): 363–382.

Buttle, L., A. Burrells, J. Good, P. Williams, P. Southgate and C. Burrells (2001). "The binding of soybean agglutinin (SBA) to the intestinal epithelium of Atlantic salmon, *Salmo salar* and Rainbow trout, *Oncorhynchus mykiss*, fed high levels of soybean meal." *Veterinary Immunology and Immunopathology 80*(3): 237–244.

Castillo, S. and D. M. Gatlin (2015). "Dietary supplementation of exogenous carbohydrase enzymes in fish nutrition: A review." *Aquaculture 435*: 286–292.

Chen, S., D. E. Coffin and R. F. Malone (1997). "Sludge production and management for recirculating aquacultural systems." *Journal of the World Aquaculture Society 28*(4): 303–315.

Chen, W., Q. Ai, K. Mai, W. Xu, Z. Liufu, W. Zhang and Y. Cai (2011). "Effects of dietary soybean saponins on feed intake, growth performance, digestibility and intestinal structure in juvenile Japanese flounder (*Paralichthys olivaceus*)." *Aquaculture 318*(1): 95–100.

Chikwati, E. M., F. F. Venold, M. H. Penn, J. Rohloff, S. Refstie, A. Guttvik, M. Hillestad and Å. Krogdahl (2012). "Interaction of soyasaponins with plant ingredients in diets for Atlantic salmon, *Salmo salar* L." *British Journal of Nutrition 107*(11): 1570–1590.

Cho, C. and D. Bureau (2001). "A review of diet formulation strategies and feeding systems to reduce excretory and feed wastes in aquaculture." *Aquaculture Research 32*(s1): 349–360.

Cho, C. Y. and D. P. Bureau (1997). "Reduction of waste output from salmonid aquaculture through feeds and feeding." *The Progressive Fish-Culturist 59*(2): 155–160.

Chou, C., K. Haya, L. Paon, L. Burridge and J. Moffatt (2002). "Aquaculture-related trace metals in sediments and lobsters and relevance to environmental monitoring program ratings for near-field effects." *Marine Pollution Bulletin 44*(11): 1259–1268.

Couto, A., T. Kortner, M. Penn, A. Bakke, Å. Krogdahl and A. Oliva-Teles (2014). "Effects of dietary phytosterols and soy saponins on growth, feed utilization efficiency and intestinal integrity of gilthead sea bream (*Sparus aurata*) juveniles." *Aquaculture 432*: 295–303.

Cripps, S. J. and A. Bergheim (2000). "Solids management and removal for intensive land-based aquaculture production systems." *Aquacultural Engineering 22*(1): 33–56.

D'Mello, J. F., C. M. Duffus and J. H. Duffus (1991). *Toxic substances in crop plants*. Elsevier, United Kingdom: Woodhead Publishing.

Davies, S. J. (1985). "The role of dietary fibre in fish nutrition." *Recent Advances in Aquaculture 2*: 219–249.

Deguara, S., K. Jauncey, J. Feord and J. Lopez (1999). "Growth and feed utilization of gilthead sea bream, *Sparus aurata*, fed diets with supplementary enzymes." In J. Brufau and A. Tacon (eds.), *Feed manufacturing in the mediterranean region: recent advances in research and technology* (Vol. *37*, pp. 195–215). Zaragoza, Spain: CIHEAM/IAMZ.

Denstadli, V., A. Skrede, Å. Krogdahl, S. Sahlstrøm and T. Storebakken (2006). "Feed intake, growth, feed conversion, digestibility, enzyme activities and intestinal structure in Atlantic salmon (*Salmo salar* L.) fed graded levels of phytic acid." *Aquaculture 256*(1): 365–376.

Drew, M., T. Borgeson and D. Thiessen (2007). "A review of processing of feed ingredients to enhance diet digestibility in finfish." *Animal Feed Science and Technology 138*(2): 118–136.

Dumas, A., T. Raggi, J. Barkhouse, E. Lewis and E. Weltzien (2018). "The oil fraction and partially defatted meal of black soldier fly larvae (*Hermetia illucens*) affect differently growth performance, feed efficiency, nutrient deposition, blood glucose and lipid digestibility of rainbow trout (*Oncorhynchus mykiss*)." *Aquaculture 492*: 24–34.

Elia, A. C., M. T. Capucchio, B. Caldaroni, G. Magara, A. J. M. Dörr, I. Biasato, E. Biasibetti, M. Righetti, P. Pastorino and M. Prearo (2018). "Influence of Hermetia illucens meal dietary inclusion on the histological traits, gut mucin composition and the oxidative stress biomarkers in rainbow trout (*Oncorhynchus mykiss*)." *Aquaculture 496*: 50–57.

Erdman, J. (1979). "Oilseed phytates: nutritional implications." *Journal of the American Oil Chemists' Society 56*(8): 736–741.

FAO (2017). "Aquaculture regional reviews. FAO Fisheries and Aquaculture Department [online]." Rome. Updated 8 February 2017.

Filioglou, M. and M. Alexis (1989). "Protein digestibility and enzyme activity in the digestive tract of rainbow trout fed diets containing increasing levels of carbo-seed germ meal." In De Pauw, N. et al. (eds.), European Aquaculture Society: Bredene. ISBN 90-71625-03-6. *Aquaculture: a biotechnology in progress* (Volume 1, pp. 839–843).

Finke, M. D. (2007). "Estimate of chitin in raw whole insects." *Zoo Biology: Published in Affiliation with the American Zoo and Aquarium Association 26*(2): 105–115.

Fountoulaki, E., D. Nikolopoulou, A. Vasilaki, M. N. Alexi and J. Dias (2014). "The effects of different raw materials used in diets for sea bass (*Dicentrarchus labrax*) on digestibility, faeces characteristics and faecal waste production; Preliminary results." European Aquaculture Society Donostia-San Sebastian, 14–17 October 2014.

Francis, G., H. P. S. Makkar and K. Becker (2001). "Antinutritional factors present in plant-derived alternate fish feed ingredients and their effects in fish." *Aquaculture 199*(3–4): 197–227.

Galasso, H. L., M. D. Callier, D. Bastianelli, J.-P. Blancheton and C. Aliaume (2017). "The potential of near infrared spectroscopy (NIRS) to measure the chemical composition of aquaculture solid waste." *Aquaculture 476*: 134–140.

Gasco, L., M. Henry, G. Piccolo, S. Marono, F. Gai, M. Renna, C. Lussiana, E. Antonopoulou, P. Mola and S. Chatzifotis (2016). "*Tenebrio molitor* meal in diets for European sea bass (*Dicentrarchus labrax* L.) juveniles: Growth performance, whole body composition and in vivo apparent digestibility." *Animal Feed Science and Technology 220*: 34–45.

Gestetner, B., Y. Birk and Y. Tencer (1968). "Soybean saponins. Fate of ingested soybean saponins and the physiological aspect of their hemolytic activity." *Journal of Agricultural and Food Chemistry 16*(6): 1031–1035.

Glencross, B. (2009). "The influence of soluble and insoluble lupin non-starch polysaccharides on the digestibility of diets fed to rainbow trout (*Oncorhynchus mykiss*)." *Aquaculture 294*(3–4): 256–261.

Glencross, B., N. Rutherford and N. Bourne (2012). "The influence of various starch and non-starch polysaccharides on the digestibility of diets fed to rainbow trout (*Oncorhynchus mykiss*)." *Aquaculture 356–357*: 141–146.

Gong, Y., H. A. D. S. Guterres, M. Huntley, M. Sørensen and V. Kiron (2018). "Digestibility of the defatted microalgae *Nannochloropsis* sp. and *Desmodesmus* sp. when fed to Atlantic salmon, *Salmo salar*." *Aquaculture Nutrition 24*(1): 56–64.

Gosavi, S. M., C. R. Verma, S. S. Kharat, M. Pise and P. Kumkar (2019). "Structural adequacy of the digestive tract supports dual feeding habit in catfish *Pachypterus khavalchor* (Siluriformes: Horabagridae)." *Acta Histochemica 121*(4): 437–449.

Gowen, R. (1994). "Managing eutrophication associated with aquaculture development." *Journal of Applied Ichthyology 10*(4): 242–257.

Han, X., R. Rosati and J. Webb (1996). *Correlation of particle size distribution of solid waste to fish feed composition in an aquaculture recirculation system*, MSc thesis, Illinois State University, United States of America Normal, Illinois.

Hansen, J. Ø., M. Penn, M. Øverland, K. D. Shearer, Å. Krogdahl, L. T. Mydland and T. Storebakken (2010). "High inclusion of partially deshelled and whole krill meals in diets for Atlantic salmon (*Salmo salar*)." *Aquaculture 310*(1–2): 164–172.

Hardy, R. W. (2010). "Utilization of plant proteins in fish diets: effects of global demand and supplies of fishmeal." *Aquaculture Research 41*(5): 770–776.

Hemaiswarya, S., R. Raja, R. Ravi Kumar, V. Ganesan and C. Anbazhagan (2011). "Microalgae: a sustainable feed source for aquaculture." *World Journal of Microbiology and Biotechnology 27*(8): 1737–1746.

Hendriks, H., T. Van den Ingh, Å. Krogdahl, J. Olli and J. Koninkx (1990). "Binding of soybean agglutinin to small intestinal brush border membranes and brush border membrane enzyme activities in Atlantic salmon (*Salmo salar*)." *Aquaculture 91*(1–2): 163–170.

Henry, M., L. Gasco, G. Piccolo and E. Fountoulaki (2015). "Review on the use of insects in the diet of farmed fish: Past and future." *Animal Feed Science and Technology 203*: 1–22.

Hetland, H., M. Choct and B. Svihus (2004). "Role of insoluble non-starch polysaccharides in poultry nutrition." *World's Poultry Science Journal 60*(4): 415–422.

Hossain, M., U. Focken and K. Becker (2001). "Galactomannan-rich endosperm of Sesbania (*Sesbania aculeata*) seeds responsible for retardation of growth and feed utilisation in common carp, *Cyprinus carpio* L." *Aquaculture 203*(1): 121–132.

Hossain, M. and K. Jauncey (1993). "The effects of varying dietary phytic acid, calcium and magnesium levels on the nutrition of common carp, *Cyprinus carpio*." *Colloques de l'INRA (France).*

Ikeda, M., H. Kakizaki and M. Matsumiya (2017). "Biochemistry of fish stomach chitinase." *International Journal of Biological Macromolecules 104*: 1672–1681.

Ikeda, M., Y. Kondo and M. Matsumiya (2013). "Purification, characterization, and molecular cloning of chitinases from the stomach of the threeline grunt *Parapristipoma trilineatum*." *Process Biochemistry 48*(9): 1324–1334.

Iwama, G. K. (1991). "Interactions between aquaculture and the environment." *Critical Reviews in Environmental Science and Technology 21*(2): 177–216.

Iwashita, Y., N. Suzuki, H. Matsunari, T. Sugita and T. Yamamoto (2009). "Influence of soya saponin, soya lectin, and cholyltaurine supplemented to a casein-based semipurified diet on intestinal morphology and biliary bile status in fingerling rainbow trout *Oncorhynchus mykiss*." *Fisheries Science 75*(5): 1307–1315.

Johnson, I., J. M. Gee, K. Price, C. Curl and G. Fenwick (1986). "Influence of saponins on gut permeability and active nutrient transport in vitro." *The Journal of Nutrition 116*(11): 2270–2277.

Kakizaki, H., M. Ikeda, H. Fukushima and M. Matsumiya (2015). "Distribution of chitinolytic enzymes in the organs and cDNA cloning of chitinase isozymes from the stomach of two species of fish, chub mackerel (*Scomber japonicus*) and silver croaker (*Pennahia argentata*)." *Open Journal of Marine Science 5*(04): 398.

Kihara, M. and T. Sakata (1997). "Fermentation of dietary carbohydrates to short-chain fatty acids by gut microbes and its influence on intestinal morphology of a detritivorous teleost tilapia (*Oreochromis niloticus*)." *Comparative Biochemistry and Physiology Part A: Physiology 118*(4): 1201–1207.

Knudsen, K. E. B. (1997). "Carbohydrate and lignin contents of plant materials used in animal feeding." *Animal Feed Science and Technology 67*(4): 319–338.

Kokou, F. and E. Fountoulaki (2018). "Aquaculture waste production associated with antinutrient presence in common fish feed plant ingredients." *Aquaculture 495*: 295–310.

Kokou, F., G. Rigos, M. Henry, M. Kentouri and M. Alexis (2012). "Growth performance, feed utilization and non-specific immune response of gilthead sea bream (*Sparus aurata* L.) fed graded levels of a bioprocessed soybean meal." *Aquaculture 364*: 74–81.

Kokou, F., G. Rigos, M. Kentouri and M. Alexis (2016). "Effects of DL-methionine-supplemented dietary soy protein concentrate on growth performance and intestinal enzyme activity of gilthead sea bream (*Sparus aurata* L.)." *Aquaculture International 24*(1): 257–271.

Kokou, F., A. Samartzis, A. Vasilaki, D. Kyriazis, G. Kokkalenios, G. Rigos, M. Kentouri and M. Alexi (2012). "Nutrient digestibility and growth performance of gilthead sea bream (*Sparus aurata*) after dietary administration of antinutritional factors present in soybean meal." XV International Symposium on Fish Nutrition and Feeding, Molde, Norway, 4–7 June 2012.

Kraugerud, O. F., M. Penn, T. Storebakken, S. Refstie, Å. Krogdahl and B. Svihus (2007). "Nutrient digestibilities and gut function in Atlantic salmon (*Salmo salar*) fed diets with cellulose or non-starch polysaccharides from soy." *Aquaculture 273*(1): 96–107.

Krogdahl, A., A. Bakke-McKellep, K. Roed and G. Baeverfjord (2000). "Feeding Atlantic salmon *Salmo salar* L. soybean products: effects on disease resistance (furunculosis), and lysozyme and IgM levels in the intestinal mucosa." *Aquaculture Nutrition 6*(2): 77–84.

Krogdahl, Å., T. B. Lea and J. J. Olli (1994). "Soybean proteinase inhibitors affect intestinal trypsin activities and amino acid digestibilities in rainbow trout (*Oncorhynchus mykiss*)." *Comparative Biochemistry and Physiology Part A: Physiology 107*(1): 215–219.

Krogdahl, Å., M. Penn, J. Thorsen, S. Refstie and A. M. Bakke (2010). "Important antinutrients in plant feedstuffs for aquaculture: an update on recent findings regarding responses in salmonids." *Aquaculture Research 41*(3): 333–344.

Langeland, M., A. Vidakovic, J. Vielma, J. Lindberg, A. Kiessling and T. Lundh (2016). "Digestibility of microbial and mussel meal for Arctic charr (*Salvelinus alpinus*) and Eurasian perch (*Perca fluviatilis*)." *Aquaculture Nutrition 22*(2): 485–495.

Leatherland, J. (1994). "Reflections on the thyroidology of fishes: From molecules to humankind." *Guelph Ichthyological Reviews, 2*. TFH Publ., Neptune City, NJ.

Leenhouwers, J., D. Adjei-Boateng, J. Verreth and J. Schrama (2006). "Digesta viscosity, nutrient digestibility and organ weights in African catfish (*Clarias gariepinus*) fed diets supplemented with different levels of a soluble non-starch polysaccharide." *Aquaculture Nutrition 12*(2): 111–116.

Leenhouwers, J., G. Santos, J. Schrama and J. Verreth (2004). Digesta viscosity and nutrient digestibility in European sea bass in a fresh water and marine environment. Conference on'Biotechnologies for quality', Barcelona, Spain, 20-23 October 2004.

Letelier-Gordo, C. O., J. Dalsgaard, K. I. Suhr, K. S. Ekmann and P. B. Pedersen (2015). "Reducing the dietary protein:energy (P:E) ratio changes solubilization and fermentation of rainbow trout (*Oncorhynchus mykiss*) faeces." *Aquacultural Engineering 66*: 22–29.

Li, S., H. Ji, B. Zhang, J. Tian, J. Zhou and H. Yu (2016). "Influence of black soldier fly (*Hermetia illucens*) larvae oil on growth performance, body composition, tissue fatty acid composition and lipid deposition in juvenile Jian carp (*Cyprinus carpio* var. Jian)." *Aquaculture 465*: 43–52.

Li, S., H. Ji, B. Zhang, J. Zhou and H. Yu (2017). "Defatted black soldier fly (*Hermetia illucens*) larvae meal in diets for juvenile Jian carp (*Cyprinus carpio* var. Jian): Growth performance, antioxidant enzyme activities, digestive enzyme activities, intestine and hepatopancreas histological structure." *Aquaculture 477*: 62–70.

Li, Y., T. M. Kortner, E. M. Chikwati, I. Belghit, E.-J. Lock and Å. Krogdahl (2020). "Total replacement of fish meal with black soldier fly (*Hermetia illucens*) larvae meal does not compromise the gut health of Atlantic salmon (*Salmo salar*)." *Aquaculture 520*: 734967.

Liener, E. I. (1980). *Toxic constituents of plant foodstuffs*. New York: Academic.

Lim, C., P. Klesius and D. Higgs (1998). "Substitution of canola meal for soybean meal in diets for channel catfish *Ictalurus punctatus*." *Journal of the World Aquaculture Society 29*(2): 161–168.

Luo, Y., Q. Ai, K. Mai, W. Zhang, W. Xu and Y. Zhang (2012). "Effects of dietary rapeseed meal on growth performance, digestion and protein metabolism in relation to gene expression of juvenile cobia (*Rachycentron canadum*)." *Aquaculture 368*: 109–116.

Maas, R. M., M. C. Verdegem, S. Debnath, L. Marchal and J. W. Schrama (2021). "Effect of enzymes (phytase and xylanase), probiotics (*B. amyloliquefaciens*) and their combination on growth performance and nutrient utilisation in Nile tilapia." *Aquaculture 533*: 736226.

Maenz, D. D., G. G. Irish and H. L. Classen (1999). "Carbohydrate-binding and agglutinating lectins in raw and processed soybean meals." *Animal Feed Science and Technology 76*(3): 335–343.

Magill, S. H., H. Thetmeyer and C. J. Cromey (2006). "Settling velocity of faecal pellets of gilthead sea bream (*Sparus aurata* L.) and sea bass (*Dicentrarchus labrax* L.) and sensitivity analysis using measured data in a deposition model." *Aquaculture 251*(2): 295–305.

Mai, K., Y. Zhang, W. Chen, W. Xu, Q. Ai and W. Zhang (2012). "Effects of dietary soy isoflavones on feed intake, growth performance and digestibility in juvenile Japanese flounder (*Paralichthys olivaceus*)." *Journal of Ocean University of China* (English Edition) *11*(4): 511–516.

Maitra, S. and A. Ray (2003). "Inhibition of digestive enzymes in rohu, *Labeo rohita* (Hamilton), fingerlings by tannin: an in vitro study." *Aquaculture Research 34*(1): 93–95.

Makkar, H. P. and K. Becker (2009). "Jatropha curcas, a promising crop for the generation of biodiesel and value-added coproducts." *European Journal of Lipid Science and Technology 111*(8): 773–787.

Mandal, S. and K. Ghosh (2010). "Inhibitory effect of Pistia tannin on digestive enzymes of Indian major carps: an in vitro study." *Fish Physiology and Biochemistry 36*(4): 1171–1180.

Marono, S., G. Piccolo, R. Loponte, C. Di Meo, Y. A. Attia, A. Nizza and F. Bovera (2015). "In vitro crude protein digestibility of *Tenebrio molitor* and *Hermetia illucens* insect meals and its correlation with chemical composition traits." *Italian Journal of Animal Science 14*(3): 3889.

Mawson, R., R. Heaney, Z. Zdunczyk and H. Kozlowska (1994). "Rapeseed meal-glucosinolates and their antinutritional effects Part 3. Animal growth and performance." *Molecular Nutrition & Food Research 38*(2): 167–177.

Mawson, R., R. Heaney, Z. Zdunczyk and H. Kozlowska (1994). "Rapeseed meal-glucosinolates and their antinutritional effects Part 4. Goitrogenicity and internal organs abnormalities in animals." *Molecular Nutrition & Food Research 38*(2): 178–191.

Mawson, R., R. Heaney, Z. Zduńczyk and H. Kozłowska (1995). "Rapeseed meal-glucosinolates and their antinutritional effects Part 7. Processing." *Molecular Nutrition & Food Research 39*(1): 32–41.

Mazur, W. M., J. A. Duke, K. Wähälä, S. Rasku and H. Adlercreutz (1998). "Isoflavonoids and lignans in legumes: nutritional and health aspects in humans." *The Journal of Nutritional Biochemistry 9*(4): 193–200.

Messina, M. J. (1999). "Legumes and soybeans: overview of their nutritional profiles and health effects." *The American Journal of Clinical Nutrition 70*(3, Supplement): 439s–450s.

Mukhopadhyay, N. and A. Ray (1999). "Effect of fermentation on the nutritive value of sesame seed meal in the diets for rohu, *Labeo rohita* (Hamilton), fingerlings." *Aquaculture Nutrition 5*(4): 229–236.

Naczk, M., R. Amarowicz, A. Sullivan and F. Shahidi (1998). "Current research developments on polyphenolics of rapeseed/canola: a review." *Food Chemistry 62*(4): 489–502.

Nasseri, A., S. Rasoul-Amini, M. Morowvat and Y. Ghasemi (2011). "Single cell protein: production and process." *American Journal of Food Technology 6*(2): 103–116.

Nogales-Mérida, S., P. Gobbi, D. Józefiak, J. Mazurkiewicz, K. Dudek, M. Rawski, B. Kierończyk and A. Józefiak (2019). "Insect meals in fish nutrition." *Reviews in Aquaculture 11*(4): 1080–1103.

Oliva-Teles, A. and P. Gonçalves (2001). "Partial replacement of fishmeal by brewers yeast (*Saccaromyces cerevisae*) in diets for sea bass (*Dicentrarchus labrax*) juveniles." *Aquaculture 202*(3–4): 269–278.

Olli, J. J., K. Hjelmeland and Å. Krogdahl (1994). "Soybean trypsin inhibitors in diets for Atlantic salmon (*Salmo salar*, L): effects on nutrient digestibilities and trypsin in pyloric caeca homogenate and intestinal content." *Comparative Biochemistry and Physiology Part A: Physiology 109*(4): 923–928.

Øverland, M. and A. Skrede (2017). "Yeast derived from lignocellulosic biomass as a sustainable feed resource for use in aquaculture." *Journal of the Science of Food and Agriculture 97*(3): 733–742.

Pavasovic, A., A. J. Anderson, P. B. Mather and N. A. Richardson (2007). "Effect of a variety of animal, plant and single cell-based feed ingredients on diet digestibility and digestive enzyme activity in redclaw crayfish, *Cherax quadricarinatus* (Von Martens 1868)." *Aquaculture 272*(1–4): 564–572.

Pelissero, C. and J. Sumpter (1992). "Steroids and "steroid-like" substances in fish diets." *Aquaculture 107*(4): 283–301.

Pérez, Ó., E. Almansa, R. Riera, M. Rodriguez, E. Ramos, J. Costa and Ó. Monterroso (2014). "Food and faeces settling velocities of meagre (*Argyrosomus regius*) and its application for modelling waste dispersion from sea cage aquaculture." *Aquaculture 420–421*: 171–179.

Refstie, S., B. Svihus, K. D. Shearer and T. Storebakken (1999). "Nutrient digestibility in Atlantic salmon and broiler chickens related to viscosity and non-starch polysaccharide content in different soyabean products." *Animal Feed Science and Technology 79*(4): 331–345.

Reid, G., M. Liutkus, S. Robinson, T. Chopin, T. Blair, T. Lander, J. Mullen, F. Page and R. Moccia (2009). "A review of the biophysical properties of salmonid faeces: implications for aquaculture waste dispersal models and integrated multi-trophic aquaculture." *Aquaculture Research 40*(3): 257–273.

Renna, M., A. Schiavone, F. Gai, S. Dabbou, C. Lussiana, V. Malfatto, M. Prearo, M. T. Capucchio, I. Biasato, E. Biasibetti and M. De Marco (2017). "Evaluation of the suitability of a partially defatted black soldier fly (*Hermetia illucens* L.) larvae meal as ingredient for rainbow trout (*Oncorhynchus mykiss* Walbaum) diets." *Journal of Animal Science and Biotechnology 8*(1): 1–13.

Richardson, N. L., D. A. Higgs, R. M. Beames and J. R. Mcbride (1985). "Influence of dietary calcium, phosphorus, zinc and sodium phytate level on cataract incidence, growth and histopathology in juvenile chinook salmon (*Oncorhynchus tshawytscha*)." *The Journal of Nutrition 115*(5): 553–567.

Riche, M. and P. B. Brown (1996). "Availability of phosphorus from feedstuffs fed to rainbow trout, *Oncorhynchus mykiss*." *Aquaculture 142*(3): 269–282.

Riche, M. and D. Garling (2004). "Effect of phytic acid on growth and nitrogen retention in tilapia *Oreochromis niloticus* L." *Aquaculture Nutrition 10*(6): 389–400.

Roy, S. S. and R. Pal (2015). Microalgae in aquaculture: a review with special references to nutritional value and fish dietetics. *Proceedings of the Zoological Society 68*(1): 1–8, Springer India.

Rumsey, G., S. Hughes, R. Smith, J. Kinsella and K. Shetty (1991). "Digestibility and energy values of intact, disrupted and extracts from brewer's dried yeast fed to rainbow trout (*Oncorhynchus mykiss*)." *Animal Feed Science and Technology 33*(3–4): 185–193.

Rumsey, G. L., S. G. Hughes and R. A. Winfree (1993). "Chemical and nutritional evaluation of soya protein preparations as primary nitrogen sources for rainbow trout (*Oncorhynchus mykiss*)." *Animal Feed Science and Technology 40*(2–3): 135–151.

Rust, M. (2002). "Nutritional physiology." In J. E. Halver and R. W. Hardy (eds.), *Fish Nutrition* (pp. 368–446). New York: The Academic Press.

Sajjadi, M. and C. Carter (2003). "Dietary phytase supplementation and the utilisation of phosphorus by Atlantic salmon (*Salmo salar* L.) fed a canola meal-based diet." *Asia Pacific Journal of Clinical Nutrition 12*: *60*.

Sánchez-Muros, M.-J., F. G. Barroso and F. Manzano-Agugliaro (2014). "Insect meal as renewable source of food for animal feeding: a review." *Journal of Cleaner Production 65*: 16–27.

Santigosa, E., M. Á. Saénz de Rodrigáñez, A. Rodiles, F. G. Barroso and F. J. Alarcón (2010). "Effect of diets containing a purified soybean trypsin inhibitor on growth performance, digestive proteases and intestinal histology in juvenile sea bream (*Sparus aurata* L.)." *Aquaculture Research 41*(9): e187–e198.

Sarker, P. K., M. M. Gamble, S. Kelson and A. R. Kapuscinski (2016). "Nile tilapia (*Oreochromis niloticus*) show high digestibility of lipid and fatty acids from marine *Schizochytrium* sp. and of protein and essential amino acids from freshwater *Spirulina* sp. feed ingredients." *Aquaculture Nutrition 22*(1): 109–119.

Schiavone, A., M. Cullere, M. De Marco, M. Meneguz, I. Biasato, S. Bergagna, D. Dezzutto, F. Gai, S. Dabbou, L. Gasco and A. Dalle Zotte (2017). "Partial or total replacement of soybean oil by black soldier fly larvae (*Hermetia illucens* L.) fat in broiler diets: effect on growth performances, feed-choice, blood traits, carcass characteristics and meat quality." *Italian Journal of Animal Science 16*(1): 93–100.

Schneider, O., A. K. Amirkolaie, J. Vera-Cartas, E. H. Eding, J. W. Schrama and J. A. J. Verreth (2004). "Digestibility, faeces recovery, and related carbon, nitrogen and phosphorus balances of five feed ingredients evaluated as fishmeal alternatives in Nile tilapia, *Oreochromis niloticus* L." *Aquaculture Research 35*(14): 1370–1379.

Sealey, W. M., T. G. Gaylord, F. T. Barrows, J. K. Tomberlin, M. A. McGuire, C. Ross and S. St-Hilaire (2011). "Sensory analysis of Rainbow Trout, *Oncorhynchus mykiss*, fed enriched Black Soldier Fly Prepupae, *Hermetia illucens*." *Journal of the World Aquaculture Society 42*(1): 34–45.

Shiau, S.-Y., H.-L. Yu, S. Hwa, S.-Y. Chen and S.-I. Hsu (1988). "The influence of carboxymethylcellulose on growth, digestion, gastric emptying time and body composition of tilapia." *Aquaculture 70*(4): 345–354.

Shiau, S., C. Kwok, C. Chen, H. Hong and H. Hsieh (1989). "Effects of dietary fibre on the intestinal absorption of dextrin, blood sugar level and growth of tilapia, *Oreochromis niloticm× O. aureus*." *Journal of Fish Biology 34*(6): 929–935.

Sinha, A. K., V. Kumar, H. P. Makkar, G. De Boeck and K. Becker (2011). "Non-starch polysaccharides and their role in fish nutrition–A review." *Food Chemistry 127*(4): 1409–1426.

Sørensen, M., G. M. Berge, K. I. Reitan and B. Ruyter (2016). "Microalga *Phaeodactylum tricornutum* in feed for Atlantic salmon (*Salmo salar*) —Effect on nutrient digestibility, growth and utilization of feed." *Aquaculture 460*: 116–123.

Sørensen, M., M. Penn, A. El-Mowafi, T. Storebakken, C. Chunfang, M. Øverland and Å. Krogdahl (2011). "Effect of stachyose, raffinose and soya-saponins supplementation on nutrient digestibility, digestive enzymes, gut morphology and growth performance in Atlantic salmon (*Salmo salar*, L)." *Aquaculture 314*(1): 145–152.

Spinelli, J., C. R. Houle and J. C. Wekell (1983). "The effect of phytates on the growth of rainbow trout (*Salmo gairdneri*) fed purified diets containing varying quantities of calcium and magnesium." *Aquaculture 30*(1–4): 71–83.

Storebakken, T. (1985). "Binders in fish feeds: I. Effect of alginate and guar gum on growth, digestibility, feed intake and passage through the gastrointestinal tract of rainbow trout." *Aquaculture 47*(1): 11–26.

Storebakken, T. and E. Austreng (1987). "Binders in fish feeds: II. Effect of different alginates on the digestibility of macronutrients in rainbow trout." *Aquaculture 60*(2): 121–131.

Storebakken, T., K. Shearer and A. Roem (1998). "Availability of protein, phosphorus and other elements in fish meal, soy-protein concentrate and phytase-treated soy-protein-concentrate-based diets to Atlantic salmon, *Salmo salar*." *Aquaculture 161*(1): 365–379.

Summerfelt, S. T., P. R. Adler, D. M. Glenn and R. N. Kretschmann (1999). "Aquaculture sludge removal and stabilization within created wetlands." *Aquacultural Engineering 19*(2): 81–92.

Surendra, K. C., R. Olivier, J. K. Tomberlin, R. Jha and S. K. Khanal (2016). "Bioconversion of organic wastes into biodiesel and animal feed via insect farming." *Renewable Energy 98*: 197–202.

Sveier, H., B. Kvamme and A. Raae (2001). "Growth and protein utilization in Atlantic salmon (*Salmo salar* L.) given a protease inhibitor in the diet." *Aquaculture Nutrition 7*(4): 255–264.

Tanaka, Y., S.-i. Tanioka, M. Tanaka, T. Tanigawa, Y. Kitamura, S. Minami, Y. Okamoto, M. Miyashita and M. Nanno (1997). "Effects of chitin and chitosan particles on BALB/c mice by oral and parenteral administration." *Biomaterials 18*(8): 591–595.

Teuling, E., J. W. Schrama, H. Gruppen and P. A. Wierenga (2017). "Effect of cell wall characteristics on algae nutrient digestibility in Nile tilapia (*Oreochromis niloticus*) and African catfish (*Clarus gariepinus*)." *Aquaculture 479*: 490–500.

Tomasso, J. (1994). "Toxicity of nitrogenous wastes to aquaculture animals." *Reviews in Fisheries Science 2*(4): 291–314.

Tran-Tu, L. C., T. T. T. Hien, R. H. Bosma, L. T. N. Heinsbroek, J. A. J. Verreth and J. W. Schrama (2018). "Effect of ingredient particle sizes and dietary viscosity on digestion and faecal waste of striped catfish (*Pangasianodon hypophthalmus*)." *Aquaculture Nutrition 24*(3): 961–969.

Turcios, A. E. and J. Papenbrock (2014). "Sustainable treatment of aquaculture effluents—what can we learn from the past for the future?" *Sustainability 6*(2): 836–856.

Tyagi, P. K., P. K. Tyagi and S. Verma (1998). "Phytate phosphorus content of some common poultry feedstuffs." *Indian Journal of Poultry Science 33*(1): 86–88.

Ugbogu, E. and O. Ugbogu (2016). "A review of microbial protein production: prospects and challenges." *FUW Trends in Science and Technology Journal 1*: 182–185.

Van der Meer, M. B., H. van Herwaarden and M. C. J. Verdegem (1997). "Effect of number of meals and frequency of feeding on voluntary feed intake of *Colossoma macropomum* (Cuvier)." *Aquaculture Research 28*(6): 419–432.

Venou, B., M. N. Alexis, E. Fountoulaki and J. Haralabous (2006). "Effects of extrusion and inclusion level of soybean meal on diet digestibility, performance and nutrient utilization of gilthead sea bream (*Sparus aurata*)." *Aquaculture 261*(1): 343–356.

Venou, B., M. N. Alexis, E. Fountoulaki, I. Nengas, M. Apostolopoulou and I. Castritsi-Cathariou (2003). "Effect of extrusion of wheat and corn on gilthead sea bream (*Sparus aurata*) growth, nutrient utilization efficiency, rates of gastric evacuation and digestive enzyme activities." *Aquaculture 225*(1–4): 207–223.

Verdegem, M. C. J. (2013). "Nutrient discharge from aquaculture operations in function of system design and production environment." *Reviews in Aquaculture 5*(3): 158–171.

Vidakovic, A., D. Huyben, H. Sundh, A. Nyman, J. Vielma, V. Passoth, A. Kiessling and T. Lundh (2020). "Growth performance, nutrient digestibility and intestinal morphology of rainbow trout (*Oncorhynchus mykiss*) fed graded levels of the yeasts *Saccharomyces cerevisiae* and *Wickerhamomyces anomalus*." *Aquaculture Nutrition 26*(2): 275–286.

Vielma, J., S. P. Lall, J. Koskela, F.-J. Schöner and P. Mattila (1998). "Effects of dietary phytase and cholecalciferol on phosphorus bioavailability in rainbow trout (*Oncorhynchus mykiss*)." *Aquaculture 163*(3): 309–323.

Yamamoto, T., T. Goto, Y. Kine, Y. Endo, Y. Kitaoka, T. Sugita, H. Furuita, Y. Iwashita and N. Suzuki (2008). "Effect of an alcohol extract from a defatted soybean meal supplemented with a casein-based semi-purified diet on the biliary bile status and intestinal conditions in rainbow trout *Oncorhynchus mykiss* (Walbaum)." *Aquaculture Research 39*(9): 986–994.

Zhang, J.-X., L.-Y. Guo, L. Feng, W.-D. Jiang, S.-Y. Kuang, Y. Liu, K. Hu, J. Jiang, S.-H. Li and L. Tang (2013). "Soybean β-conglycinin induces inflammation and oxidation and causes dysfunction of intestinal digestion and absorption in fish." *PloS One 8*(3): e58115.

Zhao, L., W. Wang, X. Huang, T. Guo, W. Wen, L. Feng and L. Wei (2017). "The effect of replacement of fish meal by yeast extract on the digestibility, growth and muscle composition of the shrimp *Litopenaeus vannamei*." *Aquaculture Research 48*(1): 311–320.

Zheng, J., P. Yang, J. Dai, G. Yu, W. Ou, W. Xu, K. Mai and Y. Zhang (2020). "Dynamics of intestinal inflammatory cytokines and tight junction proteins of turbot (*Scophthalmus maximus* L.) during the development and recovery of enteritis induced by dietary β-conglycinin." *Frontiers in Marine Science 7*: 198.

6 The Real Meaning of Ornamental Fish Feeds in Modern Society

The Last Frontier of Pet Nutrition?

Benedetto Sicuro

CONTENTS

INTRODUCTION

Ornamental aquaculture is an emerging sector of aquaculture, developed and traditionally practiced in Asia, that is gaining popularity in western countries, such as Europe and North America, with noticeable economic value. It has been estimated that ornamental aquaculture has a value between 15 and 6 billion USD (Monticini 2010; Olivotto et al. 2011; Prathvi et al. 2013; Rhyne et al. 2009; Smith et al. 2012; Tlusty 2002). Ornamental aquaculture is a promising sector of animal farming with great economic potential; it has been estimated that ornamental aquaculture generates an economic income at the global level of about 15 billion USD (Moorhead and Zeng 2010). Ornamental aquaculture is an important economic source for some rural areas, such as Brazil (Prang 1996) or Sri Lanka (Bartley 2000). From the domestication point of view, ornamental rearing represents the last invention of the human mind in the context of the relationship between humans and animals. The human feeling with aquaria has an evolutionary key of interpretation. As Theodosius Dobzhansky said, "Nothing in biology makes sense except in the light of evolution"; keeping an aquarium represents the aquatic version of anthropological relation between humans and wild animals. The pleasant sensation of keeping a colorful bird in a cage represents an unconscious remembering of our evolutionary past in the forests; similarly, we probably like aquaria as they evoke images and sensations that are unconsciously present in our inner mind and also because, in the modern urbanized society, an aquarium is a kind of artificial window on uncontaminated nature (Filan and Llewellyn-Jones 2006). Keeping ornamental aquatic fish responds almost completely to the aesthetic criteria. This is much more evident if we consider the plethora of aquarium invertebrates, such as colorful corals. It is likely that a great number of aquarium owners do not even know that corals are animals and not plants because they are appreciated only for their aesthetic appearance. Corals and other sessile invertebrates kept in the aquaria are not perceived as animals but rather as flowers. Wild populations of captured species are heavily exploited, natural ecosystems where they live are threatened (Calado 2006; Lecchini et al. 2006), and new varieties are continuously created by artificial

DOI: 10.1201/9780429331664-6

selection. Theories on the restorative value of nature (Clements et al. 2019) indicate that the home aquarium in modern society has acquired a double function: an ideal window to a lost paradise (as a coral reef or an equatorial river in South America) but also a fascinating home hobby. Some species that escape from captivity become invasive (Magalhaes and Jacobi 2013; Savini et al. 2010); furthermore, the transferring of live animals can cause the spread of fish pathologies (Whittington and Chong 2007), thus negatively impacting the natural environment. Ornamental aquaculture is a small amount if compared with subcategory of conventional aquaculture, and it is interesting not only from an "internal" point of view, such as technologies and innovations, but particularly from an "external" point of view, that is, how it is perceived by humans. An anthropological perspective of ornamental aquaculture is surprisingly explicative with respect to its current situation (Sicuro 2017). Ornamental aquaculture is probably more similar to ornamental bird-rearing than it is to conventional aquaculture. While fish feeds used in this sector of aquaculture primarily derive from conventional aquaculture, they have changed enormously during the evolution of ornamental aquaculture, even if the reared species have not changed in this transition. Given these peculiarities of ornamental fish feeding, do the modern ornamental fish feeds reflect the physiological fish needs or the owner expectations? What are the driving forces that rule the current situation of aquarium fish feeds? Can the nutritional needs of ornamental fish justify the feed prices? It is a biological or commercial matter?

The aim of this chapter is to give a perspective of ornamental fish feeding by adopting a comparative approach, thus comparing ornamental aquaculture with conventional aquaculture and the companion feeds industry.

THE THREE PILLARS OF ORNAMENTAL AQUACULTURE: FISH, FISH OWNER, AND FISH FEED

Ornamental aquaculture originated almost 2000 year ago when Chinese fish farmers first started to take care of a colorful fish, the koi carp. By carefully examining the ornamental aquaculture evolution, it is evident that not only food fish species are outnumbered by ornamental fish species, but also food fish feed users (i.e. fish farmers) are outnumbered by fish feed users (i.e. aquarium owners), and conventional fish feeds are outnumbered by aquarium fish feed typologies. The emergence of ornamental aquaculture from conventional aquaculture can be considered a kind of evolutionary radiation that has involved fish, fish users (farmers or aquarium owners), and fish feeds.

Ornamental Fish (and Their Allies)

Ornamental species probably represent the most recent product of animal domestication and can be categorized into two main groups: aquarium and pond species. The main ornamental pond fish are koi carp (*Cyprinus carpio koi*) and goldfish (*Carassius auratus*), which are traditionally raised with conventional carp feeds. Carp and salmon are two of the few successfully farmed species in conventional aquaculture, having rapidly emerged as dominant, and are currently reared all over the world (Sicuro 2021). On the one side, in ornamental farming, there are some dominant species that are imported from tropical countries or locally farmed, such as guppies (*Poecilia reticulate*), platy (*Xiphohporus gladius*), goldfish, and fighting fish (*Betta spelendes*), but there is also an impressive number of species that are wild captured. The relative simplicity of introducing new species between those interesting for aquaria has launched ornamental aquaculture into the realm of potentially uncontrolled expansion. To be considered a suitable species in conventional aquaculture, fish must first be farmed for at least one phase of their biological cycle (as eels, tuna, or tropical groupers) (Lovatelli and Holthus 2008), so the emergence of a new species is a rare and quite complicated event. On the other side in ornamental aquaculture, the emergence of a new species is much easier to realize. Currently, there are almost 1800 species and varieties that are commercialized every year worldwide (Rhyne et al. 2012; Thornhill 2012). Moreover, ornamental aquaculture regards not only fish but also invertebrates, such as crustaceans, mollusks, polychaetes, echinoderms, and cnidarians, as well as amphibians and aquatic reptiles (Murray and Watson 2014; Pomeroy et al. 2006). Almost all these latter species are exclusively collected from the wild. Marine fish and invertebrates sold in the pet industry in Europe accounted for a total of ca. 135 million € from 2000 to 2011 (Leal et al. 2015). To be rigorous, the definition of farmed species doesn't coincide in these two sectors of aquaculture, as several ornamental species are solely translocated species from the natural environment to aquaria.

For these reasons, the species diversity in these two sectors is inherently different, and the richness of ornamental species will presumably promote diversification in ornamental aquaculture feeds. The collection of wild captured fish and invertebrates is still a source of economic sustainment in some developing and tropical countries, but this activity impacts on the natural environment and on the fish. The fish mortality caused by airplane travel of thousands of kilometers and the destruction of original natural ecosystems especially in the Southeast Asian (Monticini 2010) and Caribbean regions, is not currently accounted for as an indirect price or externality. Moreover, the interest of aquarists (such as any ornamental species keeper) exponentially increases for rarer species that reach impressive prices, thus threatening the extinction of these species (Stuart et al. 2006).

Fish Owner

The differences between a fish farmer and an aquarium owner are quite obvious but unexpectedly useful to understanding the differences in fish nutrition in these two sectors. In whatever species, feeding is the principal and foremost form of animal farming, and the constant increase of ornamental species has caused a formidable proliferation of ornamental fish feeds (Sicuro 2017). The scope of farming, its finality, is substantially different between conventional and ornamental aquaculture, and this fact in turn influences the relation between human and ornamental fish that are often considered pets.

In conventional aquaculture, a fish is a product for the human consumer; in ornamental aquaculture, a fish is a companion animal (Clements et al. 2010). The main targets of ornamental fish feed are fish welfare and optimal health conditions, while the target for food fish feed is fish growth. For a fish farmer, fish represent a source of economic subsistence; for an aquarium owner, the fish is a companion animal, and it is the recipient of special attention and care. While in food fish farming the final user is the fish consumer, for ornamental aquaculture the final user is the aquarist. This makes a great difference for the fish feed producers. There are millions of aquarists around the world (Stevens et al. 2017), and it is clear that aquafeed companies make every effort to satisfy aquarists expectations more than they do the actual fish. Moreover, it is likely that ornamental aquafeed companies also create new aquarist expectations as any other expression of consumeristic modern society. This is similar to what already happens in the pet industry, where there is a formidable variety of dog and cat feeds (Cohen 2002). In fact, similarly to what happens with dog and cats, research with aquaria owners shows that they feel an emotional bond with the fish, and this attachment may play a role in the beneficial effects of human–fish interaction (Clements et al. 2019).

Humans are naturally attracted to some animals and have special feelings for companion animals (Martens et al. 2016). Dogs, cats, and parrots are considered family members (Anderson 2014; Cohen 2002). Therefore, feeding a companion animal becomes an answer to a psychological human impulse, that eventually pushes humans to feed their companions animals as themselves. The proximity between humans and ornamental fish has both beneficial and dangerous consequences for fish. The more evident consequences are the increase of obesity and longevity of ornamental fish (Butterwick 2015; Raubenheimer et al. 2015). In other words, aquarium fish are getting older and fatter with us humans; in fact, the oldest known vertebrate is a koi carp, called "Hanako", who reached the age of 226 years. These phenomena will certainly influence the future of aquarium fish feeds, and it is expected that they will be integrated with ingredients that would balance these situations.

Fish Feeds

As a response to the increasingly number of fish species, ornamental fish feeds are quite diversified, not only for different species but more often for categories. There are fish feeds for freshwater versus marine species, aquarium fish versus pond fish, carnivorous versus herbivorous, vertebrates versus invertebrates, monospecific versus multi-species, surface feeder fish versus bottom feeder fish. Looking at the numerous ornamental feed company catalogs, it appears that aquarium feeds currently can be estimated on the order of hundreds, including a heterogeneous and quite imaginative list of ingredients, including carrots, parsley, peppers, spinach, nettles, zucchini, and many other ingredients (Sicuro 2017). Diversification is also related to the physical aspects of feeds that can be produced as pellets, flakes, or granules. Information on ornamental aquaculture feeds is scarcely published, in order to protect industry confidentiality, or alternatively, generic information is spread by aquarium feed fish companies (Barton et al. 2015). The conventional aquafeed industry is worth millions of dollars (Monticini 2010), but ornamental fish represents only a fraction of that. Ornamental fish feeds are usually sold worldwide in specialized pet shops, particularly in western countries. In the US, for instance, ornamental fish aquaculture accounts only for 7% of the industry (Tlusty 2002), but it is considered an expansive, luxury industry. However, some simple deductions can elucidate this sector: some ornamental species derive directly from conventional aquaculture as common carp and koi carp; several ornamental species belong to the same family of farmed fish, such as ciprinids; fish physiology of digestive systems and natural fish alimentation are largely known. In fact, fish feed ingredients used in ornamental aquaculture derive from those used in conventional aquaculture, and the main ingredient of ornamental fish feeds is fish meal (Tlusty 2002;Velasco-Santamaría and Corredor-Santamaría, 2011).

Even considering these intuitive premises, the principal difference between ornamental and conventional fish feeds is the price. Currently, the cost of aquaculture fish feed ranges between 800 and 1300 €/ton, depending on the species being fed, while ornamental fish feed prices vary between 14 and 74 US$/kg (Tamaru and Ako 2000; Nekoubin et al. 2012) that corresponds to 14000–74000 US$/kg. Fish feed packaging, transport, and commercialization in small quantities for ornamental fish use are considered the principal reasons for these differences; in fact, the ingredients in fish feeds only determine 10–20% of the final price, while the packaging and commercialization account for the remainder (Sicuro 2017). Fish feed prices increase proportionally with a reduction of the amount commercialized. In conventional aquaculture, fish feeds are commercialized in tons (pallets made up of 25 kg bags), while ornamental fish feeds are commercialized in quantities of hundreds of grams. Some ornamental fish feeds can reach prices several times higher compared to others, following an extremely diversified commercial offer. Of course, packaging and commercialization in small quantities greatly influences retail prices, but the suspect of a commercial speculation is largely diffused between aquarium owners.

Some simple considerations can be made on this point: it is known that aquarium fish, before they arrive in the home aquaria, in the ornamental fish farms or in ornamental fish stocking farms, are often fed with conventional aquaculture feeds (Gooley and Gavine 2003; Mosig 2007; Tamaru et al. 2001; Thilakaratne et al. 2003). Tamaru and Ako (2000) showed that Angelfish (*Pterophyllum scalare*) fed with conventional fish feeds (10 times less expensive than ornamental feed) showed the same productive results as when fed ornamental feeds. Angelfish in experimental conditions were normally fed with commercial aquaculture feeds (de Azevedo et al. 2014; Kouba et al. 2013), while Kipouros et al. (2011) fed Siamese fighting fish (*Betta splendens*) with rainbow trout feed; Gunasundari et al. (2013) fed true clownfish (*Amphiprion percula*) fish meal based on extruded feeds; Mosig (2007) fed goldfish (*Carassius auratus*) a mixture containing dried fish. Discus fish (*Symphysodon haraldi*), a popular species, was fed beef heart (Zhang et al. 2021). Dollar fish (*Metynnis orinocensis*) were fed a casein-based diet (Garzon and Gutierrez-Espinosa 2020). Concerning ornamental pond fish, two principal species, koi carp and goldfish, have been farmed for centuries, presumably with conventional fish feeds (Balon 2004); therefore, the same feeds can be used in modern home ornamental ponds, with the eventual inclusion of pigments normally at 1–3%. Considering that the greater part of ornamental fish and invertebrates come from underdeveloped countries, such as Brazil, Vietnam, Sri Lanka, Indonesia, Philippines, and the Caribbean (Avella et al. 2007; Murray et al. 2012; Murray and Watson 2014), it seems unlikely that in these countries ornamental fish would be reared with same expensive and sophisticated feeds that we (western countries) use in our home aquaria.

There is also another human psychological aspect to be considered in ornamental fish nutrition, that can be exemplified with the following sentence: Special animals need special feeds. The irrational impulse to feed exotic animals and pets with exorbitant ingredients is not a recent passion of humans. In the nineteenth century, the Chinese empress Cixi was advised to feed her Pekingese dogs shark fins and eggs of exotic birds (Francis 2015). The act of consuming a special food for ourselves or for our loved ones has again trespassed the boundary of our home and transferred to companion animals.

Moreover, some fish species as sturgeons, crucian carp (*Carassius carassius*), iridescent shark (*Pangasius hypophthalmus*) (Hoseini et al. 2015), albinic wels catfish (*Silurus glanis*) (Zatkova et al. 2011), and striped gourami (*Colisa fasciatus*) (Chakrabarty et al. 2010; Shafaet Hossen et al. 2014), can be reared either as food fish or as ornamental, therefore, again, conventional feeds can be used for the ornamental scope. Moreover, it bears repeating that the first ornamental farmed fish were koi carp and goldfish, originally raised for conventional aquaculture (Balon 2004; Snellgrove and Alexander 2011; Souto et al. 2013;), and they were presumably fed the same kind of fish feed. Finally, it should be remembered that this diverse panorama of aquarium feeds must be combined with their practical utilization in everyday life, in home aquaria, that represent a multi-species and non-functional assemblage of aquatic species, solely an expression of the owner's aesthetic taste (Forsman et al. 2012; Velasco-Santamaria et al. 2011). However, the previous considerations are not meant to frustrate the efforts of ornamental aquaculture nutrition. The diversity of ornamental fish feeds has a scientific basis and specific reasons, of course, that mirror the extreme diversity of farmed animals. Pigments are relevant in ornamental fish feeds (Corcoran and Roberts-Sweeney 2014; Gouveia and Rema 2005; Moorhead and Zeng 2015) for the importance of skin color that can be efficiently improved by these compounds, together with probiotics and new ingredients (Tamaru et al. 2011). Soybean meal (Velasco-Santamaría and Corredor-Santamaría 2011; Chong et al. 2003) and insect meals (Ganguly et al. 2014) are the principal fish meal substitutes in ornamental fish nutrition.

CONCLUSIONS

The ornamental fish, together with corals and any other animal species kept in aquaria, corresponds to the human taste of beauty, related to the aquatic environment. Therefore, ornamental feeds represent our possibility (as humans) to maintain and protect this beauty in our home aquaria, in the best possible condition. Ornamental aquaculture is doomed to an increase in species diversity, and this fact will presumably cause an increase of fish feed diversity in the future. But we should be rational in respect to this expected diversity and diffuse a more sustainable idea of ornamental fish nutrition, thus including rational perspectives and growth for the ornamental aquafeed industry. Despite the size of the ornamental industry, knowledge of fish welfare is often diffused by gray literature and hobbyist magazines. But this information should nevertheless be included in the scientific sector to provide rational directions for the future (Stevens et al. 2017). The challenge of disseminating correct information can seem almost insurmountable. Scientific information on correct nutrition is rarely published in the world of aquaria. Dominated by internet sites and blogs of passionate aquarists constantly in search of the ideal feed, they never ask themselves where their "Nemo" was kept one month before arriving in their home.

The proliferation of ornamental fish feeds will presumably continue, as a consequence of new species and new owner expectations (both real and induced by consumerism). According to the "Red queen principle", new feeds will evolve together with new species. Diversification of fish feeds is a great opportunity, of course, but it should be addressed toward sustainable targets, such as reducing or eliminating fish meal in ornamental feed (Sicuro 2017) or including natural compounds that are beneficial for fish. Other forms of diversification related to useless exotic ingredients must be completely abandoned.

It is paradoxical, but the main driving force in ornamental fish feeding is not fish physiological needs, but rather owner expectations. This means that the ornamental feed industry is more intent on satisfying owner expectations than the fish needs, just as the pet feed company considers my satisfaction more important than my cats' needs. The only reason possibly explaining this situation is a psychological "transfer" from the owner to their aquarium The owner (incredibly!) supposes that because a varied human diet is optimal for himself, then it must be his fish as well. This distortion represents a paradox of perception that is largely fed and even encouraged by ornamental fish feed companies and has been transferred in ornamental aquaculture by pet nutrition. The future direction of the ornamental aquaculture industry will be more probably anticipated by companion animal feeds than fish feeds companies, as an effect of the humanization of ornamental fish.

Ornamental fish feeds are currently based on fish meal (Sicuro 2017), and this fact is ethically not acceptable as ornamental species have only an aesthetic scope. The future of nutrition of ornamental aquaculture must be oriented to alternative feedstuffs to fish meal, in order to diffuse an ethically sustainable idea of modern ornamental aquaculture, thus suggesting an alternative form of ornamental aquaculture based on locally reared species. The diffusion of locally reared ornamental fish is universally considered a sustainable alternative to species captured in the wild; similarly, fish feed based on locally available foodstuffs, alternative to fish meal, must be diffused in the future.

REFERENCES

Anderson PK (2014) Social dimensions of the human–avian bond: parrots and their persons. *Anthrozoös* 27: 371–387.

Avella MA, Olivotto I, Gioacchini G, Maradonna F, & Carnevali O (2007) The role of fatty acids enrichments in the larviculture of false percula clownfish Amphiprion ocellaris. *Aquaculture* 273: 87–95.

Balon EK (2004) About the oldest domesticates among fishes. *Journal of Fish Biology* 65: 1–27.

Bartley DM (2000) Responsible ornamental fisheries. *FAO Aquaculture Newsletter* 24: 10–14.

Barton JA, Willis BL, & Hutson KS (2015) Coral propagation: a review of techniques for ornamental trade and reef restoration. *Reviews in Aquaculture*: 1–19.

Butterwick FR (2015) Impact of nutrition on ageing the process. Bridging the gap: the animal perspective. *British Journal of Nutrition* 113(Supplement): 23–25.

Calado R. (2006) Marine ornamental species from European waters: a valuable overlooked resource or a future threat for the conservation of marine ecosystems? *Scientia Marina* 70(3): 389–398.

Chakrabarty D, Das SK, Das MK, & Bag MP (2010) Low cost fish feed for aquarium fish: a test case using Colisa fasciata. *Spanish Journal of Agricultural Research* 8: 312–316.

Chong A, Hashim R, & bin Ali A (2003) Assessment of soybean meal in diets for discus (Symphysodon aequifasciata Heckel) farming through a fishmeal replacement study. *Aquaculture Research* 34: 913 –922.

Clements H, Valentin S., Jenkins N., Rankin J., Baker JS, Gee, N, Snellgrove, D, & Sloman, K (2019) The effects of interacting with fish in aquariums on human health and well-being: A systematic review. *PloS One* 14(7): e0220524. doi:10.1371/journal.pone.0220524

Cohen, SP (2002) Can pets function as family members? *Western Journal of Nursing Research* 24: 621–638.

Corcoran M, & Roberts-Sweeney H (2014) Aquatic animal nutrition for the exotic animal practitioner. *Veterinary Clinics of North America: Exotic Animal Practice* 17: 333–346.

de Azevedo Silva Ribeiro F, Firmino Diogenes A, Silva Cacho JC, Lima de Carvalho T, & Kochenborger Fernandes JB (2014) Polyculture of Freshwater Angelfish Pterophyllum scalare and Pacific white shrimp Litopenaeus vannamei in low-salinity water. *Aquaculture Research* 45: 637–646.

Francis R (2015) *Domesticated: Evolution in a man-made world* (p. 345). New York : W.W. Norton & Company.

Filan SL, & Llewellyn-Jone RH (2006) Animal-assisted therapy for dementia: a review of the literature. *International Psychogeriatrics* 18(4): 597–611 doi:10.1017/S1041610206003322

Forsman ZH, Kimokeo BK, Bird CE, Hunter CI, & Toonen RJ (2012) Coral farming: effects of light, water motion and artificial foods. *Journal of the Marine Biological Association of the United Kingdom* 92: 721–729.

Ganguly A, Chakravorty R, Sarkar A, Mandal DK, Haldar P et al. (2014) A preliminary study on Oxya fuscovittata (Marschall) as alternative nutrient supplement in the diets of Poecillia sphenops (Valenciennes). *PLoS One* 9: e111848.

Garzon, JSV, & Gutierrez-Espinosa, MC (2020) Protein requirement of juvenile fish Metynnis orinocensis (Characiformes: Characidae) *Revista de Biologia Tropical* 68(1): 40–49.

Gooley GJ, & Gavine FM (2003) *Integrated agri-aquaculture systems – A resource handbook*. Publication No. 03/012. Project No. MFR-2A (p. 189).

Gouveia L, & Rema P (2005) Effect of microalgal biomass concentration and temperature on ornamental goldfish (Carassius auratus) skin pigmentation. *Aquaculture Nutrition* 11: 19–23.

Gunasundari V, Kumar TTA, Ghosh S, & Kumaresan S (2013) An ex vivo loom to evaluate the brewer's yeast Saccharomyces cerevisiae in clownfish aquaculture with special reference to Amphiprion percula (Lacepede, 1802). *Turkish Journal of Fisheries and Aquatic Sciences* 13: 389–395.

Hoseini SM, Rajabiesterabadi H, & Tarkhani R (2015) Anaesthetic efficacy of eugenol on iridescent shark, Pangasius hypophthalmus (Sauvage, 1878) in different size classes. *Aquaculture Research* 46: 405–412.

Kipouros K, Paschos I, Gouva E, Ergolavou A, & Perdikaris C (2011) Masculinization of the ornamental Siamese fighting fish with oral hormonal administration. *Science Asia* 37: 277–280.

Kouba A, Sales J, Sergejevova M, Kozak P, & Masojıdek J (2013) Colour intensity in angelfish (Pterophyllum scalare) as influenced by dietary microalgae addition. *Journal of Applied Ichthyology* 29: 193–199.

Leal CM, Carraro Melo Vaz M, Puga J, Miranda Rocha RJ, Brown C, Rosa R et al. (2015) Marine ornamental fish imports in the European Union: an economic perspective. *Fish and Fisheries* 17: 459–468.

Lecchini D, Polti S, Nakamura Y, Mosconi P, Tsuchiya M, Remoissenet G, & Planes S (2006) New perspectives on aquarium fish trade. *Fisheries Science* 72: 40–47.

Lovatelli A, & Holthus PF (2008) *Capture-based aquaculture. Global overview.* FAO Fisheries Technical Paper. No. 508 (p. 298). Rome: FAO.

Magalhaes ALB, & Jacobi CM (2013) Asian aquarium fishes in a Neotropical biodiversity hotspot: impeding establishment, spread and impacts. *Biological Invasions* 15:2157–2163.

Monticini P (2010) *The ornamental fish trade production and commerce of ornamental fish: technical – Managerial and legislative aspects* (Vol. 102, p. 134). GLOBEFISH Research Programme, Rome: FAO.

Moorhead JA, & Zeng C (2015) Development of captive breeding techniques for marine ornamental fish: a review. *Reviews in Fisheries Science* 18: 315–343.

Moorhead JA, & Zeng C (2010) Development of captive breeding techniques for marine ornamental fish: a review. *Reviews in Fisheries Science* 18(4):315–343.

Mosig J (2007) Queensland hatchery moves with the market to focus on ornamentals. *Austasia Aquaculture* 21: 14–18.

Murray JM, & Watson GJ (2014) A critical assessment of marine aquarist biodiversity data and commercial aquaculture: Identifying gaps in culture initiatives to inform local fisheries managers. *PLoS ONE* 9(9): e105982. doi:10.1371/journal.pone.0105982

Murray JM, Watson GJ, Giangrande A, Licciano M, & Bentley MG (2012) Managing the marine aquarium trade: revealing the data gaps using ornamental polychaetes. *PLoS One* 7: e29543.

Nekoubin H, Hatefi S, Javahery S, & Sudgar M (2012) Effects of symbiotic (Biominlmbo) on growth performance, survival rate, reproductive parameters of angelfish (Pterophyllum scalare). *Walailak Journal of Science and Technology (WJST)* 9: 327–332.

Olivotto I, Planas M, Simões N, Holt GJ, Avella MA, & Calado R (2011) Advances in breeding and rearing marine ornamentals. *Journal of the World Aquaculture Society* 42: 135 –166.

Pomeroy RS, Parks JE, & Balboa CM (2006) Farming the reef: is aquaculture a solution for reducing fishing pressure on coral reefs? *Marine Policy* 30: 111–130.

Prathvi R, Sheela I, Aananthan PS, Ojha SN, Kumar NR, & Krishnam M (2013) Export performance of Indian ornamental fish – an analysis of growth, destination and diversity. *Indian Journal of Fishery* 60: 81–86.

Raubenheimer D, Machovsky-Capuska GE, Gosbya AK, & Simpson S (2015) Nutritional ecology of obesity: from humans to companion animals. *British Journal of Nutrition* 113(Supplement): 26–39.

Rhyne A, Rotjan R, Bruckner A, & Tlusty M (2009) Crawling to collapse: ecologically unsound ornamental invertebrate fisheries. *PLoS One* 4: e8413.

Savini D, Occhipinti–Ambrogi A, Marchini A, Tricarico E, Gherardi F, & Olenin S, Gollasch S (2010) The top 27 animal alien species introduced into Europe for aquaculture and related activities. *Journal of Applied Ichthyology* 26(2): 1–7.

Shafaet Hossen M, Mohsinul Reza AHM, Ferdewsi S, Takahashi K, & Hossain Z (2014) Effects of polyunsaturated fatty acids (PUFAs) on gonadal maturation and spawning of striped gourami, Colisa fasciatus. *International Aquatic Research* 6: 65–78.

Sicuro B (2017) Nutrition in ornamental aquaculture: the raise of anthropocentrism in aquaculture? *Reviews in Aquaculture* 10: 791–799.

Smith KF, Schmidt V, Rosen GE, & Amaral-Zettler L (2012) Microbial diversity and potential pathogens in ornamental fish aquarium water. *PLoS One* 7: e39971.

Snellgrove DL, & Alexander LG (2011) Whole-body amino acid composition of adult fancy ranchu goldfish (Carassius auratus). *British Journal of Nutrition* 106(Supplement 1): 110–112.

Souto CN, Antunes de Lemos MV, Pessoa Martins G, Gomes Araújo J, Ludovico de Almeida Martinez Lopes K, & Gomes Guimarães I (2013) Protein to energy rations in goldfish (Carassius auratus) diets. *Ciência e Agrotecnologia Lavras* 37: 550–558.

Stevens CH, Croft DP, Paull GC, & Tyler CR (2017) Stress and welfare in ornamental fishes: What can be learned from aquaculture? *Journal of Fish Biology* 91: 409–428. First published: 09 July 2017. doi:10.1111/jfb.13377

Stuart BL, Rhodin AGJ, Grismer LL, & Hansel T (2006) Scientific description can imperil species. *Science* 312(5777): 1137. doi:10.1126/science.312.5777.1137b

Tamaru CS, & Ako H (2000) Using commercial feeds for the culture of freshwater ornamental fishes in Hawaii. In C Tamaru, CS Tamaru, JP McVey, & K Ikuta (Eds.). *Spawning and maturation of aquatic species*, UJNR Technical Report No. 28 (pp. 109–120). Honolulu, HI: University of Hawaii Sea Grant College Program.

Tamaru CS, Cole B, Bailey R, Brown C, & Ako H (2001) *A manual for commercial production of the swordtail, Xiphophorus helleri*. Honolulu, HI: University of Hawaii Sea Grant Extension Service, CTSA Publication number 128. Available from http://www.ctsa.org/upload/publication/CTSA 12863167286 1652747584.pdf (2001).

Thilakaratne IDSIP, Rajapaksha G, Hewakopara A, Rajapakse RPVJ, & Faizal ACM (2003) Parasitic infections in freshwater ornamental fish in Sri Lanka. *Disease of Aquatic Organisms* 54: 157–162.

Thornhill DJ (2012) *Ecological impacts and practices of the coral reef wildlife trade* (p. 179). Washington, DC: Defenders of Wildlife.

Tlusty M (2002) The benefits and risks of aquacultural production for the aquarium trade. *Aquaculture* 205: 203–219.

Velasco-Santamaría Y, & Corredor-Santamaría W (2011) Nutritional requirements of freshwater ornamental fish: a review. *Revista MVZ Cordoba* 16: 2458–2469.

Whittington RJ, & Chong R (2007) Global trade in ornamental fish from an Australian perspective: The case for revised import risk analysis and management strategies. *Preventive Veterinary Medicine* 81: 92–116.

Zatkova I, Sergejevová M, Urban J, Vachta R, Štys D, & Masojídek J (2011) Carotenoid enriched microalgal biomass as feed supplement for freshwater ornamentals: albinic form of wels catfish (Silurus glanis). *Aquaculture Nutrition* 17: 278–286.

Zhang, Y , Wen, B , Meng, LJ , Gao, JZ, & Chen, ZZ (2021) Dynamic changes of gut microbiota of discus fish (Symphysodon haraldi) at different feeding stages. *Aquaculture* 531: 2–9.

7 Life Cycle Assessment for Sustainable Improvement of Aquaculture Systems

Patricia Gullón, Gonzalo Astray, Sara García-González, Fotini Kokou, and José Manuel Lorenzo

CONTENTS

INTRODUCTION

Eating fish and seafood is traditionally associated with health benefits because of their nutritional composition, which is rich in protein and micronutrients (Lund, 2013). Accordingly, fish consumption per capita has dramatically risen to more than 20 kg per year (FAO, 2016), which, together with recent increases of yields from capture fisheries, have constituted an overexploitation of sea resources, reducing wild fish stocks (Iribarren et al., 2012a) and continuing a non-sustainable exploitation of fishery resources (Hilborn et al., 2015). However, concerns regarding the sustainability of increased fish consumption figures have been extensively debated for years, with the provision of this nutritional food presenting potential problems. Although several scientific discussions on the meaning of sustainability in the fishery sector can be identified in the literature (Roheim et al., 2011;

DOI: 10.1201/9780429331664-7

FAO, 2010; Jennings et al., 2014; FAO, 2018), there is no consistency with regard to the definition of sustainable seafood (Hilborn et al., 2015), even when considering social, economic, and environmental elements of sustainability.

In this sense and taking into account the performance of the last 30 years, it is very possible that the increase in the fisheries sector comes mainly from aquaculture, which has recently incorporated the farming of 580 aquatic species (Troell et al., 2014; FAO, 2019). In fact, the aquaculture is considered the world's fastest developing food production sector (Song et al., 2019). This rapid growth is linked to its potential to revert the trend towards depletion, since this sector is considered a solution to the aforementioned ecological problems. Thus, in 2016, world aquaculture production was around 90.9 Mt (FAO, 2018; OECD, 2019), representing 47% of total fish consumption. In addition, it is projected that aquaculture will supply 60% by 2030 (Song et al., 2019), becoming the main supply of aquatic animal food for humans, compensating deteriorated supplies from capture fisheries. This issue also involves substantial investments in many regions of the world (Troell et al., 2014) with economic and social consequences.

The issue is that this fast growth of aquaculture contributes significant food security and employment benefits, especially in developing countries; however, this growth is also associated with an increase of the exploitation of ecosystem services and, therefore, an increase of environmental impacts. Aquaculture has traditionally been associated with several negative environmental consequences, such as climate change, aquatic eutrophication, loss of biodiversity, as well as causing large-scale land use changes (Ottinger et al., 2016; Bohnes and Laurent, 2019) because of its dependence on terrestrial crops and wild fish feeds. In this situation, it is necessary to identify and implement policies and practices intended to reduce impacts on the environment and improve environmental sustainability (Mungkung et al., 2014).

The sustainability assessment of aquaculture has expanded its scope to ecosystem-based approaches, incorporating economic, social, and local and global environmental impacts. At the same time, for over two decades, Life Cycle Assessment (LCA) has been widely applied as a management technique to evaluate environmental and global-scale impacts of seafood products and provide new information about the environmental burden of the different stages involved in this activity (Ziegler et al., 2016). LCA practice is a powerful and standardized analytical method (ISO 14040, 2006) that can be used to evaluate the possible environmental impacts of a product or service (including resources and pollutants used in all the stages of the life cycle) "from cradle to grave" (from raw materials acquirement, processing, and usage to waste management, recycling, or final dumping). The use of LCA in aquaculture allows comparison of the different production processes for one specie and/or evaluates the hot spots or the main activities that contribute to the whole impact during the farming of one specie (Mendoza-Beltrán et al., 2018). This approach comprises four interrelated steps: (i) goal and scope definition, (ii) inventory analysis, (iii) impact assessment, and (iv) results interpretation (ISO, 2006; Abdou et al., 2017). Figure 7.1 shows the phases of LCA as well as the relationships between them.

Modern aquaculture is a highly complex and heterogeneous activity that includes a great variety of production methods, technologies, stages and processes, and a great amount of farmed species, including finfish, shellfish, and seaweeds (Cao et al., 2013). Production systems range from simple traditional subsistence aquaculture to industrial aquaculture with intensive facilities that can be located in different places: freshwater, brackish, and marine environments. In the case of farming technologies, it can be said that there are multiple variants embracing monoculture and polyculture systems, freshwater ponds, land-based tanks, and/or open water culture systems – cages, pens, poles, etc., according to the fish specie (Pelletier et al., 2008). All these factors influence the environmental impacts of aquaculture (Ayer et al., 2009) and have to be considered in the LCA to ensure a full life-cycle. Therefore, it is a multidisciplinary approach to integrate, in a coherent way, all aspects related to aquaculture systems with the purpose of guaranteeing a holistic assessment of sustainability.

Taking into account the state of the art about the sustainability assessment of aquafeed production systems, this chapter covers aspects related to aquaculture sustainability, such as current

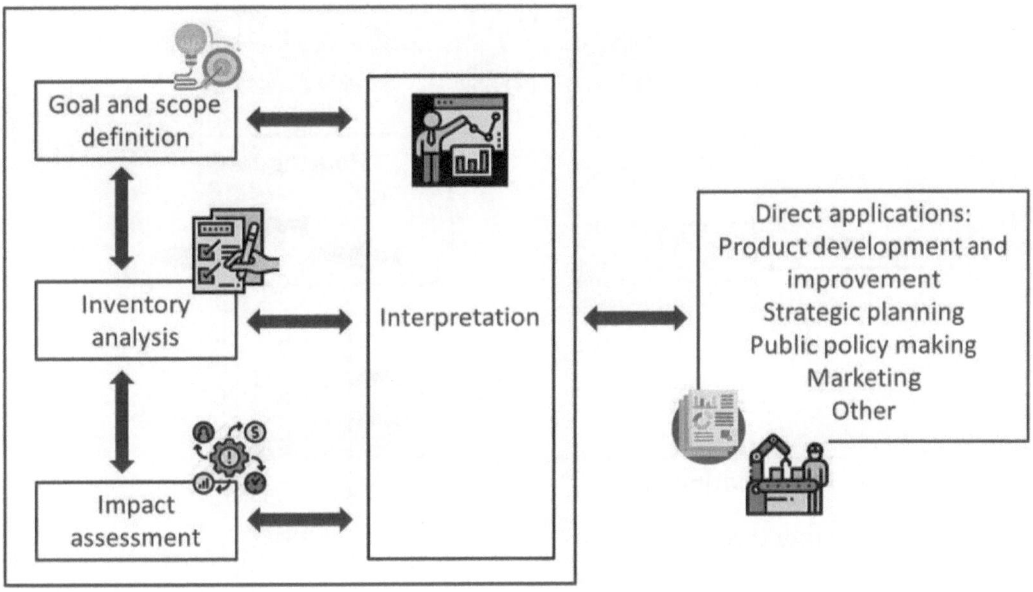

FIGURE 7.1 Phases of a life cycle assessment (LCA).

aquaculture production systems, identification of impacts caused by aquaculture activity, and the life cycle assessment of these food production systems. The purpose is to detect hot spots where the introduction of system improvements might mitigate the impacts and enable more sustainable production and consumption of seafood.

AQUACULTURE PRODUCTION SYSTEMS

Important aspects in the evaluation of the impacts of aquaculture systems on the three dimensions of sustainability (economic, social, and environmental) are the different technologies and culture types, intensities of the systems, as well as feed and energy requirements of the farmed species. In this section, we describe the configurations of the food aquatic systems to give an overview of the natural and manmade resources necessary to perform this activity and therefore, to understand and identify the potential environmental, social, and economic impacts derived from it.

Types of Aquaculture Production Systems

Based on the levels of intervention in terms of the basic requirements for fish survival such as temperature, oxygen, food, and waste removal, there are three main types of aquaculture systems: open, semi-closed, and closed systems (Tidwell, 2012a). In Figure 7.2, there is a general description of the configuration of these systems.

Open production systems rely on natural resources to fulfil their major functions, and as a consequence, they require minimal energy supply. They utilize natural water that is being stocked for aquaculture purposes. Sufficient oxygen is provided through natural processes, sourced by diffusion and/or photosynthesis by natural algal communities. Biomass densities are usually low enough in these systems (Tidwell, 2012a). According to the authors, water temperatures are ambient, while waste products are disposed of by natural processes (breakdown from heterotrophic bacteria and fungi or algae). Natural water movement (tides or currents) can be utilized to move waste products away from the farm and renew the water (Tidwell, 2012a). A negative aspect of these systems is that they are more susceptible to illness and are more vulnerable to predators. Production practices that function in an open system environment involve cages and net pens (Tidwell, 2012a).

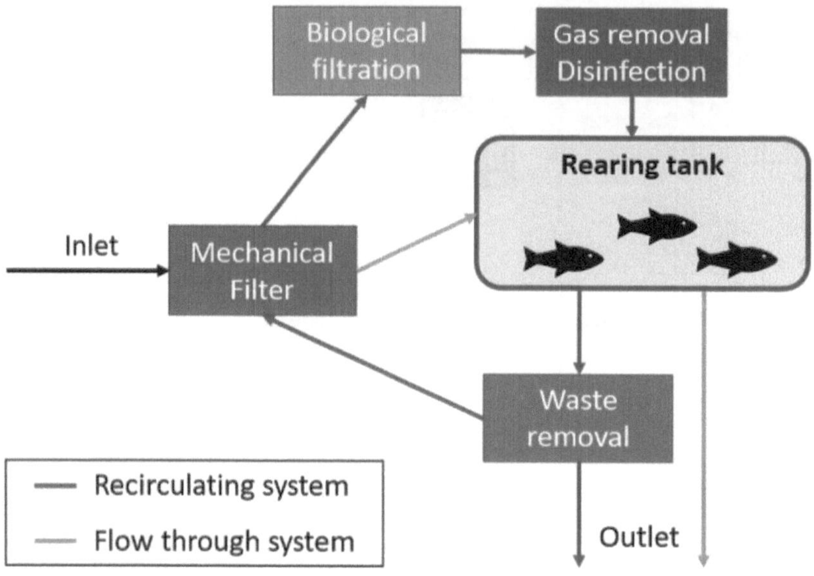

FIGURE 7.2 General configuration of flow-through and recirculating aquaculture systems. The process through the different components is highlighted in gray and black, respectively.

Semi-closed systems rely largely on natural processes to provide their basic functions; however, the production units are mainly manmade (Tidwell, 2012a). According to the authors, in these production systems, the water originates from natural resources such as from wells, rain, river or streams, and pumped into specially designed units, where the water can be cleaned and reoxygenated by natural processes. The same authors claimed that in comparison to the open systems, the semi-closed has a higher production rate (around 1000 times greater), due to a stricter control of the inputs and their physical parameters (dissolved oxygen, temperature, water quality). On the other hand, construction and equipment costs can be significantly higher in this type of system, management requirements for monitoring and intervention are necessary, energy and feed inputs are higher, and there is a higher incidence of diseases (Tidwell, 2012a). Production systems that function in a semi-closed environment include ponds and raceways (Tidwell, 2012a). Ponds are the most common aquaculture system and raceways allow farming of both cold and warm water fish.

In a closed system, the water is reused, so there is less water dependence, but this system requires high levels of intervention for all basic processes. The main advantage is full control of the environmental variables. For example, the temperature can be maintained at optimum levels for the animals, thus enhancing growth and feed efficiency, while it offers great flexibility and a large range of species to grow. Water is disinfected using ultraviolet light or ozone, and feed distribution and consumption can be accurately monitored. Closed systems utilize filters, both mechanical and biological, to get rid of the solid and soluble waste. Overall, they require a higher amount of energy compared to the other systems. Production systems of this type are the recirculating aquaculture systems (RAS) and the biofloc-based systems.

Intensity of Aquaculture Systems

Intensive culture utilizes very high densities of culture organisms (for example, 200 000–300 000 animals per hectare) and is entirely dependent on a supply of manmade feeds. Semi-intensive systems use densities lower than intensive systems (for example, 50 000–100 000 animals per hectare) and need some supplementary feeding. Extensive systems use low stocking densities (for example,

5 000–10 000 animals per hectare) and do not require supplemental feeding, although enhancements on the growth and production of natural food in the water may be achieved using fertilization.

Classification of Aquaculture Species Based on Their Natural Environment and Feeding Habits

There are several classifications to group aquaculture fish, the most common ones being based on their natural environment, temperature or salinity, and their dietary habits. Therefore, depending on the needs of each specie, appropriate systems must be developed.

Fish are generally characterized as cold water or warm water species. According to Tidwell (2012b), they are poikilothermic animals, which means that the internal environmental temperature is controlled by the external conditions (Somero and Hochachka, 1971); thus it is fundamental that the culture system provides an appropriate temperature for successful body enzymatic functions (Tidwell, 2012b).

Based on the salinity of their natural water environment, fish are characterized as freshwater, marine, or diadromous fish. Freshwater fish spend all or most of their lives in salinities of less than 1%, such as in rivers and lakes, while marine fish grow in salinities of around 3.5%. Examples of freshwater fish are tilapia and carp, while marine fish are sea bream and sea bass. Diadromous, anadromous, and catadromous fish live part of their life cycles in freshwater and part in saltwater. Anadromous fishes spend most of their adult lives at sea but return to freshwater to spawn (e.g. salmonids), while catadromous fish spend most of their juvenile life in freshwater but return to seawater to spawn (e.g. eels).

According to their nutritional requirements and habits, fish can be divided into carnivorous, omnivorous, and herbivorous. The main nutrient groups metabolized by the body to obtain the required energy for the vital physiological functions are proteins, carbohydrates, and lipids. According to their natural feeding habits, there is species variation in the ability to use these nutrients; thus, there is a relationship between natural feeding habits and dietary protein necessities (Banrie, 2013). According to Banrie, some carnivorous species demand more dietary protein than herbivorous and omnivorous species (NRC, 1993). Carnivorous species present a high ability to use dietary protein and lipids to obtain energy but a low capacity to metabolize dietary carbohydrates (Banrie, 2013). Banrei claimed that efficient use of protein for energy is largely ascribed to the mechanism of excretion of the ammonia from deaminated protein that is accomplished through the gills and associated with minimum energy expense. The diets of carnivorous species should contain fewer carbohydrates, since they use this nutrient less efficiently, due to the lack of proper digestive enzymes (Banrie, 2013).

Food composition and high-quality feed ingredients are very important for proper digestion, absorbance, and feed utilization by the fish. Moreover, meeting the minimum dietary protein requirements, or an adjusted blend of amino acids, is critical for adequate growth and health (Banrie, 2013). On the other hand, Banrei said that protein is the priciest dietary ingredient, so levels excessively high are not tenable either economically or environmentally, while an overage of dietary protein increases the excretion of nitrogenous residues. According to Banrei, herbivorous and omnivorous fish usually require a diet with 25–35% of crude protein, while carnivorous species around 40–50% (Wilson et al., 2002). Therefore, commercial feeds should be carefully formulated to guarantee that protein and amino acid requirements are fulfilled (Banrie, 2013).

Feed composition is highly related to feeding efficiency. Nutrient imbalances or the inclusion of indigestible components in the aquafeeds can greatly impact waste production. Therefore, it is a necessary component that needs great attention from both economic and environmental aspects, since it represents around 50% of aquaculture costs. Nevertheless, aquafeed production relies still on fisheries, due to the main ingredients, the fish meal and fish oil, which are becoming scarce and pricy (Papatryphon et al., 2004a), so improving this aspect will improve the aquaculture impact due to feeding consumption. Moreover, it has been reported by Bohnes et al. (2018) that the impact of aquafeed on the environment stems from the dependence on the raw material production, because of intensive combustible usage of fisheries vessels (Iribarren et al., 2010a) or the low yield of processing plants (Fréon et al., 2017). The replacement of fish meal and oil by other substitutes, mainly of plant origin, has been the subject of many recent studies. However, such replacement often causes impairment in digestion and assimilation of nutrients, and thus, the environmental impact can also be quite significant due to waste production (Kokou and Fountoulaki, 2018).

A different approach to the standard systems to utilize feed and water more efficiently are the implementation of the Integrated Multitrophic Aquaculture (IMTA) systems, which combine the production of species from several trophic levels. These systems can be a good solution to decrease waste emissions and the use of water and feeds. Aquaponic systems are a type of IMTA, which combines fish farming with hydroponics to grow plants without soil. The main environmental issues with these systems continue to be energy usage and feed production (Boxman et al., 2017; Forchino et al., 2017), despite efforts to optimize management practices (increasing feed and energy efficiency, optimizing water pumping).

IDENTIFICATION OF ENVIRONMENTAL, SOCIAL, AND ECONOMIC IMPACTS ASSOCIATED WITH AQUACULTURE SYSTEMS

As described in the previous section, aquaculture has become an important food-producing sector (Jerbi et al., 2012) and involves different activities, such as breeding and growing fish, among others (Denham et al., 2015), and can cause environmental impacts.

One of the phases with high impact is the production of feed, due to the energy required, the ingredients used, and the collection of fish used for food (Denham et al., 2015). This fact contrasts with the traditional aquaculture that is more environmentally compatible because it used wastes, by-products, or natural food before the use of pelleted feed in modern aquaculture (Edwards, 2015).

At the global level, coastal aquaculture, and – according to Ahmed and Glaser (2016) – particularly shrimp farming, has been subjected to huge criticism for its environmental impacts. Martinez-Porchas and Martinez-Cordova (2012) reported that aquaculture has been accused, with or without arguments, different environmental, economic, social, and even aesthetic impacts. Studies on specific impacts, such as eutrophication, acidification, or energy use, among others (Aubin et al., 2009), can be found in the literature. On the other hand, the environmental and social costs of the expanding, in this case, shrimp industry, are closely inter-related and affect natural resources, ecosystems, and local livelihoods (Barraclough and Finger-Stich, 1996). According to Lluch-Cota et al. (2007), intensive aquaculture can have a negative effect on the coastal topography, the different biological communities in the area, and the environmental conditions. Due to all of these, the increase in aquaculture production results in resource constraints owing to freshwater and land limitations (Ahmed et al., 2019).

All these impacts can be summarized following the classification of Martinez-Porchas and Martinez-Cordova (2012) (with slight modifications). In this review, the following impact classification is proposed (Figure 7.3).

NATURAL ECOSYSTEM DESTRUCTION

In the last few decades, huge changes in land use have been observed due to the conversion of agriculture to aquaculture systems (because of its profitability) (Ottinger et al., 2016). These changes have caused the degradation of different areas, such as coastal lakes, mangroves, and reefs, among other areas (Ottinger et al., 2016).

According to Ahmed et al. (2019), there is a usual practice in many Asian countries to convert the wetlands and the low-lying rice fields in ponds for aquaculture purposes, which affects a great

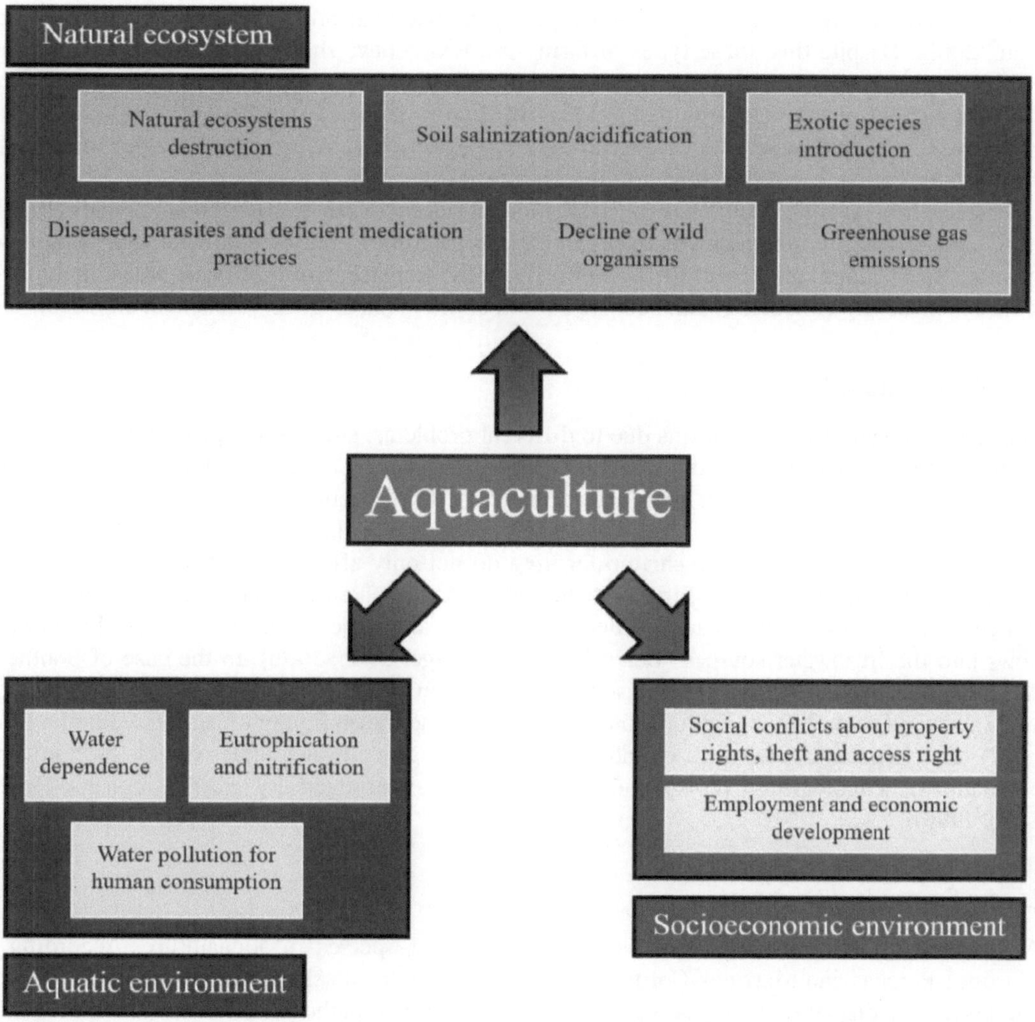

FIGURE 7.3 Impacts due to the aquaculture production cycle. Classification adapted from the proposal of Martinez-Porchas and Martinez-Cordoba (2012).

variety of flora and fauna. Another impact of aquaculture is the destruction of mangrove forests, which have great importance for the prevention of adverse phenomena (Xie et al., 2013; Walters et al., 2008).

Coastal aquaculture, where shrimp farming is located, is responsible for the mangrove deforestation process (Ahmed et al., 2019). This fact is reinforced by Martinez-Porchas and Martinez-Cordova (2012) and Ahmed et al. (2019). According to Barraclough and Finger-Stich (1996), the mangrove destruction has been sped up by commercial shrimp farming because it requires large areas. This fact can be contrasted with different research carried out by various authors. Rajitha et al. (2007) reported that the issues of shrimp culture are mainly related to the removal of mangroves because operators cut the mangrove forest.

On the other hand, other authors do not find a clear relationship between the loss of this type of forest and shrimp aquaculture. In this sense, the report prepared by DeWalt et al (2002), for: the World Bank, Network of Aquaculture Centres in Asia-Pacific (NACA), the World Wildlife Fund (WWF) and Food and the Agriculture Organization of the United Nations (FAO) Consortium Program on Shrimp Farming and the Environment, concludes there is little evidence of mangrove destruction (probably because environmental policy in Mexico affords special protections) (DeWalt et al., 2002). Despite this, these types of farms in Mexico have shown to be the most important threat to water quality (DeWalt et al., 2002). Another interesting study about mangrove deforestation was carried out by Richards and Friess (2016), who reported that between 2000 and 2012, mangrove forests in Southeast Asia were lost with an average rate of around 0.18% per year; nevertheless, authors concluded that aquaculture pressure was lower than expected.

Besides these, shrimp culture has caused the loss of rice lands due to conversion to ponds (Rajitha et al., 2007). The use of canals and dykes in shrimp farming increased sedimentation and causes erosion (Barraclough and Finger-Stich, 1996). With the transformation of these areas, mangroves reduce their fishery production and become carbon emitters (Rodríguez-Valencia et al., 2010).

Soil Salinization/Acidification

The abandonment of farming farms due to different problems, such as low productivity or animal health problems can cause varied impacts (Rodríguez-Valencia et al., 2010). In the case of the abandonment phase of shrimp farms, it has been observed soil salinization, acidification, a barrier effect due to the ponds and an erosion increase (González-Ocampo et al., 2006).

Salinization processes due to shrimp farming do not only affect the soil but also the sources of freshwater (Barraclough and Finger-Stich, 1996). In fact, the use of groundwater aquifers as a source of water for the ponds causes a decrease in the water table which implies that the seawater flows into the freshwater sources (Barraclough and Finger-Stich, 1996). In the case of Southeast Asia, the intensive shrimp farming caused the salinization of adjacent land (for example rice lands) and waterways as well as the fall of groundwater levels and the emptying of aquifers (Rajitha et al., 2007). The impacts of salinization and the descending groundwater tables affect the bordering populations, the biological diversity, and the overall biomass productivity (Barraclough and Finger-Stich, 1996).

Exotic Species Introduction

An important environmental threat is the use of exotic species in aquaculture. According to Martinez-Porchas and Martinez-Cordova (2012), the main drawbacks of the introduction of exotic species are the displacement of native species, transmission of pathogens, and new competition for food and habitat. In particular, in Mexican shrimp aquaculture some Pacific Coast species are used in the Gulf Coast (DeWalt et al., 2002). The use of exotic species should be controlled: even this is a reason for the denial of farm construction (DeWalt et al., 2002). This fact has been reported by DeWalt et al. (2002), who report that it was one of the reasons why the construction of a shrimp farm was denied in the natural protected area of Laguna de Términos (Campeche) (DeWalt et al., 2002).

It is important to keep in mind that displaced species or strains may carry exotic diseases that can spread and destroy indigenous wild populations; moreover, these species can escape and become established and take the place of wild populations (Xie et al., 2013). The same authors report studies where some numbers of animals that have escaped are estimated, such as rainbow trout, salmon, or Chinese carp.

Diseases, Parasites, and Deficient Medication Practices

The rapid development of fish farms, the poor water quality, and the presence of toxins can cause disease such as black spot and gill diseases, among others, that can cause serious mortality in shrimp farms (Ahmed et al., 2019). In addition to economic losses on farms, the intensive aquaculture practices can create hospitable environmental conditions for parasite growth and transmission to wild fish (Ahmed et al., 2019). To achieve a high population of shrimp and maximum efficiency, antibiotics are used (Rajitha et al., 2007.

According to Xie et al. (2013), the overuse of antibiotics is responsible for the presence of antibiotic residues in aquaculture products and the aquatic environment. In Mexico, shrimp farms have tested different methods such as antibiotics, vitamin C, among others, to prevent or treat shrimp diseases (DeWalt et al., 2002). After harvests, the producers dry the farm ponds and apply different products, such as nitrogen fertilizer, lime, or antibiotics (oxytetracycline, enrofloxacin, or chloramphenicol) (DeWalt et al., 2002). According to Martinez-Porchas and Martinez-Cordoba (2012), the environmental implications of the use of hormones in aquaculture have barely been studied.

Finally, workers employed on shrimp farms can handle potentially dangerous chemical products and be exposed to poor sanitary conditions, which can present health hazards (Barraclough and Finger-Stich, 1996).

Decline of Wild Organisms

Aquaculture can be a solution for overfishing but, nevertheless, can cause the decline of wild fish populations (Ahmed et al., 2019). Local fishing is affected due to the pumping of seawater to the farming pools. The pumping sucks larvae of various organisms, resulting in reduced fish production (Rodríguez-Valencia et al., 2010).

This fact is reported, for example, in shrimp farming where wild penaeid postlarvae were captured to stock shrimp ponds (Páez-Osuna, 2001). According to Páez-Osuna (2001), some indicators suggest that this practice reduced the penaeid species number in the affected regions. Likewise, during the collection, different organisms, such as fish fry or zooplankton, among others, are also captured and destroyed (see the numbers reported by Páez-Osuna in their research).

Greenhouse Gas (GHG) Emissions

As is known, most GHG emissions from terrestrial livestock supply chains are from feed production, enteric fermentation, and manure (Robb et al., 2017). According to Ahmed et al. (2019), aquaculture has caused an increment in GHG emissions. This is likely due to rapid aquaculture growth in the last 20 years to meet the global demand for seafood (Robb et al., 2017). The estimated value, taking into account the change in land use, was, depending on the species and location, between 1.61 and 2.12 kg CO_2e/kg live weight (CO_2e is a carbon dioxide equivalent) (Ahmed et al., 2019; Robb et al., 2017) with feed production being the largest source of GHG emissions (due to the need for the production and transport of the raw materials and the feed) (Robb et al., 2017). Besides this, mangrove deforestation releases blue carbon (due to the clearing process to build shrimp farms) and reduces storage capacity (Ahmed et al., 2019).

Water Dependence

This impact category, proposed by Aubin et al. (2009), refers, at farm level, to the needed water input relative to the farm fish biomass production.

Pollution of Water for Human Consumption

Probably this is the most critical effect of aquaculture in the environment. For example, the shrimp aquaculture has a water exchange rate of around 5–20% per day for most operations (DeWalt et al., 2002). This important demand on water induces competition between shrimp enterprises and other water resource users (Barraclough and Finger-Stich, 1996). This fact, in conjunction with the overutilization of feed, added to the ponds, presents a potential impact on the environment (DeWalt et al., 2002). The use of antibiotics, fertilizers, and hormones in intensive aquaculture can affect the water quality and cause water pollution (Ahmed et al., 2019). The different chemical products used in shrimp culture can be classified as disinfectants, compounds for soil or water treatment, plankton growth inducers (fertilizers and minerals), among others (Rajitha et al., 2007). Their improper use results in a highly polluted discharge into the sea or creeks, among others (Rajitha et al., 2007.

These impacts can be clearly seen in shrimp aquaculture, where effects are experienced as alteration of water drainage patterns, water quality deterioration, and contamination of groundwater aquifers, among others (Páez-Osuna, 2001).

Eutrophication and Nitrification

According to Purcell et al. (2007), eutrophication is, nowadays, one of the major global pollution problems. This process is due to the nutrient increase, nutrient ratio alterations, and turbidity increase in those coastal areas where humans carry out their activities (Purcell et al., 2007). The eutrophication increase in natural water can have undesirable ecological consequences (Rajitha et al., 2007).

Xie et al. (2013) report that the effluents from aquaculture ponds commonly present abundant suspended organic solids, phosphorus, nitrogen, and carbon that have their origin primarily from unconsumed feed and fecal material. It is also noted that to achieve high populations (in this case shrimp) and achieve maximum efficiency, different artificial foods, chemical additives, and antibiotics are used (Rajitha et al., 2007). Overstocking and overfeeding can produce increases in organic matter and wastes in the water, which reduce water quality (Ahmed et al., 2019). Besides this, the use of chemical fertilizers and organic manure causes an increase of phosphorus and nitrogen in the effluents, resulting in eutrophication and increasing the risk of red tides (Ahmed et al., 2019).

Intensive shrimp farming causes large amounts of sediments that present serious elimination problems: putrefaction inside and outside the ponds cause awful odors, hypernutrification, and eutrophication, among other undesirable processes (Barraclough and Finger-Stich, 1996). Likewise, intensive salmon farming in specific areas can cause ocean floor fouling, which contaminates the ecosystem (Jack14, 2015).

Social Conflicts about Property Rights, Theft, and Access Rights

Perhaps one of the most important social conflicts is about property rights. This effect has been observed in Mexico, where several factors complicate the use of the land (DeWalt et al., 2002): on the one hand, the federal maritime zone is not sufficiently demarcated; on the other, private owners claim lands; and finally, cooperatives and ejidos want to exploit the land and sea resources of the area. Due to all this, property rights are not well-defined (DeWalt et al., 2002) and many conflicts and economic costs appear.

Conflicts due to property rights are not the only social conflicts that have been seen in shrimp aquaculture areas. It has been reported by DeWalt et al. (2002) that concerns about robbery at shrimp farms exist throughout the country. In the state of Nayarít, robberies of shrimp and equipment are common, and the affected companies estimate losses in the range of 2–3 MT of shrimp per cycle (DeWalt et al., 2002). These crimes are carried out by organized bands (municipality of San Blas) or by individuals who are known to the workers (in Mazatlán, a farm employed the thief as a farm guard, and losses have fallen substantially) (DeWalt et al., 2002).

Finally, conflicts related to access rights occur when farms cut off traditional access pathways to coastal resources, preventing the locals from easy access to different locations such as coastline, lagoons, or pasture areas (DeWalt et al., 2002).

Employment and Economic Development

Despite all the negative effects, aquaculture, in the case of shrimp farming, can present some positive effects on the employment and economic development of rural and coastal villages (Rajitha et al., 2007).

Commercial aquaculture produces an important disruption in the traditional production and exchange systems, and the power relations changed in many areas (Barraclough and Finger-Stich, 1996).

During the operation phase of an aquaculture system, in this case shrimp, aquaculture presents positive socioeconomic impacts due to higher-educated employment offers and the increase of quality of life (González-Ocampo et al., 2006). Aquaculture generates employment, profits for producers, and multiplier effects for the industries and commercial companies around the sector (DeWalt et al., 2002). In general, people related to the shrimp aquaculture (farmers, state government, among others) are all in accord that it has been a positive economic impact in coastal regions (DeWalt et al., 2002). Related to the employment conditions DeWalt et al. (2002) reported that the permanent jobs in Mexico around the shrimp aquaculture are reasonably well paid and the salary is higher compared to the Mexican minimum wage.

Finally, just as aquaculture impacts the environment, this activity can also be impacted by different processes. For instance, climate change can impact on aquaculture, mostly on the survival, growth, and production of fish (Ahmed et al., 2019). Within the phenomena that can impact on aquaculture are global warming, droughts, or sea-level rise, among others (Ahmed and Diana, 2016; Ahmed et al., 2019; De Silva and Soto, 2009).

LIFE CYCLE ASSESSMENT AND AQUACULTURE

In regard to the information collected above, it is obvious that aquaculture production systems present high complexity with multiple impacts associated not only on the environment but also on the society and economy; therefore, it is crucial to assess the sustainability of this sector to guarantee its sustainable development (Bohnes and Laurent, 2019).

In this sense, measuring, understanding, and improving the environmental profile of aquaculture-based products takes part of the seafood industry's labors, with the aim of improving its environmental profiles, developing government regulations, and promoting its sustainability. In order to achieve a better understanding of the environmental impacts of aquaculture, the LCA methodology – an internationally standardized method for assessing and quantifying the environmental aspects throughout a products' life cycle (ISO 14040, 2006) – has been considered in the literature in numerous studies as a common tool since the early 2000s (Bohnes and Laurent, 2019). Thus, all stages performed throughout the aquaculture production system are taken into account – that is, from raw materials and energy production, manufacturing, and distribution to use and final disposal. Accordingly, mass and energy flows are identified to each stage. Therefore, the LCA methodology is considered a valuable procedure to identify strategies for improvement with the aim of reducing the environmental impact profile. The combination of this approach with economic and social studies could help to establish the concept of aquaculture sustainability (González-García et al., 2018).

In the last two decades, multiple reports are found in the literature focused on the identification of environmental impacts of aquaculture products from an LCA perspective, although some of them pay special attention to methodological practices considered and others are examples of case studies. In some of them, innovative technologies have been incorporated in the breeding of farmed species, feed practices, and rearing systems as a result of findings from an LCA.

Henriksson et al. (2012) assessed differences in results and main findings considering different methodological practices. Discrepancies have been identified in the selection of the functional unit – to report the results, the definition of system boundaries, or the allocation procedure selected, all of them considerably affecting the results together with the data quality. Accordingly, the main challenge should be focused on increasing transparency of LCA studies in order to facilitate comparison between products.

Regarding case studies, the first LCA identified in the literature analyzed the aquaculture feeds considering different compositions (Papatryphon et al., 2004a), paying special attention to the feed for rainbow trout production. This study remarked on the effect on the environment from the production of feed, mainly due to derived nutrient emissions at the farm as well as its dependence on fish oil and fish meal obtained from wild fish stocks and terrestrial crops (Aubin et al., 2009; Troell et al., 2014). This issue has also been remarked on in other related aquaculture studies regardless of the fish species (Aubin et al., 2009; Cao et al., 2011; Iribarren et al., 2012a; Ottinger et al., 2016; El-Sayed et al., 2015). In this sense, improvements have been developed limiting releases on non-ingested feed and nutrients and attention has been paid to aquafeed formulations that incorporate probiotics, which considerably contribute to achieving not only environmental but also economic and operational benefits (Iribarren et al., 2012a).

Papatryphon et al. (2004a) completed the study previously mentioned regarding rainbow trout farming activities in France (Papatryphon et al., 2004b; 2005), indicating the role that control and prevention of nutrients emission from fish farming plays on the environmental profile associated with this fish species. Rainbow trout has also been analyzed in other studies considering different countries and production systems (Seppala et al., 2001, Aubin et al., 2009; d'Orbcastel et al., 2009; Chen et al., 2015; Dekamin et al., 2015; Grönroos et al., 2006; Samuel-Fitwi et al., 2013a, 2013b).

Turbot (Aubin et al., 2006; 2009; Iribarren et al., 2012a), tilapia (Pelletier and Tyedmers, 2010; Mungkung et al., 2013; Yacout et al., 2016), carp (Mungkung et al., 2013), salmon (Arismendi et al., 2009; Ayer et al., 2016; Ellingsen and Aanondsen, 2006; Ayer and Tyedmers, 2009; Pelletier et al., 2009; Boissy et al., 2011; Cashion et al., 2016; Ford et al., 2012; McGrath et al., 2015; Song et al., 2019), sea bass (Aubin et al., 2009; Jerbi et al., 2012), and the Arctic char (Summerfelt et al., 2006; Smárason et al., 2017) are examples of species also analyzed from an LCA perspective. Concerning shellfish, shrimp (Mungkung et al., 2006; Cao et al., 2011; Jonell and Henriksson, 2015) and mussels (Iribarren et al., 2010b; 2010c; Aubin et al., 2014; Lourguioui et al., 2017; Lozano et al., 2010) have been also analyzed following the LCA principles.

According to LCA studies identified in the literature, cultivation practices, that is, culture environments and farming intensities (from extensive to hyper-intensive conditions), constitute an important and relevant issue on the assessment of environmental impacts derived from aquaculture species. The use of fertilizers, pesticides, antibiotics, disinfectants, and other feed additives is extensively practiced in aquaculture and applied in great amounts to control insects, diseases, water weeds, and plant illness as well as in cleaning activities (Ottinger et al., 2016), with the corresponding environmental consequences. This issue is also connected with the lack of consensus regarding which impact categories must be considered mandatory when assessing impacts from aquaculture. The presence of derived compounds in marine environments constitutes a threat for bacteria, fungi, and microalgae (Kümmerer, 2009). Accordingly, more research should be conducted in this area to develop toxicity impact methodologies (and thus, corresponding characterization factors) that consider this type of derived compounds (Bohnes and Laurent, 2019).

CONCLUSION

LCA can be considered a valuable tool to also propose improved production actions with the aim of decreasing energy consumption on farming activities as well as reducing nutrient discharge, both issues that require special attention according to numerous LCA studies. Current research is focused on new production systems that include more efficient water filtering technologies in order to lower water demands, reducing nutrients and solids release into the environment (Blancheton, 2000; Ebeling et al., 2006; Aubin et al., 2009). Recirculating systems are being designed as a potential option to decrease water usage and to control water quality. However, attention should be paid to research needs in LCA for aquaculture, mainly based on the quality-of-life cycle inventory data and consistency on system boundaries definition (Bohnes and Laurent, 2019; Henriksson et al., 2012). Large and complex system boundaries are required in aquaculture, which are not always included in LCA studies, excluding post-farming activities which can have remarkable contributions to the global profile (Bohnes and Laurent, 2019). Moreover, efforts should also be conducted towards broadening current knowledge about the contributions from infrastructure, since it could play a key role (Henriksson et al., 2012). Finally, there is a lack of databases specifically for developing countries, which occupy a leading position in the aquaculture sector. In addition, consistency is required regarding the sourcing of background data in order to minimize uncertainty.

ACKNOWLEDGMENTS

Dr. G. Astray thanks the University of Vigo for his contract supported by "Programa de retención de talento investigador da Universidade de Vigo para o 2018" budget application 0000 131H TAL 641. Dr. S. González-Garcia would like to express her gratitude to the Spanish Ministry of Economy and Competitiveness for financial support (Grant references RYC-2014-14984).

REFERENCES

Abdou, K., Aubin, J., Romdhane, M.S., Le Loc'h, F., and Ben Rais Lasrama, F. 2017. Environmental assessment of seabass (*Dicentrarchus labrax*) and seabream (*Sparus aurata*) farming from a life cycle perspective: A case study of a Tunisian aquaculture farm. *Aquaculture* 471:204–212.

Ahmed, N., Thompson, S., and Glaser, M. 2019. Global aquaculture productivity, environmental sustainability, and climate change adaptability. *Environmental Management* 63:159–172.

Ahmed, N., and Diana, J.S. 2016. Does climate change matter for freshwater aquaculture in Bangladesh? *Regional Environmental Change* 16:1659–1669.

Ahmed, N., and Glaser, M. 2016. Coastal aquaculture, mangrove deforestation and blue carbon emissions: Is REDD+ a solution? *Marine Policy* 66:58–66.

Arismendi, I., Soto, D., Penaluna, B., Jara, B., Leal, C., and León-Muñoz, J. 2009. Aquaculture, non-native salmonid invasions and associated declines of native fishes in northern Patagonian lakes. *Freshwater Biology* 54:1135–1147.

Aubin, J., and Fontaine, C. 2014. Impacts of producing bouchot mussels inMont-Saint-Michel Bay (France) using LCA with emphasis on potential climate change and eutrophication. In Proceedings of the 9th International Conference on Life Cycle Assessment in the Agri-Food Sector Environmental, 64–69.

Aubin, J., Papatryphon, E., van der Werf, H.M.G., and Chatzifotis, S. 2009. Assessment of the environmental impact of carnivorous finfish production systems using life cycle assessment. *Journal of Cleaner Production* 17:354–361.

Aubin, J., Papatryphon, E., van der Werf, H.M.G., Petit, J., and Morvan, Y.M. 2006. Characterisation of the environmental impact of a turbot (*Scophthalmus maximus*) re-circulating production system using life cycle assessment. *Aquaculture* 261:1259–1268.

Ayer, N., Martin, S., Dwyer, R.L., Gace, L., and Laurin, L. 2016. Environmental performance of copper-alloy net-pens: life cycle assessment of Atlantic salmon grow-out in copper-alloy and nylon net-pens. *Aquaculture* 453:93–103.

Ayer, N.W., and Tyedmers, P.H. 2009. Assessing alternative aquaculture technologies: life cycle assessment of salmonid culture systems in Canada. *Journal of Cleaner Production* 17:362–373.

Banrei. 2013. Principles of fish nutrition. Available in https://thefishsite.com/articles/principles-of-fish-nutritio n, 2020.

Barraclough, S., and Finger-Stich, A. 1996. *Discussion paper: Some ecological and social implications of commercial shrimp farming in Asia*. United Nations Research Institute for Social Development, 1–71.

Blancheton, J.P. 2000. Developments in recirculation systems for Mediterranean fish species. *Aquacultural Engineering* 22:17–31.

Bohnes, F.A., and Laurent, A. 2019. LCA of aquaculture systems: methodological issues and potential improvements. *The International Journal of Life Cycle Assessment* 24:324–337.

Bohnes, F.A., Hauschild, M. Z., Schulundt, J., and Laurent, A. 2018. Life cycle assessments of aquaculture systems: a critical review of reported findings with recommendations for policy and system development. *Reviews in Aquaculture* 11:1061–1079.

Boissy, J., Aubin, J., Drissi, A., van der Werf, H.M.G., Bell, G.J., and Kaushik, S.J. 2011. Environmental impacts of plant-based salmonid diets at feed and farm scales. *Aquaculture* 321:61–70.

Boxman, S.E., Zhang, Q., Bailey, D., and Trotz, M.A. 2017. Life cycle assessment of a commercial-scale freshwater aquaponic system. *Environmental Engineering Science* 34:299–311.

Cao, L., Diana, J.S., and Keoleian, G.A. 2013. Role of life cycle assessment in sustainable aquaculture. *Reviews in Aquaculture* 5:61–71.

Cao, L, Diana, J.S., Keoleian, G.A., and Lai, Q. 2011. Life cycle assessment of Chinese shrimp farming systems targeted for export and domestic sales. *Environmental Science Technology* 45:6531–6538.

Cashion, T., Hornborg, S., Ziegler, F., Hognes, E.S., and Tyedmers, P. 2016. Review and advancement of the marine biotic resource use metric in seafood LCAs: a case study of Norwegian salmon feed. *The International Journal of Life Cycle Assessment* 21:1106–1120.

Chen, X., Samson, E., Tocqueville, A., and Aubin, J. 2015. Environmental assessment of trout farming in France by life cycle assessment: using bootstrapped principal component analysis to better define system classification. *Journal of Cleaner Production* 87:87–95.

Dekamin, M., Veisi, H., Safari, E., Liaghati, H., Khoshbakht, K., and Dekamin, M.G. 2015. Life cycle assessment for rainbow trout (*Oncorhynchus mykiss*) production systems: a case study for Iran. *Journal of Cleaner Production* 91:43–55.

De Silva, S.S., and Soto, D. 2009. Climate change and aquaculture: potential impacts, adaptation and mitigation. In K. Cochrane, C. De Young, D. Soto et al. (Eds.), *Climate change implications for fisheries and aquaculture: overview of current scientific knowledge*. FAO Fisheries and Aquaculture Technical Paper. No. 530. Rome: FAO, 151–212.

Denham, F.C., Howieson, J., Solah, V.A., and Biswas, W.K. 2015. Environmental supply chain management in the seafood industry: past, present and future approaches. *Journal of Cleaner Production* 90:82–90.

De Walt, B.R., Ramirez-Zavala, J.R., Noriega, L., and González, R.E. 2002. Shrimp Aquaculture, the People and the Environment in Coastal Mexico. Report prepared under the World Bank, NACA, WWF and FAO Consortium Program on Shrimp Farming and the Environment. Work in Progress for Public Discussion. Published by the Consortium. 73 pages.

d'Orbcastel, E.R., Blancheton, J.P., and Aubin, J. 2009. Towards environmentally sustainable aquaculture: comparison between two trout farming systems using Life Cycle Assessment. *Aquacultural Engineering* 40:113–119.

Ebeling, J.M., Welsh, C.F., and Rishel, K.L. 2006. Performance evaluation of an inclined belt filter using coagulation/flocculation aids for the removal of suspended solids and phosphorus from microscreen backwash effluent. *Aquacultural Engineering* 35:61–77.

Edwards, P. 2015. Aquaculture environment interactions: Past, present and likely future trends. *Aquaculture* 447:2–14.

El-Sayed, A.-F.M., Dickson, M.W., and El-Naggar, G.O. 2015. Value chain analysis of the aquaculture feed sector in Egypt. *Aquaculture* 437:92–101.

Ellingsen, H., and Aanondsen, S.A. 2006. Environmental impacts of wild caught cod and farmed salmon-a comparison with chicken. *The International Journal of Life Cycle Assessment* 11:60–65.

FAO 2010 – Food and Agriculture Organisation 2010. *The state of world fisheries and aquaculture*. Rome, 197.

FAO 2016. Global per capita fish consumption rises above 20 kilograms a year. Available in http://www.fao.org/news/story/en/item/421871/icode/

FAO 2018. *The state of world fisheries and aquaculture 2018 – Meeting the sustainable development goals*. Rome. Licence: CC BY-NC-SA 3.0 IGO.

FAO 2019. Aquaculture. Available in http://www.fao.org/aquaculture/en/

Forchino, A., Lourguioui, H., Brigolin, D., and Pastres, R. 2017. Aquaponics and sustainability: The comparison of two different aquaponic techniques using the Life Cycle Assessment (LCA). *Aquacultural Engineering* 77:80–88.

Ford, J.S., Pelletier, N.L., Ziegler, F., et al. 2012. Proposed local ecological impact categories and indicators for life cycle assessment of aquaculture: a Salmon aquaculture case study. *Journal of Industrial Ecology* 16:254–26.

Fréon, P., Durand, H., Avadí, A., Huaranca, S., and Moreyra, R. O. 2017. Life cycle assessment of three Peruvian fishmeal plants: Toward a cleaner production. *Journal of Cleaner Production* 145:50–63.

González-García, S., Villanueva-Rey, P., Feijoo, G., and Moreira, M.T. 2018. Estimating carbon footprint under an intensive aquaculture regime. In F. Hai, C. Visvanathan, and B. Ramaraj (Eds.), *Sustainable aquaculture*. Switzerland: Springer Zurich, 249–263.

González-Ocampo, H.A., Beltrán Morales, L.F., Cáceres-Martínez, C., et al. 2006. Shrimp aquaculture environmental diagnosis in the semiarid coastal zone in Mexico. *Fresenius Environmental Bulletin* 15:659–669.

Grönroos, J., Seppälä, J., Silvenius, F., and Mäkinen, T. 2006. Life cycle assessment of Finnish cultivated rainbow trout. *Boreal Environment Research* 11:401–414.

Henriksson, P.J.G., Guinée, J.B., Kleijn, R., and De Snoo, G.R. 2012. Life cycle assessment of aquaculture systems-a review of methodologies. *The International Journal of Life Cycle Assessment* 17:304–313.

Hilborn, R., Fulton, E.A., Green, B.S., Hartmann, K., Tracey, S.R., and Watson, R.A., 2015. When is a fishery sustainable? *Canadian Journal of Fisheries and Aquatic Sciences* 72:1433–1441.

Iribarren, D., Moreira, M.T., and Feijoo, G. 2012a. Life cycle assessment of aquaculture feed and application to the turbot sector. *International Journal of Environmental Research* 6:837–848.

Iribarren, D., Dagá, P., Moreira, M.T., and Feijoo, G. 2012b. Potential environmental effects of probiotics used in aquaculture. *Aquaculture International* 20:779–789.

Iribarren, D., Moreira, M. T., and Feijoo, G. 2010a. Life Cycle Assessment of fresh and canned mussel processing and consumption in Galicia (NW Spain). *Resources, Conservation and Recycling* 55:106–117.

Iribarren, D., Moreira, M.T., and Feijoo, G. 2010b. Revisiting the life cycle assessment of mussels from a sectorial perspective. *Journal of Cleaner Production* 18:101–111.

Iribarren, D., Moreira, M.T., and Feijoo, G. 2010c. Implementing by-product management into the life cycle assessment of the mussel sector. *Resources, Conservation and Recycling* 54:1219–1230.

Jack14. 2015. Review: The environmental and economic impacts of salmon aquaculture. https://marinebiology co, 2019 (12/27).

Jennings, S., Smith, A.D.M., Fulton, E.A., and Smith, D.C. 2014. The ecosystem approach to fisheries: management at the dynamic interface between biodiversity conservation and sustainable use. *Annals of the New York Academy of Sciences* 1322:48–60.

Jerbi, M.A., Aubin, J., Garnaoui, K., Achour, L., and Kacem, A. 2012. Life cycle assessment (LCA) of two rearing techniques of sea bass (*Dicentrarchus labrax*). *Aquacultural Engineering* 46:1–9.

Jonell, M., and Henriksson, P.J.G. 2015. Mangrove-shrimp farms in Vietnam comparing organic and conventional systems using life cycle assessment. *Aquaculture* 447:66–75.

Kokou, F., and Fountoulaki, E. 2018. Aquaculture waste production associated with antinutrient presence in common fish feed plant ingredients. *Aquaculture* 495:295–310.

Kümmerer, K. 2009. Antibiotics in the aquatic environment-a review-part I. *Chemosphere* 75:417–434.

Lluch-Cota, S.E., Aragón-Noriega, E.A., Arreguín-Sánchez, F., et al. 2007. The Gulf of California: Review of ecosystem status and sustainability challenges. *Progress in Oceanography* 73:1–26.

ISO (International Organization for Standardization). (2006). *ISO 14040: Environmental management-Life cycle assessment-Principles and framework*. Geneva, Switzerland.

Lourguioui, H., Brigolin, D., Boulahdid, M., and Pastres, R. 2017. A perspective for reducing environmental impacts of mussel culture in Algeria. *The International Journal of Life Cycle Assessment* 22:1266–1277.

Lozano, S., Iribarren, D., Moreira, M.T., and Feijoo, G. 2010. Environmental impact efficiency in mussel cultivation. *Resources Conservation and Recycling* 54:1269–1277.

Lund, E.K. 2013. Health benefits of seafood; Is it just the fatty acids? *Food Chemistry* 140:413–420.

Martinez-Porchas, M., and Martinez-Cordova, L.R. 2012. World aquaculture: Environmental impacts and troubleshooting alternatives. *The Scientific World Journal*, 2012:1–9.

McGrath, K.P., Pelletier, N.L., and Tyedmers, P.H. 2015. Life cycle assessment of a novel closed-containment salmon aquaculture technology. *Environmental Science & Technology* 49:5628–5636.

Mendoza Beltrán, A., Chiantore, M., Pecorino, D., et al. 2018. Accounting for inventory data and methodological choice uncertainty in a comparative life cycle assessment: the case of integrated multi-trophic aquaculture in an offshore Mediterranean Enterprise. *The International Journal of Life Cycle Assessment* 23:1063–1077.

Mungkung , R., Phillips, M., Castine, S., et al. 2014. Exploratory analysis of resource demand and the environmental footprint of future aquaculture development using Life Cycle Assessment. WorldFish, Penang, Malaysia. White Paper: 2014–31.

Mungkung, R., Aubin, J., Prihadi, T.H., Slembrouck, J., van der Werf, H.M.G., and Legendre, M. 2013. Life cycle assessment for environmentally sustainable aquaculture management: a case study of combined aquaculture systems for carp and tilapia. *Journal of Cleaner Production* 57:249–256.

Mungkung, R., Udo de Haes, H., and Clift, R. 2006. Potentials and limitations of life cycle assessment in setting ecolabelling criteria: a case study of Thai shrimp aquaculture product. *The International Journal of Life Cycle Assessment* 11:55–59.

NRC, N. R. C. 1993. *Nutrient requirements of fish*. Washington DC: National Academy Press.

OECD 2019. Aquaculture production (indicator). doi:10.1787/d00923d8-en (Accessed on 04 December 2019).

Ottinger, M., Clauss, K., and Kuenzer, C. 2016. Aquaculture: relevance, distribution, impacts and spatial assessments – a review. *Ocean & Coastal Management* 119:244–266.

Papatryphon, E., Petit, J., Hayo, V., Kaushik, S.J., and Claver, K. 2005. Nutrient balance modelling as a tool for environmental management in aquaculture: the case of trout farming in France. *Environmental Management* 35:161–174.

Papatryphon, E., Petit, J., Kaushik, S. J., and van der Werf, H. M. 2004a. Environmental impact assessment of salmonid feeds using life cycle assessment (LCA). *AMBIO: A Journal of the Human Environment* 33:316–323.

Papatryphon, E., Petit, J., and Van der Werf, H.M.G. 2004b. The development of Life Cycle Assessment for the evaluation of rainbow trout farming in France. In Proceedings of the 4th International Conference on: Life Cycle Assessment in the Agri-feed Sector, October 6–8, 2003, Horsens, Denmark, 73–80.

Páez-Osuna, F. 2001. The environmental impact of shrimp aquaculture: Causes, effects, and mitigating alternatives. *Environmental Management* 28:131–140.

Pelletier, N., and Tyedmers, P. 2010. Life cycle assessment of frozen tilapia fillets from indonesian lake-based and pond-based intensive aquaculture systems. *Journal of Industrial Ecology* 14:467–481.

Pelletier, N., Tyedmers, P., Sonesson, U., et al. 2009. Not all salmon are created equal: life cycle assessment (LCA) of global salmon farming systems. *Environmental Science and Technology* 43:8730–8736.

Pelletier, N., and Tyedmers, Æ. P. 2008. Life cycle considerations for improving sustainability assessments in seafood awareness campaigns. *Environmental Management* 42:918–937. doi:10.1007/s00267-008-9148-9

Purcell, J.E., Uye, S., and Lo, W. 2007. Anthropogenic causes of jellyfish blooms and their direct consequences for humans: a review. *Marine Ecology Progress Series* 350:153–174.

Rajitha, K., Mukherjee, C.K., and Vinu Chandran, R. 2007. Applications of remote sensing and GIS for sustainable management of shrimp culture in India. *Aquacultural Engineering* 36:1–17.

Richards, D.R., and Friess, D.A. 2016. Rates and drivers of mangrove deforestation in Southeast Asia, 2000–2012. *Proceedings of the National Academy of Sciences of the United States of America* 113:344–349.

Robb, D.H.F., MacLeod, M., Hasan, M.R., and Soto, D. 2017. *Greenhouse gas emissions from aquaculture: a life cycle assessment of three Asian systems.* FAO Fisheries and Aquaculture Technical Paper No. 609. Rome, FAO, 609:1–110.

Rodríguez-Valencia, J.A., Crespo, D., and López-Camacho, M. 2010. La camaronicultura y la sustentabilidad del Golfo de California, 13 p. Available in http://www.wwf.org.mx. , 2019, 1–13.

Roheim, C.A., Asche, F., and Santos, J.I. 2011. The elusive price premium for ecolabelled products: Evidence from seafood in the UK Market. *Journal of Agricultural Economics* 62:655–668.

Samuel-Fitwi, B., Nagel, F., Meyer, S., Schroeder, J.P., and Schulz, C. 2013a. Comparative life cycle assessment (LCA) of raising rainbow trout (*Oncorhynchus mykiss*) in different production systems. *Aquacultural Engineering* 54:85–92.

Samuel-Fitwi, B., Schroeder, J.P., and Schulz, C. 2013b. System delimitation in life cycle assessment (LCA) of aquaculture: striving for valid and comprehensive environmental assessment using rainbow trout farming as a case study. *The International Journal of Life Cycle Assessment* 18:577–589.

Seppala, J., Silvenius, F., Gronroos, J., Makinen, T., Silvo, K., and Storhammar, E. 2001. *Rainbow trout production and the Environment.* Helsinki: Finnish Environmental Institute, 164.

Smárason, B.Ö., Ögmundarson, Ó., Árnason, J. Björnsdóttir, R., and Davíðsdóttir, B. 2017. Life cycle assessment of Icelandic Arctic char fed three different feed types. *Turkish Journal Fisheries and Aquatic Sciences* 17:79–90.

Somero, G. N., and Hochachka, P. W. 1971. Biochemical adaptation to the environment. *American Zoologist* 11:159–167.

Song, X., Liu, Y., Pettersen, J.B., Brandão, M., Stian, X.M., and Frostell , R.B. 2019. Life cycle assessment of recirculating aquaculture systems A case of Atlantic salmon farming in China. *Journal of Industrial Ecology* 23:1077–1086.

Summerfelt, S.T., Wilton, G., Roberts, D., Rimmer, T., and Fonkalsrud, K., 2006. Developments in recirculating systems for Arctic char culture in North America. *Aquacultural Engineering* 30:31–71.

Tidwell, J.H. 2012a. Characterization and categories of aquaculture production systems. In *Aquaculture Production Systems*, pp. 64–78.

Tidwell, J.H. 2012b. Functions and characteristics of all aquaculture systems. In *Aquaculture Production Systems*, pp. 51–63.

Troell, M., Naylor, R.L., Metian, M., et al. 2014. Does aquaculture add resilience to the global food system? *Proceedings of the National Academy of Sciences* 111:13257–13263.

Walters, B.B., Rönnbäck, P., Kovacs, J.M., et al. 2008. Ethnobiology, socio-economics and management of mangrove forests: A review. *Aquatic Botany* 89:220–236.

Wilson, R., Halver, J., and Hardy, R. 2002. *Fish nutrition.* San Diego, CA: Academic Press, Inc, 143–179.

Xie, B., Qin, J., Yang, H., Wang X., Wang, Y.H., and Li, T.Y. 2013. Organic aquaculture in China: A review from a global perspective. *Aquaculture* 414–415:243–253.

Yacout, D.M.M., Soliman, N.F., and Yacout, M.M. 2016. Comparative life cycle assessment (LCA) of Tilapia in two production systems: semiintensive and intensive. *The International Journal of Life Cycle Assessment* 21:806–819.

Ziegler, F., Hornborg, S., Green, B.S., et al. 2016. Expanding the concept of sustainable seafood using Life Cycle Assessment. *Fish and Fisheries* 17:1073–1093.

8 Innovative Protein Sources in Aquafeeds

Fernando G. Barroso, Cristina E. Trenzado, Amalia Pérez-Jiménez, Eva E. Rufino-Palomares, Dmitri Fabrikov, and Maria José Sánchez-Muros

CONTENTS

DOI: 10.1201/9780429331664-8

INTRODUCTION

One of the most important goals in aquaculture nutrition research is to find a protein source with adequate nutritive properties to replace fish meal (FM) in aquafeed. In the past century, economic and environmental problems were foreseen for using this protein source, which would be caused by the deterioration and overexploitation of the marine environment, and by rising demand due to increasing aquaculture and its use to feed other livestock species (Sanchez Muros et al., 2014). The search for alternative sources recently resulted in a significant drop in fish in-fish out (FIFO) from 0.63 in 2000 to 0.22 in 2015 (http://www.iffo.net/fish-fish-out-fifo-ratios-conversion-wild-feed, consulted 7/11/2018), of which soya is the mostly widely used. Nevertheless, soya also involves environmental problems, such as the deforestation of areas with high biological value (Carvalho, 1999; Osava, 1999), considerable water use (Steinfeld et al., 2006), the utilization of pesticides and fertilizers (Carvalho, 1999), and transgenic varieties (Garcia and Altieri, 2005), which lead to significant environmental deterioration (Osava, 1999). Other sources have been checked and show different problems in relation to nutritive value, such as anti-nutritional factor presence, inadequate balance between essential amino acids/non-essential amino acids, amino acid bioavailability, which occurs in most vegetal origin sources, or inadequate fatty acid (FA) profile characteristics of animal sources. Price is another handicap, which should be competitive and include manufacturing and transporting, but affects the source as protein concentrates that must be submitted to transformation processes. Availability must also be considered, as many studies have been doing with local sources with good results but with low local production. Food safety is another factor to take into account. The protein source must be free of organic, inorganic, and biological toxins or pollutants, and this restriction limits the inclusion of animal meal in animal feed. Finally, alternative protein sources cannot compete with human food as with soya, which is important in animal-to-human feeding.

Then the innovative alternative protein concept for aquafeed must include not only the nutritional quality of the source but must also bear in mind availability, price, food safety, human competition, and sustainability.

This chapter studies the potential of some protein sources that have aroused much expectation because they are promising matches for the above-mentioned requirements to be considered an alternative protein source.

MARINE INVERTEBRATES

Marine invertebrates include different animal families that include crustacea, Mollusca, copepods, polychaetes, rotifers, and many other interesting species in aquaculture feeding.

Many of these species are suitable for aquaculture. In fact, lots of them are cultured for human or aquaculture feeding.

Crustacea

The crustacea form part of the natural diet of wild fish, and some crustacea species are cultured as human food (i.e. shrimp) or fish feed (i.e. artemia). The culture technology is well-known for these two species and can be used as the basis for culturing other crustacea species.

Crustacea are rich in protein, with a lipid content high in EPA (eicosapentaenoic acid) and DHA (docosahexaenoic acid) (Chapelle, 1977) that depends on the species; for example, artemia possess a low DHA level (Vismara et al., 2003). Krill larvae are rich in EPA, DHA, and 16:0, while in the adult stages, FA 14:0, 16:0, and 18:1n-9 are dominant (Hagen et al., 2001). Nutritional condition, developmental mode (planktotrophy vs. lecithotrophy) and clade also affect proximate biochemical composition (Anger, 1998).

From the alternative protein sources point of view, this variability is positive because it allows the nutritive values of crustacea to be manipulated for aquaculture feed, clade choice, and to feed a rearing system to obtain adequate nutritional composition to replace FM.

Nowadays, artemia and copepods are cultivated to be used as feed for larvae fish, but there are other interesting species, such a krill.

Krill

Euphausia superba and *Euphausia pacifica* are two of the most abundant species on earth, with an estimated biomass of around 500 million tons. Gross postlarval production is estimated at 342–536 million tons/yr^{-1} (Atkinson et al., 2009). This is a vast quantity that allows it to be used without limitation. Nevertheless, krill is essential for supporting the primary production system with an estimated predator consumption of 128–470 Mt/yr^{-1}. These data reveal the need for the precautionary management of developing krill fisheries (Atkinson et al., 2009). The Commission for the Conservation of Antarctic Marine Living Resources (CCAMLR) restricts harvesting to 1 percent of the total biomass (Burri and Nunes, 2016).

Nutritional value: Krill is an excellent source of vitamins, minerals, essential amino acids, n-3 polyunsaturated FA, natural carotenoid pigments, nucleotides, and organic acids (Lee and Meyers, 1997; Everson, 2000). Krill products are known to be excellent feed attractants in the fish diet.

Whole krill is a high protein food whose protein content is estimated into fall within the 60–65% range (Nicol, 2000) and has a higher amino acid content than trout or salmon (Tou et al., 2007). Lipid content ranges from 12% to 50% on a dry weight basis. Differences are attributed to sampling occurring during different seasons (Saether et al., 1986).

The FA profile is low (26.1%) in both saturated fatty acids (SFAs) and (24.2%) monounsaturated fatty acids (MUFAs), but is high (48.5%) in polyunsaturated fatty acids (PUFAs). Palmitic acid (16:0) is the predominant SFA, oleic acid (18:1n-9) is the predominant MUFA, and PUFAs consist mainly of n-3 FA. Kolakowska et al. (1994) reported that n-3 PUFAs accounted for approximately 19% of the total FA in Antarctic krill caught in winter, while EPA and DHA were particularly abundant. Lipid content also varies with Northern krill species, which are particularly rich in lipids (182 g/kg^{-1}) (Suontama et al., 2007).

The main phospholipid in krill meal is phosphatidylcholine (Tou et al., 2007), which delivers omega-3 FA and choline. Choline is an essential vitamin that must be added to aquafeed (Gong et al., 2000). Phospholipids are also involved in cholesterol uptake and distribution, which is an essential nutrient in shrimp feed (Gong et al., 2000).

A characteristic of Arthropoda is the presence of an exoskeleton. Chitin forms part of the exoskeleton and consists in b-1,4-linked N-acetylglucosamine. It requires the action of enzymes chitinase

(EC 3.2.1.14) and chitobiase (EC 3.2.1.30) for *in vivo* degradation. The presence of chinase in fish digestive secretion has been discussed. Most examined fish seem to possess some chitin-degrading enzymes, such as chitinases and/or chitobiases, in their digestive tract (Danulat and Kausch, 1984; Lindsay, 1987; Lindsay and Gooday, 1985; Danulat, 1986; Rehbein et al., 1986; Kono et al., 1987; Sabapathy and Teo, 1993; Moe and Place, 1999; Gutowska et al., 2004).

Some data also indicate that chitinase activity in those fish feeding on chitin-rich prey is greater than in other fish (Gutowka et al., 2004; Karasuda et al., 2004; Fines and Holt, 2010) and that feeding chitin-rich diets increases enzyme activity (Danulat, 1986). However, current data are rather incomplete and, to some extent, contradictory. The current discussion is about whether the origin of chitinase activity is endogenous or due to digestive microbiota.

In the Atlantic salmon (*Salmo salar* L.), feed in which FM is replaced with krill meal seems to have no influence on the apparent digestibility coefficients (ADCs) of dry matter and protein, while chitin is not utilized to a great extent (Olsen et al., 2006). Nevertheless, the level of inclusion affects digestibility, while lipid digestion does not reduce at 60% krill FM replacement (Suontama et al., 2007). At a high Antarctic krill inclusion level (>80% of diet proteins), lipid digestibility lowers (Olsen et al., 2006).

Advantages and Constraints

One beneficial effect of chitin on the fish immune system has been described. The fish fed a diet supplemented with chitin displayed high total haemocyte counts (THCs) and marked prophenoloxidase and superoxide dismutase activities (Zhu et al., 2010; Gopalakannan and Arul, 2006). The chintin immuno-stimulating system effect depends on the administration channel (Esteban et al., 2000), the inclusion level (Esteban et al., 2001), or the size of chitin particles (Cuesta et al., 2003). Krill also affects adherent distal intestine microbiota and enterocytes, shown to be replete with numerous irregular vacuoles (Ringø et al., 2006).

The main inconvenience of krill meal is fluoride content at around 1,000–6,000 mg/kg. The European Union has set a maximum fluoride level in feed at 150 mg/kg dry feed (Council Directive, 1999) for its potential accumulation in organs, especially bone. Studies have related krill meal inclusion and fluoride accumulation in many fish species, such as Atlantic salmon *Salmo salar* (Julshamn et al., 2004), Atlantic cod *Gadus morhua*, and Atlantic halibut *Hippoglossus hippoglossus* (Moren et al., 2007). Fluoride accumulation provokes reduced growth. Indeed 30% krill FM replacement with meal reduces the growth of rainbow trout due to fluoride accumulation in vertebral bones (Yoshitomi et al., 2006). The same authors (Yoshitomi et al., 2007) obtained good results with no negative effects on growth, survival, or nutritive indices with 100% low-fluoride krill meal replacement. However, the fluoride effect depends on both salinity (Julshamn et al., 2004) and species.

Studies in replacing FM with krill meal generally report good results for Russian sturgeon at 30% replacement (Gong et al., 2016) and for gilthead sea bream at 9%, which enhances gilthead sea bream growth and reduces both lipid accumulation and hepatocyte damage (Saleh et al., 2018), as well as 60% substitution for Atlantic halibut (*Hyppoglossus hyppoglossus*) (Suontama et al., 2007) and 40% substitution in juvenile spotted halibut (*Verasper variegatus*) (Yan et al., 2018). The total replacement of FM with low-fluoride krill in the diet is successful with no defects in growth performances for *Oncorhynchus mykiss* (Yoshitomi et al., 2007) and with normal krill Atlantic salmon (*Salmo salar*) (Olsen et al., 2006).

From the environmental point of view, the main constraint lies in krill being a very abundant, but finite, source that supports the primary production system. Hence sustainable harvesting has been established at 1 percent of total biomass (Burri and Nunes, 2016).

Crabs

Dean et al. (1992) studied the inclusion of not only blue crab for the fingerling channel (*Ictalurus punciatus)* diet, but also the 10% inclusion of blue crab without carapace. These authors reported a similar weight gain and feed efficiency to the fish feed in an FM diet. Nevertheless, under their production conditions, the caged channel catfish fed the control diet or the Atlantic herring diet displayed greater daily gain and net production than those fed the crab diet.

Copepods

The utilization of copepods in larvae and juvenile feeding started a long time ago, but has not been used as a protein source alternative.

Cultured copepods have good nutritive composition; 6.9–22.5% DW (dry weight) of lipids with EPA and DHA of 8.3–24.6%, and 13.9–42.3%, respectively. Protein amounted to 32.7–53.6% (determined as protein-bound amino acids) with a stable fraction of indispensable amino acids (37.3–43.2% of PAA). Abundant astaxanthin has been detected in copepods (413–1422 μg/g DW), as have vitamin C (38–1232 μg/g DW), vitamin E (23–209 μg/g DW) thiamine (3.5–46.0 μg/g DW) and riboflavin (23.2–35.7 μg/g DW) (van der Meeren et al., 2008). The nutritive benefits in larvae nutrition of copepods have been well demonstrated (van der Meeren et al., 2008).

Nevertheless, their use as an alternative protein source of FM has not yet been checked. Nowadays, calanus (gen. *Calanus*) is considered a potential source of n-3 highly unsaturated FA (HUFA) (Olsen et al., 2004) that can help to reduce dependence on marine fish oils. The increased use of calanus oil has probably led to increased interest in calanus nutritive values or in nutritional characteristics of oil production waste.

CONCLUSIONS

Currently, the more promising crustacean as a protein source is krill because it is an abundant source with a high protein percentage. Nevertheless, more studies are needed to establish the chitin effect on digestibility and the immune system. Fluoride accumulation is well-studied and the use of low-fluoride meal reports good results in the studied fish species.

INSECTS

Introduction

If the environmental and economic sustainability of aquaculture are to be ensured, the contribution of FM as fish feed must be lower. Interest in insects is currently growing as they are one of the most promising protein sources for feed production (Gasco et al., 2020).

Interest in insects as feed has grown mainly in developed countries over the last decade. In the market scenario, the insect business continues to grow, with companies being founded worldwide, especially those that perform the mass breeding of the black soldier fly (*Hermetia illucens*, HI), whose world production was 14,000 tons (wet weight) in 2016, and was 7,000 – 8,000 tons in 2014/15 (Sogari et al., 2019). From a scientific-academic point of view, so many projects and publications have been produced in these years that it is extremely complex to synthesize all the knowledge being produced. However, to look in-depth at some aspects related to insects, such as animal feed, readers can consult different reviews (Barroso et al., 2014; Gasco et al., 2020; Gómez et al., 2019; Govorushko et al., 2019; Kenis et al., 2014; Makkar et al., 2014; Sánchez-Muros et al., 2014, among others).

Reasons for Using Insects as Feed

Insect meals appear to be safe, cheap, and sustainable. Compared to other feed or food sources, insect breeding offers several environmental benefits. Indicators that provide insight into the sustainability of insect production can be included: (1) greenhouse gas emissions are much lower than other animal production as insects only consume lignin (termites and cockroaches) and produce methane (Govorushko, 2019); (2) much less water and space are needed to reproduce and raise insects than with livestock (Tabassum et al., 2016); (3) insects offer higher feed conversion efficiencies, are poikilotherms, and do not invest energy to maintain body temperature (Oonincx et al., 2010); (4) insects can transform abundant low-cost organic waste into protein-rich animal biomass for use in animal nutrition (Ramos-Elorduy, 1999). A socio-economic advantage lies in insect breeding requiring low capital and technology investment, which could be developed by the most disadvantaged population in society (Govorushko et al., 2019).

In line with nutritional value, overall, insects are generally rich in proteins (30–68% on a dry matter (DM) basis), although less protein-rich than FM (Barroso et al., 2014), but have well-balanced amino acid profiles (Finke, 2015; Gasco et al., 2018; Koutsos et al., 2019). Insects have no anti-nutritional factors as regards vegetable ingredients (Spranghers et al., 2017).

Use of Insect Meals in Aquaculture

As insects are included in the diet of many fish species in their natural environments, we consider that their use as feed in aquaculture can be a very interesting option. The European Commission (Annexe II of Regulation 2017/893 of 24th May 2017) has recently authorized the use of insect-processed animal proteins that derive from seven species of insects farmed for aquaculture purposes. These species include two flies (HI; *Musca domestica*), two mealworms (*Tenebrio molitor*, TM; *Alphitobius diaperinus*), and three crickets (*Acheta domesticus*; *Gryllodes sigillatus*; *Gryllus assimilis*).

This chapter does not intend to be an exhaustive review, as more and more studies address the partial substitution of FM for insect meals in fish. Among other experiments, insect meals have been successfully tested:

- With HI larvae meal in Atlantic salmon (*Salmo salar* L.) (Lock et al., 2016; Belghit et al., 2018), gilthead sea bream (*Sparus aurata*) (Fabrikov et al., 2020), Nile tilapia (*Oreochromis niloticus* L.) (Devic et al., 2018), Jian carp (*Cyprinus carpio* var. Jian) (Li et al., 2017; Zhou et al., 2018), rainbow trout (*Oncorhynchus mykiss* Walbaum) (Renna et al., 2017; Elia et al., 2018; Fabrikov et al., 2020), sea bass (*Dicentrarchus labrax* L.) (Abdel-Tawwab et al., 2020; Magalhães et al., 2017), tench (*Tinca tinca*) (Fabrikov et al., 2020) and tilapia (*Oreochromis* sp.) (Bondari and Sheppard, 1981)
- With TM larvae meal in African catfish (*Clarias gariepinus*) (Ng et al., 2001), gilthead sea bream (*Sparus aurata*) (Fabrikov et al., 2020), rainbow trout (*Oncorhynchus mykiss* Walbaum) (Chemello et al., 2020; Fabrikov et al., 2020), rockfish (*Sebastes schlegeli*) (Khosravi et al., 2018), sea trout (*Salmo trutta m. trutta*) (Hoffmann et al., 2020), tench (*Tinca tinca*) (Fabrikov et al., 2020), and tilapia (*Oreochromis niloticus*)(Sánchez-Muros et al., 2016)
- With cricket meal (*Gryllus bimaculatus*) in African catfish (*Clarias gariepinus*) (Taufek et al., 2018)

The most frequently used insects in aquaculture are TM and HI. As Kenis et al. (2014) point out, this could be due to the possibility of these species being mass-reared in small production units at both the community and industrial levels and can be fed waste or by-products.

Generally, when the degree of inclusion of insect meals was below 25% of diet (regardless of the degree of FM substitution), no negative effects on fish growth performance have been observed. However, when higher inclusion rates have been evaluated, the results were not as positive. In meagre (*Argyrosomus regius*), Guerreiro et al. (2020) found that nutritional indices linearly lowered with increasing dietary HI levels. High FM substitution levels (more than 50%) make production rates worse (Reyes et al., 2020) in sea bass and, even if inclusion is very high (100% FM substitution and up to 75% feed), rejection occurs in Siberian sturgeon (*Acipenser baerii* Brandt) (Caimi et al.,

2020). With high FM replacement levels (60%) in Pacific white shrimp (*Litopenaeus vannamei*), not only is growth lower, but pathological changes appear in the hepatopancreas (Cao et al., 2012).

These results coincide with a recent meta-analysis by Hua (2020) on the effect of including insect meals on fish growth performance. This researcher concluded that moderate levels of insect meals can cause comparable growth performance to FM diets. However, when large proportions of insect meal are used, growth reduces, and the effect depends on the employed insect species. This author indicates that TM is better tolerated at high levels than HI, which usually leads to greater declining fish production rates.

The possible reasons for worse growth with higher insect meal inclusion levels are:

- Essential amino acid (EAA) deficiencies and EAA/non-essential amino acids (NEAA) imbalances are among the most important reasons behind these negative results (Cummins et al., 2017). Panini et al. (2017) found in shrimp that, if supplemented with methionine, 100% FM could be replaced with TM meal. Although insect meals are rich in EAA (lysine, methionine, leucine) (Caimi et al. 2020) compared to FM, insect meal is deficient in lysine and tryptophan and is limited in threonine and sulphur AA (Makkar et al., 2014; Sánchez-Muros et al., 2014). Barroso et al. (2014) compared the amino acids profile of different insect species with FM and found that the profile of amino acids was related to the taxonomic group. According to these authors, the order Diptera has the most similar amino acid profile to FM.
- Another limitation could lie in the cuticle (exoskeleton) of insects, as it contains chitin fibers, a polysaccharide of glucosamine and N-acetylglucosamine, both of which contain N atoms (Jonas-Levi and Martinez, 2017). As the N-factor for meat is 6.25, and based on the idea that proteins contain approximately 16% of nitrogen (Merrill and Watt, 1973), estimating the protein content (calculated as nitrogen × 6.25) could prove misleading in insect meal (Barker et al., 1998). Therefore, Janssen et al. (2017) considered that in order to estimate the protein content of whole larvae, a conversion factor of 4.76 should be used. However, Finke (2013) studied the nutrient content of several insect species and discovered that only a small amount of nitrogen was associated with chitin.
- It would appear that the crude protein digestibility of insects is affected by chitin content. Marono et al. (2015) found that the digestibility of crude protein from HI and TM correlated negatively with their fiber and chitin contents. Defatting insect meals also seems to affect digestibility. In an experiment with sea bass and feed with 20% FM substitution for several insect meals, Basto et al. (2020) revealed that the apparent digestibility coefficients of crude proteins were high in defatted TM (93%), intermediate in defatted HI (87%) and TM (89%), and moderate in HI (76%). Yet despite insects being defatted, in sturgeon Caimi et al. (2020) found that feeds with high inclusion HI rates showed lower apparent digestibility coefficients of crude protein compared to 100% FM feeds.
- We must also consider that the nutritional value of insects varies with age. Aniebo and Owen (2010) found that the fat content in *Musca domestica* larvae increased with age, and this was inversely related to protein content. It would seem that insects, with a complex metamorphosis, contain more fat and fiber (chitin) and less protein when they approach the pupal stage.
- Finally, the nutritional value of insect meals varies according to the processing followed during their manufacture. According to Hoffmann et al. (2020), the key to improving protein retention in the digestive system of insect meals may be the hydrolysis of their protein. Currently, data on the hydrolysis of insect material are limited, but these authors used diets with 20% TM in trout and found that hydrolyzed and unprocessed TM had similar effects. Another common treatment in insect processing is defatting, which can provide a meal that is easily used as an ingredient in aquaculture feed. With defatting, meals with higher percentages of crude protein that are more resistant to degradation can be obtained

(Chemello et al., 2020). In addition, the high proportion of fat in larvae, which sometimes does not have a suitable FA profile for fish, can be reduced. Chemello et al. (2020) evaluated progressive FM substitution (up to 100%) with increasing concentrations of a partially defatted TM meal in the diet of rainbow trout. These authors observed no negative effect on fish growth. As a final example of the importance of processing methods on the nutritional value of insects, Kinyuru et al. (2010) devised a method of toasting and drying grasshopper samples, which affected their vitamin content and significantly decreased their protein digestibility.

It is also necessary to correctly evaluate whether fillet quality can be affected by insect meal inclusion. This quality is primarily affected by the fatty acid profile, which depends on the quality of fat in diet (Sanchez-Muros et al., 2014). Terrestrial insects contain mainly n-6 FA and small amounts of n-3 PUFA, which could represent a limit in animal nutrition (Barroso et al., 2014). As FM is increasingly replaced with insect meal in feed, there is generally a proportional reduction in n-3 PUFA in the n-3:n-6 ratio and in the unsaturation rate in fish fillets. This has been observed, among others, in blackspot sea bream (*Pagellus bogaraveo*) (Iaconisi et al., 2017), meagre (*Argyrosomus regius*) (Guerreiro et al., 2020), Pacific white shrimp (Panini et al., 2017), and rainbow trout (Belforti et al., 2015; Stadtlander et al., 2017). In some experiments, 100% FM has been replaced with insect meal with no significant differences in the FA profile, e.g. in Atlantic salmon (Bruni et al. 2020). However, this could be due to the additional fish oil present in insect-containing diets. The FA profile also significantly affects lipid digestibility (Hua and Bureau, 2009). To avoid these disadvantages, larvae can be defatted or their FA profile can be modified by the substrates used for their feeding. Different experiments (Barroso et al., 2017, 2019; Liland et al., 2017; St-Hilaire et al., 2007) have increased omega-3 fatty acid in larvae by including components rich in these FA (fishery waste or seaweed) in their diet (Figure 8.1)

Regarding fillet quality, Bruni et al. (2020) found that complete dietary FM substitution with HI meal did not impair the physicochemical quality of Atlantic salmon fillets. By taking into account sensory aspects, Bondari and Sheppard (1981) ran an experiment with channel catfish (*Ictulurus punctatus*) and blue tilapia (*Tilapiu urea*) fed a diet with HI (50% inclusion). They found that it did not affect their taste.

FIGURE 8.1 *Hermetia illucens* larvae rearing in fish discards (Courtesy of F Barroso, University of Almería, Spain).

In a sensory analysis in rainbow trout using two HI types (normal and fish offal-enriched) with an FM substitution degree up to 50%, an untrained consumer panel found no significant differences between the trout fed different diets (Sealey et al., 2011).

Critical Points to Be Solved in the Future

- *Legislation*: Some legislative barriers must be overcome. For example, the EU currently considers that insects, as food in aquaculture, can be fed only with the animal raw material listed in Regulation (EU) 2017/1017. Therefore, insects cannot be fed manure, waste, former foodstuffs containing meat, or waste fish or food from restaurants or catering establishments (Gasco et al., 2020). We believe that this legislation should become more flexible in forthcoming years, as it limits the potential of insects as sustainable food. One of the biggest advantages of insects is precisely that they can be raised with waste from catering establishments or the food industry as they do not compete with humans for this food resource. Furthermore, by breeding insect by-products, their production becomes more economical, and, as Van Huis (2015) points out, they can alleviate waste disposal. However, the microbiological content of insect meals must be controlled both during processing and storage to guarantee their hygienic and sanitary quality, as with any other raw material.
- *Market price*: although marked insect production growth is expected for the food and feed market in forthcoming years, it still remains on a small industrial scale. It is difficult to find the market price because demand is still limited, and companies adapt their price according to the size of orders (Gasco et al., 2020). Therefore, mass insect breeding is not yet sufficiently developed to obtain a competitive price in relation to other protein sources.
- *Heterogeneity*: Nutrient content varies widely between not only insect species, but even within the same insect species. Nutrient characteristics depend on life stage, environment, diet, processing or slaughter methods, etc. This limits their use in the feed industry, as they need to include an availability of raw material of homogeneous and stable quality.
- *Cultural acceptance*: Despite the production advantages of insect meal, its direct consumption clashes with cultural barriers in more developed countries. However, these barriers can be overcome with correct information on the sustainability of their production and with nutritional advantages over other foods. Several studies have shown that consumers have neophobia when faced with direct insect consumption (food) (Sogari et al., 2019), but will readily accept eating insect-fed animals (feed) (Verbeke et al., 2015; Mancuso et al., 2016; Ferrer Llagostera et al., 2019). Specifically, Mancuso et al. (2016) studied consumer acceptance of farmed fish fed insect meals and obtained very positive consumer attitudes because almost 90% were prepared to eat fish that were fed insects.

CONCLUSIONS

In summary, we believe there is still plenty of work to be done.

Although insect meals can hardly replace FM satisfactorily, if we wish to promote the commercialization of insects as a valued raw material in aquaculture, we must assess not only their proximate composition, but also aspects like digestibility or anti-nutritional components, the effect of insect breeding methods, and feed production technologies (method of drying, processing methods, etc.) of meals on nutritional value. We must also determine the optimal levels of insects (or combinations of different insect types) to adapt to these needs according to the nutritional needs of each fish species.

YEAST

Introduction

Yeasts are potential sustainable ingredients in aquafeeds given their ability to convert low-value lignocellulosic biomass into high-value feed with limited dependence on land, water, and climate conditions (Øverland et al., 2013).

Yeast has been utilized as a nutritional supplement in animal feed for more than 70 years. In aquaculture, it has been well-studied as probiotic systems, immunostimulants, and live feed (Manoppo et al., 2011; Murthy et al., 2009; Gatesoupe, 2007; Gopalakannan and Arul, 2010; Jones et al., 2020). The use of yeast as an FM substitute in aquaculture systems has recently drawn considerable attention (Øverland et al., 2017; Montoya-Camacho et al., 2019; Guo et al., 2019). Yeast acts as a health promoter in fish given the presence of peptides, free nucleotides, and mannan oligosaccharide (Rawling et al., 2019).

Nutritive Values

Yeast proximal composition varies depending on the used species. Table 8.1 shows the different crude protein (CP), crude fat (CF), and fiber (F) percentages in several yeast species employed in aquaculture feeds.

Although CP varies from one specie to another, it also varies within the same species. CP varies in yeast species in accordance with the substrate used to cultivate the yeast (Ritala et al., 2017). As yeasts are single-cell organisms, their CP content is strongly influenced by the nitrogen present in the genome, which represents 6–12% of total nitrogen (Halasz and Lasztity, 1991, 2017). The EAA profile of the main yeasts used in aquaculture (*Saccharomyces. cerevisiae*, *Candida utilis*, and *Kluyveromyces marxianus*) show similar values to FM, although methionine levels are lower than those in FM (Øverland and Skrede, 2017). Lipids represent a low percentage of the proximal composition of yeasts, but genetically modified *Yarrowia lipolytica* produces up to 20.3% of CF with 30% EPA of total lipids. Brown et al. (1996) analyzed different marine yeast FA, of which the main FA are palmitic acid (16:0), oleic acid (18:1n9), and linoleic acid (18:2n6). Yeast is also a source of minerals like phosphorus, calcium, sodium, zinc, iron, copper, manganese, and selenium (Chanda and Chakrabarti, 1996; Cheng et al., 2004).

Use of Yeast Meals in Aquaculture

Hatlen et al. (2012) carried out an experiment with genetically modified *Y. lipolytica* to be included in the diet of *S. salar*. Although growth performance was not affected by yeast inclusion, digestibility

TABLE 8.1
Proximal Composition of Principal Yeast Used as Feed in Aquaculture

Specie	CP (%)	CF (%)	F (%)	Reference
Y. lipolytica	29.8	20.3	n.a.	Hatlen et al. (2012)
S. cerevisiae	32.0	4.0	10	Zerai et al. (2008)
S. cerevisiae	44.2	2.9	0.3	Pongpet et al. (2016), 2016)
C. Utilis	56.0	0.3	3.7	Øverland et al. (2013)
C. Utilis	39.0	2.1	n.a.	Hansen et al. (2018)
C. Utilis	41.0	n.a.	n.a.	Gamboa-Delgado et al. (2015, 2016)
K. marxianus	51.0	0.8	0.8	Øverland et al. (2013)
K. marxianus	42.0	1.3	n.a.	Ribeiro et al. (2014)
R. mucilaginosa	17.0	–	–	Chen et al. (2019)

decreased when yeast inclusion was higher. The experiment (Zerai et al., 2008) carried out with *S. cerevisiae* at different substitution levels (25%, 50%, 75%, 100%) in *Oreochromis niloticus* showed substitution-dependent growth performance, but it was possible to replace 50% FM without compromising growth. The inclusion of *K. marxianus* in the *O. niloticus* (Ribeiro et al., 2014) diet during different seasons resulted in reduced growth performance mainly in spring. However, lipid content was not altered by yeast diet, and CP was higher in the muscles from the fish fed yeast diets. *S. cerevisiae* inclusion in *Dicentrarchus labrax* led to increased growth performance for 30% yeast substitution, and CP rose in fish that were fed yeast diets (Oliva-Teles and Gonçalves, 2001). Øverland et al. (2013) substituted FM for 40% *S. cerevisiae* in *S. salar*, which negatively affected the fish growth parameters. Similar results have been reported for *Oncorhynchus mykiss* (Cheng et al., 2004), Atlantic salmon (Øverland et al., 2013), and Arctic char (*Salvelinus alpinus*) (Langeland et al., 2013).

This lower protein digestibility is due to tough cell walls and their negative effect (Øverland et al., 2013). In Arctic char, greater energy and amino acid digestibility were found for disrupted *S. cerevisiae* than for intact cells, with no significant differences in Eurasian perch (*Perca fluviatilis*) (Langeland et al., 2016).

Moreover, the digestibility of *S. cerevisiae* depends on fish species, which increased in gilthead sea bream depending on FM substitution levels (10% and 20% substitutions) (Salnur et al., 2009). In pacu (*Piaractus mesopotamicus*), lipid digestibility increased, while protein digestibility remained unaffected (Ozório et al., 2010). These differences in digestion efficiency among fish species can be related to different digestive enzyme activity levels (Langeland et al., 2013).

Digestibility can also be affected by yeast species. Substituting FM for *C. Utilis*, *K. marxianus*, or *S. cerevisiae* in *Salmo salar* feed led to a similar CP digestibility to FM for *C. utilis* and *K. Marxianus*, while low CP digestibility was observed for *S. cerevisiae* (Øverland et al., 2013).

Yeast is often used in aquaculture as a growth promoter and immunostimulant in functional feeds due to various bioactive components. Positive health effects are well-documented in several fish species, such as salmonids (Tukmechi et al., 2014; Refstie et al., 2010), *Ictalurus punctatus* (Welker et al., 2012), *Lateolabrax japonicas*, (Yu et al., 2014), hybrid *Morone saxatilis* (Li and Gatlin, 2003; Li and Gatlin, 2004), *Sparua aurata* (Rodríguez et al., 2003), hybrid tilapia (He et al., 2011), Cyprinus carpio (Gopalakannan and Arul, 2006, 2010), and *Labeo rohita* (Tewary and Patra, 2011). Yeast inclusion increases the total gut weight of fish, possibly due to the high content of nucleic acid converted into nucleotides and acting as a growth promoter of intestinal epithelial cells. Morphological studies on the distal intestine have indicated no adverse effect for *C. utilis* and even report diminishing the possible adverse effects of high inclusion levels of vegetal protein (Grammes et al., 2013; Hansen et al., 2018).

The manipulation and preparation of yeast cells are most important because these processes influence CP digestibility in fish. The tough cell walls of yeasts can inhibit access to the nutrients inside cells. This can be seen in the different digestibility observed during fish trials with several fish species (Nazzaro et al., 2021). The cell wall represents 26–32% of the cell's total dry weight. Nesseri et al. (2011) suggested that mechanical (high-pressure homogenization, wet milling, sonication) and enzymatic methods can be used to disrupt the cell wall. Enzymatic disruption has several advantages, given selectivity only in the cell wall, but this process is slow compared to mechanical methods. Asenjo and Dunnill (1981) conducted a study to combine the enzymatic and mechanical cell wall disruption.

Yeast extract is obtained by removing the cell wall material with higher protein contents than in whole or hydrolyzed yeast. However, the cell wall fraction is rich in bioactive and immunostimulant compounds like -glucan and mannan oligosaccharides, which are a very interesting feed ingredients for combining properties as a source of nutrients and bioactive components (Overland and Skrede, 2017) that can be lost in yeast extract.

CONCLUSIONS

As stated before, yeasts have been used in aquaculture for different purposes. They can be employed in dietary compounds of diet as health promoters to improve gut microbiota, to act as an immunosuppressor, and to promote growth. As aquaculture rapidly grows, the industry cannot meet the protein demand for aquaculture feeds. Hence yeasts have been tested in fish trials by partially replacing FM with similar growth rates to FM diets. Yeast can be an alternative protein source for aquaculture, but further studies are necessary to improve the production and processes to obtain yeasts with better chemical compositions and to improve digestibility.

BIOFLOCS

Introduction

Biofloc technology (BFT) has emerged as an alternative aquaculture system based on the limitation of water exchange and the culture of microorganisms for feeding purposes, whose origin is established in France in the 1970s, with several studies performed on various penaeid species (Dauda, 2020). However, it was not until the beginning of the 21st century when this technology became very important thanks to the numerous advantages that it provides from both the economic and environmental points of view (de Schryver et al., 2008; Khanjani and Sharifinia, 2020; Robles-Porchas et al., 2020). So initial studies confirmed the beneficial influence of BFT on water quality and growth performance of fish and crustacea, such as Nile tilapia (*Oreochromis niloticus*) and Pacific white shrimp (*Litopenaeus vannamei*), respectively (Azim and Little, 2008; Haveman et al., 2009).

BFT is a zero-water exchange system. It improves the feed conversion ratio of reared species due to the production of microbial protein, which becomes an important food source that, in the long term, can efficiently decrease dietary protein content, even in intensive and super-intensive cultures (Walker et al., 2020) (Figure 8.2). Moreover, BFT can be a source of compounds with different bioactivities to help to improve the health status of cultured species by enhancing their immune system and antioxidant status (Haridas et al., 2017; Aguilera-Rivera et al., 2019). BFT has been successfully applied to the culture of many crustacea, but is more limited in fish species (Robles-Porchas et al., 2020). Notwithstanding, BFT is not free of disadvantages, such as the slightly higher initial cost for modernizing aquaculture facilities, limited species available to be cultivated in this system, or the

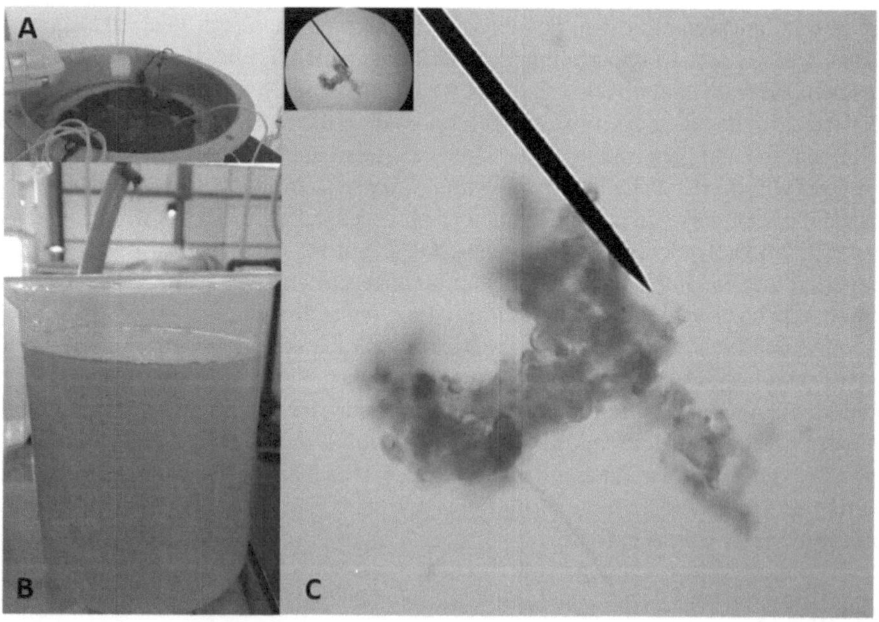

FIGURE 8.2 Experimental tank with BFT system (A), beaker with water sample containing bioflocs in suspension (B) and morphology of bioflocs under the microscope (C) (Courtesy of iMare Natural S.L. (2020)).

high dependence of constant aeration. However, its outstanding advantages, such as being mid and long-term cost-effective and its enviro-friendly technology, make BFT a firm option for a sustainable aquaculture future (Crab et al., 2012; Khanjani and Sharifinia, 2020).

Biofloc Development and Composition

BFT is based on a culture system of microorganisms that recycle waste nitrogen to proliferate (Dauda, 2020). The massive accumulation of these microorganisms, combined with other components, makes up amorphous structures of variable sizes called bioflocs, which range from several micrometres to millimetres (de Schryver et al., 2008). The start of biofloc formation takes place in the first week of culture, whereas its maturation, denoted by a change in culture color from green to brown and constant composition, can last between weeks and months depending on the factors conditioning this process (Ahmad et al., 2017; Martínez-Córdova et al., 2018; Robles-Porchas et al., 2020).

Bioflocs are heterogeneous aggregates that are composed of 30–40% inorganic material and 60–70% organic material (Chu and Lee, 2004). Among the organic components, chemoautotrophic and heterotrophic bacteria and cyanobacteria stand out for their abundance, mainly including phyla Proteobacteria, Bacteroidetes, and Actinobacteria, among others (Robles-Porchas et al., 2020; Zhao et al., 2012). Additionally, other microorganisms adhere to the organic matrix to form these aggregates, including viruses, microalgae, yeasts, and fungi, as well as invertebrates like rotifers, protozoa, amoebas, copepods, cladocera, ostracods, annelids, and nematodes (Ahmad et al., 2017; Ju et al., 2008; Martínez-Córdova et al., 2018). Other components that make up bioflocs include feces remains, uneaten food, dead cells, organic polymers, colloids, salts, and trace minerals (Azim and Little, 2008).

However, the final composition of the organisms present in bioflocs is conditioned by several parameters, among which the C:N ratio and type of carbon source are highlighted as the most important factors that can modify the relative content and type of different microorganisms (Liu et al., 2018a; Kim et al., 2021; Minabi et al., 2020; Tinh et al., 2021). Furthermore, as reviewed extensively by Dauda (2020), other aspects, like salinity levels, temperature, pH, dissolved oxygen, mixing intensity, light or cultured species are also key in biofloc composition. Controlling all the factors that determine the final biofloc composition, mainly regarding its microbial community, is indispensable as this fact conditions its nutritional value (Ahmad et al., 2017). Although biochemical composition is fixed, other aspects, like biofloc particle size or digestibility, define the nutritional value for a given species (Khanjani and Sharifinia, 2020).

Nutritive Value

In nutritional terms, widely variable protein levels ranging from 7.7% to 50%, and lipid levels between less than 0.1 and 9.9 (both on a DM basis), characterize bioflocs (Table 8.2). Regarding essential nutrients, different studies have indicated that bioflocs can also contain significant quantities of omega-3 and omega-6 FA, and several EAA. For essential FA, docosahexaenoic acid (DHA), EPA, linolenic acid (ALA), arachidonic acid (ARA) and linoleic acid (LA) have been found (Castro et al., 2021). The most representative group of EAA in bioflocs comprise arginine, isoleucine, leucine, phenylalanine, threonine, and valine, although other amino acids, like histidine, tryptophan, and methionine have been observed, although in limited amounts (Castro et al., 2021). In the same way, bioflocs can also include non-essential amino acids (i.e. aspartate, glutamate, serine, glycine, alanine, proline), minerals (i.e. calcium, phosphorous, sodium, potassium, magnesium, zinc, iron), vitamins (i.e. thiamine B1, riboflavin B2, niacin B3, vitamin B12, vitamin E) and other bioactive compounds like phytosterols, carotenoids, or chlorophylls (Castro et al., 2021; Wei et al., 2016).

Given the nutritional value of bioflocs, several studies have demonstrated that they can be used as an effective food source which, among others, enhances growth performance and feed utilization

TABLE 8.2
Nutritional Composition of Bioflocs

Authors	Protein Range (%)	Lipid Range (%)	Ash (%)
Azim and Little (2008)	50.0	2.5	7.0
Ju et al. (2008)	30.5–36.9	1.9–5.9	14.8–38.9
Kuhn et al. (2010)	38.8–40.5	< 0.1–1.1	11.8–24.7
Xu and Pan (2014a)	21.3–32.1	1.6–2.8	43.4–61.4
Dantas et al. (2016)	24.7	0.4	36.6
Wei et al. (2016)	31.5–41.2	4.2–8.5	12.4–15.2
Da Silva et al. (2018)	14.2–21.3		53.2–60.1
Ekasari et al. (2019)	18.1–33.8	2.4–9.9	3.4–9.4
Mabroke et al. (2019)	24.5	5.0	16.0
Castro et al. (2021)	7.7–18.1	0.9–2.1	53.5–62.4

of fish and crustacea, and this improvement is achieved in different ways. Hence bioflocs can be simultaneously utilized with artificial diet in aquaculture production by being a natural food that is directly ingested from the culture water that supplements diets (Tierney and Ray, 2018). In the same way, biofloc intake can save dietary protein as protein content can be reduced in diets without negatively affecting growth (Ebrahimi et al., 2020), while the use of alternative protein sources, like plant or insect meals, to replace FM improves when BFT is applied (Jatobá et al., 2017; Tubin et al., 2020). Finally, bioflocs can also be included directly as a dietary ingredient to replace FM in both fish and crustacean species (Ekasari et al., 2019; Khatoon et al., 2016).

Use of Bioflocs in Aquaculture

Bioflocs as a Natural Complementary Feeding Source

In situ bioflocs are an important nutrient source as they are available 24 h/day as a fresh and constantly renewed food. Tacon et al. (2002) ran a feeding trial in Pacific white shrimp to investigate the effects of indoor running water culture system *vs.* an outdoor zero-water exchange culture system. They observed that this system achieved the best growth performance and feed utilization. Final body weight (FBW) was nearly 3-fold heavier than for those animals cultivated in the indoor running water system. These results were attributed to the possibility of obtaining additional nutrients from the organisms present in outdoor "green water". Following the line of these initial studies performed in Pacific white shrimp, other authors have observed the same positive effects of BFT on the culture of this species. Khanjani et al. (2015) obtained feed conversion ratio (FCR) values that ranged between 1.52 and 1.29 in a clear water system (CW) *vs.* BFT, respectively, and better animal growth performance. Subsequent studies that focused on the effect of carbon sources or probiotic supplementation on Pacific white shrimp performance have demonstrated that, once again, any of the tested BFT systems were better for growth and food utilization than the control groups without bioflocs (Khanjani et al., 2017; Liu et al., 2018b).

In other crustacean species, Cardona et al. (2015) studied the effect of CW, BFT, and BFT without external feed on blue shrimp (*Litopenaeus stylirostris*) culture. They observed that growth was about 4.4-fold better in BFT than in CW, plus survival improved. Moreover, the natural food consumed by the shrimps reared in BFT contributed to shrimp growth and increased the activity of digestive enzymes alpha-amylase and trypsin, which demonstrate this species' effective use of bioflocs (Cardona et al., 2015). In tiger shrimp (*Penaeus monodon*), Arnold et al. (2009) studied the effect of culture density (2,500 and 5,000 shrimp m^{-3}) and the addition, or not, of artificial substrates that could increase biofloc production. The results showed that, with the addition of an artificial substrate, BFT significantly enhanced animal growth performance regardless of stocking density, but especially impacted those animals cultivated at 5,000 shrimp m^{-3}. At this density, the final tiger shrimp weight doubled (from 0.2 to 0.4 g) with the artificial substrate, mainly due to higher nutritional source availability but also due to the improved water quality (lower concentrations of total ammonia nitrogen and nitrite) obtained with this treatment. Moreover, at this high density, the FCR significantly improved and obtained lower values (from 1.49 to 2.5) with added substrate compared to the treatment with no artificial substrate (Arnold et al., 2009). Similar results have been reported for the Pacific white shrimp cultured in CW and BFT with the addition of artificial substrates (de Morais et al., 2020; Olier et al., 2020). Olier et al. (2020) determined that other than the addition of a vertical substrate improving the growth performance and feed utilization of Pacific white shrimp, it can also save dietary protein without negatively affecting production and the chemical aspects of those shrimp reared in BFT.

In fish species, BFT has also drawn considerable attention, with initial studies carried out in Nile tilapia and are being extended to other species. Ekasari et al. (2015) observed that Nile tilapia larvae did not improve significantly in growth performance when they were cultivated in a BFT system, which indicates that nutritional requirements were met independently of the culture system, while additional biofloc consumption did not contribute significantly to fish growth. Although the BFT effect was not shown on animal growth, such growth seemed more uniform in the BFT group compared to the control, and larvae survival of BFT (90–98%) was significantly higher than in the control group (67–75%) (Ekasari et al., 2015). These authors challenged fish to pathogenic infection with either bacterium *Streptococcus agalactiae* or salinity stress tests. They observed that the BFT group presented significantly higher survival than the control group in both challenges.

In other fish species, such as rohu (*Labeo rohita*) and Jayanti rohu (genetically improved *Labeo rohita*), it has been demonstrated that BFT enhances growth performance and feed utilization compared to the control groups (Kamilya et al., 2017; Vadhel et al., 2020). The growth performance of Jayanti rohu reared in BFT increased by around 60% (specific growth rate (SGR) from 0.54 to 0.84, and weight gain rose from 38.5 to 65.9%), and FCR improved 40%, which went from 1.56 to 0.91, whereas survival also increased from 90% to 100% (Vadhel et al., 2020). For Jayanti rohu, values also significantly improved, and were better in the genetically improved rohu with increases of around 90% (SGR from 0.37 to 0.38 and weight gain from 25.29% to 50.93%) for growth parameters and 50% for FCR (from 2.37 to 1.17) when animals were maintained in BFT. In both cases, the authors indicated that the enhanced parameters and welfare determined by diverse parameters could be attributed to the better quality and maintenance of the water in the BFT system (Vadhel et al., 2020). In golden carp (*Carassius auratus*) juveniles, Yu et al. (2020) tested five C:N ratios in BFT systems and found that the bioflocs produced higher C:N ratios (20:1 and 25:1), significantly enhanced weight gain, SGR, and protein efficiency ratio (PER) than the control group because of better water quality and the increase in the parameters related to liver protease, lipase and amylase enzymes, and different immunological and antioxidant factors.

Notwithstanding, although almost all the studies conducted in crustacean and fish species have demonstrated the positive effect of BFT compared to CW in different situations (Arnold et al., 2009; de Morais et al., 2020; Khanjani et al., 2015, 2017; Khanjani and Sharifinia, 2020; Olier et al., 2020; Vadhel et al., 2020; Yu et al., 2020), others authors have found no effect or only negative effects (Esparza-Leal et al., 2015; Ray et al., 2017; Tierney and Ray, 2018). Esparza-Leal et al. (2015) found that Pacific white shrimp (0.009 initial body weight, IBW) grew better in a CW system than in a BFT system, independently of culture density (from 1,500 to 9,000 orgs m^{-3}). However, the same authors indicated that the obtained results could be due to a problem with the alkalinity of the water employed in the BFT system, which was lower than that recommended for the species and could negatively influence good biofloc development (Esparza-Leal et al., 2015). Similar results have also been observed in shrimps of 0.42 g of the same species stocked in all the tanks at 250 m^{-3} and grown in CW *vs.* BFT, with the best results obtained for growth and FCR in CW systems (Ray et al., 2017). In fish, Fleckenstein et al. (2018) determined the effect of CW, BFT, or a hybrid treatment on tilapia with an IBW 0.17 g, and observed that the hybrid treatment was the best system to obtain the best growth performance, although no differences appeared between BFT and CW, and the FCR values worsened when the BFT system was employed. These authors indicated that, as in other cases, inferior water quality conditions could be responsible for diminished performance in BFT (Fleckenstein et al., 2018). These studies revealed the importance of carrying out effective BFT system management to positively contribute to production. Controlling the high diversity of the parameters that condition not only biofloc formation but also water quality can improve or worsen performance based on such management.

Dietary Protein Sparing Effect of Biofloc Technology

Regarding the use of FM in the diet, biofloc studies in aquatic species have clearly demonstrated the possibility of reducing protein content in the diet with no detriment to production and/or improving results related to growth performance and feed utilization when alternative meals are used. Thus, Pacific white shrimp is the most widely studied species in all its development stages. In 10-day-old postlarvae (≈1 mg), Correia et al. (2014) did not find any differences in survival (82% *vs.* 84%) and protein efficiency ratios (3.89 *vs.* 3.28) when shrimp were fed two different dietary protein levels (30 and 40% CP, respectively) in a BFT system. These authors concluded that substituting high protein (40%) for low protein (30%) feed in the nursery phase of Pacific white shrimp in a BFT system is a possible sustainable alternative to shrimp production with cheaper (lower protein) feed and reduced environmental impact, besides obtaining improved water quality. In the juveniles of this shrimp species, with an IBW of 1.5 g, Panigrahi et al. (2019) demonstrated that protein content in feed can be reduced from 40% in a CW system to 24% in a BFT system without affecting animal production and welfare. Among BFT treatments (24%, 32%, and 40% of dietary protein level), 32% was the lowest protein level in feed to give the best results. Hence the improvement in the CW system with 40% protein ranged between 32.6 and 52.6% for productivity, 22% and 27.6% for average body weight (ABW), 8.7% and 19.6% for survival, 10% and 31% for the FCR, and 32% and 83% for the PER when Pacific white shrimp were reared in BFT with different protein levels (Panigrahi et al., 2019). Similar results for this species, but with heavier IBW (5.3–6.5 g), have been obtained in other studies, which indicated that protein levels can lower to 25% or 30% when bioflocs are used as a nutritional supplement, with similar growth performance, feed utilization, and survival to those shrimps fed diets containing protein levels that come close to those in commercial diets (Jatobá et al., 2014; Panigrahi et al., 2020; Xu and Pan, 2014b).

Likewise, optimum results have been reported for other shrimp species, such as tiger shrimp and Indian white shrimp (*Penaeus indicus*). Megahed et al. (2018) carried out a study with low dietary protein levels (20%, 22%, 24%) and a 35% commercial level in Indian white shrimp (IBW 0.52 g). They observed no significant differences in final weight and the SGR and FCR between the control and low protein diet groups. However, the shrimps that were fed a low protein diet (86.6%) compared to the control (66.6%) had better survival results. In a later study, Panigrahi et al. (2020) evaluated BFT systems with different dietary protein levels (2%, 30%, 35%) against a control group without biofloc in the same shrimp species (IBW 0.75 g). These authors reported a significant improvement in all the growth and survival indices of the groups reared in a biofloc system, with the best results observed for those shrimp feds 30% protein compared to the other biofloc groups and controls. With tiger shrimp, Kumar et al. (2017) conducted a study to evaluate the effect of two dietary protein levels (32% and 40%) and two different carbon sources (rice flour and molasses), but without carbohydrates (control). Their results indicated better animal performance in all the experimental groups compared to the control, with optimum growth values and immune responses with the rice flour addition. This revealed adding rice flour at the 32% protein level could replace 40% protein feed.

Studies performed with alternative meals to replace FM in diet with Pacific white shrimp have reported optimum results when applying BFT systems. Jatobá et al. (2017) tested different soya protein concentrate levels (0%, 33%, 66%, and 100% FM replacement) in shrimps (IBW 4 g) reared in a super-intensive biofloc system. They found that growth performance was not negatively affected up to 33% replacement. In a similar study, Tesser et al. (2019) observed that FM and fish oil can be substituted for up to 75% soya protein concentrate and soya bean oil without negatively affecting Pacific white shrimp development (IBW 2.93 g) when reared in a BFT system.

Other protein sources, such as spent brewer's yeast, have been tested to replace FM in the diet for giant freshwater shrimp (*Macrobrachium rosenbergii*) with an IBW of 6.7 g when reared in a BFT *vs.* a CW system (Nguyen et al., 2019). Thus, experimental diets with 35% protein were formulated with 0%, 20%, 40%, and 60% FM replacement for spent brewer's yeast. The results revealed that neither growth nor survival was affected when increasing brewer's yeast levels in any system. In general, FCR was better in BFT than in CW, and the BFT group fed 60% spent brewer's yeast presented the best growth results (Nguyen et al., 2019).

Additionally, BFT can be improved by introducing other components into the system. In Pacific white shrimp with an IBW of 0.23 g, Brito et al. (2018) studied the effect of lowering dietary protein from 40% to 30% in a biofloc monoculture or a biofloc integrated system by using seaweed *Gracilaria birdiae.* Their results revealed that the integrated system improved the growth performance of shrimps by increasing weight gain by 21% and 5% in the shrimps fed 30% and 40% dietary protein, respectively. Moreover, growth was similar when comparing treatments with 40% protein, independently of the BFT system with 32% in the integrated BFT system. This study indicates that dietary protein content can be lowered without affecting zootechnical parameters when seaweed is employed as a supplemental food in an integrated BFT system (Brito et al. 2018).

In crustacea, dietary protein levels can also be lowered in fish species cultivated with BFT. In tilapia, one of the most widely studied fish species, da Silva et al. (2018) evaluated the effect of different protein contents (between 17% and 33%) on the diet to find that 10 g and 60 g of IBW protein can be lowered to 28% and 22%, respectively, when rearing tilapia juveniles in BFT. Likewise, Hisano et al. (2020) observed that different dietary protein levels (36%, 32%, and 28% CP), which were all supplied in diets from plant sources, did not influence growth performance, FCR, or health status (IBW 6 g) of tilapia reared in BFT systems. Therefore, a reduction up to 28% of dietary protein in this fish species would be profitable from the production cost and environmental impact points of view.

In a later study, Klanian et al. (2020) investigated the influence of two different protein levels, which were also lowered during the experiment. The Nile tilapia fingerlings (IBW 2 g), cultivated at two stock densities (40 and 80 fish m-3) using BFT and CW technologies, were fed commercial pellets containing 45% (high protein group) or 35% (low protein group) protein for the initial 7 weeks and 35% (high protein group) or 25% (low protein group) protein for the final 9 weeks. Their results showed that BFT microbial proteins compensated the dietary protein restriction without harming survival, growth, feed utilization, or fish health status. Moreover, the BFT results were better when compared to CW systems when low protein was used in the diet, independently of stock density (Klanian et al., 2020).

In the fingerlings of Genetically Improved Farmed Tilapia (GIFT) (IBW 0.99 g), Sgnaulin et al. (2020) studied the combination of different dietary digestible protein levels (22%, 26%, 30%) and digestible energy (3,000, 3,150, 3,300 kcal/kg^{-1}) levels on the growth performance of this fish reared in BFT systems. These authors found that neither FCR nor survival was affected by several treatments, and weight gain was no different between 26% and 30%, but was lower at 22% compared to 30% digestible protein. These results suggest that the optimum combination of digestible protein and energy levels for GIFT fingerlings reared in biofloc systems was 26% and 3,000 kcal/kg^{-1} (Sgnaulin et al., 2020). Another study with hybrid tilapia (*Oreochromis aureus x O. niloticus*) has suggested that diets formulated by supplementing the first four limiting amino acids (Lys, Met, Thr, Ile) could include low digestible protein levels from 32.3% to 27.7% without adversely affecting productivity (Green et al., 2019).

Similar results to those observed for tilapia have also been obtained for several carp species. In gibel carp (*Carassius auratus gibelio*; IBW 250.8 g), three protein levels in the diet (25.7%, 30.1%,

35.3%) were tested in a BFT system against a control without bioflocs (Li et al., 2018). The results demonstrated that a low protein diet was recommended for the gibel carp culture in BFT systems compared to a high protein diet, as growth was similar for both 25.7% and 30.1% BFT compared to a 35.3% CW system, and the FCR was better in the 30.1% diet than in the 35.3% one in BFT (Li et al., 2018). In common carp (*Cyprinus carpio*) juveniles (IBW 30.5 g), Ebrahimi et al. (2020) evaluated the effect of two dietary protein levels (20% and 30%) with different carbon sources for biofloc culture (sugarcane molasses, rice bran, and their combination, plus a control) on several parameters, including fish performance. Their results indicated that 30% protein in diet and the use of rice bran increased weight gain, the SGR, and survival, and improved the FCR, compared to the control and the group with only molasses. Once again, it demonstrated that bioflocs spare dietary protein and are closely related to BFT system management (Ebrahimi et al., 2020).

By contemplating the use of alternative meals to substitute FM in the diet similarly to Pacific white shrimp, Nhi et al. (2018) tested in Nile tilapia the use of spent brewer's yeast at different substitution levels (0%, 30%, 600%, 100%). Compared to the CW system, tilapias (IBW 29 g) reared in BFT obtained higher growth performance indices and better FCR, PER, and survival. The use of brewer's yeast did not affect tilapia growth, feed utilization, or survival at any of the tested substitution levels. Thus, these authors concluded that brewer's yeast is a potential substitute for FM in tilapia diet, especially when fish are reared by BFT. Another recent alternative to FM is to use insect meals in diet. Tubin et al. (2020) utilized mealworm meal (*Tenebrio molitor*) at the 0% (control), 5%, 10%, 15%, and 20% inclusion levels to feed Nile tilapia (IBW 2.1 g) reared in a biofloc system. The overall results indicated that it is possible to include this meal up to 10% in diet without negatively affecting production.

Bioflocs as a Feedstuff Ingredient in Dietary Formulations

It should be taken into account that, in order to apply BFT, aquatic species must possess several abilities and characteristics as morphologically specialized structures that allow them to feed with bioflocs (Walker et al., 2020). As numerous available studies have shown, Pacific white shrimp and tilapia are the most successfully cultivated species in BFT systems, but its application to other aquatic species, including Indian white shrimp, blue shrimp, tiger shrimp, giant freshwater shrimp, rohu, and several carp species has been demonstrated (Arnold et al., 2009; Cardona et al., 2015; Ebrahimi et al., 2020; Kamilya et al., 2017; Kumar et al., 2017; Megahed et al., 2018; Nguyen et al., 2019; Panigrahi et al., 2020; Vadhel et al., 2020). However, many other species, mainly fish species, are unable to use this technology. In such cases and/or in order to improve biofloc use efficiency, an alternative application is their direct inclusion in diet as a feed ingredient. For this purpose, bioflocs can be produced in reactors specially designed for this purpose, or collected directly by the BFT system when produced in excess (Walker et al., 2020). After their processing, bioflocs can be used as an alternative protein source to FM replacement (Walker et al., 2020).

A study by Kuhn et al. (2010) used bioflocs obtained in not only sequencing batch reactors with sucrose supplementation, but also in a membrane biological reactor without carbon supplementation. After treatment, the dried bioflocs were used to feed Pacific white shrimp postlarvae (< 1 mg). The experiment consisted in a control diet (without bioflocs) and diets were supplemented with 10%, 15%, 21%, and 30% bioflocs from both production systems by replacing FM and/or soya bean protein. The results showed that despite survival (92.9 to 100%) or harvest biomass (536 – 574 g m^{-2}) not being affected, growth performance was better in the diet with biofloc inclusion and confirmed that the external bioflocs included in diet could replace FM and soya bean protein. In other studies by Valle et al. (2015) and Dantas et al. (2016), performed with the same shrimp species but with a postlarvae 2 mg IBW, different FM replacement levels, along with a combination of biofloc meal and fish protein hydrolysate at the 1:1 ratio, or with only biofloc meal, were respectively evaluated. After 42 days, the studies showed that FM replacement with biofloc meal was possible and even improved optimum growth performance and feed utilization compared to the control diet. Valle et al. (2015) established the optimal FM replacement level with fish protein hydrolysate and biofloc to be between 15.2% and 16.5%, but also indicated that diets should be supplemented with methionine when biofloc meal was used. Dantas et al. (2016) observed the best weight gain, SGR, and PER results when employing 30% biofloc meal with no differences in survival, which exceeded 91% in all treatments. Other authors have observed similar positive results for the same shrimp species (Ju et al., 2008; Shao et al., 2017; Khanjani et al., 2020). Therefore, according to the results of all these studies, biofloc meal can clearly be used as an alternative protein source to replace FM in the Pacific white shrimp diets.

In fish species, Prabhu et al. (2018) conducted a trial in which biofloc meal was supplemented at the 0%, 20%, 30%, and 40% inclusion levels in GIFT tilapia diet (2 g IBW). Their results showed that with up to 20% biofloc meal inclusion, both growth performance and feed utilization improved *vs.* the control diet. In tilapia with 12.1 g IBW, biofloc meal was evaluated to replace soya bean meal at three replacement levels (0%, 25%, and 50%) to obtain up to 25% soya bean meal substitution that had no negative effects on tilapia growth performance (Mabroke et al., 2019). With African catfish (*Clarias gariepinus*), Ekasari et al. (2019) studied the effect of biofloc meal as a feed ingredient on juveniles (1.4 g IBW). They included 0%, 5%, 10%, and 20% of biofloc meal by replacing soya bean meal, and concluded that the dietary 10% and 20% biofloc meal inclusion levels induced better results for feed intake, PER, FCR, and weight gain, besides improving animals' health status. Finally, in other aquatic species like sea cucumber (*Apostichopus japonicus*), biofloc meal has been used to replace macroalgae *Sargasswn thunbergii*. The results showed that the optimum dietary replacement level was between 27.74% and 30.75%, which improved the SGR, FCR, and PER (Chen et al., 2018).

CONCLUSIONS

In conclusion, BFT and the use of biofloc meal seem an eco-friendly and cost-effective alternative for developing sustainable aquaculture given the possibility of reducing dietary protein content and decreasing the dependence on mainly FM but also other alternative plant-protein sources that are beginning to generate environmental problems, such as soya bean meal. They all accomplish optimum growth performance and feed utilization of species besides offering the beneficial effects that bioflocs have on health status. Notwithstanding, in order to successfully apply this technology in aquaculture, it is essential to pay special attention to the management and composition of biofloc production because, if they are not properly performed and do not adapt to each species' requirements, the opposite effect to that sought could come into play. Today, continuing to study and improve the use of bioflocs in aquaculture still remains a challenge.

ALGAE

INTRODUCTION

Algae constitute a group of organisms that is present in fresh or seawater which, despite having different phylogenetic origins, share autotrophic and photosynthetic characters. In algae, we find multicellular groups that have developed tissues and acquired a macroscopic character (macroalgae) but which, unlike plants, have simple reproductive structures. Although they lack roots, stems, leaves and vascular tissues, these algae have a structure known as a thallus. Some algae have a microscopic character (unicellular or filamentous microalgae) with a wide variety, of which some 30,000 species have been studied. However, only some 100 genera have been cultivated on the laboratory scale, and studies have focused on about 20 for their potential beneficial use, while fewer than 10 have been produced in an industrial context (Mobin and Alam, 2017; Gaignard et al., 2019).

In this section of the chapter, an updated description is provided of how both microalgae and macroalgae increasingly play an important role as an alternative to animal protein sources in aquaculture feed.

Microalgae

Microalgae are generally classified as four groups: Rhodophyta (red algae), Chlorophyta (green algae), and Chromophyta (all other algae) and are eukaryotes, and prokaryotes Cynaobacteria (blue-green algae). In the last few years, the industrial use of microalgae has increased in the areas of food production (Figure 8.3), pharmaceuticals, and biofuels. This is because algal biomass is a rich source of nutrients, such as proteins, n-3 FA, and carbohydrates, as well as vitamins, minerals, and other bioactive compounds, like antioxidants. The fact that many species can be used as protein sources has not only promoted their use for human nutrition but also for animal nutrition, including aquaculture feeds (Becker et al., 2007; Mobin and Alam, 2017; Raja et al., 2018).

The importance of employing microalgae arises from the high growth rate and their ability to accumulate nutrients of interest under not very complex culture conditions. A suitable nitrogen, phosphorus, and carbon dioxide concentration in medium, as well as pH, temperature, and light conditions, are specific to cultivated genera or species. These aspects are key for achieving adequate yields and the suitable use of the molecules of interest provided in microalgae cultures (Khan et al., 2018; Suparmaniam et al., 2019; De Morais et al., 2020).

In order to make algal biomass production economically viable, high-value by-product extraction is paramount. Obtaining microalgae extracts for commercial purposes requires separating organisms from the culture medium by means of filtration, centrifugation, sedimentation, flocculation, etc. Later, treatment will differ depending on the ultimate purpose. For instance, microalgae can be processed into algae paste (microalgae cells dispersed in liquid media) or may go through drying processes to obtain powders or freeze-dried cubes. This latter format is most useful for aquaculture feed purposes (Raja et al., 2018). In order to extract cellular components of interest, a physical, chemical, enzymatic, or mechanical treatment is required to rupture the cell wall. Finally, a selective extraction of the molecules of interest is carried out, regardless of whether they are proteins, lipids, carbohydrates, antioxidants, etc. (De Morais et al., 2020).

Microalgae and Aquaculture

Traditionally, microalgae use has been associated with one of the initial steps in aquaculture: zooplankton culture for later hatchery and nursery feeds. Their microscopic size renders them ideal for feeding rotifers, copepods, and brine shrimp (*Artemia* sp.) which, in turn, is employed to feed juvenile finfish and shellfish, including crustacea. Cultivated microalgae are also employed as direct

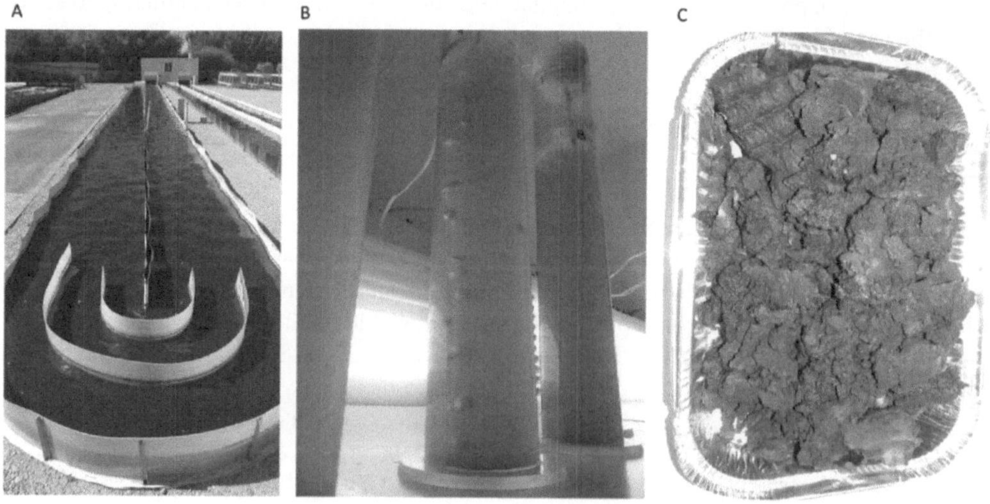

FIGURE 8.3 A and B, different technology systems for microalgae culture. C, Biomass of microalgae (Courtesy of F.G. Acien University of Almeria, Spain).

feeds for marine bivalve molluscs (oysters, clams, scallops), and the larvae of some marine gastropods (abalone and conch), sea urchins, shrimps, and some marine fish (Raja et al., 2018; Dineshbabu et al., 2019).

In recent years, the incorporation of microalgae into the market as formulated feed ingredients has gained relevance. Thus, certain types of microalgae are very promising for the industry because they present components with a high added value. *Chlorella* spp., *Pavlova lutheri*, *Haemotococcus pluvialis*, and *Phaeodactylum tricornutum* are particularly useful as algal meal. Others are reported as a source of n-3 FA (*Nitzschia* spp., *Phaeodactylum tricornutum*, and *Schizochytrium limacinum*) or pigments for fish, such as salmonids or crustacea (*Dunaliella* spp., *Haemotococcus pluvialis*) (Yarnold et al., 2019). Hence the variety of commercial microalgae preparations are intended to feed molluscs, crustacea, zooplankton, etc., covering nutritional requirements throughout their life cycle (Dineshbabu et al., 2019). It is important to bear in mind that the main biochemical components in algal cells not only varies according to species, but also to culture conditions, growth phase, and physiological status and that culture condition is critical to obtaining certain nutrients (Tibbetts et al., 2015; Madeira et al., 2017).

The use of microalgae as a source of high-quality proteins for single-cell proteins (SCPs) emerged some years ago and has been revealed as an alternative to classic protein sources of animal origin and even as an equivalent to convectional plants, while producing feed for aquaculture (Becker, 2007). In addition, greenhouse gas emissions from aquaculture have increased in recent years, while a drop in FM and fish oil production would be relevant to improve sustainable production (Hasan and Soto, 2017).

Regarding protein levels, not all microalgae are suitable for this purpose as the proximal analysis of different genera reveals a wide variability in their protein percentages (Table 8.3). The protein microalgae biomass can generally reach values between 40% and 70%. This is interesting because they are not only higher in protein than many plant sources, such as soya bean (38%), rice (approx. 10%), pea (2.8%), or even animal sources like milk (4%) or eggs (13%), but because they also present a suitable amino acid profile due to the relevance of EAA that are absent in many traditional plants (Torres-Tiji et al., 2020) (Table 8.3).

Many microalgae genera possess such high protein levels that they can sometimes represent up to half of the biomass weight (Table 8.3). Among the most remarkable ones in chlorophytes, we find *Dunaliella*, *Chlamydomonas*, *Nannocloropsis*, and *Chlorella*, and *Arthrospira phylum* in cyanobacteria (blue-green algae), commercially known as *Spirulina*. Cyanobacteria can reach 40–70% protein due to phycobiliprotein pigments. Some chorophytes show values above 50% (Niccolai et al., 2019), such as *Dunaliella*, with protein values over 80% in dry weight compared to other green microalgae like *Chlamydomonas* (48%) or the blue-green alga *Arthrospira* (46–70%) (Dineshbabu et al., 2019; De Morais et al., 2020).

However, despite the fact that some genera may present very high protein values, this is no guarantee for their suitability as a protein source because assessing its quality is essential. *Arthospira* (*Spirulina*) is one of the filamentous cyanobacteria whose culture accounts for almost one third of the world microalgae production. It has been recognized by the World Health Organization (WHO) for its high protein content and EAA profile, which make this genus, together with *Chlorella* (Table 8.3), suitable for animal feed (Mitra and Mishra, 2019; Niccolai et al., 2019; Lafarga et al., 2020). Finally, it is worth mentioning the role of certain red microalgae, such as *Porphyridium*, which, in addition to high-value compounds like pigments or FA, have recently gained relevance for their high-quality protein content and suitable EAA profile compared to conventional vegetable sources (Gaignard et al., 2019).

Another interesting aspect when considering microalgae as a protein source is the possible overestimation of protein content. In most cases, protein determination is based on nitrogen concentration, which can come from other constituents, like nucleic acids, amines, glucosamides, and nitrogen-containing cell wall materials. An overestimated protein content between 6% and 12% has been reported for *Scenedesmus*, *Arthrospira*, and *Dunaliella* (Becker, 2007).

TABLE 8.3
Protein Content and Amino Acids Profile of Soybean and Different Microalgae with Applications in Aquaculture

		Chlorophyta						*Cyanobacteria*			*Rhodophyta*	*Others*	
Amino acids (mg/100 mg protein)	*Soybean*	*Chlorella vulgaris*	*Chlorella protothecoides*	*Dunaliella bardawil*	*Scenedesmus obliquus*	*Chlamydomonas reinhardtii*	*Haematoccocus* sp.	*Arthrospira maxima*	*Arthrospira platensis*	*Aphanizonemon* sp.	*Porphyridium aerugineum*	*Nannochloropsis granulata*	*Euglena gracilis*
Alanine	5.0	7.6	6.2	7.3	9.0	8.8	11.8	6.8	9.5	4.7	8.4	7.1	15.8
Arginine	7.4	6.0	13.4	7.3	7.1	7.2	nd	6.5	7.3	3.8	8.6	7.4	3.4
Aspartic acid	1.3	8.4	7.1	10.4	8.4	9.7	11.8	11.6	11.8	4.7	15.0	11.4	7.1
Cysteine	1.9	1.4	1.6	1.2	0.6	nd	nd	0.4	0.9	0.2	2.2	1.6	0.2
Glutamic acid	19.0	10.7	10.3	12.7	10.7	11.3	13.7	12.6	10.3	7.8	15.6	14.1	9.5
Glycine	4.5	5.5	5.5	5.5	7.1	5.7	6.3	4.8	5.7	2.9	7.0	7.5	7.0
Histidine	2.6	1.9	3.0	1.8	2.1	2.3	5.8	1.8	2.2	0.9	1.9	2.3	2.2
Isoleucine	5.3	3.8	3.7	4.2	3.6	4.4	4.9	6.0	6.7	2.9	7.1	5.6	0.2
Leucine	7.7	8.3	5.6	11.0	7.3	9.8	6.4	8.0	9.8	5.2	11.9	11.0	3.7
Lysine	6.4	6.8	4.9	7.0	5.6	6.6	4.1	4.6	4.8	3.5	8.0	8.2	4.9
Methionine	1.3	2.2	2.1	2.3	1.5	2.7	nd	1.4	2.5	0.7	3.7	3.5	0.0
Phenylalanine	5.0	4.9	5.5	5.8	4.8	5.6	4.9	4.9	5.3	2.5	6.3	6.2	0.9
Proline	5.3	4.5	5.6	3.3	3.9	5.6	nd	3.9	4.2	2.9	5.0	11.2	0.0
Serine	5.8	3.7	5.1	5.0	3.8	4.3	5.9	4.2	5.1	2.9	7.0	5.6	10.6
Threonine	4.6	4.4	4.9	5.4	5.1	5.1	5.6	4.6	6.2	3.3	5.8	5.4	4.5
Tryptophan	1.4	1.5	0.5	0.7	0.3	2.8	nd	1.4	0.3	0.7	3.3	2.8	1.7
Tyrosine	3.7	3.4	4.7	3.7	3.2	4.3	2.7	3.9	5.3	nd	5.8	4.2	0.7
Valine	5.3	5.4	5.2	5.8	6.0	6.5	7.5	6.5	7.1	3.2	7.3	7.1	8.0

Bold letters: essential amino acids. nd: not determined. Reported by Becker, 2007; Szabo et al., 2013; Kim et al., 2015; Tibbetts et al., 2015; Asiedu et al., 2018; Dineshbabu et al., 2019; de Morais et al., 2020; Niccolai et al., 2019; Torres-Tiji et al., 2020

Digestibility is one of the most important and determinative aspects when estimating protein availability in microalgae as aquafeed. The presence of a robust cell wall, which represents 10% of dry algal matter, is often a problem that hinders the action of digestive enzymes to access cell content, especially for carnivorous fish with a short digestion phase (e.g. salmonids). The analysis of different microalgae has revealed a range of digestibility values among genera, with higher values in cyanobacteria (*Arthrospira*, *Nostoc*) compared to chlorophycea (Table 8.4). This aspect seems to correlate with the chemical composition and structural organization of the cell wall (Niccolai et al., 2019). Something similar occurs in red microalgae, *Porphyridium* and *Rhodella*, which highlights their content in very digestible proteins given the presence of a polysaccharide mucilage instead of the classic cell wall (Gaignard et al., 2019).

Differences observed in the protein digestibility of microalgae become evident when running an analysis of nutritional indices associated with protein assimilation. Thus indices, like biological value (BV, nitrogen retained for growth or maintenance according to absorbed nitrogen), net protein utilization (NPU, nitrogen retained according to nitrogen intake), digestibility coefficient (DC, measuring both protein digestibility and the biological value of absorbed amino acids), and the protein efficiency ratio (PER, weight gain according to intake), confirm that some genera have suitable values, namely *Chlorella*, *Spirulina*, and *Scenedesmus* compared to, for example, egg proteins (Becker, 2007) (Table 8.4).

There is a variety of methods for cellular wall treatment to increase microalgae digestibility: enzymatic (cellulases), chemical (organic solvents or acids), and physical/mechanical (bead milling, high-pressure homogenization, or micro-fluidization) (Jones et al., 2020). In general terms, physical and mechanical methods are the most widely used because chemical and enzymatic methods can affect intracellular nutrients. Enzymatic treatment seems effective in *Chlorella* and *Nannochloropsis* by improving protein digestibility in fish species, like Atlantic salmon (*Salmo salar*) and juvenile Nile tilapia (*Oreochromis niloticus*), respectively (Tibbetts et al., 2017; Teuling et al., 2019), which may also result in additive or synergistic effects on nutrient utilization (Madeira et al., 2017). Technically speaking, the aquafeeds within the above compositional ranges can be formulated entirely from microalgae, but other factors to be considered, including feed attractiveness (e.g., smell, taste) and accessibility (e.g., cell/pellet size, buoyancy).

Different studies have reported employing microalgae as a protein source in feed for aquaculture (Macias-Sancho et al., 2014; Teuling et al., 2019; Ansari et al., 2020, Table 8.5). In any case, it is still difficult to reach 100% FM replacement or other protein sources of animal origin. Indeed, a recent comparative analysis performed in Nile tilapia (*O. niloticus*) revealed maximum FM substitution levels for *Nannochloropsis*, *Chlorella*, *Scenedesmus*, and *Spirulina* with 35%,

TABLE 8.4
Digestibility of Microalgae Biomass of Interest as a Food Source

Microalgae	Dry Matter Digestibility (%)	Crude Protein Digestibility (%)
Chlorella vulgaris	> 60	76
Tetraselmis sp.	> 60	62–70
Arthrospira sp.	78	81
Nostoc sphaeroides	> 60	82
Porphyridium purpureum	47	62–70
Nannochloropsis oceanica	> 60	50
Tisochrysis lutea	> 60	62–70

Adapted from data reported by Niccolai et al. (2019).

TABLE 8.5
Studies Focused on Fishmeal Replacement for Algal Biomass in Aquaculture

Microalgae	Aquaculture Species	Challenge	Reference
Chlorella vulgaris	**Atlantic salmon** (*Salmo salar*)	Digestibility assays	Tibetts et al. (2017)
Nannochloropsis gaditana	**Nile tilapia** (*Oreochromis niloticus*)	Digestibility assays	Teuling et al. (2019)
Scenedesmus obliquus	**Nile tilapia** (*Oreochromis niloticus*)	Fishmeal replacement	Ansari et al. (2020)
Haematococcus pluvialis	**Yellow perch** (*Perca flavescens*) **Pacific white shrimp** (*Litopenaeus vannamei*)	Fishmeal replacement	Jiang et al. (2019, 2019) Ju et al. (2012)
Dunaliella tertiolecta	**Sea Urchin** (*Strongylocentrotus intermedius/ S. nudus*)	Dietary protein source	Qi et al. (2018)
Arthrospira (*Spirulina platensis*)	**Pacific white shrimp** (*Litopenaeus vannamei*)	Fishmeal substitution	Macias-Sancho et al. (2014)

50%, 50%, and 75%, respectively, which are similar to those observed for plants (Yarnold et al., 2019). Another study in the same species has shown that, although the maximum substitution values for *Scenedesmus obliquus* were around 7.5%, a significant improvement in growth and a higher PER were promoted. This study also revealed the profitability and sustainability of an integrated algae aquaculture system because the use of defatted algae after lipid extraction provided valuable source algae for FM replacement, while lipids were utilized for biodiesel synthesis purposes (Ansari et al., 2020).

The effect of partial FM protein replacement with defatted microalgae (*Haematococcus pluvialis*) meal has been evaluated as a protein ingredient in the diet of Pacific white shrimp (*Litopenaeus vannamei*), with a significantly higher growth rate and a lower feed conversion ratio at 12.5% replaced protein, which reveals a valuable alternative protein and pigmentation ingredient in shrimp feed (Ju et al., 2012).

At present, the challenge of using microalgae as a protein source in animal feed production is opening up a research area in the biotechnology field. However, it is true that today substitution levels are far from reaching 100% substitution because improvements in protein quality, suitable amino acid profile, digestibility indices, palatability, etc., are all still required.

A recent review by Hua et al. (2019), in which alternative protein sources to FM were evaluated according to the biological capacity of microalgae, reported positive research findings for its replacement efficacy in different aquaculture species, which suggests the high potential for employing microalgae as a protein source. However, this potential is affected by the technical, biological, and economic difficulties in the continuous production of high-quality microalgae biomass and in its downstream processing and subsequent scaling.

Major advances are required, including the establishment of extending algal collections for breeding purposes, genome sequencing, and optimal species identification (Yarnold et al., 2019). Indeed, the selection of breeds with a phenotype based on high protein content and an EAA profile is one of the proposals linked with improving their organoleptic qualities. Apart from all this, improving culture technology to reduce production costs and increase the use of microalgae remains a major challenge, for the near future, in the effort to support a sustainable 'circular' aquaculture industry (Torres-Tiji et al., 2020).

Macroalgae

Marine algae, referred to by the generic term "seaweeds", and colloquially known as macroalgae, are one of the most prominent primary marine photosynthetic producers (Hong et al., 2014). They are divided into three major groups according to their photosynthetic pigments: Chlorophyta (green algae), Phaeophyta (brown algae) and Rhodophyta (red algae). In recent years, they have been considered an alternative resource for sustainable biomass to produce biofuels, biochemicals, and food. Similar to microalgae, seaweeds are an outstanding source of proteins, polysaccharides, minerals, vitamins, and a series of biologically active substances. The culture and harvesting of algae are straightforward, and a significant amount of biomass can be obtained compared to microalgae (Kraan, 2013).

Within green algae, one of the most notable genera is *Ulva* spp., commonly known as sea lettuce. The brown algae group includes genera like *Ascophyllum nodosum*, *Laminaria ochroleuca* (kombu), *Undaria pinnatifida* (Japanese wakame), and *Himanthalia elongata*. Finally, red algae comprise genera such as *Eucheuma*, *Palmaria palmata*, *Gracilaria* spp., *Chondrus crispus*, *Porphyra* spp. (Japanese nori), and *Pyropia* spp., of which some are of interest for human consumption (Martínez-Hernández et al., 2018; Sudhakar et al., 2018; Hua et al., 2019; Yarnold et al., 2019).

In general terms, and compared to terrestrial biomass, macroalgae stand out for their high carbohydrate content (25–60%), with protein values between 3–47% and low lipid contents (1–3%; dry weight) (Sudhakar et al., 2018). The wide range of protein levels depends on the seaweed type, with higher values reported for red seaweeds (10–47%), moderate ones for green algae (9–26 %) and the lowest values for brown seaweeds (3–19.5%), except *U. pinnatifida*. An analysis of the dried extract of different algae by a more exhaustive methodology (total amino acid residues, TAA) to minimize the overestimation associated with non-protein nitrogen, gave very high protein values in *U. pinnatifida* (51.6%) compared to *Ulva*, *Palmaria*, or *Chondrus* (18–37%), and even values of microalgae with a marked protein character, such as *Chlorella* or *Spirullina* (32–41%) (Martínez-Hernández et al., 2018).

Culture conditions are a key factor for determining protein levels. Under growth conditions, overestimations can arise in the assayed ranges because seaweeds have lower protein contents to ranges between 10–30% under non-limiting nutrient conditions (Vieira et al., 2018; Hua et al., 2019).

Regarding total amino acid content in seaweeds, generally the percentage is around 5.5%, which is lower than for other protein sources of plant origin, such as soya bean meal (22.3%), or animal sources, like fish (31.2%) (Angell et al., 2016). The analysis of the amino acid profile in some brown and red algae species generally revealed a higher EAA concentration than soya bean protein or even FM (Martínez-Hernández et al., 2018; Hua et al., 2019). Tryptophan, methionine, and leucine were the limiting EAAs in different seaweed species and, on the contrary, high lysine concentrations have been reported in red and green seaweeds. Accordingly, green and red algae seem to be interesting and suitable sources of protein for animal nutrition (Vieira et al., 2018).

The protein nutritional value is not only determined in terms of amino acid profiles, but also by protein digestibility. The digestibility of algae proteins often appears limited by present anti-nutritional compounds, such as polysaccharides or trypsin inhibitors, whose effect can be attenuated by a series of physicochemical enzymatic treatments (Fleurence et al., 2018).

Macroalgae and Aquaculture

Some aquaculture examples appear in which macroalgae have been traditionally used for feeding, and their use in mollusks such as abalone (*Haliotis* spp.) is noteworthy. With this species and a traditional diet of fresh algae, diets supplemented with macroalgae as a protein source have been assayed. A recent study on the greenlip abalone (*Haliotis laevigata*) has shown that replacing concentrates from green algae *Ulva* spp. in diet with 10% red algae *Gracilaria cliftonii* improves animal growth performance and survival (Duong et al., 2020).

Regarding fish feed, studies of the European sea bass (*Dicentrarchus labrax*) demonstrated that *Gracilaria* and *Ulva* can be incorporated into the diet as a protein source by up to 10% without affecting growth performance (Valente et al., 2006). In tilapia, reports reveal the positive effects of red algae *Gracilaria* on growth performance, provided the FM substitution level does not exceed 20% (Younis et al., 2018).

The profitability of using algae in aquaculture does not seem relevant because, despite their high protein content, it is not readily available, which entails raising costs to obtain necessary amounts. In general, given the marked presence of polysaccharides, the substitution percentage is limited to a 10% protein source in aquafeed formulations. Even so, their employment can be useful in herbivorous, or even omnivorous, species. Their use with carnivorous species as a functional additive is more highly recommended (Hua et al., 2019).

CONCLUSIONS

In summary, employing algae as an alternative protein source for aquafeed has revealed the more prominent role of microalgae compared to macroalgae because the former generally have a higher protein content and better tolerance to replacement percentages in diet, as reported for some species like tilapia (Yarnold et al., 2019). Microalgae can provide the most suitable bulk feed in fish diets, while macroalgae might be more suitable for terrestrial livestock and lower trophic-level aquaculture species (Shields and Lupatsch, 2012). Replacing FM with macroalgae in aquaculture feeding requires biomass processing to obtain more concentrated protein than can compete with other land and freshwater crops used as protein sources (Kim et al., 2013).

VEGETABLE PROTEIN SOURCES

Introduction

Fish oil and FM have traditionally been considered the most important standard dietary protein and lipid sources for many fish species owing to their high protein content, balanced EAA profile, and considerable quantities of lipid and minerals. Moreover, they offer excellent nutrient digestibility and are low in anti-nutritional factors (ANFs) (Daniel, 2018). These characteristics explain the predominance of FM over other raw materials in aquaculture feed. The production of both FM and fish oil is based on marine fish waste that does not reach commercial sizes or has no market value (Naylor et al., 2009). However, these products are available only in limited quantities; they are usually expensive, and are considered unsustainable from the environmental and ecological points of view. For these reasons, societal and economic pressures are placed on aquaculture production to search for alternative protein sources to FM (Figure 8.4). In omnivorous fish species, the elimination of FM from the diet has been more easily achieved than in carnivorous fish and crustacea, where it is more difficult to implement (Turchini et al., 2019). In general, the main and more widely studied alternative to FM in aquafeed is vegetable protein sources. It has been shown that up to 50% FM protein can be replaced with vegetable proteins in carnivorous fish diets without negatively affecting fish growth or welfare (Hardy, 2010).

Raw materials of vegetable origin abound. The use of these vegetable meals offers a series of advantages and disadvantages. They are less expensive than those of animal origin and, therefore, feed manufacturers substitute them because they are cost-effective. They are normally endowed with a certain binding power, associated with the presence of digestible substances. They are also a source of group B vitamins. However, the growth performance obtained with vegetable meal is inferior to that of the fish fed FM-based diets as they do not meet fish nutritional requirements (Gajardo et al., 2017). Their n-3 HUFA content is zero, and they are less palatable and poorly digestible because they contain significant amounts of ANFs, such as phytate, saponins, lectins, pectins,

FIGURE 8.4 Examples of vegetables used as a protein source in aquafeed: potatoes, coffee, soybean, palm kernel, canola, green pea, lupin, Lemna, corn, faba bean, and peanut (Courtesy of E. Rufino, University of Granada, Spain).

hemicellulose, pentosans, cellulose, lignin, among others (Lall and Anderson, 2005). Furthermore, starch is the main energy source in vegetable meal, which is not always well tolerated by fish.

Many studies have explored the possibility of vegetable proteins being used to replace FM (Caruso, 2015). Each vegetable protein source has its own characteristics and FM substitution efficiency depends on each species' nutritional needs. When choosing one or another, it is necessary to bear in mind that food must be available all year round at a low cost for growers; handling, transport, and processing requirements prior to feeding must be minimal; nutritional value has to be high in protein and carbohydrates and low in fiber; and they must be well accepted by the species that they intend to feed (Kaushik et al., 1995).

The present work reviews information about the innovative vegetable protein sources that allow FM to be replaced without affecting farmed fish growth and quality.

LEGUMES

The most frequently used vegetable protein source to replace FM is legumes: soya bean, bean, green peas, and lupins (Pereira and Oliva-Teles, 2002; Thiessen et al., 2003). Legumes have low fat content and are an excellent source of protein, dietary fiber, and a variety of micronutrients, as well as other compounds like phytochemicals (Messina, 1999).

Soya Bean

Among legumes, the importance of soya bean meal (SBM) is worth highlighting, which is the most important protein source as feed for fish feeding and as partial or entire FM replacement. SBM is the by-product left after removing oil from soya beans (glycine max) (Storebakken et al., 2000). It provides a high protein content, a good amino acid profile, and good digestibility. All this, together with adequately processing its seeds to eliminate ANFs, guarantees a sufficient availability of amino acids to achieve acceptable development (Oliveira et al., 1999). Moreover, the cost of SBM is lower compared to FM. Hence many studies have reported that partially substituting FM for SBM affects the performance of many fish species, such as trout (*Oncorhynchus mykiss*) (Harlioğlu and Yilmaz, 2011), common carp (*Cyprinus carpio*) (Uran et al., 2008), Japanese sea bass *Lateolabrax japonicus* (Zhang et al., 2014), juvenile tench *Tinca tinca L.* (García et al., 2015), gilthead sea bream *Sparus aurata L.* (Kokou et al., 2012), Nile tilapia *Oreochromis niloticus* (Abdel-Warith et al., 2013), Atlantic salmon (*Salmo salar*) (Furuya et al., 2004), shrimp (Molina-Poveda et al., 2013). All this reveals that adequate SBM inclusion levels in diet differs depending on the studied species. It is known that SBM inclusion over 50% decreases fish growth. Therefore, acceptable substitutions lie between 40% and 50%, depending on the species. The use of SBM involves several limitations due to ANFs and low methionine levels, which could lead to intestinal problems in some carnivorous fish (Gatlin et al., 2007). Furthermore, SBM can have an environmental impact, including deforestation, water pollution, and pesticide use, among others (Sánchez-Muros et al., 2014).

One source that derives from SBM is soya protein concentrate (SPC). It is produced by SBM fractionation, which is a highly refined ingredient as most of the ANFs present in SBM are removed during processing (González-Rodríguez et al., 2015). Mambrini et al. (1999) reported that the 50% supplementation of dietary protein using SPC as an FM replacement showed good amino acid digestibility in rainbow trout. According to Paripatananont et al. (2001), 8% SPC inclusion could replace 22% FM in *Penaeus monodon* diet. In carp, SPC can be incorporated into diet up to 40% without affecting survival and growth (Escaffre et al., 1997). According to Swick et al. (1995) 40–50% FM can be replaced with 25–30% SBM in shrimp without compromising animal growth. Furthermore, González-Rodríguez et al. (2015) studied the effects of substituting FM for SPC in tench juveniles. The fish fed between 0% and 45% replacement diets obtained significantly lower feed conversion ratios and higher protein productive values than those fed diets with higher replacement levels.

Lupin

Other legumes represent alternative protein sources to be used for aquaculture feed. Lupin (*Lupinus albus*) is a legume that belongs to the same family of plants as peas and soya beans. Lupin seeds have been described for aquaculture diet for their high protein content (30–40 g/100 g), availability, and low cost. Lupin seeds have a CP content between 31% and 42%, which is higher than the content of most other grain legumes (Rajeev and Bavitha, 2015). In all aquaculture species for which a nutritional assessment has been made on the value of lupins, they are an acceptable and nutritionally useful ingredient. Lupin seed meal could be a good alternative vegetable protein of high nutritive quality when used up to 30% or 40% in rainbow trout diet (Glencross et al., 2004). The nutritive quality of trout diets, including lupin seed meal, at 10%, 20%, 30%, and 40% dietary protein content has been studied by de la Higuera et al. (1988). These authors concluded that 30% raw lupin seed meal could be included in trout diet. The studies by Molina-Poveda et al. (2013) found that the body weight gain in shrimp gradually decreased with increasing lupin meal concentrations in diet, and displayed excellent growth at the 50% FM replacement level with defatted and dehulled lupin meal (Glencross et al., 2001). The results reported by Anwar et al. (2020) showed that SPC can be replaced with up to 25% white lupin meal in the carp diet with no reduction in growth performance, feed utilization, body composition, gut integrity, or health status.

Green Pea

Green pea (GP) (*Pisum sativum* L.) is another legume that has been used in livestock feeds for a long time as a source of energy and protein but has only been recently evaluated in feed for aquatic species (Sonesson et al., 2005). Early studies by Kaushik et al. (1995) demonstrated that dehulling, extrusion, and milling improved GP digestibility. GP has been studied as an alternative protein source for FM and impacts growth performance, feed utilization, and phosphorus excretion for Asian sea bass, *Lates calcarifer*. The obtained results revealed that GP can replace FM at the 10% level in diets for sea bass without adverse effects on growth, feed utilization or body composition, which may also contribute to environmental protection and to lower feed costs to sustain aquaculture (Ganzon-Naret, 2013). Likewise, Borlongan et al. (2003) confirmed that GP meal could be used as a dietary feed ingredient and can replace up to 20% total dietary protein in the milkfish (*Chanos chanos*) diet.

Faba Bean

Faba bean (FB) (*Vicia faba*) is a widespread and relatively unexploited legume crop in Europe with potentially favorable characteristics, including low ANF levels – with the highest concentration in the seed coat, which can be removed during the dehulling process (El-Shemy et al., 2000). FB can be successfully used as an FM replacement because of its relatively high protein content (20–40%) and good amino acid profile (El-Shemy et al., 2000). A study by Ouraji et al. (2013) demonstrated that rainbow trout fingerlings could tolerate up to 30% FB inclusion in diet and that 15% inclusion levels were beneficial for growth performance. Similar studies have been performed in *Atlantic salmon*, in which the inclusion rates of 35% FB protein concentrate were well accepted and had no negative impact on either growth or immune capacity (De Santis et al., 2015).

Lemna and Peanut

Lemna sp. are widely used as a model plant to treat wastewater (Nasar et al., 2014) and to partially replace FM, with good Indian major carp growth performance (Kaur et al., 2012). Peters et al. (2009) evaluated the nutritional quality of duckweed meal (*Lemna obscura*) as an ingredient for preparing food for red tilapia. They concluded that up to 25% can be included in the diet of fingerlings without affecting growth, as long as it is combined with other high protein ingredients. Solid residue from peanut, obtained after oil extraction, is known as peanut meal. It contains 48% protein, but it is deficient in amino acids, like methionine and lysine. Liu et al. (2011) suggested that 50% FM replacement with peanut meal (diet with 14% peanut meal) could be included in practical *L. vannamei* diets. The potential of peanut leaves has also been studied, as has their utilization by fish for their high protein content (22.3%), which was comparable to that of FM, and can replace the latter by up to 20% without negative effects on growth performance (Yue et al., 2012).

Corn Gluten Meal

Corn gluten meal (CGM) is another important alternative protein source in aquafeed production. CGM is considered the major protein portion obtained from the wet milling process to separate starch, germ, protein, and fiber components from corn (Anderson and Lamsal, 2011). Compared to other vegetable protein sources, CGM is a cost-effective alternative protein source for aquafeed, given its high available protein content (60–70% of DM), low fiber and ANF content, competitive price, and steady supply (Glencross, 2016). CGM has been shown to successfully replace more than half the FM protein used in the diets of juvenile cobia, Japanese seabass, and sea bream and has no negative effect on growth performance (Luo et al., 2012; Pereira et al., 2003). However, in turbot, CGM negatively affects gut health by inducing enteritis and by decreasing intestinal immunity and antioxidant capacity (Bai et al., 2019).

Canola

The tested seed meal of Canola (*Brassica napus*) has experimentally achieved success. Canola is a vegetable oil deriving from rapeseed and is rich in the marine fatty acid DHA. It has high protein content and is also used as a feed ingredient in animal nutrition (Chakraborty et al., 2019). Canola meal has been used in the diet of several species, such as salmon, trout, carp, catfish, sea bass, tilapia, perch, sea bream, and shrimp, with similar results to those found in SBM (Enami, 2011). The first studies by Hardy and Sullivan (2011) suggested 20% canola meal inclusion in the rainbow trout (*Salmo gairdneri*) diet had no effect on growth. Similar results with canola have been observed for rainbow trout (Shafaeipour et al., 2008). According to Webster et al. (1997), incorporating canola meal into channel catfish (*Ictalurus punctatus*) diet is cost-effective compared to other vegetable protein ingredients used in commercial catfish feed. Buchanan et al. (1997) observed how enzyme-treated canola meal in the diets of juvenile *Penaeus monodon* led to better fish conversion ratios and greater weight gain compared to others. Furthermore, Soares et al. (2001) suggested 35–40% canola meal inclusion to replace 48.17% of soya bean meal in Nile tilapia diet.

Potato Protein Concentrate

Potato protein concentrate (PPC) isolate, obtained by a thermal coagulation process from processing waste potato juice (a by-product of the potato starch manufacturing industry), is a promising FM substitute candidate for its high crude protein content (more than 750–800 g/kg per weight) and nutritionally balanced EAA (Refstie and Storebakken, 2001). Early studies into aquaculture with PPC diets gave a slightly negative result for use and showed that solanine and chaconine, anti-nutritional compounds, influenced diet quality (Tacon and Jackson, 1985; Xie and Jokumsen, 1998). Similarly, the inclusion of up to 26% in rainbow trout feed implies no adverse effect on growth (Tusche et al., 2012). The utility of PPC has also been detailed in Atlantic salmon (*Salmo salar*), and there are reports that formulating feed to contain 21% PPC has no negative effect on growth (Refstie and Storebakken, 2001). Xie et al. (2001) showed that protein digestibility reached 93% in Gibel carp (*Carassius gibelio*) that were fed a diet comprising 32% PPC, which is the equivalent to that of FM. Experiments with *Tilapia zilli* revealed that up to 15% sweet potato leaf meal can be included in diets without compromising either growth or feed efficiency (Adewolu et al., 2008). Other results obtained by Takakuwa et al. (2020) showed that PPC can replace up to 20% FM, which is originally contained at 60% in the diet of greater amberjack (*Seriola dumerili*), and with no effect on growth or feed efficiency. However, the protein digestibility of PPC is lower, and the viscera of the fish fed PPC tend to be small.

Palm Kernel Meal

Palm kernel meal (PKM) is another alternative in aqua feed ingredients. It is cultivated mainly for its high oil content. PKM is the by-product that forms as a result of the palm kernel oil extraction process. Its incorporation as fish feed is restricted for its lower protein content (< 20%) (Chakraborty et al., 2019). Ng and Cheng (2002) reported that 20% PKM dietary inclusion in tilapia led to better growth performance. The Nile tilapia fingerlings fed 60% PKM displayed similar growth performance to that of the fish fed with an FM-based diet (Omoregie et al., 1993). Studies by Ng et al. (2002) revealed that commercial feed enzyme-treated PKM incorporated into the diet of red tilapia resulted in better growth performance than the fish fed raw PKM. The results obtained in that study indicated that the inclusion of enzyme-treated PKM up to 30% did not involve any significant decrease in fish feed utilization and growth. Souza et al. (1997) reported how pacu fish (*Piaractus mesopotamicus*) fed diets containing 70% PKM showed less digestibility (54%) compared to copra meal (a coproduct of coconut oil production). Studies performed in hybrid catfish have demonstrated that 20% PKM could be incorporated and have no negative impacts on growth performance (Ng et al., 2002).

Other Vegetable Protein Sources

Other vegetable protein sources have been investigated. Castillo et al. (2002) studied employing coffee pulp to feed tilapia fingerlings, and concluded that this input can be included in diet up to 25% without affecting production indices (Bayne et al., 1976). It also highlights that those diets, in which coffee pulp has been used, are cheaper than conventional diets. Likewise, Delgado et al. (2006) evaluated diets for tilapia containing four roatan banana flour levels (10%, 20%, 30%, 40%). These authors concluded that diet with 10% banana flour presented the best results for weight gain, SGR, and feed conversion index. Diets prepared by combining 30% fish offal meal, 5% FM, and 24% mulberry leaf meal for herbivorous *H. fossilis* fingerlings achieved 75% FM substitution (Mondal et al., 2011).

CONCLUSIONS

In conclusion, the results obtained by the current studies performed with vegetable products are promising and encourage further experimentation to meet seafood production and environmental sustainability. Currently, good results are being obtained when replacing FM and combining several plant sources and when vegetable protein sources are supplemented with amino acids.

REFERENCES

Abdel-Tawwab, M., Khalil, R. H., Metwally, A. A., Shakweer, M. S., Khallaf, M. A., and H. M. R Abdel-Latif. 2020. Effects of black soldier fly (*Hermetia illucens* L.) larvae meal on growth performance, organs-somatic indices, body composition, and hemato-biochemical variables of European sea bass, *Dicentrarchus labrax. Aquaculture* 522: 735136.

Abdel-Warith, A.-W. A.-M., Younis, E., and N. A. Al-Asgah. 2013. Influence of dietary inclusion of full-fat soybean meal and amino acids supplementation on growth and digestive enzymes activity of nile tilapia, *Oreochromis niloticus. Turkish Journal of Fisheries and Aquatic Sciences* 13: 69–77.

Adewolu, M., Ogunsanmi, A. O., and A. Yunusa. 2008. Studies on growth performance and feed utilization of two Clariid catfish and their hybrid reared under different culture systems. *European Journal of Scientific Research.* 23: 252–260.

Aguilera-Rivera, D., Escalante-Herrera, K., and Gaxiola, G., et al. 2019. Immune response of the Pacific white shrimp, Litopenaeus vannamei, previously reared in biofloc and after an infection assay with *Vibrio harveyi. Journal of the World Aquaculture Society* 50: 119–136.

Ahmad, I., Rani, A. M. B., Verma, A. K., and M. Maqsood. 2017. Biofloc technology: An emerging avenue in aquatic animal healthcare and nutrition. *Aquaculture International* 25: 1215–1226.

Anderson, T. and B. Lamsal. 2011. Zein extraction from corn, corn products, and coproducts and modifications for various applications: A review. *Cereal Chemistry* 88: 159–173.

Angell, A. R., Angell, S. F., de Nys, R., and N. A. Paul. 2016. Seaweed as a protein source for mono-gastric livestock. *Trends in Food Science & Technology* 54: 74–84.

Anger, K. 1998. Patterns of growth and chemical composition in decapod crustacean larvae. *Invertebrate Reproduction & Development* 33: 159–176.

Anwar, A., Wan, A., Omar, S., El-Haroun, E., and S. Davies. 2020. The potential of a solid-state fermentation supplement to augment white lupin (*Lupinus albus*) meal incorporation in diets for farmed common carp (*Cyprinus carpio*). *Aquaculture Reports* 17: 100348.

Aniebo, A. O. and O. J. Owen. 2010. Effects of age and method of drying on the proximate composition of housefly larvae (Musca domestica Linnaeus) meal (HFLM). *Pakistan Journal of Nutrition* 9: 485–487.

Ansari, F. A., Nasr, M., Guldhe, A., Gupta, S. K., Rawat, I., and F. Bux. 2020. Techno-economic feasibility of algal aquaculture via fish and biodiesel production pathways: A commercial-scale application. *Science of the Total Environment* 704: 1–11.

Arnold, S. J., Coman, F. E., Jackson, C. J., and S. A. Groves. 2009. High-intensity, zero water-exchange production of juvenile tiger shrimp, *Penaeus monodon*: An evaluation of artificial substrates and stocking density. *Aquaculture* 293: 42–48.

Asenjo, J. A. and P. Dunnill. 1981. The Isolation of Lytic Enzymes FromCytophaga and Their Application to the Rupture of Yeast Cells. *Biotechnology and Bioengineering* 23(5): 1045–1056. Accessed 1 October 2020. doi:10.1002/bit.260230512

Asiedu, A., Ben, S., Resurreccion, E. and S. Kumar. 2018. Techno-economic analysis of protein concentrate produced by flash hydrolysis of microalgae. *Environmental Progress & Sustainable Energy* 37: 881–890.

Atkinson, A., Siegel, V., Pakhomov, E. A., Jessopp, M. J., and V. Loeb. 2009. A re-appraisal of the total biomass and annual production of Antarctic krill. *Deep Sea Research Part I: Oceanographic Research Papers* 56: 727–740.

Azim, M. E. and D. C. Little. 2008. The biofloc technology (BFT) in indoor tanks: Water quality, biofloc composition, and growth and welfare of Nile tilapia (*Oreochromis niloticus*). *Aquaculture* 283: 29–35.

Bai, N., Gu, M., Liu, M., Jia, Q., Pan, S., and Z. Zhang. 2019. Corn gluten meal induces enteritis and decreases intestinal immunity and antioxidant capacity in turbot (*Scophthalmus maximus*) at high supplementation levels. *PLoS One* 14: e0213867.

Barker, D., Marianne, P., Fitzpatrick, D., and E. S Dierenfeld. 1998. Nutrient composition of selected whole invertebrates. *Zoo Biology* 17: 123–134.

Barroso, F. G., de Haro, C., Sánchez-Muros, M. J., Venegas, E., Martínez-Sánchez, A. and C. Pérez-Bañón. 2014. The potential of various insect species for use as food for fish. *Aquaculture* 422–423: 193–201.

Barroso, F. G., Sánchez-Muros, M. J., Rincón, M. Á., et al. 2019. Production of n-3-rich insects by bioaccumulation of fishery waste. *Journal of Food Composition and Analysis* 82: 103237.

Barroso, F. G., Sánchez-Muros, M. J., Segura, M., et al. 2017. Insects as food: Enrichment of larvae of Hermetia illucens with omega 3 fatty acids by means of dietary modifications. *Journal of Food Composition and Analysis* 62: 8–13.

Basto, A., Matos, E., and L. M. P. Valente. 2020. Nutritional value of different insect larvae meals as protein sources for European sea bass (*Dicentrarchus labrax*) juveniles. *Aquaculture* 521: 735085.

Bayne, R. D., David, D. D., and R. C. García. 1976. Supplemental FEEDS containing coffee pulp for rearing Tilapia in central América. *Aquaculture* 7: 133–146.

Becker, E. W. 2007. Micro-algae as a source of protein. *Biotechnology Advances* 25: 207–210.

Belforti, M., Gai, F., Lussiana, C., et al. 2015. Tenebrio Molitor meal in rainbow trout (Oncorhynchus Mykiss) diets: Effects on animal performance, nutrient digestibility and chemical composition of fillets. *Italian Journal of Animal Science* 14: 4170.

Belghit, I., Liland, N. S., Waagbø, R., et al. 2018. Potential of insect-based diets for Atlantic salmon (Salmo salar). *Aquaculture* 491: 72–81.

Bondari, K. and D. C. Sheppard. 1981. Soldier fly larvae as feed in commercial fish production. *Aquaculture* 24: 103–109.

Borlongan, I. G., Eusebio, P. S., and T. Welsh. 2003. Potential of feed pea (*Pisum sativum*) meal as a protein source in practical diets for mikfish (*Chanos chanos Forsskal*). *Aquaculture* 225: 89–98.

Brito, L. O., Junior, L. C., Abreu, J. L., Severi, W., Moraes, L. B. S., and A. O. Gálvez. 2018. Effects of two commercial feeds with high and low crude protein content on the performance of white shrimp *Litopenaeus vannamei* raised in an integrated biofloc system with the seaweed *Gracilaria birdiae*. *Spanish Journal of Agricultural Research* 16: e0603. doi:10.5424/sjar/2018161-11451

Brown, M. R., Barrett, S. M., Volkman, J. K., Nearhos, S. P., Nell, J. A., and G. L. Allan. 1996. Biochemical composition of new yeasts and bacteria evaluated as food for bivalve aquaculture. *Aquaculture* 143: 341–60. doi:10.1016/0044-8486(96)01286-0

Bruni, L., Belghit, I., Lock, E. J., Secci, G., Taiti, C., and G. Parisi. 2020. Total replacement of dietary fish meal with black soldier fly (*Hermetia illucens*) larvae does not impair physical, chemical or volatile composition of farmed Atlantic salmon (*Salmo salar* L.). *Journal of the Science of Food and Agriculture* 100: 1038–1047.

Buchanan, J., Sarac, H. Z., Poppi, D., and R. T. Cowan. 1997. Effects of enzyme addition to canola meal in prawn diets. *Aquaculture* 151: 29–35.

Burri, L. and A. J. Nunes. 2016. Benefits of including krill meal in shrimp diets. *World Aquaculture*: 47: 19–23.

Caimi, C., Renna, M., Lussiana, C., et al. 2020. First insights on Black Soldier Fly (Hermetia illucens L.) larvae meal dietary administration in Siberian sturgeon (*Acipenser baerii* Brandt) juveniles. *Aquaculture* 515: 734539.

Cao, J. M., Yan, J., Wang, G. X., Huang, et al. 2012. Effects of replacement of fish meal with housefly maggot meal on digestive enzymes, transaminases activities and hepatopancreas histological structure of *Litopenaeus vannamei*. *South China Fisheries Science* 8: 72–79.

Cardona, E., Lorgeoux, B., Geffroy, C., et al. 2015. Relative contribution of natural productivity and compound feed to tissue growth in blue shrimp (*Litopenaeus stylirostris*) reared in biofloc: Assessment by C and N stable isotope ratios and effect on key digestive enzymes. *Aquaculture* 448: 288–297.

Caruso, G. 2015. Use of plant products as candidate fish meal substitutes: An emerging issue in aquaculture productions. *Fisheries and Aquaculture Journal* 6: e123.

Carvalho, R. 1999. A Amazônia rumo ao 'ciclo da soja'. In: Amazônia Papers No. 2. São Paulo, Brazil: Programa Amazônia, Amigos da Terra, p. 8. URL: http://www. amazonia.org.br.

Castillo, E., Acosta, Y., Betancourt, N., Lidia, E., Mildred, A., Cobos, V., and M. Jover. 2002. Utilización de la pulpa de café en la alimentaciónde alevines de tilapia roja. *AquaTIC*,16.

Castro, L. F., Pinto, R. C. C., and A. J. P. Nunes. 2021. Nutrient value and contribution of microbial floc to the growth performance of juvenile shrimp, *Litopenaeus vannamei*, fed fatty acid and amino acid-restrained diets under a zero-water exchange intensive system. *Aquaculture* 531: 735–789.

Chakraborty, P., Mallik, A., Sarang, N., and S. Lingam. 2019. A review on alternative plant protein sources available for future sustainable aqua feed production. *International Journal of Chemical Studies* 7: 1399–1404.

Chanda, S. and S. Chakrabarti. 1996. Plant origin liquid waste: A resource for single cell protein production by yeast. *Bioresource Technology* 57(1): 51–54.

Chapelle, S. 1977. Lipid composition of tissues of marine crustaceans. *Biochemical Systematics and Ecology* 5: 241–248.

Chemello, G., Renna, M., Caimi, C., et al. 2020. Partially defatted *tenebrio molitor* larva meal in diets for grow-out rainbow trout, *oncorhynchus mykiss* (Walbaum): Effects on growth performance, diet digestibility and metabolic responses. *Animals* 10: 229. doi:10.3390/ani10020229

Chen, J. H., Liu, P., Li, Y. Q., Li, M., and B. Xia. 2018. Effects of dietary biofloc on growth, digestibility, protein turnover and energy budget of sea cucumber *Apostichopus japonicus* (Selenka). *Animal Feed Science and Technology* 241: 151–162.

Chen, X. Q., Zhao, W., Xie, S., et al. 2019. Effects of dietary hydrolyzed yeast (*Rhodotorula mucilaginosa*) on growth performance, immune response, antioxidant capacity and histomorphology of juvenile Nile Tilapia (*Oreochromis Niloticus*). *Fish and Shellfish Immunology* 90: 30–39. Accessed 1 October 2020. doi:10.1016/j.fsi.2019.03.068

Cheng, Z. J., Hardy, R. W. and N. J Huige. 2004. Apparent digestibility coefficients of nutrients in brewer's and rendered animal by-products for rainbow trout (*Oncorhynchus mykiss* (Walbaum)). *Aquaculture Research* 35: 1–9. doi:10.1111/j.1365-2109.2004.00941.x

Chu, C. P., and D. J. Lee. 2004. Multiscale structures of biological flocs. *Chemical Engineering Science* 59: 1875–1883.

Correia, E. S., Wilkenfeld, J. S., Morris, T. C., Wei, L. Z., Prangnell, D. I., and T. M. Samocha. 2014. Intensive nursery production of the Pacific white shrimp *Litopenaeus vannamei* using two commercial feeds with high and low protein content in a biofloc-dominated system. *Aquacultural Engineering* 59: 48–54.

Council Directive. 1999. On the undesirable substance and production animal nutrition. *European Journal of Communication*. Council Directive 29/EC, L115/32–L115/46.

Crab, R., Defoirdt, T., Bossier, P., and W. Verstraete. 2012. Biofloc technology in aquaculture: Beneficial effects and future challenges. *Aquaculture* 356: 351–356.

Cuesta, A., Esteban, M. A., and J. Meseguer. 2003. In vitro effect of chitin particles on the innate cellular immune system of gilthead seabream (*Sparus aurata* L.). *Fish Shellfish Immunology* 15, 1–11.

Cummins, V. C., Rawles, S. D., Thompson, K. R., et al. 2017. Evaluation of black soldier fly (Hermetia illucens) larvae meal as partial or total replacement of marine fish meal in practical diets for Pacific white shrimp (*Litopenaeus vannamei*). *Aquaculture* 473: 337–344.

da Silva, M. A., de Alvarenga, E. R., Alves, G. F. D., et al. 2018. Crude protein levels in diets for two growth stages of Nile tilapia (*Oreochromis niloticus*) in a biofloc system. *Aquaculture Research* 49: 2693–2703.

Daniel, D. 2018. A review on replacing fish meal in aqua feeds using plant protein sources. *International Journal of Fisheries and Aquatic Studies* 6: 164–179.

Dantas, E. M., Valle, B. C. S., Brito, C. M. S., Calazans, N. K. F., Peixoto, S. R. M., and R. B. Soares. 2016. Partial replacement of fishmeal with biofloc meal in the diet of postlarvae of the Pacific white shrimp *Litopenaeus vannamei*. *Aquaculture Nutrition* 22: 335–342.

Danulat, E. 1986. The effects of various diets on chitinase and beta-glucosidase activities and the condition of cod, *Gadus morhua* (L). *Journal of Fish Biology* 28: 191–197.

Danulat, E. and H. Kausch. 1984. Chitinase activity in the digestive tract of the cod, *Gadus morhua* (L). *Journal of Fish Biology* 24: 125–133.

Dauda, A. B. 2020. Biofloc technology: A review on the microbial interactions, operational parameters and implications to disease and health management of cultured aquatic animals. *Reviews in Aquaculture* 12: 1193–1210.

De la Higuera, M., García-Gallego, M., Sanz, A., et al. 1988. Evaluation of lupin seed meal as an alternative protein source in feeding of rainbow trout (*Salmo gairdneri*). *Aquaculture* 71: 37–50.

De Morais, A. P. M., Abreu, P. C., Wasielesky, W., and D. Krummenauer. 2020. Effect of aeration intensity on the biofilm nitrification process during the production of the white shrimp *Litopenaeus vannamei* (Boone, 1931) in Biofloc and clear water systems. *Aquaculture* 514: 734516.

De Santis, C., Ruohonen, K., Tocher, D. R., et al. 2015. Atlantic salmon (*Salmo salar*) as a model to predict the optimum inclusion of air classified faba bean protein concentrate in feeds for seawater salmon. *Aquaculture* 444: 70–78.

De Schryver, P., Crab, R., Defoirdt, T., Boon, N., and W. Verstraete. 2008. The basics of bio-flocs technology: The added value for aquaculture. *Aquaculture* 277: 125–137.

Dean, J. C., Nielsen, L. A., Helfrich, L. A., and Jr, D. L. Garling. 1992. Replacing fish meal with seafood processing wastes in channel catfish diets. *The Progressive Fish-Culturist* 54(1): 7–13.

Delgado, V. F. K., Piñón, R. D. A., and P. C. L. Cuevas. 2006. Evaluación de dietas para tilapia (*Oreochromis niloticus*, L. 1758) con inclusión de harina de plátano roatán (Musa sapientum L.). *XIV Congreso Nacional de Oceanografía. Manzanillo, Colima, México. Memorias*: 504–507.

Devic, E., Leschen, W., Murray, F., and D. C. Little. 2018. Growth performance, feed utilization and body composition of advanced nursing Nile tilapia (*Oreochromis niloticus*) fed diets containing Black Soldier Fly (*Hermetia illucens*) larvae meal. *Aquaculture Nutrition* 24: 416–423.

Dineshbabu, G., Goswami, G., Kumar, R., Sinha, A., and D. Das. 2019. Microalgae–nutritious, sustainable aqua-and animal feed source. *Journal of Functional Foods* 62: 1–14.

Duong, D. N., Stone, D. A., Qin, J. G., Hoang, T. H., Bansemer, M. S., and J. O. Harris. 2021. An improvement of ingested food and somatic growth energy in greenlip abalone (*Haliotis laevigata* Donovan) fed different types and levels of macroalgae meal supplementation. *Aquaculture* 531: 735816.

Ebrahimi, A., Akrami, R., Najdegerami, E. H., Ghiasvand, Z., and H. Koohsari. 2020. Effects of different protein levels and carbon sources on water quality, antioxidant status and performance of common carp (*Cyprinus carpio*) juveniles raised in biofloc based system. *Aquaculture* 516: 734639.

Ekasari, J., Rivandi, D. R., Firdausi, A. P., et al. 2015. Biofloc technology positively affects Nile tilapia (*Oreochromis niloticus*) larvae performance. *Aquaculture* 441: 72–77.

Ekasari, J., Setiawati, R., Ritonga, F. R., Setiawati, M., and M. A. Suprayudi. 2019. Growth and health performance of African catfish *Clarias gariepinus* (Burchell 1822) juvenile fed with graded levels of biofloc meal. *Aquaculture Research* 50: 1802–1811.

El-Shemy, H., Abdel-Rahim, E., Shaban, O., et al. 2000. Comparison of nutritional and antinutritional factors in soybean and faba bean seeds with or without cortex. *Soil Science and Plant Nutrition* 46: 515–524.

Elia, A. C., Capucchio, M. T., Caldaroni, B., Magara, G., Dörr, A. J. M., Biasato, I., Biasibetti, E., Righetti, M., Pastorino, P., Prearo, M., Gai, F., Schiavone, A., and L. Gasco. 2018. Influence of Hermetia illucens meal dietary inclusion on the histological traits, gut mucin composition and the oxidative stress biomarkers in rainbow trout (Oncorhynchus mykiss). *Aquaculture* 496: 50–57.

Enami, H. 2011. A review of using canola/rapeseed meal in aquaculture feeding. *Journal of Fisheries and Aquatic Science* 6: 22–36.

Escaffre, A. M., Infante, J. Z., Cahu, C. L., et al. 1997. Nutritional value of soy protein concentrate for larvae of common carp (*Cyprinus carpio*) based on growth performance and digestive enzyme activities. *Aquaculture* 153: 63–80.

Esparza-Leal, H. M., Cardozo, A. P., and W. Wasielesky. 2015. Performance of *Litopenaeus vannamei* postlarvae reared in indoor nursery tanks at high stocking density in clear-water versus biofloc system. *Aquacultural Engineering* 68: 28–34.

Esteban, M. A., Cuesta, A., Ortuño, J., and J. Meseguer. 2001. Immunomodulatory effects of dietary intake of chitin on gilthead seabream (*Sparus aurata* L.) innate immune system. *Fish Shellfish Immunology* 11: 303–315.

Esteban, M. A., Mulero, V., Cuesta, A., Ortuño, J., and J. Meseguer. 2000. Effects of injecting chitin particles on the innate immune response of gilthead seabream (*Sparus aurata* L.). *Fish Shellfish Immunology* 10: 543–554.

Everson, I. (Ed.). 2000. *Krill: Biology, ecology, and fisheries*. Oxford, UK: Blackwell Science.

Fabrikov, D., Sánchez-Muros, M. J., Barroso, F. G., et al. 2020. Comparative study of growth performance and amino acid catabolism in *Oncorhynchus mykiss*, *Tinca tinca* and *Sparus aurata* and the catabolic changes in response to insect meal inclusion in the diet. *Aquaculture* 529: 735731.

Ferrer Llagostera, P., Kallas, Z., Reig, L., and D. Amores de Gea. 2019. The use of insect meal as a sustainable feeding alternative in aquaculture: Current situation, Spanish consumers' perceptions and willingness to pay. *Journal of Cleaner Production* 229: 10–21.

Finke, M. D. 2015. Complete nutrient content of four species of commercially available feeder insects fed enhanced diets during growth. *Zoo Biology* 34: 554–564.

Finke, M. D. 2013. Complete Nutrient Content of Four Species of Feeder Insects. *Zoo Biology* 32: 27–36.

Fines, B. C. and G. J. Holt. 2010. Chitinase and apparent digestibility of chitin in the digestive tract of juvenile cobia, *Rachycentron canadum*. *Aquaculture* 303: 34–39. doi:10.1016/j.aquaculture.2010.03.010

Fleckenstein, L. J., Tierney, T. W., and A. J. Ray. 2018. Comparing biofloc, clear-water, and hybrid recirculating nursery systems (Part II): Tilapia (*Oreochromix niloticus*) production and water quality dynamics. *Aquacultural Engineering* 82: 80–85.

Fleurence, J., Morançais, M., and J. Dumay. 2018. Seaweed proteins. In R. Yada (Ed.), *Proteins in food processing* (pp. 245–262). Cambridge: Woodhead Publishing.

Furuya, W. M., Pezzato, L. E., Barros, M. M., et al. 2004. Use of ideal protein concept for precise formulation of amino acid levels in fish-meal-free diets for juvenile Nile tilapia (*Oreochromis niloticus* L.). *Aquaculture Research* 35: 1110–1116.

Gaignard, C., Gargouch, N., Dubessay, P., et al. 2019. New horizons in culture and valorization of red microalgae. *Biotechnology Advances* 37: 193–222.

Gajardo, K. Jaramillo-Torres, A., Kortner, T. M., et al. 2017. Alternative protein sources in the diet modulate microbiota and functionality in the distal intestine of Atlantic salmon (*Salmo salar*). *Applied and Environmental Microbiology* 83: e02615–e02616.

Gamboa-Delgado, J., Fernández-Díaz, B., Nieto-López, M., and L. E. Cruz-Suárez. 2016. Nutritional contribution of torula yeast and fish meal to the growth of shrimp *Litopenaeus vannamei* as indicated by natural nitrogen stable isotopes. *Aquaculture* 453: 116–121.

Ganzon-Naret, E. 2013. The use of green pea (*Pisum sativum*) as alternative protein source for fish meal in diets for Asian sea bass, *Lates calcarifer. AACL Bioflux* 6: 399–406.

Garcia, M. A. and M. A. Altieri. 2005. Transgenic crops: Implications for biodiversity and sustainable agriculture. *Bulletin Science and Technology Society* 25: 335e353.

García, V., Celada, J. D., González, R., Carral, J. M., Sáez-Royuela, M., and Á. González. 2015. Response of juvenile tench (*Tinca tinca* L.) fed practical diets with different protein contents and substitution levels of fish meal by soybean meal. *Aquaculture Research* 46: 28–38.

Gasco, L., Acuti, G., Bani, P., et al. 2020. Insect and fish by-products as sustainable alternatives to conventional animal proteins in animal nutrition. *Italian Journal of Animal Science* 19: 360–372.

Gasco, L., Gai, F., Maricchiolo, G., et al. 2018. *Fishmeal Alternative Protein Sources for Aquaculture Feeds, Feeds for the Aquaculture Sector: Current Situation and Alternative Sources.* Cham: Springer International Publishing, pp. 1–28.

Gatesoupe, F. J. 2007. Live yeasts in the gut: Natural occurrence, dietary introduction, and their effects on fish health and development. *Aquaculture* 267(1–4): 20–30.

Gatlin, D. M., Barrows, F. T., Brown, P., et al. 2007. Expanding the utilization of sustainable plant products in aquafeeds: A review. *Aquaculture Research* 38: 551–579.

Glencross, B., Blyth, D., Irvin, S., Bourne, N., Campet, M., Boisot, P., and N. M. Wade. 2016. An evaluation of the complete replacement of both fishmeal and fish oil in diets for juvenile Asian seabass, Lates calcarifer. *Aquaculture* 451: 298–309.

Glencross, B. D. (Ed.). 2001. *Feeding lupins to fish: A review of thenutritional and biological value of lupins in aquaculture feeds* (pp. 1–12). Western Australia: Department of Fisheries.

Glencross, B., Evans, D., Hawkins, W., et al. 2004. Evaluation of dietary inclusion of yellow lupin (*Lupinus luteus*) kernel meal on the growth, feed utilisation and tissue histology of rainbow trout (*Oncorhynchus mykiss*). *Aquaculture* 235: 411–422.

Gómez, B., Munekata, P. E. S., Zhu, Z., et al. 2019. Challenges and opportunities regarding the use of alternative protein sources: Aquaculture and insects. *Advances in food and nutrition research* 89: 259–295.

Gong, Y. Y., Huang, Y. Q., Gao, L. J., Lu, J. X., and Huang. 2016. Substitution of Krill meal for Fish Meal in Feed for Russian Sturgeon, Acipenser gueldenstaedtii. *The Israeli Journal of Aquaculture-Bamidgeh* 16. http://hdl.handle.net/10524/54968

Gong, H., Lawrence, A. L., Jiang, D. H., Castille, F. L., and D. M. Gatlin. 2000. Lipid nutrition of juvenile *Litopenaeus vannamei.* Dietary cholesterol and de-oiled control diet lecithin requirements and their interaction. *Aquaculture* 190: 305–324.

González-Rodríguez, Á., Celada, J. D., Carral, J. M., et al. 2015. Evaluation of pea protein concentrate as partial replacement of fish meal in practical diets for juvenile tench (*Tinca tinca L.*). *Aquaculture Research* 47: 2825–2834.

González-Rodríguez, Á., Celada, J. D., Carral, J. M., et al. 2015. Evaluation of pea protein concentrate as partial replacement of fish meal in practical diets for juvenile tench (*Tinca tinca L.*) *Aquaculture Research* 47: 2825–2834.

Gopalakannan, A. and V. Arul. 2006. Immunostimulatory effects of dietary intake of chitin, chitosan and levamisole on the immune system of *Cyprinus carpio* and control of *Aeromonas hydrophila* infection in ponds. *Aquaculture* 255: 179–187.

Gopalakannan, A. and V. Arul. 2010. Enhancement of the innate immune system and disease-resistant activity in cyprinus carpio by oral administration of β-glucan and whole cell yeast. *Aquaculture Research* 41: 884–892. doi:10.1111/j.1365-2109.2009.02368.x

Govorushko, S. 2019. Global status of insects as food and feed source: A review. *Trends in Food Science & Technology* 91: 436–445.

Grammes, F., Reveco, F. E., Romarheim, O. H., Landsverk, T., Mydland, L. T., and M. Øverland. 2013. *Candida utilis* and *Chlorella vulgaris* counteract intestinal inflammation in Atlantic salmon (*Salmo salar L.*). *PloS one* 8: e83213.

Green, B. W., Rawles, S. D., Schrader, K. K., Gaylord, T. G., and M. E. McEntire. 2019. Effects of dietary protein content on hybrid tilapia (*Oreochromis aureus* x *O. niloticus*) performance, common microbial off-flavor compounds, and water quality dynamics in an outdoor biofloc technology production system. *Aquaculture* 503: 571–582.

Guerreiro, I., Castro, C., Antunes, B., et al. 2020. Catching black soldier fly for meagre: Growth, whole-body fatty acid profile and metabolic responses. *Aquaculture* 516: 734613.

Guo, J., Qiu, X., Salze, G., and D. A. Davis. 2019. Use of high-protein brewer's yeast products in practical diets for the Pacific white shrimp *Litopenaeus vannamei*. *Aquaculture Nutrition* 25(3): 680–690.

Gutowska, M. A., Drazen, J. C., and B. H. Robison. 2004. Digestive chitinolytic activity in marine fishes of Monterey Bay, California. *Comparative Biochemistry and Physiology* 139A: 351–358.

Hagen, W., Kattner, G., Terbruggen, A., and E. S. Van Vleet. 2001. Lipid metabolism of the Antarctic krill *Euphausia superba* and its ecological implications. *Marine Biology*. 139: 95–104.

Halasz, A. and R. Lasztity. 1991. *Use of Biomass in Food Production*. Boca Raton, FL: CRC Press.

Halász, A. and R. Lásztity. 2017. *Use of Yeast Biomass in Food Production Use of Yeast Biomass in Food Production*. CRC Press. doi:10.1201/9780203734551

Hansen, K. G., Aviram, N., Laborenz, J., Bibi, C., Meyer, M., Spang, A., et al. 2018. An ER surface retrieval pathway safeguards the import of mitochondrial membrane proteins in yeast. *Science* 361(6407): 1118–1122.

Hardy, R. W. 2010. Utilization of plant proteins in fish diets: Effects of global demand and supplies of fishmeal. *Aquaculture Research* 41: 770–776.

Hardy, R. and C. Sullivan. 2011. Canola meal in Rainbow trout (*Salmo gairdneri*) production diets. *Canadian Journal of Fisheries and Aquatic Sciences* 40: 281–286.

Haridas, H., Verma, A. K., Rathore, G., Prakash, C., Sawant, P. B., and A. M. B. Rani. 2017. Enhanced growth and immuno-physiological response of Genetically Improved Farmed Tilapia in indoor biofloc units at different stocking densities. *Aquaculture Research* 48: 4346–4355.

Harlioğlu, A. G. and Ö. Yilmaz. 2011. Fatty acid composition, cholesterol and fat-soluble vitamins of wild-caught freshwater spiny eel, *Mastacembelus simack* (Walbaum, 1792). *Journal of Applied Ichthyology* 27: 1123–1127.

Hasan, M. R. and S. Soto. 2017. *Improving Feed Conversion Ratio and Its Impact on Reducing Greenhouse Gas Emissions in Aquaculture. FAO Non-Serial Publication*. Rome: FAO, 33 pp.

Hatlen, B., Berge, G. M., Odom, J. M., Mundheim, H., and B.Ruyter. 2012. Growth performance, feed utilisation and fatty acid deposition in Atlantic salmon, *Salmo salar* L., fed graded levels of high-lipid/high-EPA *Yarrowia lipolytica* biomass. *Aquaculture* 364: 39–47.

Haveman, J., Venero, J. A., Lewis, B. L., et al. 2009. Effect of photoautotrophic and heterotrophic biofloc communities on productivity of Pacific white *whrimp Litopenaeus Vannamei* red a plant-based diet in superintensive, zero-water exchange systems. *Journal of Shellfish Research* 28: 702–702.

He, S., Zhou, Z., Meng, K., Zhao, H., Yao, B., Ringø, E., & I. Yoon. 2011. Effects of dietary antibiotic growth promoter and *Saccharomyces cerevisiae* fermentation product on production, intestinal bacterial community, and nonspecific immunity of hybrid tilapia (*Oreochromis niloticus* female× *Oreochromis aureus* male). *Journal of Animal Science* 89: 84–92.

Hisano, H. J., Parisi, I. L. Cardoso, G. H. Ferri, and P. M. F. Ferreira. 2020. Dietary protein reduction for Nile tilapia fingerlings reared in biofloc technology. *Journal of the World Aquaculture Society* 51: 452–462.

Hoffmann, L., Rawski, M., Nogales-Merida, S. and J. Mazurkiewicz. 2020. dietary inclusion of *Tenebrio molitor* meal in sea trout larvae rearing: Effects on fish growth performance, survival, condition, and git and liver enzymatic activity. *Annals of Animal Science* 20: 579–598.

Hong, I. K., Jeon, H. and S. B. Lee. 2014. Comparison of red, brown and green seaweeds on enzymatic saccharification process. *Journal of Industrial and Engineering Chemistry* 20: 2687–2691.

Hua, K. 2020. A meta-analysis of the effects of replacing fish meals with insect meals on growth performance of fish. *Aquaculture* 530: 735732. doi:10.1016/j.aquaculture.2020.735732

Hua, K. and D. P. Bureau. 2009. Development of a model to estimate digestible lipid content of salmonid fish feeds. *Aquaculture* 286: 271–276.

Hua, K., Cobcroft, J. M., Cole, A., et al. 2019. The future of aquatic protein: Implications for protein sources in aquaculture diets. *One Earth* 1: 316–329.

Iaconisi, V., Marono, S., Parisi, G., et al. 2017. Dietary inclusion of *Tenebrio molitor* larvae meal: Effects on growth performance and final quality treats of blackspot sea bream (*Pagellus bogaraveo*). *Aquaculture* 476: 49–58.

Janssen, R. H., Vincken, J. P., Van Den Broek, L. A. M., Fogliano, V., and C. M. M. Lakemond. 2017. Nitrogen-to-protein conversion factors for three edible insects: *Tenebrio molitor, Alphitobius diaperinus, and Hermetia illucens. Journal of Agricultural and Food Chemistry* 65: 2275–2278.

Jatobá, A., da Silva, B. C., da Silva, J. S., et al. 2014. Protein levels for *Litopenaeus vannamei* in semi-intensive and biofloc systems. *Aquaculture* 432: 365–371.

Jatobá, A., Vieira, F. D., da Silva, B. C., Soares, M., Mourino, J. L. P., and W. Q. Seiffert. 2017. Replacement of fishmeal for soy protein concentrate in diets for juvenile *Litopenaeus vannamei* in biofloc-based rearing system. *Revista Brasileira De Zootecnia-Brazilian Journal of Animal Science* 46: 705–713.

Jiang, M., Zhao, H. H., Zai, S. W., Shepherd, B., Wen, H. and D. F. Deng. 2019. A defatted microalgae meal (*Haematococcus pluvialis*) as a partial protein source to replace fishmeal for feeding juvenile yellow perch *Perca flavescens. Journal of Applied Phycology* 31: 1197–1205.

Jonas-Levi, A. and J.-J. I. Martinez. 2017. The high level of protein content reported in insects for food and feed is overestimated. *Journal of Food Composition and Analysis* 62: 184–188.

Jones, S. W., Karpol, A., Friedman, S., Maru, B. T., and B. P. Tracy. 2020. Recent advances in single cell protein use as a feed ingredient in aquaculture. *Current Opinion in Biotechnology* 61: 189–197.

Ju, Z. Y., Deng, D. F., and W. Dominy. 2012. A defatted microalgae (*Haematococcus pluvialis*) meal as a protein ingredient to partially replace fishmeal in diets of Pacific white shrimp (*Litopenaeus vannamei*, Boone, 1931). *Aquaculture* 354: 50–55.

Ju, Z. Y., Forster, I., Conquest, L., and W. Dominy. 2008. Enhanced growth effects on shrimp (*Litopenaeus vannamei*) from inclusion of whole shrimp floc or floc fractions to a formulated diet. *Aquaculture Nutrition* 14: 533–543.

Julshamn, K., Malde, M. K., Bjorvatn, K., and P. Krogdal. 2004. Fluoride retention of Atlantic salmon (Salmo salar) fed krill meal. *Aquaculture Nutrition* 10: 9–13.

Kamilya, D., Debbarma, M., Pal, P., Kheti, B., Sarkar, S., and S. T. Singh. 2017. Biofloc technology application in indoor culture of *Labeo rohita* (Hamilton, 1822) fingerlings: The effects on inorganic nitrogen control, growth and immunity. *Chemosphere* 182: 8–14.

Karasuda, S., Yamamoto, K., Kono, M., Sakuda, S., and D Koga. 2004. Kinetic analysis of a chitinase from red sea bream, Pagrus major. *Bioscience, Biotechnology, and Biochemistry* 68: 1338–1344.

Kaushik, S. J., Cravedi, J. P., Lalles, J. P., et al. 1995. Partial or total replacement of fish meal by soybean protein on growth, protein utilization, potential estrogenic or antigenic effects, cholesterolemia and flesh quality in rainbow trout, *Oncorhynchus mykiss. Aquaculture* 133: 257–274.

Kaur, V. I., Ansal, M. D., and A. Dhawan. 2012. Effect of feeding duckweed (Lemna minor) based diets on the growth performance of rohu, Labeo rohita (Ham.). *Indian Journal of Animal Nutrition* 29(4): 406–409.

Kenis, M., Koné, N., Chrysostome, C. A. A. M., Devic, E., Koko, G. K. D., and V. A. Clottey. 2014. Insects used for animal feed in West Africa. *Entomologia* 2(218): 107–114.

Khan, M. I., Shin, J. H., and J. D. Kim. 2018. The promising future of microalgae: Current status, challenges, and optimization of a sustainable and renewable industry for biofuels, feed, and other products. *Microbial Cell Factories* 17: 1–21.

Khanjani, M. H. and M. Sharifinia. 2020. Biofloc technology as a promising tool to improve aquaculture production. *Reviews in Aquaculture* 12: 1836–1850.

Khanjani, M. H., Alizadeh, M., and M. Sharifinia. 2020. Rearing of the Pacific white shrimp, *Litopenaeus vannamei* in a biofloc system: The effects of different food sources and salinity levels. *Aquaculture Nutrition* 26: 328–337.

Khanjani, M. H., Sajjadi, M., Alizadeh, M., and I. Sourinejad. 2015. Effect of different feeding levels on water quality, growth performance and survival of western white shrimp (*Litopenaeus vannamei* Boone, 1931) post larvae with application of biofloc technology. *Iranian Scientific Fisheries Journal* 24: 13–28.

Khanjani, M. H., Sajjadi, M., Alizadeh, M., and I. Sourinejad. 2017. Nursery performance of Pacific white shrimp (*Litopenaeus vannamei* Boone, 1931) cultivated in a biofloc system: The effect of adding different carbon sources. *Aquaculture Research* 48: 1491–1501.

Khatoon, H., Banerjee, S., Yuan, G. T. G., et al. 2016. Biofloc as a potential natural feed for shrimp postlarvae. *International Biodeterioration and Biodegradation* 113: 304–309.

Khosravi, S., Kim, E., Lee, Y. S. and S. M. Lee. 2018. Dietary inclusion of mealworm (*Tenebrio molitor*) meal as an alternative protein source in practical diets for juvenile rockfish (*Sebastes schlegeli*). *Entomological Research* 48: 214–221.

Kim, J. H., Affan, M. A., Jang, J., et al. 2015. Morphological, molecular, and biochemical characterization of astaxanthin-producing green microalga *Haematococcus* sp. KORDI03 (*Haematococcaceae*, Chlorophyta) isolated from Korea. *Journal of Microbiology and Biotechnology* 25: 238–246.

Kim, Y.-S., Kim, S.-E., Kim, S.-J., et al. 2021. Effects of wheat flour and culture period on bacterial community composition in digestive tracts of *Litopenaeus vannamei* and rearing water in biofloc aquaculture system. *Aquaculture* 531: 735908.

Kim, Y., Kim, D., Kim, T., et al. 2013. Use of red algae, Ceylon moss (*Gelidium amansii*), hydrolyzate for clostridial fermentation. *Biomass and Bioenergy* 56: 38–42.

Kinyuru, J. N., Kenji, G. M., Njoroge, S. M., and M. Ayieko. 2010. Effect of Processing Methods on the In Vitro Protein Digestibility and Vitamin Content of Edible Winged Termite (*Macrotermes subhylanus*) and Grasshopper (*Ruspolia differens*). *Food and Bioprocess Technology* 3: 778–782.

Klanian, M. G., Díaz, M. D., Solis, M. J. S., Aranda, J., and P. M. Moral. 2020. Effect of the content of microbial proteins and the poly-beta-hydroxybutyric acid in biofloc on the performance and health of Nile tilapia (*Oreochromis niloticus*) fingerlings fed on a protein-restricted diet. *Aquaculture* 519: 734872.

Kolakowska, A., Kolakowski, E., and M. Szcygielski. 1994. Season krill (*Euphausia superba* Dana) as a source of n-3 polyunsaturated fatty acids. *Die Nahrung* 38: 128–134.

Kokou, F., Henry, M., and M. Alexis. 2012. Growth performance, feed utilization and non-specific immune response of gilthead sea bream (*Sparus aurata L.*) fed graded levels of a bioprocessed soybean meal. *Aquaculture* 364: 74–81.

Kono, M., Matsui, T., and C. Shimizu. 1987. Chitin decomposing bacteria in the digestive tract of red sea bream and Japanese eel. *Nippon Suisan Gakkaishi* 53: 305–310.

Koutsos, L., McComb, A., and M Finke. 2019. Insect composition and uses in animal feeding applications: A brief review. *Annals of the Entomological Society of America* 112: 544–551.

Kraan, S. 2013. Mass-cultivation of carbohydrate rich macroalgae, a possible solution for sustainable biofuel production. *Mitigation and Adaptation Strategies for Global Change* 18: 27–46.

Kuhn, D. D., Lawrence, A. L., Boardman, G. D., Patnaik, S., Marsh, L., and G. J. Flick. 2010. Evaluation of two types of bioflocs derived from biological treatment of fish effluent as feed ingredients for Pacific white shrimp, *Litopenaeus vannamei*. *Aquaculture* 303: 28–33.

Kumar, S., Anand, P. S. S., De, D., et al. 2017. Effects of biofloc under different carbon sources and protein levels on water quality, growth performance and immune responses in black tiger shrimp *Penaeus monodon* (Fabricius, 1978). *Aquaculture Research* 48: 1168–1182.

Lafarga, T., Fernández-Sevilla, J. M., González-López, C., and F. G. Acién-Fernández. 2020. *Spirulina* for the food and functional food industries. *Food Research International* 137: 1–10.

Lall, S. P. and S. Anderson. 2005. Amino acid nutrition of salmonids: Dietary requirements and bioavailability. *Cahiers Options Méditerranéennes* 63: 73–90.

Langeland, M., Vidakovic, A., Vielma, J., Lindberg, J. E., Kiessling, A., and T. Lundh. 2016. Digestibility of microbial and mussel meal for Arctic Charr (*Salvelinus Alpinus*) and Eurasian perch (*Perca Fluviatilis*). *Aquaculture Nutrition* 22(2): 485–95. Accessed 1 October 2020. doi:10.1111/anu.12268

Langeland, M., Lindberg, J. E. and T. Lundh. 2013. Digestive enzyme activity in Eurasian perch (*Perca fluviatilis*) and Arctic charr (*Salvelinus alpinus*). *Journal of Aquaculture Research and Development* 5: 208.

Lee, P. G. and S. P. Meyers. 1997. Chemoattraction and feeding stimulation. In: *Crustacean Nutrition* (D'Abramo, L. R., Conklin, D. E., and Akiyama, D. M. eds), pp. 292–352. Baton Rouage, LA: World aquaculture Society, Lousiana State University.

Li, H. D., Han, D., Zhu, X. M., et al. 2018. Effect of biofloc technology on water quality and feed utilization in the cultivation of gibel carp (*Carassius auratus gibelio* var. CAS III). *Aquaculture Research* 49: 2852–2860.

Li, P. and D. M. Gatlin III. 2004. Dietary brewers yeast and the prebiotic Grobiotic™ AE influence growth performance, immune responses and resistance of hybrid striped bass (*Morone chrysops*× *M. saxatilis*) to *Streptococcus iniae* infection. *Aquaculture* 231: 445–456.

Li, P. and D. M. Gatlin III. 2003. Evaluation of brewers yeast (*Saccharomyces cerevisiae*) as a feed supplement for hybrid striped bass (*Morone chrysops*× *M. saxatilis*). *Aquaculture* 219: 681–692.

Li, S., Ji, H., Zhang, B., Zhou, J. and H. Yu. 2017. Defatted black soldier fly (Hermetia illucens) larvae meal in diets for juvenile Jian carp (*Cyprinus carpio* var. Jian): Growth performance, antioxidant enzyme activities, digestive enzyme activities, intestine and hepatopancreas histological structure. *Aquaculture* 477: 62–70.

Liland, N. S., Biancarosa, I., Araujo, P., et al. 2017. Modulation of nutrient composition of black soldier fly (Hermetia illucens) larvae by feeding seaweed-enriched media. *PLoS ONE* 12: e0183188.

Lindsay, G. J. H. 1987. Seasonal activities of chitinase and chitobiase in the digestive tract and serum of Cod, *Gadus morhua* (L). *Journal of Fish Biolology.* 30: 495–500.

Lindsay, G. J. H. and G. W.Gooday. 1985. Chitinolytic enzymes and the bacterial microflora in the digestive tract of cod, *Gadus morhua*. *Journal of Fish Biolology* 26: 255–265.

Liu, W. C., Luo, G. Z., Chen, W., et al. 2018a. Effect of no carbohydrate addition on water quality, growth performance and microbial community in water-reusing biofloc systems for tilapia production under high-density cultivation. *Aquaculture Research* 49: 2446–2454.

Liu, G., Ye, Z. Y., Liu, D. Z., and S. M. Zhu. 2018b. Inorganic nitrogen control, growth, and immunophysiological response of *Litopenaeus vannamei* (Boone, 1931) in a biofloc system and in clear water with or without commercial probiotic. *Aquaculture International* 26: 981–999.

Liu, X.-H., Ye, J.-D., Wang, K., Kong, J.-H. Yang, W., and L. Zhou. 2011. Partial replacement of fish meal with peanut meal in practical diets for the Pacific white shrimp, Litopenaeus vannamei. *Aquaculture Research* 43: 745–755.

Lock, E. R., Arsiwalla, T., and R. Waagbø. 2016. Insect larvae meal as an alternative source of nutrients in the diet of Atlantic salmon (*Salmo salar*) postsmolt. *Aquaculture Nutrition* 22: 1202–1213.

Luo, Y., Ai, Q., Zhang, W., Xu, W. Zhang, Y., and Z. Liufu. 2013. Effects of dietary corn gluten meal on growth performance and protein metabolism in relation to IGF-I and TOR gene expression of juvenile cobia (*Rachycentron canadum*). *Journal of Ocean University of China* 12: 418–426.

Mabroke, R. S., El-Husseiny, O. M., El-Naem, A., Zidan, F. A., Tahoun, A. A., and A. Suloma. 2019. Floc meal as potential substitute for soybean meal in tilapia diets under biofloc system conditions. *Journal of Oceanology and Limnology* 37: 313–320.

Macias-Sancho, J., Poersch, L. H., Bauer, W. Romano, L. A., Wasielesky, W. and M. B. Tesser. 2014. Fishmeal substitution with Arthrospira (*Spirulina platensis*) in a practical diet for *Litopenaeus vannamei*: Effects on growth and immunological parameters. *Aquaculture* 426: 120–125.

Madeira, M. S., Cardoso, C., Lopes, P. A., et al. 2017. Microalgae as feed ingredients for livestock production and meat quality: A review. *Livestock Science* 205: 111–121.

Magalhães, R., Sánchez-López, A., Leal, R. S., Martínez-Llorens, S., Oliva-Teles, A., and H. Peres. 2017. Black soldier fly (*Hermetia illucens*) pre-pupae meal as a fish meal replacement in diets for European seabass (*Dicentrarchus labrax*). *Aquaculture* 476: 79–85.

Makkar, H. P. S., Tran, G., Heuzé, V., and P. Ankers. 2014. State-of-the-art on use of insects as animal feed. *Animal Feed Science and Technology* 197(0): 1–33.

Mambrini, M., Roem, A. J., Carvedi, J. P., et al. 1999. Effects of replacing fish meal with soy protein concentrate and of DL-methionine supplementation in high-energy, extruded diets on the growth and nutrient utilization of rainbow trout, *Oncorhynchus mykiss*. *Journal of animal science* 77: 2990–2999.

Mancuso, T., Baldi, L., and L. Gasco. 2016. An empirical study on consumer acceptance of farmed fish fed on insect meals: The Italian case. *Aquaculture International* 24: 1489–1507.

Manoppo, H., Djokosetiyanto, D., Sukadi, M. F., and E. Harris. 2011. Enhancement of non-specific immune response, resistance and growth of (*Litopenaeus vannamei*) by oral administration of nucleotide. *Journal Akuakultur Indonesia* 10(1): 1–7.

Marono, S., Piccolo, G., Loponte, R., et al. 2015. In vitro crude protein digestibility of *Tenebrio molitor* and *Hermetia illucens* insect meals and its correlation with chemical composition traits. *Italian Journal of Animal Science* 14: 3889.

Martínez-Córdova, L. R., Vargas-Albores, F., Garibay-Valdez, E., et al. 2018. Amaranth and wheat grains tested as nucleation sites of microbial communities to produce bioflocs used for shrimp culture. *Aquaculture* 497: 503–509.

Martínez-Hernández, G., Castillejo, N., Carrión-Monteagudo, M., Artés, F., and F. Artés-Hernández. 2018. Nutritional and bioactive compounds of commercialized algae powders used as food supplements. *Food Science and Technology International* 24: 172–182.

Megahed, M. E., Elmesiry, G., Ellithy, A., and K. Mohamed. 2018. Genetic, nutritional and pathological investigations on the effect of feeding low protein diet and biofloc on growth performance, survival and disease prevention of Indian white shrimp *Fenneropenaeus indicus*. *Aquaculture International* 26: 589–615.

Messina, M. J. 1999. Legumes and soybeans: Overview of their nutritional profiles and health effects. *American Journal of Clinical Nutrition* 70 Supplement: 439S–450S.

Merrill, A. L. and B. K. Watt. 1973. *Energy Value of Foods—Basis and Derivation*. US Dept. of Agriculture, Agriculture Handbook No. 74, 105 pp.

Minabi, K., Sourinejad, I., Alizadeh, M., Ghatrami, E. R., and M. H. Khanjani. 2020. Effects of different carbon to nitrogen ratios in the biofloc system on water quality, growth, and body composition of common carp (*Cyprinus carpio* L.) fingerlings. *Aquaculture International* 28: 1883–1898.

Mitra, M. and S. Mishra 2019. Multiproduct biorefinery from *Arthrospira* spp. towards zero waste: Current status and future trends. *Bioresource Technology* 291: 1–12.

Mobin, S. and F. Alam. 2017. Some promising microalgal species for commercial applications: A review. *Energy Procedia* 110: 510–517.

Moe, C. M. and A. R. Place. 1999. Characterization of a vertebrate gastric chitinase. *American Zoologist* 39(71A) meeting abstract: 418.

Molina-Poveda, C., Lucas, M. and M. Jover. 2013. Evaluation of the potential of Andean lupin meal (Lupinus mutabilis Sweet) as an alternative to fish meal in juvenile Litopenaeus vannamei diets. *Aquaculture* 410–411: 148–156.

Mondal, K., Kaviraj, A. and P. K. Mukhopadhyay. 2011. Introducing mulberry leaf meal along with fish offal meal in the diet of freshwater catfish. *Heteropneustes fossilis. Electronic Journal of Biology* 7: 54–59.

Montoya-Camacho, N., Marquez-Ríos, E., Castillo-Yáñez, F. J., et al. 2019. Advances in the use of alternative protein sources for tilapia feeding. *Reviews in Aquaculture* 11: 515–526.

De Morais Junior, W. G., Gorgich, M., Corrêa, P. S., Martins, A. A., Mata, T. M., and N. S. Caetano. 2020. Microalgae for biotechnological applications: Cultivation, harvesting and biomass processing. *Aquaculture* 528: 735562.

Moren, M., Malde, M. K., Olsen, R. E., Hemre, G. I., Dahl, L., Karlsen, Ø., and K. Julshamn. 2007. Fluorine accumulation in Atlantic salmon (Salmo salar), Atlantic cod (Gadus morhua), rainbow trout (Onchorhyncus mykiss) and Atlantic halibut (Hippoglossus hippoglossus) fed diets with krill or amphipod meals and fish meal based diets with sodium fluoride (NaF) inclusion. *Aquaculture* 269: 525–531.

Murthy, H. S., Li, P., Lawrence, A. L., and D. M. Gatlin. 2009. Dietary β-glucan and nucleotide effects on growth, survival and immune responses of pacific white shrimp, *Litopenaeus vannamei. Journal of Applied Aquaculture* 21: 160–68.

Nasar, A., Pushpendra Kumar Vishwakarma, I., and M. Sohaib. 2014. *In vitro* antibacterial, antifungal and phytotoxic activities of *Ficus carica* methanolic leaves extracts, *International Journal of Current Biotechnology* 2: 11–15.

Nasseri, A. T., Rasoul-Amini, S., Morowvat, M. H., and Y. Ghasemi. 2011. Single cell protein: Production and process. *American Journal of Food Technology* 6: 103–16.

Nazzaro, J., San Martin, D., Perez-Vendrell, A. M., Padrell, L., Iñarra, B., Orive, M., and A. Estévez. 2021. Apparent digestibility coefficients of brewer's by-products used in feeds for rainbow trout (Oncorhynchus mykiss) and gilthead seabream (Sparus aurata). *Aquaculture* 530: 735796.

Naylor, R. L., Hardy, R. W., Bureau, D. P., et al. 2009. Feeding aquaculture in an era of finite resources. *Proceedings of the National Academy of Sciences* 106: 15103–15110.

Ng, W. K. and M. L. Chen. 2002. Replacement of soybean meal with palm kernel meal in practical diets for hybrid Asian-African catfish, *Clarias macrocephalus* × *C. gariepinus. Journal of Applied Aquaculture* 12: 67–76.

Ng, W. K., Liew, F. L., Ang, L. P., and K. W. Wong. 2001. Potential of mealworm (*Tenebrio molitor*) as an alternative protein source in practical diets for African catfish, *Clarias gariepinus. Aquaculture Research* 3: 273–280.

Ng, W. K., Lim, H. A., Lim, S. L., et al. 2002. Nutritive value of palm kernel meal pretreated with enzyme or fermented with *Trichoderma koningii* (Oudemans) as a dietary ingredient for red hybrid tilapia (*Oreochromis* sp.). *Aquaculture Research* 33: 1199–1207.

Nguyen, N. H. Y., Trinh, L. T., Chau, D. T., Baruah, K., Lundh, T., and A. Kiessling. 2019. Spent brewer's yeast as a replacement for fishmeal in diets for giant freshwater prawn (*Macrobrachium rosenbergii*), reared in either clear water or a biofloc environment. *Aquaculture Nutrition* 25: 970–979.

Nhi, N. H. Y., C. T. Da, T. Lundh, T. T. Lan, and A. Kiessling. 2018. Comparative evaluation of Brewer's yeast as a replacement for fishmeal in diets for tilapia (*Oreochromis niloticus*), reared in clear water or biofloc environments. *Aquaculture* 495: 654–660.

Niccolai, A., Zittelli, G. C., Rodolfi, L., Biondi, N., and M. R. Tredici. 2019. Microalgae of interest as food source: Biochemical composition and digestibility. *Algal Research* 42: 1–9.

Nicol, S., Forster, I., and J. Spence. 2000. Products derived from krill. In: Everson, I., ed. *Krill: Biology, Ecology and Fisheries*. Malden, MA: Blackwell Sciences Ltd, 262–283.

Olier, B. S., Tubin, J. S. B., de Mello, G. L., Martínez-Porchas, M., and M. G. C. Emerenciano. 2020. Does vertical substrate could influence the dietary protein level and zootechnical performance of the Pacific white shrimp *Litopenaeus vannamei* reared in a biofloc system?. *Aquaculture International* 28: 1227–1241.

Oliva-Teles, A. and P. Gonçalves. 2001. partial replacement of fishmeal by brewers yeast (*Saccaromyces cerevisae*) in diets for sea bass (*Dicentrarchus labrax*) juveniles'. *Aquaculture* 202: 269–78.

Oliveira, J. T., Silveira, S. B., Vasconcelos, I. M., Cavada, B. S. and R. A. Moreira. 1999. Compositional and nutritional attributes of seeds from the multiple purpose tree Moringa oleifera Lamarck. *Journal of the Science of Food and Agriculture* 79: 815–820.

Olsen, R. E., Henderson, R. J., Sountama, J., Hemre, G., Ring, E., Melle, W., and D. R. Tocher. 2004. Atlantic salmon, *Salmo salar*, utilizes wax ester-rich oil from *Calanus finmarchicus* effectively. *Aquaculture* 240: 433–449.

Olsen, R. E., Suontama, J., Langmyhr, E., Mundheim, H., Ring, E., Melle, W., Malde, M. K., and G.-I Hemre. 2006. The replacement of fishmeal to Antarctic krill, *Euphausia superba* in diets for Atlantic salmon, *Salmo salar. Aquaculture Nutrition* 12: 280–290.

Omoregie, E. and F. I. Ogbemudia. 1993. Effect of substituting fishmeal with palm kernel meal on growth and food utilization of the Nile tilapia, *Oreochromis niloticus. Israeli Journal of Aquaculture* 45: 113–113.

Oonincx, D. G. A. B., van Itterbeeck, J., Heetkamp, M. J. W., van den Brand, H., van Loon, J. J. A., and A. van Huis. 2010. Sn exploration on greenhouse gas and ammonia production by insect species suitable for animal or human consumption. *PLoS ONE* 5: e14445.

Osava, M. 1999. ENVIRONMENT-BRAZIL: Soy production spreads, threatens Amazon. *Inter Press Service.* http://www.ipsnews.

Ouraji, H., Zaretabar, A. and H. Rahmani. 2013. Performance of rainbow trout (*Oncorhynchus mykiss*) fingerlings fed diets containing different levels of faba bean (*Vicia faba*) meal. *Aquaculture* 416–417: 161–165.

Øverland, M. and A. Skrede. 2017. Yeast derived from lignocellulosic biomass as a sustainable feed resource for use in aquaculture. *Journal of the Science of Food and Agriculture* 97: 733–742.

Øverland, M., Karlsson, A., Mydland, L. T., Romarheim, O. H., and A. Skrede. 2013. Evaluation of Candida utilis, *Kluyveromyces marxianus* and *Saccharomyces cerevisiae* yeasts as protein sources in diets for Atlantic salmon (*Salmo salar*). *Aquaculture* 402: 1–7.

Ozório, R. O. A., Turini, B. G. S., Môro, G. V., Oliveira, L. S. T., Portz, L., and J. E. P. Cyrino. 2010. Growth, nitrogen gain and indispensable amino acid retention of pacu (*Piaractus mesopotamicus*, Holmberg 1887) fed different brewers yeast (*Saccharomyces cerevisiae*) levels. *Aquaculture Nutrition* 16(3): 276–283.

Ozório, R. O., Portz, L., Borghesi, R., and J. E. Cyrino. 2012. Effects of dietary yeast (*Saccharomyces cerevisia*) supplementation in practical diets of tilapia (*Oreochromis niloticus*). *Animals* 2: 16–24.

Panigrahi, A., Sivakumar, M. R., Sundaram, M., et al. 2020. Comparative study on phenoloxidase activity of biofloc-reared pacific white shrimp *Penaeus vannamei* and Indian white shrimp *Penaeus indicus* on graded protein diet. *Aquaculture* 518: 734654.

Panigrahi, A., Sundaram, M., Saranya, C., et al. 2019. Influence of differential protein levels of feed on production performance and immune response of pacific white leg shrimp in a biofloc-based system. *Aquaculture* 503: 118–127.

Panini, R. L., Pinto, S. S., Nóbrega, R. O., Vieira, F. N., Fracalossi, D. M., Samuels, R. I., Prudêncio, E. S., Silva, C. P., and R. D. M. C. Amboni. 2017. Effects of dietary replacement of fishmeal by mealworm meal on muscle quality of farmed shrimp *Litopenaeus vannamei. Food Research International* 102: 445–450.

Paripatananont, T., Boonyaratpalin, M., Pengseng, P., and P. Chotipuntu. 2001. Substitution of soy protein concentrate for fishmeal in diets of tiger shrimp *Penaeus monodon. Aquaculture Research* 32: 369–374.

Pereira, T. G. and A. Oliva-Teles. 2002. Preliminary evaluation of pea seed meal in diets for gilthead sea bream (*Sparus aurata*) juveniles. *Aquaculture Research* 33: 183–189.

Pereira, T. G. and A. Oliva-Teles. 2003. Evaluation of corn gluten meal as a protein source in diets for gilthead sea bream (*Sparus aurata, L.*) juveniles. *Aquaculture Research* 34: 1111–1117.

Peters, R. R., Morales, E. D., Morales, N. M., and J. L. Hernández. 2009. Evaluación de la calidad alimentaria de la harina de Lemna obscura como ingrediente en la elaboración de alimento para tilapia roja (Orechromis spp.). *Revista científica* 19(3): 303–310.

Pongpet, J., Ponchunchoovong, S., and K. Payooha. 2016. Partial replacement of fishmeal by brewer's yeast (*saccharomyces cerevisiae*) in the diets of thai panga (*Pangasianodon Hypophthalmus* × *Pangasius Bocourti*)'. *Aquaculture Nutrition* 22: 575–85.

Prabhu, E., Rajagopalsamy, C. B. T., Ahilan, B., Jeevagan, J. M. A., and M. Renuhadevi. 2018. Effect of dietary supplementation of biofloc meal on growth and survival of GIFT tilapia. *Indian Journal of Fisheries* 65: 65–70.

Qi, S., Zhao, X., Zhang, W., Wang, C., et al. 2018. The effects of 3 different microalgae species on the growth, metamorphosis and MYP gene expression of two sea urchins, *Strongylocentrotus intermedius* and *S. nudus. Aquaculture* 492: 123–131.

Raja, R., Coelho, A., Hemaiswarya, S., Kumar, P., Carvalho, I. S. and A. Alagarsamy. 2018. Applications of microalgal paste and powder as food and feed: An update using text mining tool. *Beni-Suef University Journal of Basic and Applied Sciences* 7: 740–747.

Rawling, M. D., Pontefract, N., Rodiles, A., Anagnostara, I., Leclercq, E., Schiavone, M., et al. 2019. The effect of feeding a novel multistrain yeast fraction on European seabass (Dicentrachus labrax) intestinal health and growth performance. *Journal of the World Aquaculture Society* 50(6): 1108–1122.

Rajeev, R. and M. Bavitha. 2015. Lupins – An alternative protein source for aquaculture diets. *International Journal of Applied Research* 1: 04–08.

Ramos-Elorduy, J. 1999. Insects as intermediate biotransformers to obtain proteins. In: Dickinson-Bannack, F. and Garcia-Santaella, E. (Eds.), *Homo sapiens: An endangered species, towards a global strategy for survival. Proceedings of the 4th World Academic Conference on Human Ecology 1993*, Yucatan, Mexico, pp. 157–165.

Ray, A. J., Drury, T. H., and A. Cecil. 2017. Comparing clear-water RAS and biofloc systems: Shrimp (*Litopenaeus vannamei*) production, water quality, and biofloc nutritional contributions estimated using stable isotopes. *Aquacultural Engineering* 77: 9–14.

Refstie, S. and T. Storebakken. 2001. Vegetable protein sources for carnivorous fish: Potential and challenges. *Recent Advances in Animal Nutrition in Australia* 13: 195–203.

Refstie, S., Baeverfjord, G., Seim, R. R., and O. Elvebø. 2010. Effects of dietary yeast cell wall β-glucans and MOS on performance, gut health, and salmon lice resistance in Atlantic salmon (*Salmo salar*) fed sunflower and soybean meal. *Aquaculture* 305(1–4): 109–116. doi:10.1016/j.aquaculture.2010.04.005

Rehbein, H., Danulat, E., and M. Leineman. 1986. Activities of chitinase and protease and concentration of fluoride in the digestive tract of Antarctic fishes feeding on krill (*Euphausia superba Dana*). *Comparative Biochemistry and Physiology* 85A: 545–551.

Renna, M., Schiavone, A., Gai, F., et al. 2017. Evaluation of the suitability of a partially defatted black soldier fly (*Hermetia illucens* L.) larvae meal as ingredient for rainbow trout (*Oncorhynchus mykiss* Walbaum) diets. *Journal of Animal Science and Biotechnology* 8: 57.

Reyes, M., Rodríguez, M., Montes, J., et al. 2020. Nutritional and growth effect of insect meal inclusion on seabass (*Dicentrarchuss labrax*) feeds. *Fishes* 5(2): 16

Ribeiro, C. S., Moreira, R. G., Cantelmo, O. A., and E. Esposito. 2014. The use o *Kluyveromyces marxianus* in the diet of Red-Stirling tilapia (*Oreochromis niloticus*, L nnaeus) exposed to natural climatic variation: Effects on growth performance, fatty acids, and protein deposition. *Aquaculture Research* 45: 812–827. doi:10.1111/are.12023

Ringø, E., Sperstad, S., Myklebust, R., Mayhew, T. M., Mjelde, A., Melle, W., and R. E. Olsen. 2006. The effect of dietary krill supplementation on epithelium-associated bacteria in the hindgut of Atlantic salmon (*Salmo salar* L.). A microbial and electron microscopical study. *Aquaculture Ressearch* 37: 1644–1653.

Ritala, A., Häkkinen, S. T., Toivari, M., and M. G. Wiebe. 2017. Single cell protein—state-of-the-art, industrial landscape and patents 2001–2016. *Frontiers in Microbiology* 8: 2009.

Robles-Porchas, G. R., Gollas-Galván, T., Martínez-Porchas, M., Martínez-Cordova, L. R., Miranda-Baeza, A., and F. Vargas-Albores. 2020. The nitrification process for nitrogen removal in biofloc system aquaculture. *Reviews in Aquaculture* 12(4): 2228–2249. doi:10.1111/raq.12431

Rodríguez, A., Cuesta, A., Ortuño, J., Esteban, M. A., and J. Meseguer. 2003. Immunostimulant properties of a cell wall-modified whole saccharomyces cerevisiae strain administered by diet to seabream (*Sparus aurata* L.)'. *Veterinary Immunology and Immunopathology* 96: 183–192. doi:10.1016/j.vetimm.2003.07.001

Saleh, R., Burri, L., Benitez-Santana, T., Turkmen, S., Castro, P., and M. Izquierdo. 2018. Dietary krill meal inclusion contributes to better growth performance of gilthead seabream juveniles. *Aquaculture Research* 49(10): 3289–3295.

Sabapathy, U. and L. H. Teo. 1993. A quantitative study of some digestive enzymes in the rabbitfish, *Siganus canaliculatus* and the sea bass, *Lates calcarifer. Journal of Fish Biology* 42: 595–602.

Saether, O., Ellingsen, T. E., and V. Mohr. 1986. Lipids of North Atlantic krill. *Journal Lipid Research* 27: 274–285.

Salnur, S., Gultepe, N., and B. Hossu. 2009. Replacement of fish meal by yeast (Saccharomyces cerevisiae): Effects on digestibility and blood parameters for gilthead sea bream (*Sparus aurata*). *Journal of Animal and Veterinary Advances* 8: 2557–2561.

Sánchez-Muros, M. J., Barroso, F. G. and F. Manzano-Agugliaro. 2014. Insect meal as renewable source of food for animal feeding: A review. *Journal of Cleaner Production* 65: 16–27.

Sánchez-Muros, M., De Haro, C., Sanz, A., Trenzado, C. E., Villareces, S., and F. G. Barroso. 2016. Nutritional evaluation of Tenebrio molitor meal as fishmeal substitute for tilapia (Oreochromis niloticus) diet. *Aquaculture Nutrition* 22(5): 943–955.

Sealey, W. M., Gaylord, T. G., Barrows, F. T., Tomberlin, J. K., McGuire, M. A., Ross, C., and S. St-Hilaire. 2011. Sensory analysis of rainbow trout, Oncorhynchus mykiss, fed enriched black soldier Fly Prepupae, Hermetia illucens. *Journal of the World Aquaculture Society* 42: 34–45.

Sgnaulin, T., E. G. Durigon, S. M. Pinho, G. T. Jeronimo, D. L. D. Lopes, and M. G. C. Emerenciano. 2020. Nutrition of Genetically Improved Farmed Tilapia (GIFT) in biofloc technology system: Optimization of digestible protein and digestible energy levels during nursery phase. *Aquaculture* 521: 734998.

Shafaeipour, A., Yavari, V., Falahatkar, B., Maremmazi, J. and E. Gorjipour. 2008. Effects of canola meal on physiological and biochemical parameters in rainbow trout (*Oncorhynchus mykiss*). *Aquaculture Nutrition* 14: 110–119.

Shao, J. C., Liu, M., Wang, B. J., Jiang, K. Y., Wang, M. Q., and L. Wang. 2017. Evaluation of biofloc meal as an ingredient in diets for white shrimp *Litopenaeus vannamei* under practical conditions: Effect on growth performance, digestive enzymes and TOR signaling pathway. *Aquaculture* 479: 516–521.

Shields, R. and I. Lupatsch. 2012. Algae for aquaculture and animal feeds. *TATuP-Zeitschrift für Technikfolgenabschätzung in Theorie und Praxis* 21: 23–37.

Soares, C. M., Hayashi, C., Faria, A. C. E. A., and W. Furuya. 2001. Replacement of soybean meal protein by canola meal protein in diets for Nile tilapia (*Oreochromis niloticus*) in the growing phase. *Revista Brasileira de Zootecnia* 30: 1172–1177.

Sogari, G., Amato, M., Biasato, I., Chiesa, S. and L. Gasco. 2019. The potential role of insects as feed: A multi-perspective review. *Animals* 9: 119.

Sonesson, U., Antesson, F., Davis, J., and P-O. Sjödén. 2005. Home transports and wastage – environmentally relevant household activities in the life cycle of food. *Ambio* 34: 368–372.

Souza, V. L., Urbinati, E., and E. G. Oliveira. 1997. Restrição alimentar, realimentação and as alterações no desenvolvimento de juvenis de pacu (*Piaractus mesopotamicus HOLMBERG*, 1887). *Boletim do Instituto de Pesca* 24: 19–24.

Spranghers, T., Ottoboni, M., Klootwijk, C., et al. 2017. Nutritional composition of black soldier fly (*Hermetia illucens*) prepupae reared on different organic waste substrates. *Journal of the Science of Food and Agriculture* 97: 2594–2600.

St-Hilaire, S., Cranfill, K., McGuire, M. A., et al. 2007. Fish offal recycling by the black soldier fly produces a foodstuff high in omega-3 fatty acids. *Journal of the World Aquaculture Society* 38: 309–313.

Stadtlander, T., Stamer, A., Buser, A., Wohlfahrt, J., Leiber, F., and C. Sandrock. 2017. *Hermetia illucens* meal as fish meal replacement for rainbow trout on farm. *Journal of Insects as Food and Feed* 3: 165–175.

Steinfeld, H., Gerber, P., Wassenaar, T., Castel, V., Rosales, M., and C. P. R. F. De Haan. 2006. *Livestock's Long Shadow: Environmental Issues and Options*. FAO, Rome, Italy. ftp://ftp.fao.org/docrep/fao/010/a0701e/.

Storebakken, T. Refsite, S. and B. Ruyter. 2000. Soy products as fat and protein sources in fish feeds for intensive aquaculture. In J. K. Darckly (Ed.), *Federation of Animal Science Societies, Champaign IL* pp. 127–170.

Sudhakar, K., Mamat, R., Samykano, M., Azmi, W. H., Ishak, W. F. W., and T. Yusaf. 2018. An overview of marine macroalgae as bioresource. *Renewable and Sustainable Energy Reviews* 91: 165–179.

Suontama, J., Karlsen, Ø., Moren, M., et al. 2007. Growth, feed conversion and chemical composition of Atlantic salmon (*Salmo salar* L.) and Atlantic halibut (*Hippoglossus hippoglossus* L.) fed diets supplemented with krill or amphipods. *Aquaculture Nutrition* 13(4): 241–255.

Suparmaniam, U., Lam, M. K., Uemura, Y., Lim, J. W., Lee, K. T., and S. H. Shuit. 2019. Insights into the microalgae cultivation technology and harvesting process for biofuels production: A review. *Renewable and Sustainable Energy Reviews* 115: 109361.

Swick, R. A., Akiyama, D. M., Boonyaratpalin, M., et al. 1995. Use of soybean meal and synthetic methionine in shrimp feed. *ASA – Technical Bulletin* AQ43-1995.

Szabo, N. J., Matulka, R. A., and T. Chan. 2013. Safety evaluation of whole algalin protein (wap) from *Chlorella protothecoides*. *Food and Chemical Toxicology* 59: 34–45.

Tabassum, A., Abbasi, T., and S. A. Abbasi. 2016. Reducing the global environmental impact of livestock production: The minilivestock option. *Journal of Cleaner Production* 112: 1754–1766.

Tacon, A. and A. Jackson. 1985. Utilization of conventional and unconventional protein sources in practical fish feeds. In Cowey, C. B., Mackie, A. M., and Bell, J. G. (Eds.), *Nutrition and feeding in fish*. Academic Press, London, 119–145.

Tacon, A. G. J., Cody, J. J., Conquest, L. D., Divakaran, S., Forster, I. .P, and O. E. Decamp. 2002. Effect of culture system on the nutrition and growth performance of Pacific white shrimp *Litopenaeus vannamei* (Boone) fed different diets. *Aquaculture Nutrition* 8: 121–137.

Takakuwa, F., Suzuri, K., Horikawa, T., et al. 2020. Availability of potato protein concentrate as an alternative protein source to fish meal in greater amberjack (*Seriola dumerili*) diets. *Aquaculture Research* 51: 1293–1302.

Taufek, N. M., Muin, H., Raji, A. A., Md Yusof, H., Alias, Z., and S. A. Razak. 2018. Potential of field crickets meal (*Gryllus bimaculatus*) in the diet of African catfish (*Clarias gariepinus*). *Journal of Applied Animal Research* 46: 541–546.

Tesser, M. B., Cardozo, A. P., Camano, H. N., and W. Wasielesky. 2019. Replacement of fishmeal and fish oil with vegetable meal and oil in feedstuffs used in the growing phase of the Pacific white shrimp *Litopenaeus vannamei*, in biofloc systems. *Arquivo Brasileiro De Medicina Veterinaria e Zootecnia* 71: 703–710.

Teuling, E., Wierenga, P. A., Agboola, J. O., Gruppen, H., and J. W. Schrama. 2019. Cell wall disruption increases bioavailability of *Nannochloropsis gaditana* nutrients for juvenile Nile tilapia (*Oreochromis niloticus*). *Aquaculture* 499: 269–282.

Tewary, A. and B. C. Patra. 2011. Oral administration of baker's yeast (*Saccharomyces cerevisiae*) acts as a growth promoter and immunomodulator in *Labeo rohita* (Ham.). *Journal of Aquaculture Research and Development* 2: 109.

Thiessen, D. L., Campbell, G. L., and P. D. Adelizi. 2003. Digestibility and growth performance of juvenile rainbow trout (*Oncorhynchus mykiss*) fed with pea and canola products. *Aquaculture Nutrition* 9: 67–75.

Tibbetts, S. M., Mann, J., and A. Dumas. 2017. Apparent digestibility of nutrients, energy, essential amino acids and fatty acids of juvenile Atlantic salmon (*Salmo salar* L.) diets containing whole-cell or cell-ruptured Chlorella vulgaris meals at five dietary inclusion levels. *Aquaculture* 481: 25–39.

Tibbetts, S. M., Milley, J. E., and S. P. Lall. 2015. Chemical composition and nutritional properties of freshwater and marine microalgal biomass cultured in photobioreactors. *Journal of Applied Phycology* 27: 1109–1119.

Tierney, T. W., and A. J. Ray. 2018. Comparing biofloc, clear-water, and hybrid nursery systems (Part I): Shrimp (*Litopenaeus vannamei*) production, water quality, and stable isotope dynamics. *Aquacultural Engineering* 82: 73–79.

Tinh, T. H., Koppenol, T., Hai, T. N., Verreth, J. A. J., and M. C. J. Verdegem. 2021. Effects of carbohydrate sources on a biofloc nursery system for whiteleg shrimp (*Litopenaeus vannamei*). *Aquaculture* 531: 735795.

Torres-Tiji, Y., Fields, F. J., and S. P. Mayfield. 2020. Microalgae as a future food source. *Biotechnology Advances* 41: 1–13.

Tou, J. C., Jaczynski, J., and Y. C. Chen. 2007. Krill for human consumption: Nutritional value and potential health benefits. *Nutrition Reviews* 65: 63–77.

Tubin, J. S. B., Paiano, D., Hashimoto, G. S. D., et al. 2020. *Tenebrio molitor* meal in diets for Nile tilapia juveniles reared in biofloc system. *Aquaculture* 519: 734763.

Tukmechi, A. and M. Bandboni. 2014. Effects of *saccharomyces cerevisiae* supplementation on immune response, hematological parameters, body composition and disease resistance in rainbow trout, *Oncorhynchus mykiss* (Walbaum, 1792)'. *Journal of Applied Ichthyology* 30: 55–61. doi:10.1111/jai.12314

Turchini, G. M., Trushenski, J. T. and B. D. Glencross. 2019. Thoughts for the future of aquaculture nutrition: Realigning perspectives to reflect contemporary issues related to judicious use of marine resources in aquafeeds. *North American Journal of Aquaculture* 81: 13–39.

Tusche, K., Arning, S., Wuerts, S., et al. 2012. Wheat gluten and potato protein concentrate – Promising protein sources for organic farming of rainbow trout (*Oncorhynchus mykiss*). *Aquaculture* 344–349: 120–125.

Uran, P., Gonçalves, A., Taverne-Thiele, J., et al. 2008. Soybean meal induces intestinal inflammation in common carp (*Cyprinus carpio L.*). *Fish and Shellfish Immunology* 25: 751–760.

Vadhel, N., Pathan, J., Shrivastava, V., et al. 2020. Comparative study on the performance of genetically improved rohu Jayanti and native rohu, *Labeo rohita* fingerlings reared in biofloc system. *Aquaculture* 523: 735201.

Valente, L. M. P., Gouveia, A., Rema, P., Matos, J., Gomes, E. F., and I. S. Pinto. 2006. Evaluation of three seaweeds *Gracilaria bursa-pastoris*, *Ulva rigida* and *Gracilaria cornea* as dietary ingredients in European sea bass (*Dicentrarchus labrax*) juveniles. *Aquaculture* 252: 85–91.

Valle, B. C. S., Dantas, E. M., Silva, J. F. X., et al. 2015. Replacement of fishmeal by fish protein hydrolysate and biofloc in the diets of *Litopenaeus vannamei* postlarvae. *Aquaculture Nutrition* 21: 105–112.

van der Meeren, T., Olsen, R. E., Hamre, K., and H. J Fyhn. 2008. Biochemical composition of copepods for evaluation of feed quality in production of juvenile marine fish. *Aquaculture* 274: 375–397.

van Huis, A. 2015. Edible insects contributing to food security? *Agriculture & Food Security* 4: 20.

Verbeke, W., Spranghers, T., De Clercq, P., De Smet, S., Sas, B., and M. Eeckhout. 2015. Insects in animal feed: Acceptance and its determinants among farmers, agriculture sector stakeholders and citizens. *Animal Feed Science and Technology* 204: 72–87.

Vieira, E. F., Soares, C., Machado, S., et al. 2018. Seaweeds from the Portuguese coast as a source of proteinaceous material: Total and free amino acid composition profile. *Food Chemistry* 269: 264–275.

Vismara, R., Vestri, S., Barsanti, L., and P. Gualtieri. 2003. Diet-induced variations in fatty acid content and composition of two on-grown stages of Artemia salina. *Journal of Applied Phycology* 15: 477–483.

Walker, D. A. U., Suazo, M. C. M., and M. G. C. Emerenciano. 2020. Biofloc technology: Principles focused on potential species and the case study of Chilean river shrimp *Cryphiops caementarius*. *Reviews in Aquaculture* 12: 1759–1782.

Wei, Y. F., Liao, S.-A., and A.-L. Wang. 2016. The effect of different carbon sources on the nutritional composition, microbial community and structure of bioflocs. *Aquaculture* 465: 88–93.

Welker, T. L., Lim, C., Yildirim-Aksoy, M., and P. H. Klesius. 2012. Effect of short-term feeding duration of diets containing commercial whole-cell yeast or yeast subcomponents on immune function and disease resistance in channel catfish, *Ictalurus punctatus'*. *Journal of Animal Physiology and Animal Nutrition* 96: 159–171. doi:10.1111/j.1439-0396.2011.01127.x

Webster, C. D., Tiu, L. G., Tidwell, J. H., and J. M. Grizzle 1997. Growth and body composition of channel catfish (*Ictalurus punctatus*) fed diets containing various percentages of canola meal. *Aquaculture* 150: 103–112.

Xie, S. and A. Jokumsen. 1998. Effects of dietary incorporation of potato protein concentrate and supplementation of methionine on growth and feed utilization of rainbow trout. *Aquaculture Nutrition* 4: 183–186.

Xie, S., Zhu, X., Cui, Y., et al. 2001. Utilization of several plant proteins by Gibel carp (*Carassius auratus gibelio*). *Journal of Applied Ichthyology* 17: 70–76.

Xu, W. J. and L. Q. Pan. 2014a. Dietary protein level and C/N ratio manipulation in zero-exchange culture of *Litopenaeus vannamei*: Evaluation of inorganic nitrogen control, biofloc composition and shrimp performance. *Aquaculture Research* 45: 1842–1851.

Xu, W. J. and L. Q. Pan. 2014b. Evaluation of dietary protein level on selected parameters of immune and antioxidant systems, and growth performance of juvenile *Litopenaeus vannamei* reared in zero-water exchange biofloc-based culture tanks. *Aquaculture* 426: 181–188.

Yan, J., Chang, Q., Chen, S., Wang, Z., Lu, B., Liu, C., and J. Hu. 2018. Effect of dietary Antarctic krill meal on growth performance, muscle proximate composition, and antioxidative capacity of juvenile spotted halibut, Verasper variegatus. *Journal of the World Aquaculture Society* 49(4): 761–769.

Yarnold, J., Karan, H., Oey, M., and B. Hankamer. 2019. Microalgal aquafeeds as part of a circular bioeconomy. *Trends in Plant Science* 24: 959–970.

Yoshitomi, B., Aoki, M., and S. I. Oshima. 2007. Effect of total replacement of dietary fish meal by low fluoride krill (*Euphausia superba*) meal on growth performance of rainbow trout (*Oncorhynchus mykiss*) in fresh water. *Aquaculture* 266(1–4): 219–225.

Yoshitomi, B., Aoki, M., Oshima, S., and K. Hata. 2006. Evaluation of krill (*Euphausia superba*) meal as a partial replacement for fish meal in rainbow trout (*Oncorhynchus mykiss*) diets. *Aquaculture* 261: 440–446.

Younis, E. S. M., Al-Quffail, A. S., Al-Asgah, N. A., Abdel-Warith, A. W. A., and Y. S. Al-Hafedh. 2018. Effect of dietary fish meal replacement by red algae, *Gracilaria arcuata*, on growth performance and body composition of Nile tilapia *Oreochromis niloticus*. *Saudi Journal of Biological Sciences* 25: 198–20.

Yu, H. H., Han, F., Xue, M., Wang, J., Tacon, P., Zheng, Y. H., Wu, X. F., and Y. J. Zhang. 2014. Efficacy and tolerance of yeast cell wall as an immunostimulant in the diet of japanese seabass (*Lateolabrax japonicus*). *Aquaculture* 432: 217–224. doi:10.1016/j.aquaculture.2014.04.043

Yu, Z., Li, L., Zhu, R., Li, M., et al. 2020. Monitoring of growth, digestive enzyme activity, immune response and water quality parameters of Golden crucian carp (*Carassius auratus*) in zero-water exchange tanks of biofloc systems. *Aquaculture Reports* 16: 100283.

Yue, Z., Han, X., Mei, Y., Chuanzhi, Z., and X. W. Ai Qin Li. 2012. Cloning and expression analysis of peanut (*Arachis hypogaea* L.). *Electronic Journal of Biotechnology* 15: 1.

Zerai, D. B., Fitzsimmons, K. M., Collier, R. J., and G. C. Duff. 2008. Evaluation of brewer's waste as partial replacement of fish meal protein in Nile tilapia, *Oreochromis niloticus*, diets. *Journal of the World Aquaculture Society* 39(4): 556–564. doi:10.1111/j.1749-7345.2008.00186.x

Zhang, Y., Wu, Y., Jiang, D., et al. 2014. Gamma-irradiated soybean meal replaced more fish meal in the diets of Japanese seabass (*Lateolabrax japonicus*). *Animal Feed Science and Technology* 197: 155–163.

Zhao, P., Huang, J., Wang, X.-H., et al. 2012. The application of bioflocs technology in high-intensive, zero exchange farming systems of *Marsupenaeus japonicus*. *Aquaculture* 354–355: 97–106.

Zhou, J. S., Liu, S. S., Ji, H., and H. B. Yu. 2018. Effect of replacing dietary fish meal with black soldier fly larvae meal on growth and fatty acid composition of Jian carp (*Cyprinus carpio* var. Jian). *Aquaculture Nutrition* 24(1): 424–433.

Zhu, F. Z., Quan, H., Du, H., and Z. Xu. 2010. The effect of dietary chitosan and chitin supplementation on the survival and immune reactivity of crayfish, *Procambarus clarkii*. *Journal of World Aquaculture Society* 41: 284–290.

ABBREVIATIONS

ALA –: linolenic acid
ARA –: arachidonic acid
CP –: crude protein
DHA –: docosahexaenoic acid
DW –: dry weight
EPA –: eicosapentaenoic acid
LA –: linoleic acid
BFT –: Biofloc technology
CW –: clear-water system
EAA –: essential amino acids
FA –: fatty acids
FBW –: final body weight
FCR –: feed conversion ratio
FM –: fishmeal
HI –: *Hermetia illucens*
HUFA –: highly unsaturated fatty acids
IBW –: initial body weight,
MUFA –: monounsaturated
PER –: protein efficiency ratio
PKM –: palm kernel meal
PPC –: potato protein concentrate
PUFA –: polyunsaturated fatty acids
SGR –: specific growth rate
SFA –: satutared fatty acids
TM –: *Tenebrio molitor*

9 Fish Oil Sparing and Alternative Lipid Sources in Aquafeeds

Mansour Torfi Mozanzadeh, Fatemeh Hekmatpour, and Enric Gisbert

CONTENTS

INTRODUCTION

Aquaculture production, as a global industrial activity, accounts for at least 50% of all finfish consumed in the world, and this fast-growing industry has been fueled by manufactured compound aquafeeds. In 2018, the aquaculture industry produced *ca.* 114,508,041.8 t of seafood (US$263.4 billion), of which 54.3 MT were finfish (US$ 138.5 billion) production (FAO 2020). Marine-derived feed ingredients, mainly fishmeal (FM) and fish oil (FO), are the main feedstuffs for protein and lipid sources in formulated diets using by this growing industry. Marine-derived oils not only have an excellent palatability, a highly digestible energy content, but they also own the gold standard for amount and ratios of long-chain polyunsaturated fatty acids (LC-PUFA), in particular arachidonic acid (ARA; 20:4 *n*-6), eicosapentaenoic acid (EPA; 20:5 *n*-3), and docosahexaenoic acid (DHA; 22:6 *n*-3). The aquaculture segment uses a major portion of FO for fish feeding, whereas direct human

DOI: 10.1201/9780429331664-9

consumption of this ingredient is comparatively low. However, in recent years due to increasing awareness regarding the benefits of FO on cardiovascular disease and health condition, the FO market is rising worldwide. But, there is a stagnation in FO supply during recent years, with values ranging from *ca.* 0.8 to 1.0 MT (IFFO 2016). It has been reported that 73% of FO was used by the aquafeed industry, 21% for human purposes (e.g. medicine and cosmetic industries) and 6% for other usages in 2016 (IFFO 2016). In addition, the proportion of FO consumption for different farmed species is as follows: aquafeeds production in aquaculture of salmonids (58%), marine fish species (23%), tilapias (7%), crustaceans (5%), eels (2%) and other aquatic species (4%) (IFFO 2016). The importance of FO in the nutrition of farmed aquatic organisms is based on their global production and its nutritional requirements in terms of dietary lipids and essential fatty acids (EFA).

However, the uncertain supply of FO, mainly due to overexploitation and steady fisheries of small pelagic fish stocks and their susceptibility to climate change and natural impacts (e.g. El Niño events), as well as high demand for human consumption, has resulted in a progressive increase in price of this commodity (Hardy 2010; Tacon and Metian 2015). Furthermore, the market value of n-3 LC-PUFAs is estimated to be about US 57.07 billion by 2025 with the growing rate of 6% during 2018–2025 (Grand View Research 2017). According to the IFFO (2016), the worldwide market of FO in 2016 was US$ 2.22 billion and it is expected to increase up to US$ 3.69 billion by 2025. Regardless of the good nutritional qualities of FO, several drawbacks limit their use at high inclusion rates in aquafeeds including (i) limited availability of FO with regard to aquaculture production needs; (ii) marine-derived oils are susceptible to contamination and bioaccumulation of organic pollutants (e.g. dioxins and dioxin-like polychlorinated biphenyls) or mercury (Turchini et al. 2009); (iii) increasing environmental, ethical and economic concerns about "fishing down" marine food webs for providing marine-derived aquafeeds (Froese et al. 2016); (iv) low nutritional stability of FO (Ji et al. 2011) and (v) the quality of FO may fluctuate depend on the season and location of fishing (Park et al. 2018). These issues confirmed that aquafeed manufacture industry cannot more rely only on FO as the main dietary lipid source and should find alternative lipid sources (ALS).

ALTERNATIVE LIPID SOURCES

Sparing costly ingredients like FO with alternative sources of energy and EFA may reduce aquafeed costs; however, successful least-cost feed formulation requires considerable knowledge of the intended fish's nutritional demands and tolerances. Under this scenario, a large body of literature exists evaluating dietary FO sparing in many aquaculture species using many kinds of alternative lipid sources (ALS) (Turchini et al. 2009; Turchini et al. 2011b; Alhazzaa et al. 2019), which has resulted in a progressive replacement of FO with ALS in compound aquafeeds (Shepherd et al. 2017). The selection of different ALS for FO sparing in diets is generally conducted based on different principles such as their price, fatty acid (FA) composition, accessibility, and sustainable production levels and lower concentrations of organic pollutants (Turchini et al. 2009) (Tables 9.1 and 9.2). The main ALS for FO sparing in aquafeeds may be listed as vegetal oils (VO), oils derived from genetically modified oilseeds (GMO), rendered animal fats (RAF), other marine resources (e.g. Antarctic krill and lantern fishes) and microbial oils.

Vegetal Oils

Vegetal oils are rich in n-3, n-6, and n-9 polyunsaturated fatty acids (PUFA) but devoid of any LC-PUFA, especially EPA and DHA abundant in FO (Glencross 2009; Turchini et al. 2009). The global use of vegetal oils (VO) can be divided into domestic (mainly as food) and industrial (e.g. biofuel) consumption. The global production of VO exceeds 135 MT, while 80% of the production corresponds to different oils commonly used in aquafeeds like palm (42.4 MT), soybean (37.7 MT), rapeseed/canola (19.4 MT), and sunflower (10.1 MT) oils. Other VO that have been evaluated as ALS in many fish compound diets are cottonseed, groundnut, coconut, olive, corn, sesame,

TABLE 9.1
Global Production of Oils in the Last Decade (USDA, December 2019)

	Production (million metric tons)									
	2010/11	2011/12	2012/13	2013/14	2014/15	2015/16	2016/17	2017/18	2018/19	2019/20
Fish Oil			0.97	0.95	0.96	0.91	0.9			
Coconut	3.71	3.41	3.66	3.45	3.43	3.31	3.39	3.66	3.67	3.58
Cottonseed	4.96	5.21	5.21	5.14	5.2	4.30	4.43	5.18	5.12	5.22
Olive	3.27	3.2	2.38	3.03	3.19	3.13	2.48	3.26	3.09	3.36
Palm	48.84	52.11	55.97	59.56	63.29	58.86	65.18	70.63	73.90	75.69
Palm kernel	5.73	6.13	6.52	6.96	7.31	7.01	7.63	8.34	8.59	8.79
Peanut	5.31	5.3	5.49	5.58	5.55	5.42	5.70	5.90	5.79	6.00
Rapeseed	23.46	24.11	24.99	26.09	26.47	27.35	27.55	28.11	27.44	27.04
Soybean	41.29	42.6	42.9	44.61	46.88	51.56	53.81	55.15	55.73	56.73
Sunflower seed	12.43	14.87	13.48	15.59	15.44	15.38	18.15	18.51	19.71	20.65
Total	148.98	156.95	160.59	170.01	176.76	176.29	188.33	198.73	203.04	207.06

linseed, wheat germ, rice bran, and echium oils. Regardless of their origin, VO are generally classified according to their FA profiles: (i) VO rich in saturated fatty acids (SFA) like palm (CPO), coconut, and fully hydrogenated soybean oils; (ii) VO rich in monounsaturated fatty acid (MUFA) like canola/rapeseed, olive, sesame, groundnut, and high oleic sunflower oils; (iii) VO rich in n-6 PUFA, like soybean, sunflower, cottonseed, and corn oils; and (iv) VO rich in n-3 PUFA-rich VO like linseed, perilla, and camelina seed oils.

Among different VO rich in SFA, palm, coconut, and hydrogenated soybean oils are rich in palmitic (16:0, *ca.* 50%), lauric (12:0, *ca.* 40%), and stearic (18:0, ≥70%) acids, respectively. Oleic acid (OA, 18:1n-9) is the main FA in the VO rich in MUFA, whereas linoleic (LNA, 18:2n-6) and α-linolenic (ALA, 18:3n-3) acids are the main FA in the VO rich in n-6 and n-3 PUFA, respectively. In addition, the expansion of genetically modified oilseeds (GMO) has represented a valuable source of VO for human and animal nutrition and an alternative to FO. For instance, genetically modified soybean produces an oil rich in stearidonic acid, which is a significant precursor of EPA (Winwood 2015). Other modified crops are rich in n-3 LC-PUFA due to their capacity for synthetizing EFA (Betancor et al. 2016; Betancor et al. 2017; Napier et al. 2019), although their use is based upon national and international regulations and approvals. For example, genetically modified plants like canola (or rapeseed) produce oils with high DHA levels and represent a good source of n-3 LC-PUFA for aquafeeds. For instance, a canola crop oil can produce up to 1,200–1,500 kg of oil per hectare; which represents 100–150 kg of n-3 LC-PUFA (OGTR 2018). Furthermore, transgenic *Camelina sativa*, which was developed by transferring algal genes encoding for n-3 LC-PUFA biosynthetic pathways into the plant genome, showed promising properties for producing either EPA alone or both EPA and DHA (Ruiz-Lopez et al. 2014; Haslam et al. 2015).

Rendered Animal Fats

Rendered animal fats (RAF), like poultry fat, lard (rendered fat from pigs), tallow (from cattle and sheep), and yellow grease, are the processing by-products of the meat and leather industries. These ALS are characterized for their richness in SFA and MUFA with high β-oxidation potential that make them preferable as energy sources. Production of tallow, especially from cattle, is higher than other RAF such as greases, poultry fats, and lards. The physical and biochemical properties of RAF

TABLE 9.2
Fatty Acid Composition of Alternative Lipid Sources to Fish Oil (% total fatty acids)

Oils/Fats	SFA	MUFA	LNA	ALA	ARA	EPA	DHA	n-6 PUFA	n-3 PUFA	n-3/n-6 Ratio	Reference
Fish Oils											
Anchovy oil	28.8	24.9	1.2	0.8	0.1	17.0	8.8	1.3	31.2	24.0	(Turchini et al. 2011)
Capelin oil	20.0	61.7	1.7	0.4	0.1	4.6	3.0	1.8	12.2	6.8	(Turchini et al. 2011)
Menhaden oil	30.5	24.8	1.3	0.3	0.2	11.0	9.1	1.5	25.1	16.7	(Turchini et al. 2011)
Herring oil	20.0	56.4	1.1	0.6	0.3	8.4	4.9	1.4	17.8	12.7	(Turchini et al. 2011)
Cod liver oil	19.4	46.0	1.4	0.6	1.6	11.2	12.6	3.0	27.0	9.0	(Turchini et al. 2011)
Jack mackerel	26.0	37.0	4.0	1.0	1.0	2.0	8.0	5.0	26.0	5.2	(Glencross 2009)
Squid oil	28.0	35.0	2.0	2.0	1.0	11.0	12.0	3.0	31.0	10.3	(Turchini et al. 2011)
Vegetable Oils											
Crude palm oil	48.8	37.0	9.1	0.2	-	-	-	9.6	0.2	0.0	(Turchini et al. 2011)
Soybean oil	14.2	23.2	51.0	6.8	-	-	-	51.0	6.8	0.1	(Turchini et al. 2011)
Fully hydrogenated SBO	98.5	0.8	0.2	0.1	-	--	-	0.2	0.1	0.5	(Woitel et al. 2014)
Canola/ rapeseed oil	4.6	62.3	20.2	12.0	-	-	-	20.2	12.0	0.6	(Turchini et al. 2011)
Sunflower oi	10.4	19.5	65.7	-	-	-	-	65.7	-	0.0	(Turchini et al. 2011)
Cottonseed oil	30.5	17.8	51.5	0.2	-	-	-	51.5	0.2	0.0	(Turchini et al. 2011)
Groundnut oil	11.8	46.2	32.0	-	-	-	-	32.0	-	-	(Turchini et al. 2011)
Corn oil	12.7	24.2	58.0	0.7	-	-	-	58.0	0.7	0.0	(Turchini et al. 2011)
Linseed oil	9.4	20.2	12.7	53.3	-	-	-	12.7	53.3	4.2	(Turchini et al. 2011)
Olive oil	13.4	74.6	11.0	1.0	-	-	-	11.0	1.0	0.1	(Salini et al. 2015b)
Flaxeed oil											(Turchini et al. 2011)
Coconut oil	86.5	5.9	1.8	0.1	-	-	-	1.9	0.1	0.1	(Turchini et al. 2011)
Wheat germ oil	17.1	16.5	60.7	4.2	-	-	-	60.7	4.2	0.1	(Turchini et al. 2011)
Echium oil	11.0	19.0	18.0	32.0	-	-	-	18.0	46.0	2.6	(Turchini et al. 2011)
Sesame oil	15.5	42.3	42.4	0.3	-	-	-	42.0	0.3	-	(Turchini et al. 2011)
Camelina Oil											(Turchini et al. 2011)
Wiled type	8.8	13.8	17.1	31.7	-	-	-	-	-	-	(Usher et al. 2017)
EPA-B4-1	9.7	12.5	12.4	18.2	3.2	16.2	-	-	-	-	(Usher et al. 2017)
DHA-5-33	9.2	7.7	20.4	22.1	1.3	4.3	4	-	-	-	(Usher et al. 2017)
Animal Fats											(Turchini et al. 2011)
Beef tallow	47.5	40.5	3.1	0.6	0.4	-	-	3.1	0.6	0.2	(Turchini et al. 2011)
Pork lard	38.6	44.0	10.2	1.0	-	-	-	10.2	1.0	0.1	(Turchini et al. 2011)
Poultry fat	28.5	43.1	19.5	1.0	-	-	-	19.5	1	0.0	(Turchini et al. 2011)
Microalgae Oils/Meals											(Turchini et al. 2011)
***Schizochytrium* sp**	9.2	1.7	-	-	1.4	0.8	43.2	17.8	45.9	2.6	(Sarker et al. 2016a)
***Schizochytrium* sp. T18 oil**	41.9	8.1	0.3	-	-	0.8	40.9	7.8	42.3	5.4	(Tibbetts et al. 2020)
***Nannochloropsis* sp.**	15.4	48.9	5.0	0.3	0.6	3.7	4.9	6.4	9.2	1.4	(Turchini et al. 2011)
***Arthrospira* sp.**	50.6	2.1	19.7	-	-	0.3	-	38.8	0.3	0.0	(Perez-Velazquez et al. 2019)

greatly depend on the history of the diet, species, and age. The quantity of PUFA is limited in RAF and it is lower than 4% in beef tallow and in poultry fat is approximately 20% (mainly n-6 PUFA) (Turchini et al. 2009). In addition, the amount of ALA in poultry fat is higher than in FO (Emery et al. 2014). The lower cost and higher availability of RAF compared to other ALS make them attractive alternatives for FO in formulated aquafeeds (Turchini et al. 2009). However, the level of n-3 LC-PUFA is commonly reported only at trace levels in these ALS, which limits their wider use (Turchini et al. 2011b).

Alternative Marine Oil Sources

Considerable interest in underutilized marine sources and their by-products exist as ALS for aquafeed formulation. Candidate resources include marine invertebrates (copepods, euphausiids, amphipods, mussels, etc.), seal products, and other marine organisms like mesopelagic fish (Olsen et al. 2011). Alternative lipid sources from non-food marine organisms from lower trophic levels, such as zooplankton (e.g. krill or copepods), have been considered as potential ALS in aquafeeds (Olsen et al. 2004; Olsen et al. 2006; Vang et al. 2013). However, these raw ALS have the same restriction factors that incentivize the dietary FO sparing, so it is better to consider these feedstuffs as dietary supplements rather than ALS for FO replacement (Turchini et al. 2018).

Krill is reputed by its content of lipids (0.5–3.6%), but mostly by its content in large quantities of bioactive compounds such as phospholipids, pigments, and fat-soluble vitamins and minerals (Xie et al. 2019). Copepods accumulate high levels of lipids up to 50–70% of their body dry weight, and are of interest to aquaculture mainly as live prey (Vang et al. 2013), although only small exploratory-scale harvesting of *Calanus* sp. was reported (Olsen et al. 2011). Nevertheless, C-Feed AS (Trondheim, Norway) currently grow copepods on a commercial scale as live feeds for early developmental stages of marine fish, crustaceans, and other marine organisms (C-Feed 2014). However, up to now, the harvest of copepods occurred only at a small exploratory scale due to substantial technological and economic challenges. Although the primary commercial drivers for krill fishery are nutritional supplements for human nutrition due to krill oil having a uniquely high content of phospholipid-bound n-3 LC-PUFA, which are thought to be more bioavailable, as well as being involved in regulating more metabolic pathways than the triacylglycerol-bound EPA and DHA found in FO (Ulven and Holven 2015; Xie et al. 2019), new companies like Qrillaqua (AKER BIOMARINE AS, Lysaker, Norway) have promoted its use and commercialization for aquafeeds.

Mesopelagic fishes are the most abundant fishes in the oceanic waters, but they are the least exploited by mankind and their biomass has recently been estimated at 10 billion MT (St. John et al. 2016). The mesopelagic fish, especially those with diel vertical migrations, have high levels of lipid content (32% of dry mass), which contain more than 34% of n-3 LC-PUFA, with the exception of the non-migrant planktivorous fishes whose n-3 PUFA content is lower (21–29%). In addition, the DHA/EPA ratios in the migrant planktivorous range from 3.7 to 4.4, but in non-migrant planktivorous fishes, the range is from 3.0 to 3.7 (Wang et al. 2019). Thus, these fish resources can be considered potential ALS for FO in aquafeeds formulation.

Although marine mammals, such as whales and seals, may be used for the production of oil-related products, there is little likelihood that they will contribute much to marine oils used in aquafeeds. As Olsen et al. (2011) indicated, political and citizenship pressures have worked, or are working, toward bans on the import of seal products, making import difficult in many countries. For this reason, this ALS is not going to be considered in this revision.

Marine Microalgae

Many marine and freshwater microalgae species (i.e., *Spirulina* sp., *Chlorella* sp., *Nannochloropsis* sp., among others) have gained importance as renewable sources to substitute the conventional ingredients in the aquaculture and animal feeds due to their nutritional content and easy production and harvest (Yaakob et al. 2014). In this sense, microalgae are characterized by the high lipid content (30–70% dry weight) (Ward and Singh 2005), but they also have high concentrations of n-3 LC-PUFA (30–40% of total FA) among other important micro- and macro-nutrients and bioactive compounds. Microalgae can deposit large amounts of lipids during their production, especially when manipulating the culture conditions (e.g. stress). Microalgae oil yield in terms of n-3 LC-PUFA and biomass (*ca.* 300 kg/MT) compares very positively with other oil sources, for instance, n-3 LC-PUFA yield in pelagic fish is *ca.* 30 kg/MT of fish and *ca.* 49 kg/MT of vegetal oilseeds (Finco et al. 2017). In addition, microalgae have high quantities of carotenoids with antioxidant effects that can preserve LC-PUFA from auto-oxidation (Dineshbabu et al. 2019). They are also a good source for bio-macromolecules, vitamins (e.g. A, B1, B2, B6, B12, C, E, K, niacin, nicotinate, biotin, and folic acid), and minerals (e.g. calcium, phosphorous, magnesium, potassium, sodium, zinc, iron, copper, and sulphur) (Fox and Zimba 2018; Dineshbabu et al. 2019). In addition to their high nutritional values, microalgae also have immunostimulating effects and increase disease resistance in farmed aquatic species (Dineshbabu et al. 2019). Despite the above-mentioned advantages of microalgae use for dietary purpose, there are some drawbacks and challenges to using microalgae as a replacement for FM and FO in the aquaculture industry, such as their high production costs, poorly digestible cell walls, while the availability of a large amount of biomass could be affected by contamination. Thus, the improvement of technologies for mass production of microalgae and extraction of algal oils are required to reduce the cost of these ALS compared to FO price (Finco et al. 2017). As the LC-PUFA in microalgae oil extracts and algal biomass are in the phospholipid fraction of the oil (Artamonova et al. 2017), the digestibility of these LC-PUFA should be considered for application in aquafeeds (Miller et al. 2011). In this sense, many researchers have reported promising results in different aquatic species when replacing FO with microalgae raw materials (meal and oils), because these feedstuffs enhance the levels of LC-PUFA and n-3/n-6 ratio in diets (Qiao et al. 2014; Sarker et al. 2016; Kangpanich et al. 2017; Perez-Velazquez et al. 2019; Torres Rosas et al. 2019). However, in some cases, total replacement of FO with microalgal meals reduced growth performance in different fish species due to lower digestibility of cell walls of these ingredients compared to FO (Miller et al. 2007; Kousoulaki et al. 2015b; Sprague et al. 2015; Gbadamosi and Lupatsch 2018).

Microbial Oils

Microbes, especially heterotrophic algae (i.e. *Schizochitrium* sp.), yeast, or bacteria due to their exponential growth rate, high cell biomass, and oil productivity as well as accumulation of n-3 LC-PUFA are regarded as very attractive sources of feed ingredients for aquafeeds as they are known as the "Omega-3 Biotechnology" (Orozco Colonia et al. 2020). In this sense, the metabolic engineering of the yeast (*Yarrowia lipolytica*) resulted in a strain that synthetized EPA at 15% of its dry weight (Xue et al. 2013), and it was successfully used as a feedstuff for Atlantic salmon (Hatlen et al. 2012). However, for enhancing the digestibility and bioavailability of EPA, the disruption of the yeast cell walls was required (Berge et al. 2013). In addition, thraustochytrids, which are fungal protists in marine or brackish water environments, have demonstrated superior capacity in LC-PUFA synthesis under optimal heterotrophic cultivation conditions (Lewis et al. 1999; Burja et al. 2006; Orozco Colonia et al. 2020). They can deposit more than half of their dry weight as lipids, whereas DHA usually represent more than a quarter of their total FA content. In addition, because of their highly controlled fermentation processes, their oil extracts are free of the environmental pollutants that contaminate FO (Harwood 2019). Microbial species, such as the marine diatom, *Crypthecodinium* sp., as well as the thraustochytrids, *Thraustochytrium* sp. and *Ulkenia* sp., but especially *Schizochytrium* sp., have been identified due to their ease of cultivation under controlled conditions to produce high lipid biomasses rich in n-3 LC-PUFA and with high n-3/n-6 PUFA ratios (Sprague et al. 2017). For example, a novel strain of *Schizochytrium* sp. (T18) can produce a DHA-rich oil (41%) with an n-3/n-6 ratio five times higher that of other ALS or even FO and krill oil (Tibbetts et al. 2020). Moreover, it has high oxidative stability because of the great number of carotenoids (Orozco Colonia et al. 2020). Nowadays, several companies that are involved in production of food-grade DHA-rich oil products may soon be able to use new thraustochytrid strains for producing LC-PUFA-rich oils at industrial scale for large aquafeed producers (Tibbetts et al. 2017).

Aquaculture and Fisheries By-Products

Economic, environmental, and food security benefits are strong motivating factors in prioritizing the strategic management of aquaculture by-products toward the sustainable growth of the aquaculture industry (Stevens et al. 2018). In this sense, aquaculture and fisheries by-products (e.g. catfish offal oil) when considered as ALS do not only provide economic benefits for the producers, but they also may serve to set up new standards of fish by-product disposal strategy that finally could contribute to sustainability and cost-effective aquaculture production, boosting the circular economy concept (Šimat et al. 2019). Aquatic by-products (e.g. heads, viscera, bones with attached flesh, skins, fins, trimmings, blood) may constitute up to 70% of fish and shellfish after processing, and, in the case of fish fillet processing, its range of 30–50% depends on fish species (Olsen et al. 2014). The FO production from by-products accounts for 26.1% of the total global FO production (Jackson and Newton 2016). In this regard, appropriate storage, processing, and quality control of FO from these materials should be considered as producing environmentally friendly products.

Insects

Although insects have been regarded mainly as alternative sources of proteins in aquafeed, some studies have indicated that they may be an interesting source of LC-PUFA (Tzompa-Sosa et al. 2014). Up to now, the application of insect meal derived from black soldier fly, common housefly, yellow mealworm, lesser mealworm, house cricket, banded cricket, and field cricket is allowed in aquafeeds (EU commission regulation (2017/893–24/05/2017). Usually, insects during larval stages contain higher amounts of lipid with highly variable FA composition that profoundly affect feed's FA profile. Insects have good content of lipids which varies from 8% to 35%, depending on the stages of growth (larval, pupa, nympha, or imago), their growth media, and lipid extraction process (Gasco et al. 2018). The FA profile of most commonly used insect species is dominated by MUFA and PUFA (mainly LNA), but some others, such as black soldier fly (BSF) larvae (*Hermetia illucens*) contain high levels of lauric acid (12:0) (Belghit et al. 2019). In this context, Belghit et al. (2018) reported that the inclusion of the brown algae, *Ascophyllum nodosum*, in the culture media of BSF larvae improved the nutritional value of its extracted oil by introducing n-3 LC-PUFA (mainly EPA), iodine, and vitamin E. The aforementioned authors also reported the inclusion of BSF larvae extracted oil grown on a substrate enriched with marine macroalgae enhanced growth performance in Atlantic salmon compared to fish fed on extracted oil from BSF larvae grown on media containing only terrestrial organic waste. Furthermore, it has been confirmed that the plasticity of the nutritional make-up of insects' larvae allows them to bio-accumulate n-3 LC-PUFA (Liland et al. 2017). Thus, the inclusion of n-3 LC-PUFA is suggested, when using these ingredients in diet. In this regard, pilot research has demonstrated interesting findings regarding partial substitution of both FO and FM in diet with insects or worm meals in farmed fish (Belforti et al. 2016; Gasco et al. 2016), although their use, especially in terms of ALS, deserve further investigation and industrial development.

ESSENTIAL FATTY ACIDS REQUIREMENTS IN AQUATIC SPECIES

Although FO sparing is vital for global aquaculture sustainability, the gaps in knowledge regarding dietary EFA requirements of farmed aquatic species hindered the effective replacement of dietary FO with ALS (Turchini et al. 2009; Turchini et al. 2011b; Trushenski and Rombenso 2020). In this sense, the proper determination of the quantitative determination of EFA requirements for the *ca.* 600 farmed aquatic species worldwide is unrealistic (FAO 2018); therefore, dietary EFA requirements are generally evaluated in a selected fish species as a representative for larger groups of species with similar biological and ecological guilds. Then, this information may be extrapolated in a much broader range of species without the need of running nutritional species-specific trials. Finally, this information may be used for refining the nutritional requirements and validating ALS in FO replacement studies (Table 9.3). A large body of literature confirmed that as long as dietary EFA requirements of an aquatic species are met, FO can be partially or totally substituted with ALS without negative effects on fish growth performance. As reviewed by Glencross (2009), the signs of dietary EFA deficiency in fish species include reduction in feed intake and growth, reduction in immune competence and disease resistance, as well as an increase in body abnormalities at early life stages, and an increase in morbidity and mortality. Thus, dietary FO sparing needs to be conducted by considering EFA requirements of the target species in order to avoid the consequences of EFA deficiencies that may hamper fish performance, health, and welfare.

Like other vertebrates, fish are unable to synthesize medium chain PUFA (MC-PUFA), namely LNA and ALA, from SFA and MUFA precursors due to the lack of Δ12 and Δ15 desaturases (NRC 2011). Regarding the synthesis of LC-PUFA from their precursors, ALA as a substrate for Δ6 desaturase may be converted to 18:4n-3 that subsequently is elongated to 20:4n-3 followed by Δ5 desaturation to synthetize EPA. The bioconversion of LNA to ARA also requires the same conversion enzymes (Figure 9.1). Furthermore, DHA biosynthesis occurs by two elongation steps of EPA and 22:4n-6, followed by Δ6 desaturation, then a chain-shortening β-oxidation (Sprecher's shunt) (Sprecher 2000). Thus, the capacity of aquatic species to convert MC-PUFA to LC-PUFA coincides with their complement of fatty acyl desaturase (FAD) and elongase (ELOVL, elongase of very long chain fatty acids) enzymes (Monroig et al. 2013).

According to the review made by Trushenski and Rombenso (2020), MC-PUFA may primarily function as precursors for LC-PUFA synthesis and may not be physiologically essential. Furthermore, not all LC-PUFA may be essential for an aquatic species and their essentiality in diet can be plastic and reduced by diet manipulation (Trushenski and Rombenso 2020). For example, it has been suggested that SFA- and MUFA-rich ALS, due to their high β-oxidation potential and lower competition (compared to PUFA) with LC-PUFA, can reduce LC-PUFA requirements and facilitate greater FO sparing in diet formulation (Rombenso et al. 2015; Bowzer et al. 2016). Other nutritional factors, such as dietary lipid levels (Glencross 2009) as well as the interaction between MC-PUFA precursors and their LC-PUFA derivatives, and/or interactions with other FA, such as DHA/EPA (Jin et al. 2017a; Xu et al. 2016; Xu et al. 2018; Zhang et al. 2019), DHA/ARA (Araujo et al. 2019), ARA/EPA (Norambuena et al. 2016; Salini et al. 2016; Papiol and Estevez 2019) ratios, may also affect nutritional requirements for LC-PUFA.

Several factors determine the EFA requirements of aquatic species, such as their trophic level, stage of development (i.e. larva, fry, juvenile, and brooder), and environmental conditions, mainly temperature and salinity (Sargent et al. 2002; Glencross 2009; Tocher 2010; Nobrega et al. 2017; Lund et al. 2019; Trushenski and Rombenso 2020). In this regard, certain life stages of development may have high EFA requirements for development, such as brooders, for gonadal maturity, and larval stages, for somatic growth and metamorphosis (Tocher 2010). Furthermore, it has been speculated that the nutritional essentiality of MC-PUFA compared to LC-PUFA is mainly determined by the species' trophic level [the position of the organism in the food web hierarchy from primary producers (level = 1) to top carnivores (level = 5)]. In this sense, fish at low trophic levels (< 3) only require MC-PUFA in their diets, since they are capable of the elongation and desaturation of these precursors to LC-PUFA. In contrast, those species from higher trophic levels (> 4) cannot or have limited ability to bioconvert MC-PUFA to LC-PUFA as *de novo* synthesis of LC-PUFA (Trushenski and Rombenso 2020). In addition, fish in the middle trophic levels (3 < trophic level < 4) may require either MC-PUFA or LC-PUFA, depending on their ecological niche and life stage of development. Moreover, it has been suggested that trophic level cannot alone account for the capacity

TABLE 9.3
Essential Fatty Acid Requirements (g kg^{-1} diet) of Aquaculture Fish Species

Dietary Lipid Level (g kg^{-1})	Environment	Fish Weight (g)	LNA	ALA	ARA	EPA	DHA	n-3 LC-PUFA	n-3/n-6	DHA/EPA	Reference
				Red Seabream (*Pagrus major*)							
100	Marine-subtropical	7.5	-	-	-	10	5	15	-	2:1	(Takeuchi et al. 1992)
100			-	-	-	-	-	12	-	-	(Teshima et al. 1992)
											(Teshima et al. 1992)
150			-	-	-	-	-	37	-	-	(Teshima et al. 1992)
200			-	-	-	-	-	37	-	-	(Teshima et al. 1992)
				Gilthead Sea Bream (*Sparaus aurata*)							
120	Marine-brackish	1	-	-	-	5	5	9	-	1:1	(Kalogeropoulos et al. 1992)
100	Subtropical		-	-	-	12	6	18	-	2:1	(Ibeas et al. 1994)
100		11.5	-	-	-	5	5	10	-	1:1	(Ibeas et al. 1996)
				Silver Bream (*Rhabdosargus sarba*)							
	Marine-brackish-tropical	1.1	-	-	-	-	-	-	1.3	-	(Leu et al. 1994)
				Black Seabream (*Acanthopagrus schlegelii*)							
130	Marine-brackish-subtropical	8.1	-	-	-	-	-	9.4	-	-	(Ma et al. 2013)
140		9.5	-	-	-	-	4.6	5.3		1:0.9	(Jin et al. 2017a)
				Sobaity Seabream (*Sparidentex hasta*)							
150	Marine-brackish-tropical	13.3	-	-	-	-	-	8	-	1:2	(Mozanzadeh et al. 2015)
				Striped Jack (*Pseudocaranx dentex*)							
100	Marine-brackish Tropical	7.5	-	-	-	-	-	1.7	1.7	-	(Takeuchi et al. 1992)
				Turbot (*Scopthalamus maximus*)							
150	Marine-brackish Temperate	0.9	-	-	10	0	0	10	-	-	(Castell et al. 1994)
150		0.9	-	-	-	-	-	35	-	-	(Castell et al. 1994)
				Yellowtail Flounder (*Pleuronectes ferrugineus*)							
100	Marine-temperate	6.8	-	-	-	-	-	-	25	-	(Whalen 1999)
				Japanese Flounder (*Paralicthys olivaceus*)							
100	Marine-subtropical		-	-	-	-	-	14	-	-	(Takeuchi 1997)

TABLE 9.3 (CONTINUED)
Essential Fatty Acid Requirements (g kg^{-1} diet) of Aquaculture Fish Species

Dietary Lipid Level (g kg^{-1})	Environment	Fish Weight (g)	LNA	ALA	ARA	EPA	DHA	n-3 LC-PUFA	n-3/n-6	DHA/EPA	Reference
65		8.5	-	-	-	-	-	8-10	-	-	(Kim and Lee 2004)
				(Platyicthys stellatus)							
100	Marine; freshwater; brackish-polar	1.9	-	-	-	-	-	-	9	-	(Lee et al. 2003)
				Large Yellow Croaker (*Larmichthys crocea*)							
110	Marine-brackish-temperate	9.8							10	3:1	(Zuo et al. 2012b)
110		9.8							9.8	1:2	(Zuo et al. 2012a)
				Cobia (*Rachycentron canadum*)							
100	Marine-brackish-subtropical	52.1	-	-	-	-	3.9	12.6	18.4	-	(Trushenski et al. 2012)
135		5.9	-	-	5-19	3	10	-	-	-	(Araujo et al. 2019)
				Meagre (*Argyrosomus regius*)							
170	Marine-brackish-subtropical	2.8	-	-	-	-	9	10	20	-	(Carvalho et al. 2018)
				Golden Pompano (*Trachinotus ovatus*)							
122	Marine-brackish-subtropical	14.8	-	-	-	-	-	-	12.4–17.3	-	(Li et al. 2019)
				***Seriola* sp.**							
150	Marine-tropical		-	-	-	-	-	20	-	-	(Deshimaru et al. 1982)
				Milkfish (*Chanos chanos*)							
70	Marine-freshwater-brackish-tropical	8.6	-	10	-	5	5	10	-	1:1	(Borlongan 1992)
				Korean Rockfish (*Sebastes schlegeli*)							
	Marine-temperate	-	-	-	-	-	-	-	1	-	(Lee et al. 1993)
				Red Drum (*Sciaenops occeltus*)							
100	Marine-brackish-subtropical	60	-	-	-	5	5	10	-	1:1	(Lochmann and Gatlin 1993)
				Grouper (*Epinephelus* sp.)							
40	Marine-subtropical	10.3	-	-	-	2	2	4	-	-	(Lin and Shiau 2007)
				Grouper (*Epinephelus malabaricus*)							
116	Marine-brackish-tropical	11.3	9.0	29.5	1.1	4.3	3.7	-	-	-	(Wu and Chen 2012)

(Continued)

TABLE 9.3 (CONTINUED)
Essential Fatty Acid Requirements (g kg^{-1} diet) of Aquaculture Fish Species

Dietary Lipid Level (g kg^{-1})	Environment	Fish Weight (g)	LNA	ALA	ARA	EPA	DHA	n-3 LC-PUFA	n-3/n-6	DHA/ EPA	Reference
					Fat Cod (*Hexagrammos otakii*)						
93	Marine-temperate	13.2	-	-	-	-	-	12-17	-	-	(Lee and Cho 2009)
					American Catfish (*Ictalurus punctatus*)						
50	Freshwater-subtropical	122–175	-	-	-	5	5	10	-	-	(Satoh et al. 1989)
					Yellow Catfish (*Pelteobagrus fulvidraco*)						
103	Freshwater-temperate	1.5	-	-	6.6				-	-	(Ma et al. 2018a)
					Carp (*Cyprinus carpio*)						
60	Freshwater-brackish-subtropical	-	10	10					-	-	(Takeuchi and Watanabe 1977)
					Grass Carp (*Ctenopharyngodon idellus*)						
49	Freshwater-brackish-subtropical	5.7	-	-	-	0.2	2.9	2.1	-	-	(Ji et al. 2011)
					Silver Perch (*Bidyanus bidyanus*)						
100	Freshwater-subtropical	-	27	-	-	-	-	-	-	-	(Smith et al. 2004)
					Nile Tilapia (*Oreochromis niloticus*)						
60	Freshwater-brackish-tropical	2.1	-	6.3	-	-	-	-	-	-	(Chen et al. 2013)
					Atlantic Salmon (*Salmo salar*)						
80	Marine-freshwater; brackish-temperate	4	-		-	5	5	10	-	-	(Ruyter et al. 2000a)
80		4	-	10	-	-	-	-	-	-	(Ruyter et al. 2000b)
253		50	-		-	10	10	20	-	-	(Bou et al. 2017)
					Rainbow Trout (*Oncorhynchus mykiss*)						
50	Marine-freshwater; brackish-subtropical	-	-	10	-	-	-	-	-	-	(Watanabe and Takeuchi 1976)
100		-	-	20	-	-	-	-	-	-	(Watanabe and Takeuchi 1976)
50		-	-	-	-	1	1	3	-	-	(Watanabe and Takeuchi 1976)
50		-	-	-	-		3	3	-	-	(Watanabe and Takeuchi 1976)
					European Seabass (*Dicentrarchus labrax*)						
180	Marine-freshwater; brackish-subtropical	14.4	-	-	-	5	2	7	-	1.5:1	(Skalli and Robin 2004)
					Asian Seabass (*Lates calcarifer*)						
153	Marine-freshwater; brackish-tropical	176	-	-	-	7.5	11.5	19	-	-	(Williams et al. 2006)

(Continued)

TABLE 9.3 (CONTINUED)
Essential Fatty Acid Requirements (g kg^{-1} diet) of Aquaculture Fish Species

Dietary Lipid Level (g kg^{-1})	Environment	Fish Weight (g)	LNA	ALA	ARA	EPA	DHA	n-3 LC-PUFA	n-3/n-6	DHA/EPA	Reference
100		16.5	-	-	-	1		1	-	1:1	(Morton et al. 2014)
					Japanese Seabass, (*Lateolabrax japonicas*)						
140	Marine-freshwater; brackish-subtropical	9.5	-	-	-	3.6	6.7	10.6	-	-	(Xu et al. 2010)
120		9.5	-	-	-	0.8	7.1	14.8	-	-	(Xu et al. 2014)
					Black Tiger Shrimp (*Penaeus monodon*)						
75	Marine-brackish-tropical	1.9–3.0	16		-	-	-	-	-	-	(Glencross and Smith 1999)
75		1.9–3.0	11	16	-	-	-	-	-	-	(Glencross and Smith 1999)
75		1.8-2.0	11	16	-		9	9	-	-	(Glencross and Smith 2001)
75		1.8-2.0	11	16	-	3	3	6	-	-	(Glencross and Smith 2001)
					Kuruma Shrimp (*Penaeus japonicas*)						
50	Marine-brackish-subtropical	-	-	-	-	10	-	10	-	-	(Kanazawa et al. 1979)
50		-	-	-	-	20	-	20	-	-	(Kanazawa et al. 1979)
					Chinese Shrimp (*Penaeus chinensis*)						
60	Marine-brackish-subtropical	0.23	-	10	-	-	-	-	-	-	(Xu et al. 1994)
60		0.23	-		-	-	10	10	-	-	(Xu et al. 1994)
					White Shrimp (*Litopenaeus vannamei*)						
50	Marine-brackish-subtropical	0.4	-	-	-	-	2	1	3	-	(Gonzalez-Felix et al. 2002)
50		0.4	-	-	-	-		3	3	-	(Gonzalez-Felix et al. 2002))
					Freshwater Shrimp (*Macrobrachium rosenbergii*)						
60	Freshwater-brackish-subtropical	0.0064–0.0099	-	-	-	-	-	10	-	-	(D'Abramo and Sheen 1993)
60		0.0064–0.0099	-	-	10	-	-	-	-	-	(D'Abramo and Sheen 1993)

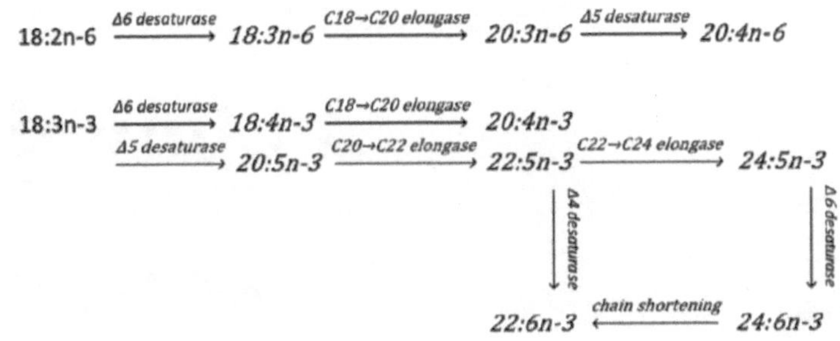

FIGURE 9.1 Pathways of selected n-3 and n-6 long-chain polyunsaturated fatty acid synthesis from C18 precursors in teleost fish. Adapted from Li et al. (2010a), NRC (2011).

of fish in biosynthesis of LC-PUFA, and other parameters like the species' phylogenic position may also be more effective when assessing species capacity for LC-PUFA biosynthesis (Garrido et al. 2019). In this sense, it has been reported that the activities of FAD2 (fatty acid desaturase 2) in goldline seabream (*Sarpa salpa*, Sparidae), which is an herbivorous species, are consistent with those reported for carnivorous Sparidae counterparts including black seabream (*Acanthopagrus schlegelii*) and gilthead sea bream (*Sparus aurata* (Garrido et al. 2019) (Figure 9.2).

A large body of literature has reported that the dietary EFA requirements of freshwater and/or warm water (> 25 °C) species as well as salmonids and/or cold water (< 18 °C) species can be met with MC-PUFA (i.e. LNA and ALA) due to their high capacity of desaturation/elongation pathway activities (Mourente et al. 2005; Glencross 2009; Tocher et al. 2010, 2015). However, EFA requirements of marine fish, including the top carnivorous fish species, are typically met only with LC-PUFA due to the apparent deficiency in the activity of one or more key enzymes that are required in the LC-PUFA biosynthesis pathway (Sargent et al. 2002; Tocher 2003; Glencross 2009; Mourente and Tocher 2009). In marine species, this limited capacity of bioconversion of MC-PUFA to LC-PUFA may be due to their adaptation to an LC-PUFA-rich environment (Sargent et al. 2002; Glencross 2009). Indeed, the lack of genes coding for fatty acyl ELOVL2 and Δ5 FAD proteins in marine fish were hypothesized as the main constraints for bioconversion of LC-PUFA from their precursors (Morais et al. 2009b; Monroig et al. 2013). The mode of action of ELOVL5 differs from ELOVL2, whose activity is centered predominantly toward MC-PUFA and C20 PUFA, while ELOVL2 activity predominantly is focused toward LC-PUFA (Morais et al. 2009 a,b Monroig et al. 2013). Thus, FO as a rich source of LC-PUFA is the key feedstuff for marine fish species, but it is not necessarily required for freshwater species, including salmonids, cyprinids, and catfishes, three of the main groups of freshwater farmed fish worldwide (Turchini et al. 2009). It should be mentioned that some marine fish species, such as the herbivorous rabbit fish (*Siganus canaliculatus*, Siganidae) and the carnivorous Senegalese sole (*Solea senegalensis*, Soleidae), have the capacity of direct desaturating 22:5n-3 to DHA by means of the Δ4 FAD (Buzzi et al. 1996; Christie 2003; Li et al. 2010b). In addition, the ELOVL4 with the capacity of elongation of 22:5n-3 to 24:5n-3 for the biosynthesis of DHA was characterized in cobia (*Rachycentron cancadum*, Rachycentridae) and Senegalese sole larvae, which may compensate the lack of ELOVL2 enzyme (Monroig et al. 2011; Morais et al. 2012). Moreover, water salinity can influence the LC-PUFA biosynthesis capacity in euryhaline fish species and affect their EFA requirements. For example, the mRNA expression of FADS2 in the liver of rabbitfish reared in 10 ppt salinity was much higher than that in fish reared in 32 ppt salinity, regardless of the diet (Li et al. 2008a). However, Reis et al. (2020) reported that increasing rearing salinities tended to enhance Δ6 FAD activity in pikeperch (*Sander lucioperca*, Percidae), but Δ5 FAD activity was not detected, indicating the inability of this fresh water carnivorous species to biosynthesize DHA. All these studies provide evidence that, when considering the EFA requirements of aquatic species, it is of vital importance to know the nutritional physiology of the target species in order to properly formulate and select feed ingredients and screen for potential ALS for FO sparing in aquaculture.

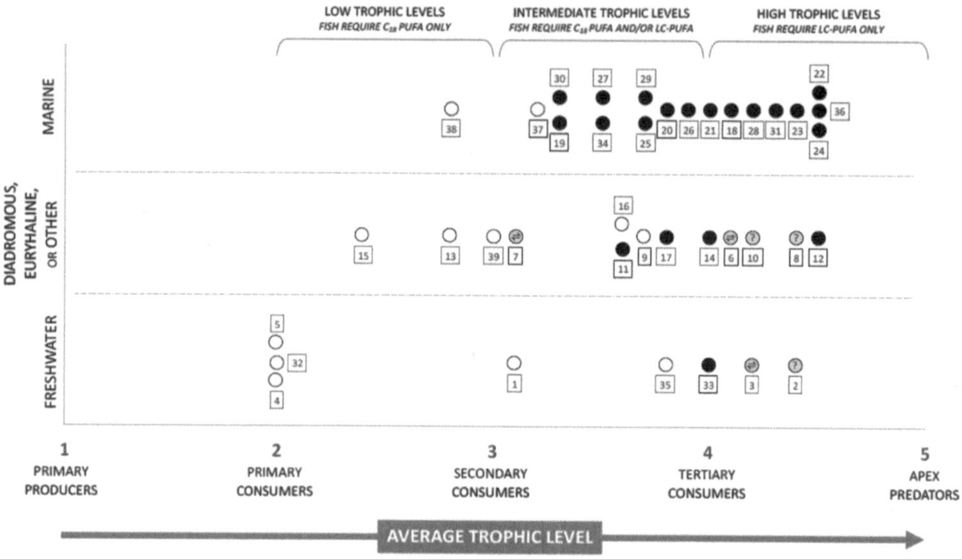

FIGURE 9.2 Dietary polyunsaturated fatty acid (PUFA) essentiality of selected fishes according to salinity and temperature preferences and average trophic level. Each circle represents a single taxon, indicated by the numeric labels as follows: 1) Common Carp *Cyprinus carpio*, 2) Wels Catfish/Sheatfish *Silurus glanis*, 3) Channel Catfish *Ictalurus punctatus*, 4) Nile Tilapia *Oreochromis niloticus*, 5) Grass Carp *Ctenopharyngodon idella*, 6) Rainbow Trout/Steelhead *Oncorhynchus mykiss*, 7) European Whitefish *Coregonus lavaretus*, 8) Arctic Charr *Salvelinus alpinus*, 9) Chum Salmon *Oncorhynchus keta*, 10) Coho Salmon *Oncorhynchus kisutch*, 11) Masu Salmon *Oncorhynchus masou*, 12) Atlantic Salmon *Salmo salar*, 13) Ayu Sweetfish *Plecoglossus altivelis*, 14) hybrid Striped Bass *Morone chrysops* × *M. saxatilis*, 15) Milkfish *Chanos chanos*, 16) Japanese Eel *Anguilla japonica*, 17) Barramundi *Lates calcarifer*, 18) Atlantic Cod *Gadus morhua*, 19) Yellowtail Flounder *Limanda ferruginea*, 20) Korean Rockfish *Sebastes schlegeli*, 21) Japanese Amberjack *Seriola quinqueradiata*, 22) Japanese Flounder/Bastard Halibut *Paralichthys olivaceus*, 23) Turbot *Scophthalmus maximus*, 24) Red Seabream *Pagrus major*, 25) Gilthead Sea Bream *Sparus aurata*, 26) White Trevally *Pseudocaranx dentex*, 27) European Seabass *Dicentrarchus labrax*, 28) Grouper *Epinephelus* spp., 29) Red Drum *Sciaenops ocellatus*, 30) Goldlined Seabream *Rhabdosargus sarba*, 31) Meagre *Agryrosomus regius*, 32) Tambaqui *Colossoma macropomum*, 33) Pikeperch *Sander lucioperca*, 34) Silvery-black Porgy *Sparidentex hasta*, 35) African Catfish *Clarias gariepinus*, 36) Common Dentex *Dentex dentex*, 37) Ballan Wrasse *Labrus bergyita*, 38) Rabbitfish *Siganus canaliculatus*, and 39) Spotted Scat *Scatophagus argus*. Average trophic level is indicated horizontally, with primarily producers on the left (trophic level = 1) and apex predators on the right (trophic level = 5). Salinity preference is indicated vertically, with marine fish at the top, freshwater fish at the bottom, and fish exhibiting other salinity preferences/tolerances (e.g., amphidromy, anadromy, catadromy, etc. in at least one life history) in the middle. The color of the numeric label border indicates temperature preference for cold water (<18°C, dark blue), cool water (18 – 25°C, light blue), or warm water (>25°C, red). PUFA essentiality is expressed in terms of dietary requirements for C18 PUFA (white-filled circles) or long-chain (LC)-PUFA (black-filled circles); yellow-filled circles indicate conflicting reports regarding C18 PUFA vs. LC-PUFA nutritional essentiality (yellow-filled circles with opposing arrow symbols) or studies with incomplete experimental designs (i.e., studies examined only C18 PUFA or LC-PUFA, but not both, yellow-filled circles with question marks). Nutritional PUFA essentiality does not adhere to any obvious patterns related to temperature or salinity preferences, but is strongly related to trophic level. Fish occupying low trophic levels (i.e., trophic level < 3) require C18 PUFA only, whereas those occupying high trophic levels (i.e., trophic level > 4) require LC-PUFA. Fish occupying intermediate trophic levels (i.e., 3 < trophic level < 4) may require either class of PUFA, depending on other aspects of their ecological niche and life history. Average trophic levels and temperature and salinity preferences were obtained from FishBase.org (Froese and Pauly 2019) and essential dietary fatty acid requirements were obtained from Carvalho et al. (2018), Ferraz et al. (2019), Kabeya et al. (2018), Li et al. (2008a), Lund et al. (2019), Mozanzadeh et al. (2015), NRC (2011), Oboh et al. (2016), Papiol and Estevez (2019), Sargent et al. (2002), Xie et al. (2016) in: Trushenski and Rombenso (2020).

GROWTH AND FEED UTILIZATION

Turchini et al. (2009) reviewed that fish performance when dealing with somatic growth, feed efficiency, and feed intake parameters is not profoundly affected by replacing a large portion of dietary FO (*ca.* 60-75%) with ALS when EFA requirements and a balanced n-3/n-6 PUFA ratio are met. Generally, ALS do not adversely affect feed intake in farmed fish, while energy requirements may be supported by ALS with different FA profiles. However, a meta-analysis of different studies conducted by Sales and Glencross (2011) confirmed that total replacement of FO with VO could compromise growth performance in most finfish species, whereas the oil type determined the extent of this negative effect. In addition, many ALS, including VO, GMO, RAF, and algal-derived oils and meals, have been tested in many nutritional studies with variable success according to the aquatic species considered, which is related to their requirement in LC-PUFA (Turchini et al. 2009; Turchini et al. 2011c; Oliva-Teles et al. 2015). When evaluating ALS, authors need to also consider the initial body EFA levels, diet composition in terms of ingredients, macro- and micro-nutrients, the amount of residual FO in FM, the duration of feeding trials, and tolerance of aquatic species to the reduction of EFA, factors that have been identified as variables that may generate potential background noise in the aforementioned nutritional studies (Subhadra et al. 2006). In this context, it is generally recommended that studies evaluating ALS should be carried out over a relatively long period in order to avoid the typical type II error when running statistics (failing to see an effect, when one is actually present). Thus, the true effect of replacement of dietary FO with ALS should be monitored throughout the production cycle, as was recommended by Glencross and Turchini (2011).

Marine Fish Species

Regarding marine fish species, the amount of ALS in feeds is generally limited by the quantity of FO that provides the minimum dietary EFA requirements for supporting fish condition, health, and welfare. As it is shown in Table 9.4, FO-deprived diets remarkably reduced the growth rate, but partial (up to 75%) FO replacement with ALS did not compromise growth in many marine fish species evaluated so far. As represented in Table 9.4 the substitution of dietary FO by VO in FM- and plant protein (PP)-based diets, for strictly carnivorous fish species like gilthead sea bream, had a certain threshold (60–70%, depending on dietary FM level) after which growth performance, nutrient composition, or health indices were compromised (Oliva-Teles et al. 2015). The use of ALS in marine fish species and its efficiency with regard to growth performance and feed utilization parameters mainly depends on the appropriate level of FM in the diet (Alhazzaa et al. 2019). As FM usually has 8–15% residual oil containing 20–35% n-3 LC-PUFA, its presence in compound feeds may provide sufficient EFA required for marine fish species; thus, efficient FO replacement may be feasible without drastic negative effects on fish performance (Bimbo 2000). On the other hand, the complete substitution of FO with ALS in PP-rich diets (70–90% of FM replaced with blends of PP) did not impair growth and feed efficiency performance in gilthead sea bream (Benedito-Palos et al. 2008), California pompano (*Trachinotus carolinus*, Carangidae; Rombenso et al. 2016), greater amberjack (*Seriola dumerili*, Carangidae; Monge-Ortiz et al. 2018), and Asian seabass (*Lates calcarifer*, Latidae, Salini et al. 2015b; Glencross et al. 2016). In contrast, Aminikhoei et al. (2014) reported that total replacement of FO with linseed and soybean oils or their blends in an FM-based diet was possible for black seabream (*Spondylosoma cantharus*, Sparidae), but the complete replacement of FO in a PP-rich diet compromised growth performance in this species (Peng et al. 2008; Jin et al. 2017b).

On the other hand, some marine fish species with omnivorous feeding habits, such as sharpsnout seabream (*Diplodus puntazzo*, Sparidae) and grey mullet (*Mugil cephalus*, Mugilidae), can tolerate total replacement of dietary FO with VO without detrimental effects on their growth performance (Argyropoulou et al. 1992; Piedecausa et al. 2007; Almaida-Pagán et al. 2007). In this context, Almaida-Pagán et al. (2007) reported that there are some regulatory mechanisms in selective

TABLE 9.4
Effects of Dietary FO Replacement with ALS on Growth and Feed Utilization of Marine Fish Species

Alternative Lipid Source	Details	Dietary Lipid Level (%)	% FO Replacement	Duration of Feeding Trial	Growth Performance					References
					IBW (g)	FBW (g)	SGR (% BW day^{-1})	FCR or FER (bold)	WG (% compared to the FO diet)	
Gilthead Sea Bream (*Sparus aurata*)										
LO	FM based diet	10	50	45 days	22.5	84.5	2.8	-	−9.8 (↓)	(El-Kerdawy and Salama 1997)
LO	*Ca.* 40% of FM was replaced with corn gluten	25	60	101 days	10	67.6	1.9	-	+16.4 (=)	(Caballero et al. 2004)
LO	*Ca.* 40% of FM was replaced with corn gluten	25	60	6 months	79.6	461.1	0.98	-	−13.9 (=)	(Caballero et al. 2004)
LO	*Ca.* 40% of FM was replaced with corn gluten	25	80	6 months	78.4	442.7	0.96	-	−28.5 (↓)	(Caballero et al. 2004)
LO	*Ca.* 40% of FM was replaced with corn gluten	25	60	101 days	10.7	67.6	1.88	1	+16.4 (=)	(Izquierdo et al. 2003)
LO	*Ca.* 40% of FM was replaced with corn gluten	25	60	159 days	85	458.8	0.67	1.43	−4.24 (=)	(Izquierdo et al. 2005)
LO	*Ca.* 40% of FM was replaced with corn gluten	25	80	159 days	85	441.0	0.66	1.46 (↓)	−19.52 (↓)	(Izquierdo et al. 2005)
LO	-	22	100	6 months	35	217.6	1.16	1.57	−62 (↓)	(Montero et al. 2008)
LO	*Ca.* 36% of FM was replaced with PP blends	16.3	100	75 days	35	106	1.54	1.2	−45.2 (↓)	(Montero et al. 2010)
SBO	-	22	100	6 months	35	215.7	1.2	1.7 (↓)	−80.3 (↓)	(Montero et al. 2008)
SBO	*Ca.* 36% of FM was replaced with PP blends	16.3	100	75 days	35	101	1.44	1.23	−67 (↓)	(Montero et al. 2010)
SBO	Casein-based diet	13	17	150 days	1.2	12.64	1.56	**0.67**	−70.8 (=)	(Kalogeropoulos et al. 1992)
SBO	Casein-based diet	13	33	150 days	1.2	13.54	1.61	**0.67**	+5 (=)	(Kalogeropoulos et al. 1992)

(Continued)

TABLE 9.4 (CONTINUED)
Effects of Dietary FO Replacement with ALS on Growth and Feed Utilization of Marine Fish Species

Alternative Lipid Source	Details	Dietary Lipid Level (%)	% FO Replacement	Duration of Feeding Trial	Growth Performance					References
					IBW (g)	FBW (g)	SGR (% BW day^{-1})	FCR or FER (bold)	WG (% compared to the FO diet)	
SBO	Casein-based diet	13	50	150 days	1.2	13.06	1.59	**0.67**	−36.7 (=)	(Kalogeropoulos et al. 1992)
SBO	Casein-based diet	13	67	150 days	1.2	11.27	1.49	**0.57**	−186.7 (↓)	(Kalogeropoulos et al. 1992)
SBO	Casein-based diet	13	83	150 days	1.2	10.5	1.44	**0.52 (↓)**	−252.5 (↓)	(Kalogeropoulos et al. 1992)
SBO	FM based diet	10	50	45 days	22.5	82.6	2.9	-	−17.8 (↓)	(El-Kerdawy and Salama 1997)
SBO	*Ca.* 40% of FM was replaced with corn gluten	25	60	101 days	10.1	63.1	1.81	-	−22.8 (=)	(Caballero et al. 2004)
SBO	*Ca.* 40% of FM was replaced with corn gluten	25	60	6 months	79.6	461.1	0.98	-	−8.6 (=)	(Caballero et al. 2004)
SBO	*Ca.* 40% of FM was replaced with corn gluten	25	80	6 months	79	432	0.94	-	−40.4 (↓)	(Caballero et al. 2004)
SBO/RO/LO (1:3:6)	*Ca.* 40% of FM was replaced with corn gluten	25	60	6 months	78.1	451.6	0.97	-	−15 (=)	(Caballero et al. 2004)
SBO	*Ca.* 40% of FM was replaced with corn gluten	25	60	101 days	10.1	63.1	1.9	1.04	−22.8 (=)	(Izquierdo et al. 2003)
SBO/RO/LO (1:3:6)	*Ca.* 40% of FM was replaced with corn gluten	25	60	101 days	10.07	66.3	1.87	1.02	+9.3(=)	(Izquierdo et al. 2003)
SBO	*Ca.* 40% of FM was replaced with corn gluten	25	60	159 days	85	460.2	0.68	1.45	−4.5 (=)	(Izquierdo et al. 2005a)
SBO	*Ca.* 40% of FM was replaced with corn gluten	25	80	159 days	85	441.1	0.62	1.55	−26.9 (↓)	(Izquierdo et al. 2005)
SBO	*Ca.* 60% of FM was replaced with PP blends	15	24	309 days	14.7	343	1	1.89	−41.7 (=)	(Martinez-Llorens et al. 2007)

(Continued)

TABLE 9.4 (CONTINUED)
Effects of Dietary FO Replacement with ALS on Growth and Feed Utilization of Marine Fish Species

Alternative Lipid Source	Details	Dietary Lipid Level (%)	% FO Replacement	Duration of Feeding Trial	Growth Performance					References
					IBW (g)	FBW (g)	SGR (% BW day^{-1})	FCR or FER (bold)	WG (% compared to the FO diet)	
SBO	*Ca.* 60% of FM was replaced with PP blends	15	48	309 days	14.7	338	0.96	1.89	−76.4 (=)	(Martinez-Llorens et al. 2007)
SBO	*Ca.* 60% of FM was replaced with PP blends	15	72	309 days	14.7	325	0.97	1.96	−166.7 (↓)	(Martinez-Llorens et al. 2007)
SBO	*Ca.* 80% of FM was replaced with PP blends	20	69	170 days	113.9	259.4	0.48	1.72	+0.3 (=)	(Fountoulaki et al. 2009)
LO/PO/RO(58/25/17%	*Ca.* 70% of FM was replaced with PP blends	21	100	8 months	16	237.4	1.1	**0.97**	−127.5(↓)	(Benedito-Palos et al. 2008)
LO/PO/RO(58/25/17%	*Ca.* 70% of FM was replaced with PP blends	21	33-66	330 days	18	294	-	-	-	(Benedito-Palos et al. 2009)
PO	*Ca.* 70% of FM was replaced with PP blends	20	69	170 days	114.34	229.41	0.75	2.14	−26.5 (↓)	(Fountoulaki et al. 2009)
EO	-	50	50	210 days	256.1	432.7	0.3	-	(=)	(Diaz-Lopéz et al. 2009)
RO	FM based diet	10	50	45 days	22.5	82.7	2.9	-	−17.8 (↓)	(El-Kerdawy and Salama 1997)
RO	*Ca.* 40% of FM was replaced with corn gluten	25	60	101 days	10.1	68.7	1.9	-	+32 (=)	(Caballero et al. 2004)
RO	*Ca.* 40% of FM was replaced with corn gluten	25	60	6 months	79.6	443.9	0.95	-	−25.3 (=)	(Caballero et al. 2004)
RO	*Ca.* 40% of FM was replaced with corn gluten	25	60	101 days	10.1	68.6	1.9	1.02	+32 (=)	(Izquierdo et al. 2003)
RO	*Ca.* 40% of FM was replaced with corn gluten	25	80	159 days	85	446.5	0.66	1.45 (↓)	−18.4 (=)	(Izquierdo et al. 2005)
RO	*Ca.* 70% of FM was replaced with PP blends	20	69	170 days	112.0	252.4	0.48	1.85	−6.7 (=)	(Fountoulaki et al. 2009)

(Continued)

TABLE 9.4 (CONTINUED)
Effects of Dietary FO Replacement with ALS on Growth and Feed Utilization of Marine Fish Species

Alternative Lipid Source	Details	Dietary Lipid Level (%)	% FO Replacement	Duration of Feeding Trial	Growth Performance					References
					IBW (g)	FBW (g)	SGR (% BW day^{-1})	FCR or FER (bold)	WG (% compared to the FO diet)	
LO	*Ca.* 40% of FM was replaced with SBM	18	50	63 days	4.2	22.88	2.5	1.77	−35.2 (=)	(Wassef et al. 2015a)
LO	*Ca.* 40% of FM was replaced with SBM	18	70	63 days	4.1	18.51	2.38	2.14 (↓)	−131.8 (↓)	(Wassef et al. 2015a)
SFO	*Ca.* 40% of FM was replaced with SBM	18	50	63 days	4.2	23.37	2.7	1.76	−24.9 (=)	(Wassef et al. 2015a)
SFO	*Ca.* 40% of FM was replaced with SBM	18	70	63 days	4.1	20.13	2.5	1.81	−86.6 (↓)	(Wassef et al. 2015a)
CSO	*Ca.* 40% of FM was replaced with SBM	18	60	63 days	4.1	22.47	2.66	1.77	−34.4(=)	(Wassef et al. 2015b)
CSO	*Ca.* 40% of FM was replaced with SBM	18	70	63 days	4.1	20.68	2.54	1.91(↓)	−78(↓)	(Wassef et al. 2015b)
TO	In FM base diet	15	63	120 days	135	271.4	0.58	0.63	−7.2 (=)	(Pérez et al. 2014)
TO/CO	Supplemented with 1% LNA in FM based diet	15.7	100	120 days	135	255.2	0.53	**0.57** (↓)	−19.2(↓)	(Pérez et al. 2014)
SFO/CSO/LO (1:1:1)	*Ca.* 36% of FM was replaced with corn gluten	18	60	140 days	131.1	340.7	0.8	1.19	−15.1 (↓)	(Wassef et al. 2009)
SFO/CSO/SBO (1:1:1)	*Ca.* 36% of FM was replaced with corn gluten	18	60	140 days	130.9	353.5	0.83	1.13	−5 (=)	(Wassef et al. 2009)
SBO/RO (1:1)	*Ca.* 70% of FM was replaced with PP	19	75	133 days	197.7	415.5	0.56	1.81	−1.3 (=)	Álvarez et al. (2020)
Sharpsnout Seabream (*Diplodus puntazzo*)										
SO	25% of FM was replaced with wheat gluten	21	100	92 days	15.0	70.6	1.7	1.15	−15.4 (=)	(Piedecausa et al. 2007)

(Continued)

TABLE 9.4 (CONTINUED)
Effects of Dietary FO Replacement with ALS on Growth and Feed Utilization of Marine Fish Species

Alternative Lipid Source	Details	Dietary Lipid Level (%)	% FO Replacement	Duration of Feeding Trial	Growth Performance					References
					IBW (g)	FBW (g)	SGR (% BW day^{-1})	FCR or FER (bold)	WG (% compared to the FO diet)	
LO	25% of FM was replaced with wheat gluten	21	100	92 days	15.11	64.9	1.58	1.21	−62.5 (=)	(Piedecausa et al. 2007)
LO	25% of FM was replaced with wheat gluten	21	100	268 days	14.9	151.3	0.86	-	+25.5 (=)	(Almaida-Pagán et al. 2007)
SBO	35% of FM was replaced with sunflower meal	20	75	84 days	36.2	105.7	1.33	2.28	−21.3 (=)	(Nogales-Mérida et al. 2017)
Lard	35% of FM was replaced with sunflower meal	20	75	84 days	33.4	96.8	1.27	2.21	−36.8 (=)	(Nogales-Mérida et al. 2011)
Black Seabream (*Acanthopagrus Schlegelii*)										
SO	55% of FM was replaced with SBM	15	60	63 days	20.3	63.6	2.61	1.11	+6 (=)	(Peng et al. 2008)
SO	55% of FM was replaced with SBM	15	80	63 days	20.3	55	2.34	1.12	−36.6 (=)	(Peng et al. 2008)
SO	55% of FM was replaced with SBM	15	100	63 days	20.3	44.9	2.03	1.32	−86.6 (↓)	(Peng et al. 2008)
LO	*Ca.* 20% FM replaced with PP blends	12	100	56 days	1.13	4.5	2.9	**0.8**	+23 (=)	(Aminikhoei et al. 2014)
Coconut oil	*Ca.* 70% of FM was replaced with alternatives	14	100	56 days	5.38	32.7	3.2	**0.71(↓)**	−118.9 (↓)	(Jin et al. 2017a)
Perilla oil	*Ca.* 70% of FM was replaced with alternatives	14	100	56 days	5.38	35.2	3.4	**0.85**	−72.2(↓)	(Jin et al. 2017b)
SFO	*Ca.* 70% of FM was replaced with alternatives	14	100	56 days	5.38	34.6	3.3	**0.79(↓)**	−83.8(↓)	(Jin et al. 2017b)
ARA-enriched oil	*Ca.* 70% of FM was replaced with alternatives	14	100	56 days	5.38	33.3	3.3	**0.77(↓)**	−107.75(↓)	(Jin et al. 2017b)

(Continued)

TABLE 9.4 (CONTINUED)
Effects of Dietary FO Replacement with ALS on Growth and Feed Utilization of Marine Fish Species

Alternative Lipid Source	Details	Dietary Lipid Level (%)	% FO Replacement	Duration of Feeding Trial	Growth Performance					References
					IBW (g)	FBW (g)	SGR (% BW day^{-1})	FCR or FER (bold)	WG (% compared to the FO diet)	
EPA+DHA enriched oil	*Ca.* 70% of FM was replaced with alternatives	14	100	56 days	5.38	37.5	3.5	**0.84**	−29.9(=)	(Jin et al. 2017b)
			Red Snapper (*Pagrus auratus*)							
SO	FM-based diet	10	25	56 days	28	88.6	2.95	1.21	+2.5 (=)	(Glencross et al. 2003)
SO	FM-based diet	10	50	56 days	28	86.7	2.92	1.24	−4.3 (=)	(Glencross et al. 2003)
SO	FM-based diet	10	75	56 days	28	86.6	2.85	1.26	−4.6 (=)	(Glencross et al. 2003)
SO	FM-based diet	10	100	56 days	28	87.9	2.91	1.30	(=)	(Glencross et al. 2003)
CAO	FM-based diet	10	25	56 days	28.3	88.1	2.95	1.2	+1(=)	(Glencross et al. 2003)
CAO	FM-based diet	10	50	56 days	28.4	89.6	2.99	1.21	+6.3 (=)	(Glencross et al. 2003)
CAO	FM-based diet	10	75	56 days	28.1	85.0	2.85	1.34	−8.8 (=)	(Glencross et al. 2003)
CAO	FM-based diet	10	100	56 days	28.6	79.7	2.62	1.36	−29.1 (↓)	(Glencross et al. 2003)
			Red Seabream (*Pagrus major*)							
CAO	*Ca.* 17% FM replaced with PP blends	15	25-70	84 days	3.61	52.3-44.1	3.07		(=)	(Huang et al. 2007)
AM	*Schizochytrium sp. in fish meal free diet*	16.4	100	84 days	8.8	42.4	1.86	1.15	−38.6(=)	(Seong et al. 2019)
			Silver Seabream (*Rhabdosargus sarba*)							

(Continued)

TABLE 9.4 (CONTINUED)
Effects of Dietary FO Replacement with ALS on Growth and Feed Utilization of Marine Fish Species

Alternative Lipid Source	Details	Dietary Lipid Level (%)	% FO Replacement	Duration of Feeding Trial	Growth Performance					References
					IBW (g)	FBW (g)	SGR (% BW day^{-1})	FCR or FER (bold)	WG (% compared to the FO diet)	
SBO	50% of FM was replaced with casein	10	100	56 day	1.1	8 ± 1.2	3.6		−100.3 (↓)	(Leu et al. 1994)
			Sobaity Seabream (*Sparidentex hasta*)							
SFO	*Ca.* 45% of FM was replaced with blends of APS	20	50	56 days	14.6	34.0	1.5	1.6	−9.3 (=)	(Mozanzadeh et al. 2016)
SFO	*Ca.* 45% of FM was replaced with blends of APS	20	100	56 days	14.5	35.9	1.6	1.6	+1.4 (=)	(Mozanzadeh et al. 2016)
CAO	*Ca.* 45% of FM was replaced with blends of APS	20	50	56 days	14.7	33.6	1.5	1.6	−9.4 (=)	(Mozanzadeh et al. 2016)
CAO	*Ca.* 45% of FM was replaced with blends of APS	20	100	56 days	14.7	34.1	1.5	1.7	−14 (=)	(Mozanzadeh et al. 2016)
T	*Ca.* 45% of FM was replaced with blends of APS	20	100	56 days	14.5	32.2	1.4	1.7	−19.7 (↓)	(Mozanzadeh et al. 2016)
T	*Ca.* 45% of FM was replaced with blends of APS	20	50	56 days	14.6	32.6	1.4	1.8 (↓)	−19 (↓)	(Mozanzadeh et al. 2016)
			Yellowfin Seabream (*Acanthopagrous latus*)							
SBO/CAO	*Ca.* 80% of FM was replaced with blends of APS	15	80	56 days	0.5	1.9	2.35	1.12	+2.1 (=)	(Abbasi et al. 2020)
			European Seabass (*Dicentrarchus labrax*)							
LO	About 40% of FM was replaced with corn gluten	25	60	89 days	8.04	29.23	1.48	1.22	−16.8 (=)	(Izquierdo et al. 2003)
LO	*Ca.* 40% of FM was replaced with corn gluten	22	60	8 months	75	358.4	0.52	1.47	−27.0 (=)	(Montero et al. 2005b)
LO	*Ca.* 40% of FM was replaced with corn gluten	22	80	8 months	75	366.0	0.51	1.49	−16.46 (↓)	(Montero et al. 2005b)

(Continued)

TABLE 9.4 (CONTINUED)
Effects of Dietary FO Replacement with ALS on Growth and Feed Utilization of Marine Fish Species

Alternative Lipid Source	Details	Dietary Lipid Level (%)	% FO Replacement	Duration of Feeding Trial	Growth Performance					References
					IBW (g)	FBW (g)	SGR (% BW day^{-1})	FCR or FER (bold)	WG (% compared to the FO diet)	
LO	*Ca.* 40% of FM was replaced with corn gluten	22	60	8 months	93.8	434.3	0.64		+8.2 (=)	(Mourente et al. 2005)
LO/PO/RO	*Ca.* 40% of FM was replaced with corn gluten	22	60	448 days	5.2	143.2				(Mourente et al. 2005)
SBO	*Ca.* 40% of FM was replaced with corn gluten	22	60	8 months	75	371.7	0.52	1.55	−8.8 (=)	(Montero et al. 2005)
SBO		16.1	50	84 days	16.2	38.1	1.0	1.4	+3.1 (=)	(Martins et al. 2006)
SBO		16.1	80	84 days	16.3	35.5	0.9	1.4	−13 (=)	(Martins et al. 2006)
SBO	*Ca.* 40% of FM was replaced with corn gluten	25	60	101 days	7.9	31.4	1.53	1.21	−7.4 (=)	(Izquierdo et al. 2003)
SBO	*Ca.* 20% of FM was replaced by corn gluten.	23	50	90 days	53.9	66.2	0.23		−4.8 (=)	(Özşahinoğlu et al. 2013)
SBO/RO/LO	(1:3:6)		60	89 days	8.09	32.03				(Izquierdo et al. 2003)
PF	*Ca.* 70% of FM was replaced with PP	9	50	112 days	19.1	67.3	1.3	1.3	+25.7 (↑)	(Campos et al. 2019)
PF	*Ca.* 70% of FM was replaced with PP	9	100	112 days	19.1	61.5	1.2	1.5	−4.7 (↓)	(Campos et al. 2019)
PF	Supplemented with 1% Soy lecithin	9	100	112 days	19.1	62.6	1.2	1.4	+1 (=)	(Campos et al. 2019)
SBO	*Ca.* 35% of FM was replaced with SBM.	12	100	166 days	94.2	201.3	0.43		+5.5 (=)	(Parpoura and Alexis 2001)
OO	*Ca.* 35% of FM was replaced with SBM	12	100	168 days	95.3	195.2	0.44		(=)	(Parpoura and Alexis 2001)
OO	*Ca.* 35% of FM was replaced with SBM	12	45	168 days	92.4	195.3	0.45		(=)	(Parpoura and Alexis 2001)

(Continued)

TABLE 9.4 (CONTINUED)
Effects of Dietary FO Replacement with ALS on Growth and Feed Utilization of Marine Fish Species

Alternative Lipid Source	Details	Dietary Lipid Level (%)	% FO Replacement	Duration of Feeding Trial	Growth Performance					References
					IBW (g)	FBW (g)	SGR (% BW day^{-1})	FCR or FER (bold)	WG (% compared to the FO diet)	
RO	*Ca.* 40% of FM was replaced with corn gluten	25	60	101 days	7.96	29.2	1.44	1.3	–36.1 (=)	(Izquierdo et al. 2003)
RO	*Ca.* 40% of FM was replaced with corn gluten	22	60	8 months	75	356.3	0.51	1.49	–29.3 (↓)	(Montero et al. 2005b)
RO	*Ca.* 40% of FM was replaced with corn gluten	22	60	8 months	93.8	430.4	0.64		–12.4 (=)	(Mourente et al. 2005)
PO			60	448 days	5.2					(Richard et al. 2006)
OO	*Ca.* 40% of FM was replaced with maze gluten	22	60	8 months	93.8	434.3	0.61		+8.2 (=)	(Mourente et al. 2005)
CAO	*Ca.* 20% of FM was replaced by corn gluten.	23	50	90 days	53.3	66.5	0.25		–4.2 (=)	(Özşahinoğlu et al. 2013)
Sesame oil	*Ca.* 20% of FM was replaced by corn gluten.	23	50	90 days	54.7	71.7	0.3		+5.3 (=)	(Özşahinoğlu et al. 2013)
PF/mammalian fats	(1:1)	19	50	114 days	20.3	63.3	1.1	1.5	–2 (=)	(Monteiro et al. 2018)
PF /mammalian fats	(1:1)	19	100	114 days	20.3	59.6	1.0	1.6	–20.2 (↓)	(Monteiro et al. 2018)
PF /mammalian fats	(1:1)	19	100	85 days	60.4	121.2	0.82		–7.7 (=)	(Silva-Brito et al. 2019)
			Asian Seabass (*Lates calcarifer*)							
LO	*Ca.* 30% of FM was replaced with blends of APS	20	100	40 days	19.1	57.8	2.8	0.7	-124 (↓)	(Raso and Anderson 2003)
SBO	*Ca.* 30% of FM was replaced with blends of APS	20	100	40 days	19.2	66.7	3.1	0.7	-77.1 (=)	(Raso and Anderson 2003)
CAO	*Ca.* 30% of FM was replaced with blends of APS	20	100	40 days	19.2	51.7	2.5	0.8	155.1 (↓)	(Raso and Anderson 2003)

(Continued)

TABLE 9.4 (CONTINUED)
Effects of Dietary FO Replacement with ALS on Growth and Feed Utilization of Marine Fish Species

Alternative Lipid Source	Details	Dietary Lipid Level (%)	% FO Replacement	Duration of Feeding Trial	Growth Performance					References
					IBW (g)	FBW (g)	SGR (% BW day⁻¹)	FCR or FER (bold)	WG (% compared to the FO diet)	
LO	FM based diet	19	100	15 days	0.6	1.9	7.8	**0.9**	–16.7 (=)	(Alhazzaa et al. 2012)
EO	FM based diet	19	100	15 days	0.6	1.7	6.8	**0.8**	–50 (↓)	(Alhazzaa et al. 2012)
PO/P flake	*Ca.* 80% of FM was replaced with blends of APS	13.5	50	56 days	47.3	231.2	3.3	1.12	–15.3 (=)	(Salini et al. 2015a)
OO	*Ca.* 80% of FM was replaced with blends of APS	13.5	50	56 days	47.1	230.9	3.3	1.09	–15.7 (=)	(Salini et al. 2015a)
PF	*Ca.* 80% of FM was replaced with blends of APS	15.1	40	84 days	211.2	546.4	1.13	1.15	–2.3(=)	(Salini et al. 2015b)
PF	*Ca.* 80% of FM was replaced with blends of APS	15.7	70	84 days	208.6	553.6	1.16	1.14	+4.4(=)	(Salini et al. 2015b)
PF	*Ca.* 80% of FM was replaced with blends of APS	16.2	100	84 days	207.5	549.7	1.16	1.12	+3.9(=)	(Salini et al. 2015b)
Rice bran oil	*Ca.* 70% of FM was replaced with blends of APS	13.3	100	56 days	154.4	354.4	-	-	+6.1 (=)	(Glencross et al. 2016)
Rice bran oil	*Ca.* 90% of FM was replaced with blends of APS	13.3	100	56 days	154.4	352.6	-	-	+4.9 (=)	(Glencross et al. 2016)
Rice bran oil	100% of FM was replaced with blends of APS	12.9	100	56 days	154.4	325.3	-	-	–12.7 (↓)	(Glencross et al. 2016)
			Japanese Seabass (*Lateolabrax japonicas*)							
SBO	*Ca.* 55% of FM was replaced with SBM and WG	13.5	50	70 days	5.9	30.6	2.84	1.02	–1.3 (=)	(Xue et al. 2006)
CO	*Ca.* 55% of FM was replaced with SBM and WG	13.5	50	70 days	5.87	30.39	2.83	1.03	–3.9 (=)	(Xue et al. 2006)

(Continued)

TABLE 9.4 (CONTINUED)
Effects of Dietary FO Replacement with ALS on Growth and Feed Utilization of Marine Fish Species

Alternative Lipid Source	Details	Dietary Lipid Level (%)	% FO Replacement	Duration of Feeding Trial	Growth Performance					References
					IBW (g)	FBW (g)	SGR (% BW day^{-1})	FCR or FER (bold)	WG (% compared to the FO diet)	
TO	*Ca.* 55% of FM was replaced with SBM and WG	13.5	50	70 days	5.87	28.89	2.74	1.07	−29.5 (=)	(Xue et al. 2006)
PF	*Ca.* 55% of FM was replaced with SBM and WG	13.5	50	70 days	5.87	31.35	2.89	0.99	+11.8 (=)	(Xue et al. 2006)
PL	*Ca.* 55% of FM was replaced with SBM and WG	13.5	50	70 days	5.87	28.49	2.72	1.05	−36 (=)	(Xue et al. 2006)
TO/CO/FO (6:2:2)	*Ca.* 55% of FM was replaced with SBM and WG	13.5	50	70 days	5.87	30.29	2.82	1.06	−7.1 (=)	(Xue et al. 2006)
PO	FM based diet	18	60	50 days	1.7	11.3	3.5	1.3	−49 (=)	(Gao et al. 2012)
PO	FM based diet	17	100	50 days	1.7	10	3.7	1.2	−124 (↓)	(Gao et al. 2012)
LO	*Ca.* 40% of FM was replaced with SBM	12	66	66 days	10.1	85.6	3.24	**0.81**	+21.8 (=)	(Xu et al. 2015)
LO	*Ca.* 40% of FM was replaced with SBM	12	100	66 days	10.1	84.7	3.22	**0.78**	+11.9 (=)	(Xu et al. 2015)
SBO	*Ca.* 40% of FM was replaced with SBM	12	66	66 days	10.1	77.2	3.08	**0.77**	−61.4 (=)	(Xu et al. 2015)
SBO	*Ca.* 40% of FM was replaced with SBM	12	100	66 days	10.1	76.0	3.06	**0.75 (↓)**	−67.9 (↓)	(Xu et al. 2015)
Malabar Grouper (*Epinephelus malabaricus*)										
CO	Casein based diet	11	100	56 days	13.3	66.0	2.9	0.91	−3 (↓)	(Lin et al. 2007)
Humpback Grouper (*Cromileptes altivelis*)										
SBO	*Ca.* 30% of FM was replaced SBM	10	100	56 days	10.6	24.4	1.6	1.68	−3.6 (=)	(Shapawi et al. 2008)
PO	*Ca.* 30% of FM was replaced SBM	10.4	100	56 days	10.6	24.4	1.6	1.52	−5.8 (=)	(Shapawi et al. 2008)

(Continued)

TABLE 9.4 (CONTINUED)
Effects of Dietary FO Replacement with ALS on Growth and Feed Utilization of Marine Fish Species

Alternative Lipid Source	Details	Dietary Lipid Level (%)	% FO Replacement	Duration of Feeding Trial	Growth Performance					References
					IBW (g)	FBW (g)	SGR (% BW day^{-1})	FCR or FER (bold)	WG (% compared to the FO diet)	
CAO	Ca. 30% of FM was replaced SBM	10	100	56 days	10.6	24.8	1.6	1.71	−7.3 (=)	(Shapawi et al. 2008)
Giant Grouper (*Epinephelis lanceolatus*)										
AM	*Schizochytrium limacinum* meal	10	40	84 days	47.77	169.3	1.5	0.95	−32.52 (=)	(Garcia-Ortega et al. 2016)
AM	*Schizochytrium limacinum* meal	10	80	84 days	45.4	109.38	1.04	1.29 (↓)	−146.56 (↓)	(Garcia-Ortega et al. 2016)
Hybrid Grouper (*Epinephelis lanceolatus* × *E. fuscoguttatus*)										
WGO	45% of FM was replaced with SBC	12	60	70 days	55.37	145.63	1.38	0.9	+1.2 (=)	(Baoshan et al. 2019)
WGO	45% of FM was replaced with SBC	12	100	70 days	55.33	144.93	1.37	0.9	−1.5 (=)	(Baoshan et al. 2019)
Orange Spot Grouper (*Epinephelus coioides*)										
Coconut oil			100	56 days	13					(Lin et al. 2007)
Flathead Grey Mullet (*Mugil cephalus*)										
SBO	Semi-purify diet	8	100	84 days	0.5	1.39	1.2	**0.33**	+31.9 (=)	(Argyropoulou et al. 1992)
CO	Semi-purify diet	8	100	84 days	0.5	1.35	1.2	**0.34**	+29.2 (=)	(Argyropoulou et al. 1992)
LO	Semi-purify diet	8	100	84 days	0.5	1.5	1.3	**0.34**	+50 (=)	(Argyropoulou et al. 1992)
Mullet (*Mugil liza*)										
AM	*Arthrospira platensis/LO* *Simultaneous FO and FM replacement*	9	50	75 days	0.48	11.71	4.32	1.64 (↑)	+1206.3 (↑)	(Torres Rosas et al. 2019)

(Continued)

TABLE 9.4 (CONTINUED)
Effects of Dietary FO Replacement with ALS on Growth and Feed Utilization of Marine Fish Species

Alternative Lipid Source	Details	Dietary Lipid Level (%)	% FO Replacement	Duration of Feeding Trial	Growth Performance					References
					IBW (g)	FBW (g)	SGR (% BW day^{-1})	FCR or FER (bold)	WG (% compared to the FO diet)	
AM	*Arthrospira platensis/LO Simultaneous FO and FM replacement*	9	100	75 days	0.47	8.59	3.89	2.2	+568 (↑)	(Torres Rosas et al. 2019)
			Atlantic Cod (*Gadus morhua*)							
Camelina oil			100	91 days	14.6	43.6	1.1	0.97 (↓)	−56.6 (↓)	(Hixson et al. 2014)
			Yellowtail (*Seriola quinqueradiata*)							
OO	FM-based diet	16	100	40 days	252	406.5	1.2	**0.52**	−3.2 (=)	(Seno-o et al. 2008)
			Yellowtail Kingfish (*Seriola lalandi*)							
CAO	At 18° C	25	100	35 days	101.2	147.1	1.13	2.13	−37.32 (↓)	(Bowyer et al. 2012b)
CAO	At 22° C	25	100	35 days	95.6	236.9	2.67	1.32	−30.32 (↓)	(Bowyer et al. 2012b)
CAO	In FM base diet	23	50	35 days	95.8	259.1	2.8	-	−6.1 (=)	(Bowyer et al. 2012a)
CAO	In FM base diet	23	100	35 days	95.5	236.8	2.6	-	−28.6 (↓)	(Bowyer et al. 2012a)
PF	In FM base diet	23	50	35 days	95.5	276.1	3.0	-	11.5 (=)	(Bowyer et al. 2012a)
PF	In FM base diet	23	100	35 days	95.4	262.1	2.9	-	−1.9 (=)	(Bowyer et al. 2012a)
			Greater Amberjack (*Seriola dumerili*)							
PO/LO (4:1)	*Ca.* 80% of FM was replaced WG	13.5	100	154 days	39.2	375	1.5	1.72	−54 (=)	(Monge-Ortiz et al. 2018)
			Longfin Yellowtail (*Seriola revoliana*)							
CAO/*Schizochytrium limacinum*	About 80% of FM was replaced with blends of *Hematococcus pluvialis* and SBC	15	85	63 days	2.5	74	5.4	0.8	−432.8 (=)	(Kissinger et al. 2016)
			Murray Cod (*Maccullochella peelii peelii*)							
LO	Casein-based diet	18	25-100	112 days	125.4	962.0	-	-	(=)	(Turchini et al. 2006)
CAO	Casein-based diet	18	100	84 days	6.5	26.9	1.69	0.8	−109.2(↓)	(Francis et al. 2007b)

(Continued)

TABLE 9.4 (CONTINUED)
Effects of Dietary FO Replacement with ALS on Growth and Feed Utilization of Marine Fish Species

Alternative Lipid Source	Details	Dietary Lipid Level (%)	% FO Replacement	Duration of Feeding Trial	Growth Performance: IBW (g)	FBW (g)	SGR (% BW day^{-1})	FCR or FER (bold)	WG (% compared to the FO diet)	References
CAO	Casein-based diet	19	25	112 days	21.4	88.9	1.27	0.76	−108.3 (=)	(Francis et al. 2007b)
CAO	Casein-based diet	19	50	112 days	21.6	90.3	1.28	0.78	−105.6 (=)	(Francis et al. 2007b)
CAO	Casein-based diet	19	75	112 days	21.4	88.3	1.26	0.83	−112 (=)	(Francis et al. 2007b)
CAO	Casein-based diet	19	100	112 days	21.2	73.8	1.1	0.88	−176.4 (↓)	(Francis et al. 2007b)
LO/PO/OO (4.5:4.3:1.2)	Casein-based diet	19	100	98 days	14.4	49.9	-	-	(=)	(Francis et al. 2007a)
Red Drum (*Sciaenops ocellatus*)										
SBO	*Ca.* 60% of FM was replaced with blends of APS	16	100	56 days	0.36	4.68	4.6	1.31	−1322.2 (↓)	(Tucker et al. 1997)
T		13	38.5							(Craig and Gatlin 1995)
AM	Arthrospira sp./*Schizochytrium* meals	9	50	42 days	2.3	15.3	4.7	**0.67**	+152.2 (=)	(Perez-Velazquez et al. 2019)
Large Yellow Croaker (*Larmichthys crocea*)										
SBO	FM-based diet	11.2	50	84 days	245.3	477.9	0.67	1.33	+31.6 (↑)	(Duan et al. 2014)
SBO	FM-based diet	11.4	100	84 days	245.3	375.7	0.43	1.63 (↓)	−10.1 (=)	(Duan et al. 2014)
PO	FM-based diet	10.9	100	84 days	245.3	448.0	0.61	1.29	+19.4 (=)	(Duan et al. 2014)
LO	*Ca.* 33% of FM was replaced with PP blends	10	100	56 days	13.8	25.1	1.92	**0.55**	−8.4 (=)	(Qiu et al. 2017)
SBO	*Ca.* 33% of FM was replaced with PP blends	10	100	56 days	13.8	26.8	2.0	**0.58**	+4 (=)	(Qiu et al. 2017)
RO	*Ca.* 33% of FM was replaced with PP blends	10	100	56 days	13.8	25.1	1.92	**0.52**	−8.5(↓)	(Qiu et al. 2017)
Peanut oil	*Ca.* 33% of FM was replaced with PP blends	10	100	56 days	13.8	26.6	1.99	**0.62**	−3 (=)	(Qiu et al. 2017)

(*Continued*)

TABLE 9.4 (CONTINUED)
Effects of Dietary FO Replacement with ALS on Growth and Feed Utilization of Marine Fish Species

Alternative Lipid Source	Details	Dietary Lipid Level (%)	% FO Replacement	Duration of Feeding Trial	Growth Performance					References
					IBW (g)	FBW (g)	SGR (% BW day^{-1})	FCR or FER (bold)	WG (% compared to the FO diet)	
			Starry Flounder (*Platichthys stellatus*)							
CO	Semi-purify diet	9.3	100	70 days	1.9	3.74	0.96		−225 (↓)	(Lee et al. 2003)
LO	Semi-purify diet	9.3	100	70 days	1.8	4.06	1.16		−196 (↓)	(Lee et al. 2003)
Olive Flound				**er (*Paralichthys olivaceous*)**						
SBO	FM based diet	8	100	91 days	3.0	12.3	1.6		−716.7 (↓)	(Kim et al. 2002)
SBO		10	50	56 days	16.5	36	1.3	1.0	−11.2 (=)	(Qiao et al. 2014)
AM	*Schizochytrium* and *Nannochloropsis* meals (2:3)	10	50	56 days	17	40.8	1.5	0.9	+10 (=)	(Qiao et al. 2014)
AM	*Schizochytrium* and *Nannochloropsis* meals (2:3)	10	100	56 days	16.5	36.5	1.3	1.1 (↓)	−7.6 (=)	(Qiao et al. 2014)
T	*Ca.* 27% of FM was replaced with blends of proteins	10	50	56 days	3.93	13.2	3.05	**1.1**	−4 (=)	(Lee et al. 2020)
T	*Ca.* 27% of FM was replaced with blends of proteins	10	100	56 days	3.93	13.6	3.13	**1.1**	+6 (=)	(Lee et al. 2020)
LO	FM based diet	10	100	119 days	296	507	0.45	**0.51 (↓)**	−3.4 (=)	(Kim et al. 2012)
SBO	FM based diet	10	100	119 days	290	523	0.5	**0.8**	+5.6 (=)	(Kim et al. 2012)
LO/SBO (1:1)	FM based diet	10	80	119 days	295	568	0.55	**0.67**	+17.8 (=)	(Kim et al. 2012)
			Senegalese Sole (*Solea senegalensis*)							
Blends of LO/SBO/RO	FM based diet	9	50	5 months	152.9	254.9	0.71	1.31	+8 (=)	(Reis et al. 2014)

(*Continued*)

TABLE 9.4 (CONTINUED)
Effects of Dietary FO Replacement with ALS on Growth and Feed Utilization of Marine Fish Species

Alternative Lipid Source	Details	Dietary Lipid Level (%)	% FO Replacement	Duration of Feeding Trial	Growth Performance: IBW (g)	FBW (g)	SGR (% BW day^{-1})	FCR or FER (bold)	WG (% compared to the FO diet)	References
Blends of LO/SBO/RO	FM based diet	9	100	5 months	152.3	247.3	0.67	1.35	+3 (=)	(Reis et al. 2014)
Blends of LO/SBO/RO	*Ca.* 50% of FM was replaced with PP blends	10	50	5 months	152.7	247.7	0.67	1.33	+3.3 (=)	(Reis et al. 2014)
LO	FM based diet	13	100	90 days	3.5	13.2	1.5	1.6	−14.3 (=)	(Montero et al. 2015)
SBO	FM based diet	13	100	90 days	3.5	11.0	1.3	1.7	−77 (↓)	(Montero et al. 2015)
SBO/LO/RO (1:1:1)	*Ca.* 50% of FM was replaced with PP blends	8	75	91 day	5	21.7	1.6	1.3	−34(=)	(Bonacic et al. 2017)
SBO/LO/RO (1:1:1)	*Ca.* 50% of FM was replaced with PP blends	18	75	91 day	5	20.7	1.57	1.3	−60(=)	(Bonacic et al. 2017)
			Turbot (*Psetta maxima*)							
SBO	FM based diet	16.5	100	91 day	579	912	-	**1.17**	−7.6 (↓)	(Regost et al. 2003)
LO	FM based diet	16.5	100	91 day	579	918	-	**1.17**	−6.6 (↓)	(Regost et al. 2003)
SBO	FM based diet contained FO	16.5	-	56 days	912	1159	-	**1.17**	−10.5 (=)	(Regost et al. 2003)
LO	FM based diet contained FO	16.5	-	56 days	918	1172	-	**1.17**	−9.5 (=)	(Regost et al. 2003)
			Atlantic Halibut (*Hippoglossus Hippoglossus*)							
FXO	*Ca.* 20% of FM was replaced with PP blends	24	40	231 days	849	1503.6	0.2	1.2	+12.9 (=)	(Martins et al. 2011)
FXO	*Ca.* 20% of FM was replaced with PP blends	24	70	231 days	849	1452.1	0.2	1.2	+6.8 (=)	(Martins et al. 2011)
			Rabbit Fish (*Siganus canaliculatus*)							
Perilla/SFO	(1:1)		100		10.4	18.1	-	-	–	(Li et al. 2008b)
			Cobia (*Rachycentron canadum*)							
SBO	About 70% of FM was replaced with corn gluten	11	50	56 days	77.9	211.0	1.8	1.5	+ 1(=)	(Woitel et al. 2014)

(Continued)

TABLE 9.4 (CONTINUED)
Effects of Dietary FO Replacement with ALS on Growth and Feed Utilization of Marine Fish Species

Alternative Lipid Source	Details	Dietary Lipid Level (%)	% FO Replacement	Duration of Feeding Trial	Growth Performance					References
					IBW (g)	FBW (g)	SGR (% BW day^{-1})	FCR or FER (bold)	WG (% compared to the FO diet)	
Partially hydrogenated SBO	About 70% of FM was replaced with corn gluten	11	50	56 days	77.3	208.9	1.8	1.5	+ 0.5(=)	(Woitel et al. 2014)
Fully hydrogenated SBO	About 70% of FM was replaced with corn gluten	11	50	56 days	77.6	211.0	2.1	1.5	+50.5 (=)	(Woitel et al. 2014)
PL	*Ca.* 70% of FM was replaced with corn gluten	11	50	56 days	77.5	248.4	1.8	1.7	+5.3 (=)	(Woitel et al. 2014)
T	*Ca.* 70% of FM was replaced with corn gluten	11	50	56 days	77.1	212.7	1.8	1.6	+8.3 (=)	(Woitel et al. 2014)
California Pompano (*Trachinotus carolinus*)										
T	*Ca.* 70% of FM was replaced with blends of protein sources	18	100	56 days	41.3	120.8	1.91	1.85	−14 (=)	(Rombenso et al. 2016)

Abbreviations: AM, algal meal; APS, alternative protein sources; CAO, canola oil; CO, corn oil; EO, echium oil; FBW, final body weight; FCR, feed conversion ratio; FER, feed efficiency ratio; FM, fish meal; FXO, flaxseed oil; IBW, initial body weight; LO, linseed oil; OO, olive oil; PF, poultry fat; PL, pork lard; PO, palm oil; SGR, specific growth rate, SPC, soy protein concentrate; SBM, soybean meal; SBO, soybean oil; SFO, sunflower oil; T, tallow; WG, wheat gluten, WFO, wheat germ oil. Semi-purified diet: diet containing a high level of casein. The difference of the measured parameters in each treatment compared with control (FO-based diet). (↑), (↓) and (=) show increasing, decreasing, and no significant differences, respectively, in each treatment compared with the control (FO-based diet).

retention of LC-PUFA and oxidation of MC-PUFA in the liver of sharpsnout seabream, even though the key enzymes in the LC-PUFA bioconversion pathway do not exist in the hepatocytes of this species.

It has been estimated that about 25% of global FO production in 2016 was used in aquafeeds for farmed marine carnivorous tropical fish (Alhazzaa et al. 2019). Regarding this group of species, total replacement of FO with poultry fat in Asian sea bass diets was successful, which may be due to the low EFA requirements of this species. In addition, it has been confirmed that tropical marine fish species that were adapted to high water temperature potentially had a higher capacity to utilize SFA- and MUFA-rich lipid sources (e.g. RAF and palm oil) for energy purposes, which resulted in LC-PUFA sparing from β-oxidation (Salini et al. 2015b; Alhazzaa et al. 2018). However, a meta-analysis conducted by Alhazzaa et al. (2019) demonstrated that using VO in diets for tropical marine carnivorous fish reduced their growth performance and feed efficiency, which coincided with a remarkable reduction of n-3 LC-PUFA content in their body composition, results that depended on the replacement levels of FO and FM in experimental diets. Thus, it is generally recommended to use a combination of ALS, such as VO and RAF, in order to improve diet digestibility and energy content, as well as providing a varied FA profile with balanced n-3/n-6 PUFA ratios (Oliva-Teles et al. 2015).

Freshwater Fish Species

Compared to marine fish species, freshwater fish species and salmonids (see next section for this group of fish) have a high capacity for bioconverting MC-PUFA to LC-PUFA (Tocher 2003; Glencross 2009; Tocher 2010). As illustrated in Table 9.5, partial or even total FO replacement with VO in freshwater fish species can be successfully achieved, since ALS are generally rich in precursors of LC-PUFA, namely LNA and ALA. Freshwater fish species cultured in warm temperatures (> 25°C), such as tilapia, carps, and catfish species, require high levels of n-6 PUFA compared to n-3 PUFA for maximal growth performance (NRC 2011). For example, tropical catfish species showed better performances when fed diets containing VO in comparison with those formulated with FO, even when using FM-free diets (Ng et al. 2003). In some cases, excessive n-3 LC-PUFA has been found to impair fish growth performance, possibly due to the effect of oxidative stress (Sargent et al. 1993a; Østbye et al. 2009), which might explain the relatively lower growth performance of fish fed FO compared to that of fish fed with ALS (Duan et al. 2014; Campos et al. 2019).

For salmonids, Turchini et al. (2009) concluded that partial, or even total substitution of FO with ALS in FM-based diets is possible throughout the production cycle (Table 9.6). However, when both FM and FO are simultaneously replaced with their vegetal counterparts, the minimum EFA requirement for salmonids should be considered. In this sense, Torstensen et al. (2008) reported that simultaneous replacement of dietary FM (80%) and FO (70%) with blends of PP and VO (palm, linseed, and rapeseed oils) sources resulted in remarkably reduced final body weight (−17%) in Atlantic salmon after one year. The above-mentioned authors revealed that the growth depression was associated with a reduction in feed intake and a decrease in digestibility of dietary starch and palmitic acid. In addition, Beheshti-Foroutani et al. (2018) reported that dietary FM could be reduced to 5% without affecting the growth of Atlantic salmon, as long as there was a minimum of 5% FO included in the diet.

Regarding crustaceans, many studies have revealed that a large portion of FO (*ca.* 60 to 100%) can be replaced with ALS without significant negative effects on their performance (Table 9.7). These findings were related to their omnivorous feeding habits, which may increase their tolerance to low concentrations of LC-PUFA in the diet. On the other hand, it has been documented those excess levels of FO in the diet suppressed growth performance in giant freshwater prawn (*Macrobrachium rosenbergii*, Palaemonidae) due to extra levels of n-3 LC-PUFA that results in oxidative stress, hepatocyte apoptosis, and hemolymph proinflammatory cytokines (Sun et al. 2020). However, it has been reported that total replacement of FO with coconut or castor oils in a PP-based

TABLE 9.5
Effects of Dietary FO Replacement with ALS on Growth and Feed Utilization of Freshwater Fish Species

Alternative lipid source	Details	Dietary Lipid Level (%)	% FO Replacement	Duration of Feeding Trial	Growth Performance: IBW (g)	FBW(g)	SGR (% BW day^{-1})	FCR or FER (bold)	WG (% compared to the FO diet)	References
				European Perch (*Perca fluviatilis*)						
LO	FM-based diet	18.6	33	76 days	43.	106.6	1.19	**0.88**	+13(=)	(Blanchard et al. 2008)
SFO	FM-based diet	18.6	33	76 days	43.6	98.7	1.07	**0.75**	–4.9(=)	(Blanchard et al. 2008)
LO/SFO (1:1)	FM-based diet	18.6	66	76 days	43.1	105.7	1.14	**0.85**	11.4 (=)	(Blanchard et al. 2008)
OO	Casein-based diet	16	100	70 days	25	33.1	-	**0.86 (↓)**	–111.7 (↓)	(Xu and Kestemont 2002)
										(Xu and Kestemont 2002)
SBO	Casein-based diet	16	100	70 days	25	35.9	-	**1.1**	–100.7 (↓)	(Xu and Kestemont 2002)
LO	*Ca.* 40% of FM was replaced with blends of PP	11.6	100	70 days	17.4	40.9	1.22	-	–46 (↓)	(Geay et al. 2015)
				Pike Perch (*Sander lucioperca*)						
LO	-	13	100	57 days	15.12	34.8	1.44	3.37	(=)	(Schulz et al. 2005)
SBO	-	13	100	57 days	14.9	34.3	1.5	3.27	(=)	(Schulz et al. 2005)
				***Climbing Perch* (*Anabas testudineus*)**						
PO	FM-based diet		100	60 days	21	32.3	-	-	(=)	(Varghese and Oommen 2000)
Coconut oil	FM-based diet		100	60 days	21	32.4	-	-	(=)	(Varghese and Oommen 2000)
				Largemouth Bass (*Micropterus salmoides*)						
LO	*Ca.* 50% of FM was replaced with SBM	7	100	84 days	15.7	93.1	2.2	1.6	–7 (↓)	(Tidwell et al. 2007)
CO	*Ca.* 50% of FM was replaced with SBM	7.9	100	84 days	15.7	91.8	2.2	1.5	–15 (↓)	(Tidwell et al. 2007)

(Continued)

TABLE 9.5 (CONTINUED)
Effects of Dietary FO Replacement with ALS on Growth and Feed Utilization of Freshwater Fish Species

Alternative lipid source	Details	Dietary Lipid Level (%)	% FO Replacement	Duration of Feeding Trial	Growth Performance					References
					IBW (g)	FBW(g)	SGR (% BW day^{-1})	FCR or FER (bold)	WG (% compared to the FO diet)	
CAO	Fat free FM	13.8	100	84 days	5	31.7	2.2	1.33	–46 (=)	(Subhadra et al. 2006)
PF	Fat free FM	13.8	50	84 days	5	33	2.2	1.37	–40 (=)	(Subhadra et al. 2006)
PF	Fat free FM	13.8	100	84 days	5	32	2.2	1.36	–40 (=)	(Subhadra et al. 2006)
Low α-linolenic acid SBO	*Ca.* 80% of FM was replaced with blends of APS	15	50	70 days	1.64	9.79	2.52	1.33	+8 (=)	(Laporte and Trushenski 2011)
Hydrogenated SBO	*Ca.* 80% of FM was replaced with blends of APS	15	50	70 days	1.64	10.23	2.58	1.27	+25 (=)	(Laporte and Trushenski 2011)
SFA-enriched SBO	*Ca.* 80% of FM was replaced with blends of APS	15	50	70 days	1.64	11.07	2.76	1.22	+109 (=)	(Laporte and Trushenski 2011)
White Bass (*Morone chrysops*)										
CO	FM-based diet	14.5	100	196	800	1200	-	-	+12.5 (=)	(Lane and Kohler 2006)
Hybrid Bass (*Morone saxatilis* × *Morone chrysops*)										
CO	FM-based diet	14.4	100	196	362	710	-	1.7	+10 (=)	(Lane et al. 2006)
Arthrospira sp./ Schizochytrium limacinum meals	*Reared in brackish water (3 ppt)*	10	50	42 days	10.5	24.4	2		23.5 (↑)	(Perez-Velazquez et al. 2019)
Channel Catfish (*Ictalurus punctatus*)										
LO	Casein-based diet	100	100	63 days	0.9	14.81	4.44		–170 (↓)	(Fracalossi and Lovell 1994)
T	Casein-based diet	7	100	112 days	3.2	10.8	1.1	-	–145 (↓)	(Fracalossi and Lovell 1995)

(Continued)

TABLE 9.5 (CONTINUED)
Effects of Dietary FO Replacement with ALS on Growth and Feed Utilization of Freshwater Fish Species

Alternative lipid source	Details	Dietary Lipid Level (%)	% FO Replacement	Duration of Feeding Trial	Growth Performance: IBW (g)	FBW(g)	SGR (% BW day^{-1})	FCR or FER (bold)	WG (% compared to the FO diet)	References
CO	Casein-based diet	7	100	112 days	3.2	13.2	1.26	-	−70 (=)	(Fracalossi and Lovell 1995)
LO	Casein-based diet	7	100	112 days	3.2	13.9	1.31	-	−47.8 (=)	(Fracalossi and Lovell 1995)
CO/T	Casein-based diet	7	66	112 days	3.2	14.4	1.34	-	−32.2 (=)	(Fracalossi and Lovell 1995)
				African Sharptooth Catfish (*Clarias gariepinus*)						
CPO	*Ca.* 40% of FM was replaced with blends of PS	13	100	63 days	7.01	53.97	-	**1.09**	–	(Lim et al. 2001)
RBDPO	*Ca.* 40% of FM was replaced with blends of PS	13	100	63 days	7.01	63.96	-	**1.06**	–	(Lim et al. 2001)
Crude PO	Casein-based diet	10	100	49 days	9.03	38.2	-	**1.07 (↑)**	107.4 (↑)	(Ng et al. 2003)
SFO	Casein-based diet	10	100	49 days	9.03	40.8	-	**1.14 (↑)**	134.4 (↑)	(Ng et al. 2003)
RBDPO	Casein-based diet	10	100	49 days	9.03	40.5	-	**1.15 (↑)**	132.1 (↑)	(Ng et al. 2003)
CPKO	Casein-based diet	10	100	49 days	9.03	39.1	-	**1.13 (↑)**	115.2 (↑)	(Ng et al. 2003)
PO	Ca. 45% of FM was with blends of PP	12	100	56 days	7.1	60.1	2.9	1.07	–	(Ng et al. 2006)
				Surubim (*Psudoplatystoma coruscans*)						
Lard	FM-based diet	18	100	63 days	3.0	36.9	4.0	1.0	–	(Martino et al. 2002)
CO	FM-based diet	18	100	63 days	2.7	28	3.7	1.1	–	(Martino et al. 2002)
SBO	FM-based diet	18	100	63 days	2.6	28	3.8	1.1	–	(Martino et al. 2002)
LO	FM-based diet	18	100	63 days	2.7	32.1	3.9	1.1	–	(Martino et al. 2002)
				Heterobranchus longifilis						
PO	-	-	100	17 days	0.002	114.5 mg	27	-	–	(Legendre et al. 1995)

(Continued)

TABLE 9.5 (CONTINUED)
Effects of Dietary FO Replacement with ALS on Growth and Feed Utilization of Freshwater Fish Species

Alternative lipid source	Details	Dietary Lipid Level (%)	% FO Replacement	Duration of Feeding Trial	Growth Performance IBW (g)	FBW(g)	SGR (% BW day^{-1})	FCR or FER (bold)	WG (% compared to the FO diet)	References
Claiias batvachus										
SFO	Casein-based diet	5	100	42 days	0.51	2.15	-	2.06	−32(=)	(Mukhopadhyay and Mishra 1998)
				River Catfish (*Mystus nemurus*)						
CO	Semi-purify diet	10	100	70 days	13.0	23.1	0.82	**0.3**	+9.5 (=)	(Ng et al. 2000)
Crude PO	Semi-purify diet	10	100	70 days	13.0	21.4	0.71	**0.27**	−3.1 (=)	(Ng et al. 2000)
RBDPO	Semi-purify diet	10	100	70 days	13.0	28.3	1.1	**0.41(↑)**	+51 (↑)	(Ng et al. 2000)
				South American Catfish (*Rhamdia quelen*)						
CO	FM-free diet	10	100	31 days	1.0	4.6	5.12	0.83	+42(=)	(Vargas et al. 2008)
LO	FM-free diet	10	100	31 days	1.0	3.9	4.53	0.83	−17(=)	(Vargas et al. 2008)
				Murray Cod (*Maccullochella peelii peelii*)						
LO	Casein-based diet	18	25-100	112 days	125.4	962.0	-	-	(=)	(Turchini et al. 2006)
CAO	Casein-based diet	18	100	84 days	6.5	26.9	1.69	0.8	−109.2(↓)	(Francis et al. 2007b)
CAO	Casein-based diet	19	25	112 days	21.4	88.9	1.27	0.76	−108.3 (=)	(Francis et al. 2007b)
CAO	Casein-based diet	19	50	112 days	21.6	90.3	1.28	0.78	−105.6 (=)	(Francis et al. 2007b)
CAO	Casein-based diet	19	75	112 days	21.4	88.3	1.26	0.83	−112 (=)	(Francis et al. 2007b)
CAO	Casein-based diet	19	100	112 days	21.2	73.8	1.1	0.88	−176.4 (↓)	(Francis et al. 2007b)
LO/PO/OO (4.5:4.3:1.2)	Casein-based diet	19	100	98 days	14.4	49.9	-	-	(=)	(Francis et al. 2007a)
				Nile Tilapia (*Oreochromis niloticus*)						**T**
Casein-based diet	7	100	84 days	2.85	14.05	-	**0.9 (↓)**	−400(↓)	(Yildirim-Aksoy et al. 2007)	**CO**
Casein-based diet	7	100	84 days	2.85	30.65	-	**1.1**	+122.8(=)	(Yildirim-Aksoy et al. 2007)	**LO**
Casein-based diet	7	100	84 days	2.85	29.45	-	**1.2**	−73.7(=)	(Yildirim-Aksoy et al. 2007)	**LO**

(Continued)

TABLE 9.5 (CONTINUED)
Effects of Dietary FO Replacement with ALS on Growth and Feed Utilization of Freshwater Fish Species

Alternative lipid source	Details	Dietary Lipid Level (%)	% FO Replacement	Duration of Feeding Trial	Growth Performance: IBW (g)	FBW(g)	SGR (% BW day^{-1})	FCR or FER (bold)	WG (% compared to the FO diet)	References
Ca. 90% of FM was with blends of PP	3.5	100	140 days	1.06	174.0	3.43	2.06	(=)	(Karapanagiotidis et al. 2007)	***Nanochloropsis salina***
FM-free diet	13	100	36 days	12.62	39.79	3.19	1.28	−19 (↓)	(Gbadamosi and Lupatsch 2018)	***Schizochytrium sp.***
Ca. 66% of FM was replaced with SBM and CG	11	100	84 days	1.6	28.8	3.5	0.9 (↑)	(↑)	(Sarker et al. 2016)	**Coconut oil**
Ca. 80% of FM was replaced with blends of PP	4.63	50	56 days	52.23	205.2	2.44	2.72	+12.8 (=)	(Apraku et al. 2017)	**Coconut oil**
Ca. 80% of FM was replaced with blends of PP	4.29	100	56 days	59.67	214.6	2.29	3.19	+14.7 (=)	(Apraku et al. 2017)	**Perilla oil**
Casein-based diet	9	100	75 days	8.93	111.26	3.35	0.94	−154.8 (=)	(Teoh and Ng 2016)	**CAO**
Casein-based diet	9	100	75 days	8.93	107.04	3.31	0.98	−202.9 (=)	(Teoh and Ng 2016)	**SFO**
Casein-based diet	9	100	75 days	8.94	100.88	3.23	0.98	−272.3 (↓)	(Teoh and Ng 2016)	**RBDPO**
Casein-based diet	9	100	75 days	8.93	109.91	3.34	1.02 (↓)	−272.3 (↓)	(Teoh and Ng 2016)	**Hybrid Tilapia (*Oreochromis* sp.)**
CPO	Ca. 40% of FM was replaced with SBM	11.4	100	140 days	31.2	436.6	1.88	1.45	−102.2 (=)	(Bahurmiz and Ng 2007b)
RBDPO	Ca. 40% of FM was replaced with SBM	11.4	100	140 days	31.2	414.3	1.84	1.44	−173.5 (=)	(Bahurmiz and Ng 2007b)
PFAD	Ca. 40% of FM was replaced with SBM	11.4	100	140 days	31.2	423.5	1.86	1.35	−145 (=)	(Bahurmiz and Ng 2007b)

(Continued)

TABLE 9.5 (CONTINUED)
Effects of Dietary FO Replacement with ALS on Growth and Feed Utilization of Freshwater Fish Species

Alternative lipid source	Details	Dietary Lipid Level (%)	% FO Replacement	Duration of Feeding Trial	Growth Performance: IBW (g)	FBW(g)	SGR (% BW day^{-1})	FCR or FER (bold)	WG (% compared to the FO diet)	References
CPO	Semi-purified diet	-	100	56 days	11.73	46.4	2.5	**1.12 (↑)**	+27.4(=)	(Ng et al. 2001)
SFO	Semi-purified diet	-	100	56 days	11.73	48.5	2.53	**1.03**	+48(=)	(Ng et al. 2001)
CPKO	Semi-purified diet	-	100	56 days	11.73	45.2	2.4	**1.01**	+20.1(=)	(Ng et al. 2001)
lard	Casein-based diet	5	100	56 days	0.83	1.86	1.44	**0.35(↓)**	−93.6 (↓)	(Chou and Shiau 1999)
CO	Casein-based diet	5	100	56 days	0.84	2.03	1.6	**0.41(↓)**	−76(↓)	(Chou and Shiau 1999)
CO/Lard	Casein-based diet	5	66	56 days	0.84	3.06	2.32	**0.69(↑)**	+46.6 (↑)	(Chou and Shiau 1999)
SBO	Casein-based diet	8	100	70 days	3	5.7	1.52	-	−10 (=)	(Huang et al. 1998)
CAO	FM based diet	14	100	140 days	40	110	0.72	3.0		(Wonnacott et al. 2004)
				Grass Carp (*Ctenopharyngodon idella*)						
CO	Casein-based diet	10	100	56 days	4.2	9.1	2.1	**0.61**	+117.4 (↑)	(Du et al. 2008)
LO/maize (1:1)	Casein-based diet	10	100	56 days	4.2	9.4	2.14	**0.61**	+125.2 (↑)	(Du et al. 2008)
				Common Carp (*Cyprinus carpio*)						
LO	*Ca.* 75% of FM was replaced with blends of PS	10	100	63 days	16.9	32.7	1	2.01	−25.1 (=)	(Nuguyen et al. 2019)
SFO	*Ca.* 75% of FM was replaced with blends of PS	10	100	63 days	16.4	33.6	1.1	2.23	−13.7 (=)	(Nuguyen et al. 2019)
				Indian Major Carp (*Catla catla*)						
SFO	Casein-based diet	5	50	56 days	0.25	1.232	-	2.52	+90.4	(Mukhopadhayay and Rout 1996)
SFO	Casein-based diet	5	100	56 days	0.25	1.075	-	2.52	+29.5	(Mukhopadhayay and Rout 1996)
SFO	FM based diet	6	50	60	0.25	0.424	-	-	–	(Dalbir et al. 2015)
SFO	FM based diet	6	75	60	0.25	0.336	-	-	–	(Dalbir et al. 2015)

(*Continued*)

TABLE 9.5 (CONTINUED)
Effects of Dietary FO Replacement with ALS on Growth and Feed Utilization of Freshwater Fish Species

Alternative lipid source	Details	Dietary Lipid Level (%)	% FO Replacement	Duration of Feeding Trial	Growth Performance					References
					IBW (g)	FBW(g)	SGR (% BW day^{-1})	FCR or FER (bold)	WG (% compared to the FO diet)	
				Gold Fish (*Carrassios auratus gibelio*)						
CO	Semi-purified diet	0-21	50	70 days	3.5	18.13-24.53	2.35-2.78	-	–	(Zhou et al. 2014)
				River Chub (*Zacco barbata*)						
SBO	Casein-based diet	0-20	50	56 days	0.26	0.62-0.71	-	2.63-3.48	(=)	(Huang et al. 2001)
				Marble Goby (*Oxyeleotris marmorata*)						
SBO	*Ca.* 30% of FM was replaced with SBM	12	90	120 days	13.7	18.5	0.3	2.8	–3.6 (=)	(Ti et al. 2019)
CAO	*Ca.* 30% of FM was replaced with SBM	12	90	120 days	13.5	16.6	0.2	6.6 (↓)	–14.9 (=)	(Ti et al. 2019)
SBO/CO (4:5)	*Ca.* 30% of FM was replaced with SBM	12	90	120 days	13.6	18.3	0.3	3.7	–3 (=)	(Ti et al. 2019)
				Java Barb (*Puntius gonionotus*)						
LO	*Ca.* 80% of FM was replaced with SBM	11.8	50	60 days	11.6	28.9	1.53	1.82	–26 (=)	(Nayak et al. 2017)
LO	*Ca.* 80% of FM was replaced with SBM	11.6	100	60 days	12.3	27.4	1.33	1.92	–53 (↓)	(Nayak et al. 2017)
				Tambaqui (*Colossoma macropomum*)						
Different blends of CO/LO/PO	PP-based diet	8	100	49 days	43.1	139.4	3.4	**0.9**	–5(=)	(Paulino et al. 2018)

Abbreviations: AM, algal meal; APS, alternative protein sources; CAO, canola oil; CO, corn oil; EO, echium oil; FBW, final body weight; FCR, feed conversion ratio; FER, feed efficiency ratio; FM, fish meal; FXO, flaxseed oil; IBW, initial body weight; LO, linseed oil; OO, olive oil; PF, poultry fat; PL, pork lard; PO, palm oil; SGR, specific growth rate, SPC, soy protein concentrate; SBM, soybean meal; SBO, soybean oil; SFO, sunflower oil; T, tallow; WG, wheat gluten, WFO, wheat germ oil. Semi-purify diet: diet contained high level of casein. The difference of the measured parameters in each treatment comparing with control (FO-based diet). (↑), (↓) and (=) show increasing, decreasing, and no significant differences, respectively, in each treatment comparing with control (FO-based diet).

TABLE 9.6
Effects of Dietary FO Replacement with ALS on Growth and Feed Utilization of Salmonids

Alternative lipid source	Details	Dietary lipid level (%)	% FO replacement	Duration of feeding trial	Growth performance: IBW (g)	FBW(g)	SGR (% BW day^{-1})	FCR or FE (bold)	WG (% compared to the FO diet)	References
Atlantic salmon (*Salmo salar*)										
LO - - 100 84 days 54 198-210 - - (=) (Bell et al. 1996)	-	30–37	50–60	12 months	120	2700	0.85		(=)	(Rosenlund et al. 2001)
LO Ca. 60% of FM was replaced with blends of PP 19 100 238 days 30.1 114.6 0.56 - -36(↓) (Tocher et al. 2000)										
LO										
LO	FM-based diet	24.1	33	350 days	120	2300	0.84	-	+166.7 (=)	(Bell et al. 2003a)
LO	FM-based diet	24.1	66	350 days	120	2300	0.84	-	+166.7 (=)	(Bell et al. 2003a)
LO	FM-based diet	24.1	100	350 days	120	2100	0.82	-	(=)	(Bell et al. 2003a)
LO	-	-	100	84 days	54	198–210	-	-	(=)	(Bell et al. 1996)
LO	Ca. 60% of FM was replaced with blends of PP	19	100	238 days	30.1	114.6	0.56	-	−36(↓)	(Tocher et al. 2000)
LO			100	112 days	132.4	412.9	-			(Leaver et al. 2008)
LO	-	29.4	100	280 days	127	1870	-	-	+63 (=)	(Bell et al. 2004)
LO	FM-based diet	36	100	84 days	220	556.1	-	-	+14 (=)	(Menoyo et al. 2005)
EO	Ca. 80% of FM was replaced with blends of APS	17	100	42 days	44.8	76.9	1.3	**0.8**	−6.5 (=)	(Miller et al. 2007)
EO	Ca. 75% of FM was replaced with blends of APS		100	84 days	106.9	236.5	-	-	(=)	(Miller et al. 2008a,b)

(*Continued*)

TABLE 9.6 (CONTINUED)
Effects of Dietary FO Replacement with ALS on Growth and Feed Utilization of Salmonids

Alternative lipid source	Details	Dietary lipid level (%)	% FO replacement	Duration of feeding trial	Growth performance					References
					IBW (g)	FBW(g)	SGR (% BW day^{-1})	FCR or FE (bold)	WG (% compared to the FO diet)	
LO/RO (1:1)	Ca. 60% of FM was replaced with blends of PP		100	175 days	21.9	34.7	0.54	-	-	(Bell et al. 1997)
LO/Arasco (98:2)	Ca. 60% of FM was replaced with blends of PP	19	100	238 days	30.8	113.3	0.55	-	−39.5(↓)	(Tocher et al. 2000)
LO/RO(1:1)	-	-	100	790 days	-	2020	-	-	-	(Tocher et al. 2003b)
LO/RO(2:1)	FM-based diet	-	100	172 days	19.3	38	-	-	-	(Jobling and Bendiksen 2003)
LO/RO (1:1)			75	760 days		2100				(Bell et al. 2005)
RO/PO/LO	*Simultaneous FM and FO replacement with VO and PP sources*	28–34	35–70	348 days	355	3280–3967	0.86–0.94	0.78–0.82	-	(Torstensen et al. 2008)
LO/SFO	FM-based diet	32	100	84 days	220	531	1.02–1.16	0.72–0.79	(=)	(Menoyo et al. 2007)
EO/RO(1:1)	Ca. 80% of FM was replaced with blends of APS	17	100	42 days	44.6	80.1	1.4	**0.8**	+3.1 (=)	(Miller et al. 2007)
Camelina/RO/PO (2:5:3)	Ca. 60% of FM was replaced with blends of PP	28	100	294 days	-	2000	-	-	-	(Petropoulos et al. 2009)
CAO	--	28	17	112 days	160.2	417.4	0.86	**0.89**	+20.6 (↑)	(Dosanjh et al. 1998)
CAO	-	28	34	112 days	168.7	379.9	0.73	**0.8**	−14.1 (=)	(Dosanjh et al. 1998)
CAO	-	28	51	112 days	169.2	406.3	0.78	**0.79**	+0.8 (=)	(Dosanjh et al. 1998)
RO	-	30–37	50–60	12 months	120	2700	0.85	-	(=)	(Rosenlund et al. 2001)
RO	FM-based diet	27	50	112 days	56.3	185.8	1.21	1.53	+39.8 (=)	(Bell et al. 2003a)

(Continued)

TABLE 9.6 (CONTINUED)
Effects of Dietary FO Replacement with ALS on Growth and Feed Utilization of Salmonids

Alternative lipid source	Details	Dietary lipid level (%)	% FO replacement	Duration of feeding trial	Growth performance					References
					IBW (g)	FBW(g)	SGR (% BW day^{-1})	FCR or FE (bold)	WG (% compared to the FO diet)	
RO	FM-based diet	27	100	112 days	52.4	177.1	1.22	1.55	+24 (=)	(Bell et al. 2003a)
RO	FM-based diet	24.1	33	350 days	120	2000	0.8	-	−83.3 (↓)	(Bell et al. 2003a)
RO	FM-based diet	24.1	66	350 days	120	2400	0.86	-	+250 (↑)	(Bell et al. 2003a)
RO	FM-based diet	24.1	100	350 days	120	2600	0.88	-	+416.7 (↑)	Bell et al. (2003b)
RO	-	30	25–100	294 days	143	1463	0.74	-	(=)	(Torstensen et al. 2004)
RO	-	28	100	56 days	85	285	1.8	1.41 (↓)	+82 (=)	(Moya-Falcón et al. 2005)
RO	*Ca. 40% of FM was replaced with oilseeds and legume meals*	35	30	84 days	1171.9	1760.3	0.52	0.87	+3.3 (=)	(Karalazos et al. 2007)
RO	*Ca. 40% of FM was replaced with oilseeds and legume meals*	36	60	84 days	1168.4	1767.7	0.53	0.85	+3.9 (=)	(Karalazos et al. 2007)
RO	-	35	60	70 days	2053	3340.2–3664.2	-	1.1	(=)	(Karalazos et al. 2011)
SBO	-	30–37	50–60	12 months	120	2700	0.85		(=)	(Rosenlund et al. 2001)
SBO	Ca. 55% of FM was replaced with maize gluten	30	50	3 months	117	317	1.1	-	−12.7 (=)	(Grisdale-Helland et al. 2002)
SBO	Ca. 55% of FM was replaced with maize gluten	30	100	3 months	119	320	1.1	-	−10.2 (=)	(Grisdale-Helland et al. 2002)
SBO	Ca. 50% of FM was replaced with APS	19	72	161 days	1429	1950	0.19	-	(↓)	(Hardy et al. 1987)
PO	-	30–37	50–60	12 months	120	2700	0.85	-	(=)	(Rosenlund et al. 2001)

(Continued)

TABLE 9.6 (CONTINUED)
Effects of Dietary FO Replacement with ALS on Growth and Feed Utilization of Salmonids

Alternative lipid source	Details	Dietary lipid level (%)	% FO replacement	Duration of feeding trial	Growth performance					References
					IBW (g)	FBW(g)	SGR (% BW day^{-1})	FCR or FE (bold)	WG (% compared to the FO diet)	
PO	Ca. 80% of FM was replaced with APS	15	100	63 days	40	78.4	1.14	-	−19.25 (=)	(Miller et al. 2007)
***Shyzochytrium* oil**	Ca. 80% of FM was replaced with APS	15	100	63 days	38.8	80	1.17	-	−9.1 (=)	(Miller et al. 2007)
Shyzochytrium oil /PO *(4:1)*	Ca. 80% of FM was replaced with APS	15	100	63 days	38.5	74.5	1.04	-	−21.7 (↓)	(Miller et al. 2007)
Scizochytrium sp	-	-	40	84 days	213	819	1.6	0.88	−2.3 (=)	(Kousoulaki et al. 2015a)
Scizochytrium sp	-	-	100	84 days	213	797	1.57	0.83	−12.7 (↓)	(Kousoulaki et al. 2015a)
SFO	-			147 days	1500	3600	0.6			(Torstensen et al. 2000)
		30	50	294 days	143	1458.6	0.75		(=)	(Torstensen et al. 2004)
RO/Shyzochytrium (5:1)	Ca. 55% of FM was replaced with blends of PP	-	100	-	1522	3170	0.58	1.40 (↓)	−2 (=)	(Sprague et al. 2015)
RO/ Shyzochytrium (2:1)	Ca. 55% of FM was replaced with blends of PP	-	100	-	1543	3030	0.53	1.42 (↓)	−14 (↓)	(Sprague et al. 2015)
Camelina oil	Fat-free FM	18.5	100	112 days	236	613	0.86	1.06	−40.7 (↓)	(Hixson et al. 2014a)
Camelina oil	Fat-free FM	20	100	112 days	231	537	0.76	1.21	−66.5 (↓)	(Hixson et al. 2014b)
Camelina oil	Ca. 40% of FM was replaced with APS	20	50	112 days	8.4	28	1.45	0.91	+15.5 (=)	(Ye et al. 2016)
Camelina oil	Ca. 40% of FM was replaced with APS	20	100	112 days	8.2	28.5	1.44	0.99	+24.4 (=)	(Ye et al. 2016)

(Continued)

TABLE 9.6 (CONTINUED)
Effects of Dietary FO Replacement with ALS on Growth and Feed Utilization of Salmonids

Alternative lipid source	Details	Dietary lipid level (%)	% FO replacement	Duration of feeding trial	Growth performance					References
					IBW (g)	FBW(g)	SGR (% BW day^{-1})	FCR or FE (bold)	WG (% compared to the FO diet)	
Camelina oil	Ca. 30% of FM was replaced with blends of PP	24	100	49 days	82.5	200.5	1.9	0.9	+4.8 (=)	(Betancor et al. 2015)
EPA-Camelina oil	Ca. 30% of FM was replaced with blends of PP	24	100	49 days	82.5	207.9	2.0	0.9	+109 (=)	(Betancor et al. 2015)
T	Ca. 50% of FM was replaced with APS	19	72	161 days	1611	2184	0.19	-	(=)	(Hardy et al. 1987)
Poultry by product/ tallow (1:2)	-	20	50	98 days	138	637.5	1.52	0.91	+3.6 (=)	(Emery et al. 2014)
Poultry by product	-	30	80	195 days	1714	3892	0.43	1.22	-	(Emery et al. 2016)
Poultry by product/ tallow (1:1)	-	30	80	195 days	1727	4016	0.44	1.3	-	(Emery et al. 2016)
RO	*Ca. 30% of FM was replaced with oilseed meals*	34.5	60	70 days	780	2180	1.34	0.98	+4.9(=)	(Pratoomyot et al. 2008)
RO/SBO	*Ca. 40% of FM was replaced with oilseed meals*	34.5	60	70 days	780	2190	1.35	0.96	(=)	(Pratoomyot et al. 2008)
SBO	*Ca. 40% of FM was replaced with oilseed meals*	34.5	60	70 days	770	2190	1.33	0.96	+ 1.3 (=)	(Pratoomyot et al. 2008)
SFO	FM-based diet	30	100	63 days	21.7	66.7		**1.36**	+11 (=)	(Bransden et al. 2003)
Chinook salmon (*Oncorhynchus tshawytscha*)										
CAO	FM-based diet	16	50	62 days	0.51	1.97	2.17		+16 (=)	(Dosanjh et al. 1988)

(Continued)

TABLE 9.6 (CONTINUED)
Effects of Dietary FO Replacement with ALS on Growth and Feed Utilization of Salmonids

Alternative lipid source	Details	Dietary lipid level (%)	% FO replacement	Duration of feeding trial	Growth performance					References
					IBW (g)	FBW(g)	SGR (% BW day^{-1})	FCR or FE (bold)	WG (% compared to the FO diet)	
CAO	FM-based diet	16	100	62 days	0.5	2.03	2.3		+28 (=)	(Dosanjh et al. 1988)
CAO	FM-based diet	19	11-54	139 days	11.0	82.9	1.15		(=)	(Grant et al. 2008)
CAO	Ca. 45% of FM was replaced with APS	22	25-72	210 days	0.79	15	1.4	**0.43**	(=)	(Huang et al. 2008)
			Rainbow trout (*Onchorhynchus mykiss*)							
LO	-	25	100	72 days	55	159	1.95		−12.8 (=)	(Geurden et al. 2009)
SBO	Ca. 50% of FM was replaced with APS	30	50	64days	252.2	751.8	1.72	0.723	−12.9 (=)	(Caballero et al. 2002)
SBO	FM-based diet	16.1	50	84 days	51.9	176.4	1.5	1.0	+10.4 (=)	(Martins et al. 2006)
SBO	FM-based diet	16.1	100	84 days	52.7	171.0	1.4	1.0	(=)	(Martins et al. 2006)
PO		20		70 days	27	82.5				(Fonseca-Madrigal et al. 2005)
OO	Ca. 50% of FM was replaced with APS	30	50	64days	242.8	743.1	1.76	0.786	−16.8 (=)	(Caballero et al. 2002)
OO			100	185 days	42	77.7				(Choubert et al. 2006)
RO	Ca. 50% of FM was replaced with APS	30	50	64days	253.5	761.7	1.73	0.73	−8.9 (=)	(Caballero et al. 2002)
RO	Semi-purified diet	19.5	25	51 days	76.9	142.7	1.9		−5.2 (=)	(Pettersson et al. 2009)
RO	Semi-purified diet	19.5	50	51 days	74.7	138.1	1.9		−11.5 (=)	(Pettersson et al. 2009)
RO	Semi-purified diet	19.5	75	51 days	74.1	142.4	2		−5.5 (=)	(Pettersson et al. 2009)
Camelina oil	-		100	84 days	48	184	1.59	0.93	−7.4 (=)	(Hixson et al. 2014)
PO	-	20–22		84 days	38					(Oo et al. 2007)
T	-	21	38							(Bureau 1997)
Lard/OO(2:1)	Ca. 50% of FM was replaced with APS	18	60	64days	242.8	743.1	1.76	0.79(↓)	−8.8(=)	(Caballero et al. 2002)

(*Continued*)

TABLE 9.6 (CONTINUED)
Effects of Dietary FO Replacement with ALS on Growth and Feed Utilization of Salmonids

Alternative lipid source	Details	Dietary lipid level (%)	% FO replacement	Duration of feeding trial	Growth performance IBW (g)	FBW(g)	SGR (% BW day^{-1})	FCR or FE (bold)	WG (% compared to the FO diet)	References
PF		18	60							(Liu et al. 2004)
T		18	45							(Bureau et al. 2008)
T	Casein-based diet	15	100	42 days	2	7.12	3.02	0.8	−7.6 (=)	(Bayraktar and Bayır 2012)
Sheep tail fat	Casein-based diet	15	100	42 days	2	7.07	3.01	0.8	−8.9 (=)	(Bayraktar and Bayır 2012)
Goose fat	Casein-based diet	15	100	42 days	2	7.31	3.06	0.79	−1.2 (=)	(Bayraktar and Bayır 2012)
LO	Ca. 30% of FM was replaced with SBM	20	100	112 days	47.6	458.5	2.04	1.03	+32.5 (=)	(Francis et al. 2014)
OO	Ca. 30% of FM was replaced with SBM	20	100	112 days	47.8	431.7	1.98	1.01	−27.9 (=)	(Francis et al. 2014)
OO	Ca. 30% of FM was replaced with SBM	20	100	112 days	47.6	423.8	1.97	1.13	−40.4 (=)	(Francis et al. 2014)
	Ca. 30% of FM was replaced with SBM	20	100	112 days	47.6	441.8	2.01	1.02	−2.3 (=)	(Francis et al. 2014)
T	Ca. 80% of FM was replaced with APS	14	100	210 days	46.2	570.8	1.1	1.5	+3(=)	(Gause and Trushenski 2013)
Manola oil	Ca. 80% of FM was replaced with APS	20	75	189 days	5.05	391.6	2.96	1.26	+28 (=)	(Turchini et al. 2013)
CAO	Ca. 80% of FM was replaced with APS	20	75	189 days	5.18	401.5	2.99	1.23	−197 (=)	(Turchini et al. 2013)
PF	Ca. 80% of FM was replaced with APS	20	75	189 days	5.23	370.9	2.88	1.29	−864 (=)	(Turchini et al. 2013)
PO	Ca. 80% of FM was replaced with APS	20	75	189 days	5.07	358.2	2.85	1.33 (↓)	−893 (↓)	(Turchini et al. 2013)

(Continued)

TABLE 9.6 (CONTINUED)
Effects of Dietary FO Replacement with ALS on Growth and Feed Utilization of Salmonids

Alternative lipid source	Details	Dietary lipid level (%)	% FO replacement	Duration of feeding trial	Growth performance					References
					IBW (g)	FBW(g)	SGR (% BW day^{-1})	FCR or FE (bold)	WG (% compared to the FO diet)	
SFO	Ca. 80% of FM was replaced with APS	20	75	189 days	5.18	402.4	2.99	1.22	−188 (=)	(Turchini et al. 2013)
High oleic SFO	Ca. 80% of FM was replaced with APS	20	75	189 days	5.11	429.6	3.08	1.13 (↑)	+463 (↑)	(Turchini et al. 2013)
SBO	Ca. 80% of FM was replaced with APS	20	75	189 days	5.08	390.2	2.96	1.2	−274 (=)	(Turchini et al. 2013)
Coconut oil	Ca. 30% of FM was replaced with WGM	11 LF	100	21 days	71.29	115	2.3		+ 1(=)	(Luo et al. 2014)
Coconut oil	Ca. 30% of FM was replaced with WGM	21 HF	100	21 days	71.38	118.6	2.3		−9.18 (↓)	(Luo et al. 2014)
CSO	Ca. 30% of FM was replaced with APS	16.5	100	84 days	15.4	116.6	2.4	**0.8**	−8.3 (↓)	(Yıldız et al. 2018)
CAO	Ca. 30% of FM was replaced with APS	16.5	100	84 days	15.7	116.1	2.4	**0.9**	−8.8 (↓)	(Yıldız et al. 2018)
CSO/CAO	Ca. 30% of FM was replaced with APS	16.5	50	84 days	15.5	120.9	2.4	**0.9**	−4 (=)	(Yıldız et al. 2018)
CSO/CAO	Ca. 30% of FM was replaced with APS	16.5	50	84 days	15.9	122.3	2.4	**0.8**	−2.6 (=)	(Yıldız et al. 2018)
Arctic charr (*Salvelinus alpinus*)										
SBO	Casein-based diet	10	100	90 days	26.3	52.3	0.76	-	−2.3 (=)	(Olsen and Henderson 1997)
EO	Ca. 15% of FM was replaced with SBM	20	80	112 days	6.3	31.1	1.4	1.5	+48(=)	(Tocher et al. 2006)
Atlantic charr (*Salvelinus fontinalis*)										
SBO	Ca. 20% of FM was replaced with WGM	18.6	100	63 days	41.5	112.1	1.6	1.11	−14 (=)	(Guillou et al. 1995)

(Continued)

TABLE 9.6 (CONTINUED)
Effects of Dietary FO Replacement with ALS on Growth and Feed Utilization of Salmonids

Alternative lipid source	Details	Dietary lipid level (%)	% FO replacement	Duration of feeding trial	Growth performance: IBW (g)	FBW(g)	SGR (% BW day^{-1})	FCR or FE (bold)	WG (% compared to the FO diet)	References
CAO	Ca. 20% of FM was replaced with WGM	18.6	100	63 days	41.5	108.3	1.52	1.16	−23 (=)	(Guillou et al. 1995)
RO	Ca. 85% of FM was replaced with APS	25	83	15 months	15–20	270.3	0.81		(=)	(Murray et al. 2014)
				Brown trout (*Salmo trutta*)						
CO	FM-based diet	29	100	110 days	1605	2610	1.87	1.13	+2 (=)	(Arzel et al. 1994)
OO	-	20	100	70 days	58.3	100.6	0.78	1.16	−6.4 (=)	(Turchini et al. 2003)
PF	-	20	100	70 days	58.1	98.9	0.76	1.1	−8.5 (=)	(Turchini et al. 2003)
Lard	-	20	100	70 days	58.5	100.9	0.78	1.05	−6.1 (=)	(Turchini et al. 2003)
CAO	-	20	100	70 days	58.4	99.9	0.77	1.32	−6.8 (=)	(Turchini et al. 2003)
Hazelnut oil	Casein-based diet	15	100	42 days	1	2.8	2.1	1.2	−30.1 (↓)	(Arslan et al. 2012)
SBO	Casein-based diet	15	100	42 days	1	3.5	2.6	1.3	+31.1 (↑)	(Arslan et al. 2012)
SBO/LSO (1:1)	Casein-based diet	15	100	42 days	1	3.1	2.3	1.3	−1.2 (=)	(Arslan et al. 2012)
Caspian brown trout (*Salmo trutta caspious*)										
CO/SBO (85:15)	FM-based diet	10	100	56 days	10.04	15.75	0.61	-	+22.3 (↑)	(Kenari et al. 2011)
CO/SBO (85:15)	FM-based diet	20	100	56 days	10.03	16.28	0.71	-	+10.5 (=)	(Kenari et al. 2011)

Abbreviations: AM, algal meal; APS, alternative protein sources; CAO, canola oil; CO, corn oil; EO, echium oil; FBW, final body weight; FCR, feed conversion ratio; FER, feed efficiency ratio; FM, fish meal; FXO, flaxseed oil; IBW, initial body weight; LO, linseed oil; OO, olive oil; PF, poultry fat; PL, pork lard; PO, palm oil; SGR, specific growth rate, SPC, soy protein concentrate; SBM, soybean meal; SBO, soybean oil; SFO, sunflower oil; T, tallow; WG, wheat gluten, WFO, wheat germ oil. Semi-purify diet: diet contained a high level of casein. The difference of the measured parameters in each treatment comparing with control (FO-based diet). (↑), (↓) and (=) show increasing, decreasing and no significant differences, respectively, in each treatment comparing with control (FO-based diet).

TABLE 9.7
Effects of Dietary FO Replacement with ALS on Growth and Feed Utilization of Crustaceans

Alternative Lipid Source	Aquatic Species	Dietary Lipid Level (%)	% FO Replacement	Duration of Feeding Trial	Growth Performance IBW (g)	FBW(g)	SGR (% BW day^{-1})	FCR	WG (% compared to the FO diet)	References
LO	*Penaeus monodon*	3.6	100	40 days	1.49	3.6	-	-	(↑)	(Deering et al. 1997)
LO	*Penaeus vannamei*	8	100	70 days	0.9	3.95	2.11	1.78	−105.5 (↓)	(Lim et al. 1997)
Schizochytrium/SBO (1:2)	*Macrobrachium rosenbergii*	10	100	60 days	3.13	8.12	1.59	1.88 (↑)	+13.39 (↑)	(Kangpanich et al. 2017)
Schizochytrium/SBO (1:1)	*Macrobrachium rosenbergii*	10	100	60 days	3.17	8.25	1.59	1.87 (↑)	+14.16 (↑)	(Kangpanich et al. 2017)
Schizochytrium/SBO (2:1)	*Macrobrachium rosenbergii*	10	100	60 days	3.12	8.76	1.72	1.65 (↑)	+35.23 (↑)	(Kangpanich et al. 2017)
Schizochytrium	*Macrobrachium rosenbergii*	10	100	60 days	3.14	7.59	1.47	2.12 (↑)	−4.6 (=)	(Kangpanich et al. 2017)
SBO	*Macrobrachium rosenbergii*	9	100	56 days	0.24	12.4	6.33	1.47 (↑)	+1103.3 (↑)	(Sun et al. 2020)
RO	*Macrobrachium rosenbergii*	9	100	56 days	0.24	11.9	6.27	1.68 (↑)	+912.8 (↑)	(Sun et al. 2020)
CO	*Cherax quadricarinatus*	8	100	84 days	4.0	35.51	2.56	1.34	(=)	(Hernández-Vergara et al. 2003)
PL	*Litopenaeus vannamei*	100	60	50 days	0.1	1.41	4.74	3.98(↑)	−1240.1 (↓)	(Zhou et al. 2007)
SBO	*Litopenaeus vannamei*	100	60	50 days	0.1	1.74	5.11	2.75(↑)	−921.7(↓)	(Zhou et al. 2007)
PN	*Litopenaeus vannamei*	100	60	50 days	0.1	1.46	4.78	3.65(↑)	-1213.2(↓)	(Zhou et al. 2007)
RO	*Litopenaeus vannamei*	100	60	50 days	0.1	1.55	4.82	3.43(↑)	−1182.7 (↓)	(Zhou et al. 2007)
SBO	*Litopenaeus vannamei*	50	60	50 days	0.1	2.4	5.68	1.95(=)	−255.8 (=)	(Zhou et al. 2007)
High linolenic acid SBO	*Litopenaeus vannamei*	7	90	58 days	1.55	16.42	4	1.18	−31 (=)	(González-Félix et al. 2010)
LO	*Litopenaeus vannamei*	7	90	58 days	1.55	16.22	4	1.22	−43.9 (=)	(González-Félix et al. 2010)
Low linolenic acid SBO	*Litopenaeus vannamei*	7	90	58 days	1.55	16.78	4.1	1.22	−7.7 (=)	(González-Félix et al. 2010)
PF	*Litopenaeus vannamei*	8	44	119 days	0.25	21.6	-	1.45	+14.3 (=)	(Soller et al. 2017)
SBO	*Litopenaeus vannamei*	8	44	119 days	0.25	18.0	-	1.43	−4.8 (=)	(Soller et al. 2017)
FXO	*Litopenaeus vannamei*	8	44	119 days	0.25	21.0	-	1.37	+11.1 (=)	(Soller et al. 2017)
Nanochloropsis oil	*Marsupenaeus japonicus*	10	27	50 days	0.46	1.86	2.61	1.47 (↑)	+76 (↑)	(Oswald et al. 2019)

Abbreviations: CO, corn oil; FBW, final body weight; FCR, feed conversion ratio; FER, feed efficiency ratio; FXO, flaxseed oil; IBW, initial body weight; LO, linseed oil; PF, poultry fat; PL, pork lard; PN, peanut oil; PO, palm oil; SGR, specific growth rate, SBO, soybean oil; WG, weight gain. The difference of the measured parameters in each treatment comparing with control (FO-based diet). (↑), (↓) and (=) show increasing, decreasing and no significant differences, respectively, in each treatment comparing with control (FO-based diet).

diet remarkably reduced growth performance in giant freshwater prawn (Muralisankar et al. 2014). The authors of above-mentioned studies suggested that at least 10% FM should be included in the diet of this species for providing sufficient n-3 LC-PUFA.

Dietary Strategies for Improving FO Sparing Efficiency

Several authors have suggested that FO replacement with ALS may be more feasible when providing only adequate dietary levels of DHA, but not EPA or total n-3 LC-PUFA. Some studies confirmed that the complete substitution of dietary FO by ALS may be possible when EFA, especially DHA in pure form, was supplemented in the feed. For example, Florida pompano (*Trachinotus carolinus*, Carangidae) that were fed a plant-based diet supplemented with algal DHA grew similarly or even better than their congeners fed a FO diet (Rombenso et al. 2017). Similar results have been reported in different fish species like olive flounder (*Paralichthys olivaceus*, Paralychtidae; Lee et al. 2020), rainbow trout (*Oncorhynchus mykiss*, Salmonidae; Betico et al. 2015), white seabass, (*Atractoscion nobilis*, Sciaenidae; Rombenso et al. 2015), *Totoaba macdonaldi* (Sciaenidae; Mata-Sotres et al. 2018), and even in crustaceans, like white leg shrimp (*Litopenaeus vannamei*, Penaeidae; Araujo et al. 2019). In addition, enhancing the endogenous elongation and desaturation of MC-PUFA into LC-PUFA by using bioactive compounds can increase the endogenous FA synthesis (Alhazzaa et al. 2019). In this context, it has been confirmed that diet supplementation with different bioactive compounds, such as flavonoids (Nichols et al. 2011; Alhazzaa and Sase 2016), sesamin (Trattner et al. 2008; Alhazzaa et al. 2012), phytosterols (Brufau et al. 2008; Alhazzaa et al. 2013), conjugated LA (Diez et al. 2007; Corl et al. 2008), polyphenols (Caro et al. 2017), genistein (Schiller Vestergren et al. 2011), butyrate (Benedito-Palos et al. 2016), L-carnitin (Chen et al. 2020), and resveratrol (Torno et al. 2018), may modulate LC-PUFA biosynthesis and lipid metabolism (e.g. an increase in the abundance or activity of the desaturases or the protection of LC-PUFA from oxidation). This dietary strategy will not only improve growth performance but also enhance the rate of n-3 LC-PUFA biosynthesis in the above-mentioned fish species fed on VO-rich diets.

EFFECTS OF DIETARY FO REPLACEMENT WITH ALS ON NUTRIENT DIGESTIBILITY

The determination of the lipid and FA digestibilities from ALS is important because their low digestibility may compromise fish growth and feed efficiency, as well as fish health and welfare. Thus, such information is essential for producing cost-effective feeds (De Silva and Anderson 1995; Allan et al. 2000; Tibbetts et al. 2006; Francis et al. 2007a; Dernekbaşi et al. 2011). The ability of an aquatic species to digest and utilize lipids is mainly determined by the lipid physical state (i.e. liquid or solid at environmental temperature), their structural form (e.g. waxes, sterols, phospholipids, triglycerols) and FA profile. In addition, the innate digestive capacity (e.g. gut morphology and digestive enzyme activities) of the taxon should be considered (Trushenski and Lochmann 2009), although it has been proved that this factor has a negligible effect on lipid digestibility (Hua and Bureau 2009).

Regarding lipid structural form, the findings of Olsen et al. (1999, 2000) confirm that the apparent digestibility coefficient (ADC) of phospholipids (PL) is greater than that of triacylglycerols (TAG) in the diet. The former authors found that high levels of TAG in VO (linseed oil) altered the structural integrity of the intestinal mucosa and, consequently, resulted in epithelial damage in Arctic charr (*Salvelinus alpinus*, Salmonidae). This damage resulted in a reduction of the intestinal epithelium absorption capacity. In fact, the emulsifying action of PL is important for increasing lipid digestion and absorption (Tocher et al. 2008; Trushenski and Lochmann 2009). However, PL may be a limiting factor for ALS, especially in refined VO (Olsen et al. 2000); thus, supplementing diets with commercial PL, such as soy lecithin, can enhance overall lipid digestion and absorption (Trushenski and Lochmann 2009).

The digestibility of an ALS is fundamentally determined by its proportions in SFA, MUFA, PUFA, and LC-PUFA in its FA composition (Bureau and Meeker 2011; Emery et al. 2014). The digestibility of different FA classes is determined by their carbon chain length and the degree of unsaturation that define the melting point of an individual FA (Caballero et al. 2002). Other biotic factors, such as the preference order of individual FA for the bile salt-activated lipase (Izquierdo et al. 2000; Caballero et al. 2002; Piedecausa et al. 2007), and the solubility of FA in the micelles (Koven et al. 1994; Olsen and Ringø 1997; Tocher 2003) can also affect the ADC of FA. In fact, SFA and long-chain MUFA (e.g. cetoleic acid, 22:1n-11) provide a smaller surface area available for digestive enzyme processes, which makes them less digestible in fish (Bureau and Meeker 2011; Monteiro et al. 2018). According to the aforementioned factors, the ADC of FA classes reported in different finfish species is as follows: PUFA > MUFA > SFA and short-chain > longer-chain of FA (Cravedi et al. 1987; Sigurgisladottir et al. 1992; Olsen et al. 1998; Johnson et al. 2000; Torstensen et al. 2000; Caballero et al. 2002; Menoyo et al. 2003; Ng et al. 2003; Turchini et al. 2005; Francis et al. 2007a; Martins et al. 2009; Castro et al. 2014). In this regard, many studies reported that using SFA and MUFA-rich ALS (e.g. RAF, palm oil, coconut oil) is associated with lower PUFA content, and PUFA/SFA ratio in the diet resulted in a reduction in lipid digestibility and absorption (Bureau et al. 2002; Caballero et al. 2002; Bahurmiz and Ng 2007b; Hua and Bureau 2009; Martins et al. 2009; Turchini et al. 2009; Karalazos et al. 2011; Mozanzadeh et al. 2015; Monteiro et al. 2018). Furthermore, different studies in which FO was replaced by SFA-rich ALS, such as RAF or palm oil, resulted in a reduction in growth performance and feed utilization in different species, such as whiteleg shrimp (Zhou et al. 2007), black seabream, (Jin et al. 2017b), sobaity seabream (*Sparidentex hasta*, Sparidae; Mozanzadeh et al. 2016) and Nile tilapia (*Oreochromis niloticus*, Cichlidae; Teoh and Ng 2016) (Table 9.8). Thus, it has been suggested that SFA may reduce the ADC of dietary energy, including lipids and FA (Caballero et al. 2002; Francis et al. 2007a; Bureau and Meeker 2011). In this context, Ng et al. (2003, 2004) observed that up to 25 and 50% replacement of dietary FO with palm oil in Atlantic salmon and rainbow trout, respectively, did not affect ADC of lipid, but total FO substitution with palm oil significantly reduced ADC of lipids in rainbow trout. The findings of previous studies suggested that the proportion of SFA at levels lower than 23% (± 1%) increased the ADC of dietary lipids (*ca.* > 90%), but above this threshold, the ADC of lipids was linearly reduced with the increasing proportion of SFA (Hua and Bureau 2009). Moreover, the position of the first double bond can also affect FA digestibility as follow: n-3 > n-6 > n-9 (Francis et al. 2007a; Eroldoğan et al. 2013). In this sense, Bandarra et al. (2011) reported that the medium ratio of n-3/n-6 PUFA in diets resulted in the highest nutrient digestibility compared to low or high n-3/n-6 PUFA ratios. But, Huguet et al. (2015) reported that dietary EPA/ARA had minor effects on overall nutrient digestibility, but it only affected the digestibility of dietary EPA and ARA.

On the other hand, Cho and Kaushik (1990) revealed that the ADC of FO and VO with a high n-3/n-6 PUFA ratio and MUFA were high (ADC = 80–95%) over a wide range of water temperatures (5–15 °C) in rainbow trout. In contrast, the ADC of RAF characterized for being rich in SFA was reduced (ADC = −8 %) at lower water temperatures, because cold-water temperature is below the melting point of SFA. Thus, it seems that partial replacement of dietary FO with RAF in cold-water species may enhance digestibility of SFA compared to total replacement. In contrast, it seems that tropical species farmed in high water temperatures (28–35°C) may have more capacity to cope with SFA- and MUFA-rich ALS (Alhazzaa et al. 2019). Regarding marine microalgal meals that provide both protein and sufficient n-3 LC-PUFA, Sarker et al. (2020) reported that ADC of macronutrients, especially n-3 LC-PUFA in *Isochrysis sp.*, was higher than that of *Nannochloropsis* sp (Table 9.8). They reported that ADC of n-3 PUFA was 63.9% vs. 92.6% in *Nannochloropsis sp.* and *Isochrysis sp.* meals, respectively. In addition, Sarker et al. (2016) reported that *Schizochytrium*-derived meal and oil had the highest ADC for lipid (total PUFA = 97.5%), n-3 LC-PUFA (97.2%), and n-6 LC-PUFA (92.4%). On the other hand, Tibbetts et al. (2020) demonstrated that the ADC of macronutrients in Atlantic salmon that were fed diets containing *Schizochytrium sp.* (T18) extracted oil were similar to those fed FO-based diets. Regarding insect oil extracts, it has been demonstrated

TABLE 9.8
Effects of Dietary FO Replacement with ALS on Nutrients Digestibility

Species	Environmental Temperature	Alternative Lipid Source	% FO Replacement	Percentage (%) of the Apparent Digestibility Coefficient Compared to FO Diet					WG (%compared to fish fed on the FO diet)	Reference
				Protein	Lipid	Dry matter	NFE/ carbohydrate	Energy		
Rainbow trout	Cold Water	MO	75	−1.8 (=)	+0.7 (=)	−1.9 (=)	−2 (=)	-	−172 (=)	(Turchini et al. 2013)
Rainbow trout	Cold Water	CAO	75	−2.5 (=)	−0.2 (=)	−1.5 (=)	−0.2 (=)	-	−197 (=)	(Turchini et al. 2013)
Rainbow trout	Cold Water	PF	75	−3.8 (=)	−2.4 (=)	−5.3 (↓)	−7.4 (↓)	-	−864 (↓)	(Turchini et al. 2013)
Rainbow trout	Cold Water	PO	75	−3.7 (=)	−1.5 (=)	−5.6 (↓)	−10.9 (↓)	-	−893 (↓)	(Turchini et al. 2013)
Rainbow trout	Cold Water	SFO	75	−0.6 (=)	+1.7 (=)	+1.4 (=)	+ 3.5 (=)	-	−188 (=)	(Turchini et al. 2013)
Rainbow trout	Cold Water	High oleic SFO	75	−0.3 (=)	+1.8 (=)	+1.3 (=)	+4.9 (=)	-	+463 (=)	(Turchini et al. 2013)
Rainbow trout	Cold Water	SBO	75	−1.6 (=)	+1.3 (=)	+0.1 (=)	+ 2.3 (=)	-	−272 (=)	(Turchini et al. 2013)
European seabass	Temperate	PF	50	(=)	−1.5 (=)	−1.2 (=)	-	−1.4 (=)	−2 (=)	(Monteiro et al. 2018)
European seabass	Temperate	PF	75	+0.4 (=)	−0.9 (=)	+0.2 (=)	-	+0.1 (=)	−1 (=)	(Monteiro et al. 2018)
European seabass	Temperate	PF	100	+0.2 (=)	−2.6 (↓)	+0.5 (=)	-	−0.4 (=)	−20.2 (↓)	(Monteiro et al. 2018)
European seabass	Temperate	PO:LO:RO (3:5:2)	97	−0.1 (=)	−0.4 (↓)	(=)	(=)	−0.6 (↓)	-	(Castro et al. 2015)
Sharpsnout seabream	Tropical	LO	100	−2.9 (↓)	−3.5 (↓)	−9.1 (↓)	-	-	−8.05 (=)	(Piedecausa et al. 2007)
Sharpsnout seabream	Tropical	SBO	100	+0.4 (=)	−0.4 (=)	+0.1 (=)	-	-	−15.4 (=)	(Piedecausa et al. 2007)
Red hybrid tilapia	Tropical	CPO	100	+1.5 (=)	−6.9 (↓)	−2.8 (=)	-	-	−102.9 (=)	(Bahurmiz and Ng 2007a)
Red hybrid tilapia	Tropical	RBDPO	100	(=)	−5.8 (↓)	−0.7 (=)	-	-	−145 (=)	(Bahurmiz and Ng 2007a)
Red hybrid tilapia	Tropical	PFAD	100	0.9 (=)	−7 (↓)	−1.2 (=)	-	-	−174.6 (=)	(Bahurmiz and Ng 2007a)
Murray cod	Tropical	OO:PO:LO (1.2:4.3:4.5)	25	−0.9 (=)	−1.2 (=)	−1.3 (=)	-	-	(=)	(Francis et al. 2007a)
Murray cod	Tropical	OO:PO:LO (1.2:4.3:4.5)	50	−0.8 (=)	−1.4 (=)	−0.5 (=)	-	-	(=)	(Francis et al. 2007a)
Murray cod	Tropical	OO:PO:LO (1.2:4.3:4.5)	75	−0.8 (=)	−3.7 (↓)	−2.1 (=)	-	-	(=)	(Francis et al. 2007a)
Murray cod	Tropical	OO:PO:LO (1.2:4.3:4.5)	100	−2.3 (↓)	−4.2 (↓)	−4.5 (↓)	-	-	(↓)	(Francis et al. 2007a)
Silvery-black porgy	Tropical	CAO	50	+0.7 (=)	+1.5 (=)	-	-	-	−14 (=)	(Mozanzadeh et al. 2016)

(Continued)

TABLE 9.8 (CONTINUED)
Effects of Dietary FO Replacement with ALS on Nutrients Digestibility

Species	Environmental Temperature	Alternative Lipid Source	% FO Replacement	Percentage (%) of the Apparent Digestibility Coefficient Compared to FO Diet					WG (%compared to fish fed on the FO diet)	Reference
				Protein	Lipid	Dry matter	NFE/ carbohydrate	Energy		
Silvery-black porgy	Tropical	SFO	50	−3.6 (=)	+1.1 (=)	-	-	-	+1.4 (=)	(Mozanzadeh et al. 2016)
Silvery-black porgy	Tropical	T	50	−1.1 (=)	−1.2 (=)	-	-	-	−19 (↓)	(Mozanzadeh et al. 2016)
Silvery-black porgy	Tropical	CAO	100	−2.4 (=)	−0.9 (=)	-	-	-	−9.4 (=)	(Mozanzadeh et al. 2016)
Silvery-black porgy	Tropical	SFO	100	+4.2 (=)	+1.7 (=)	-	-	-	−9.3 (=)	(Mozanzadeh et al. 2016)
Silvery-black porgy	Tropical	T	100	−6.3 (↓)	−2.6 (↓)	-	-	-	−19.7 (↓)	(Mozanzadeh et al. 2016)
Atlantic halibut	Cold Water	PF	100	+0.2 (=)	+0.1 (=)	(=)	-	−0.2 (=)	-	(Martins et al. 2009)
Atlantic halibut	Cold Water	FXO	100	+0.2 (=)	+5.1 (↑)	+1.6 (↑)	-	+2.5 (↑)	-	(Martins et al. 2009)
Atlantic halibut	Cold Water	CAO	100	+0.4 (↑)	+3.7 (↑)	+1.7 (↑)	-	+2.6 (↑)	-	(Martins et al. 2009)
Atlantic halibut	Cold Water	SBO	100	+0.2 (=)	+3.6 (↑)	+1.7 (↑)	-	+2.4 (↑)	-	(Martins et al. 2009)
Rainbow trout	Cold Water	SBO	50	+0.1 (=)	+0.1 (=)	+0.6 (=)	−0.6 (=)	-	(=)	(Martins et al. 2005)
Rainbow trout	Cold Water	SBO	100	+0.4 (=)	+0.3 (=)	−0.3 (=)	−0.2 (=)	-	(=)	(Martins et al. 2005)
Rainbow trout	Cold Water	*Isochrysis sp.* powder	100	−8.5 (↓)	−8.1 (↓)	−7.6 (↓)	−7.2 (↓)	-		(Sarker et al. 2020)
Rainbow trout	Cold Water	*Nannochloropsis* sp. powder	100	−1.8 (=)	−12.1 (↓)	+0.7 (=)	−3.7 (↓)	-		(Sarker et al. 2020)
European seabass	Temperate	SBO	50	+0.4 (=)	+0.3 (=)	+0.6 (=)	+0.7 (=)	-	+3 (=)	(Martins et al. 2005)
European seabass	Temperate	SBO	100	−0.1 (=)	+0.5 (=)	+0.3 (=)	(=)	-	−13 (=)	(Martins et al. 2005)
Atlantic salmon	Cold water	T/PF (2:1)	75	−1 (=)	−6 (↓)	−3.6 (=)	−6.5 (=)	−3.5 (↓)	−18.8 (=)	(Emery et al. 2014)
Atlantic salmon	Cold water	*Schizochytrium* sp. (T18) oil	33	−0.2 (=)	−0.3 (=)	−1.4 (=)	-	−1 (=)	(=)	(Tibbetts et al. 2020)
Atlantic salmon	Cold water	*Schizochytrium* sp. (T18) oil	66	−0.5 (=)	+0.7 (=)	−1.2 (=)	-	−0.8 (=)	(=)	(Tibbetts et al. 2020)
Atlantic salmon	Cold water	*Schizochytrium* sp. (T18) oil	100	−0.3 (=)	+ 1.1 (=)	−0.3 (=)	-	−0.4 (=)	(=)	(Tibbetts et al. 2020)

Abbreviations: AM, algal meal; APS, alternative protein sources; CAO, canola oil; CO, corn oil; EO, echium oil; FXO, flaxseed oil; LO, linseed oil; OO, olive oil; PF, poultry fat; PL, pork lard; PO, palm oil; SBO, soybean oil; SFO, sunflower oil; T, tallow. The difference of the measured parameters in each treatment comparing with control (FO-based diet). (↑), (↓) and (=) show increasing, decreasing and no significant differences, respectively, in each treatment comparing with control (FO-based diet).

that the ADC of most FA was remarkably reduced compared to the insect-based diets in Atlantic salmon; however, the ADC of n-3 LC-PUFA were not influenced by insect oils (extracted from black soldier fly larvae) (Belghit et al. 2019).

It should be mentioned that the specific interactions between FA classes may also affect lipid digestibility due to the physicochemical characteristics of individual FA (Olsen and Ringø 1997). In this regard, Campos et al. (2016) reported that poultry fat reduced ADC of PUFA, especially n-3 PUFA compared to FO in European seabass (*Dicentrachus labrax*, Moronidae), but the ADC of total lipid and other FA classes were not affected. Furthermore, Caballero et al. (2002) observed that the ADC of MUFA, n-6 and n-3 PUFA, in rainbow trout were slightly higher in feeds with lower SFA concentration than those with higher SFA levels. Moreover, many studies reported that the ADC of EPA and DHA were positively affected by the n-3 LC-PUFA content of the diet, whereas they were not affected by dietary VO inclusion (Caballero et al. 2002; Ng et al. 2004; Torstensen et al. 2000; Karalazos et al. 2011). However, Menoyo et al. (2003) reported that the ADC of EPA was affected by the dietary SFA, while the ADC of DHA was affected by both the SFA and n-3 LC-PUFA levels in the diet.

According to above-mentioned findings, there are different nutritional strategies which modify lipid digestibility that may be used for maximizing the use of ALS, including (i) using blends of ALS with different FA profiles can optimize the ADC of dietary lipid; (ii) the inclusion of phospholipid sources in the diet, such as soy lecithin, for increasing dietary lipid digestibility, and (iii) supplementing diets, especially PP-rich diets with additives (e.g. acidifiers, taurine, and probiotics) to maintain normal lipid digestion and absorption.

EFFECTS OF DIETARY FO REPLACEMENT ON FATTY ACID PROFILE OF CULTURED AQUATIC SPECIES

The main drawback of FO replacement in aquafeeds is the profound change in FA profile of the fillet and the significant reduction of EPA and DHA, which are absent or present only at trace levels in most ALS (except for microalgal derivatives), GMO, microbial oils, and aquaculture fisheries by-products. This is a key issue since fish are generally recommended for human consumption due to their n-3 LC-PUFA content, which are reputed for reducing cardiovascular risk and preventing diseases, such as breast cancer, as well as contributing to the prevention or treatment of mental illnesses, such as psychotic disorders, depression or Alzheimer's. Furthermore, dietary replacement of FO with ALS that are especially rich in n-6 PUFA, like VO, increase the deposition of LNA in the fish fillet, which could increase the risks of human diseases, like cardiovascular disorders, inflammatory, and autoimmune diseases (Simopoulos 2008). In addition, many studies demonstrated the fillet FA composition in carnivorous marine species with a restricted capacity of bioconversion of LNA and ALA to LC-PUFA was more affected by ALS compared to freshwater species (Table 9.9). In this sense, n-3 LC-PUFA, mainly DHA, are selectively retained in the muscular tissue of salmonids by the selective incorporation of DHA from the dietary FM and/or the contribution of bioconversion of dietary ALA into DHA. However, previous research in salmonids also proved that the bioconversion of ALA to n-3 LC-PUFA was not enough to preserve n-3 LC-PUFA or reduce the extra retention of dietary MC-PUFA, especially LNA, in the fillet (Table 9.10). In this sense, it has been reported that, due to the progressive reduction of FO in salmon diets, the levels of n-3 LC-PUFA in the fillet of farmed salmon have decreased by 50% over the last decade (Bou et al. 2017). In addition, the changes of the fillet's FA profile become more prominent when FO and FM simultaneously were replaced with their vegetal counterparts in feed formulation.

It has been proved that the tissue FA composition change in fish generally follows a simple dilution model; thus, based on this modeling, the FA profile of the fillet mainly reflects the FA profile of the feed (Jobling 2003; Robin et al. 2003). In addition, the interactions between various FA may have a profound influence on assimilation and retention of other FA from ALS, changes that are

TABLE 9.9
Effects of Dietary FO Replacement on Fatty Acid Profile of Marine Fish Species

	Duration of Feeding Trial	SFA	OA	MUFA	LA	ALA	ARA	EPA	DHA	n-6 PUFA	n-3 PUFA	LC-PUFA	n-3/n-6 PUFA Ratio
		Gilthead sea bream (*Sparus aurata*) (Izquierdo et al. 2003)											
Diet FO	101 days	25.0	11.9	25.9	7.3	3.8	0.6	13.9	8.9	9.4	31.2	–	3.3
Diet LO (60%)		17.6	15.1	25.1	13.0	23.0	0.3	7.3	5.3	14.1	38.2	–	2.7
Difference (%)		−29.6	+27	−3.1	+78.1	+505.3	−50	−47.5	−40	+50	+22.4	–	−18.2
Whole body FO		24.9	19.7	34.2	5.9	2.0	0.7	7.6	11.3	7.3	28.1	26.3	3.8
Whole body LO (60%)		20.3	25.4	35.3	10.8	14.5	0.3	4.0	5.9	12.6	30.2	15.9	2.4
Difference (%)		−18.5	+28.9	+3.1	+83	+625	−57	−47.4	−47.8	+72.6	+7.5	−39.5	−36.8
		Gilthead sea bream (*Sparus aurata*) (Fountoulaki et al. 2009)											
Diet FO	170 days	30.6	11.1	22.0	11.1	1.0	0.8	16.7	9.4	12.9	33.7	29.9	2.6
Diet PO (68%)		37.9	31.3	35.6	16.5	0.8	0.2	4.4	2.9	16.7	9.7	8.0	0.6
Difference (%)		+23.9	+182	+61.8	48.6	−25	−75	−73.7	−69.1	29.5			
Fillet FO After 170 days		26.9	16.2	30.3	9.7	–	0.9	9.9	11.8	11.3	27.0	–	2.8
Fillet PO (68%) Afetr 170 days		27.7	29.5	39.6	16.0	–	0.4	3.9	7.0	16.5	15.7	–	1.0
Difference (%)		+3	+82.1	+30.7	+64.9	–	−55.6	−60.6	−40.7	+46	−41.9	–	−64.3
		Red sea bream (*Pagrus major*) (Huang et al. 2007)											
Diet FO	84 days	22.1	17.7	44.8	4.7	0.8	0.6	8.3	10.0	5.8	22.4	20.9	3.9
Diet CAO (70%)		11.5	52.8	55.6	19.0	5.7	0.2	1.1	4.3	19.4	11.7	11.8	0.6
Difference (%)		−48	+198.3	+24.1	+304.3	+612.5	−66.7	−86.7	−57	+234.5	−100.5	−43.5	−84.6
Whole body FO		24.5	22.3	45.8	4.2	0.6	0.5	5.6	11.4	5.4	20.8	20.5	0.9
Whole body CAO (70%)		14.1	52.2	56.0	16.2	4.0	0.2	0.9	5.3	17.0	11.0	13.2	0.7
Difference (%)		−42.4	+134	+22.3	+285.7	+566.7	−60	−83.9	−53.5	+214.8	−47.1	−35.6	−22.2
		Red sea bream (*Pagrus major*) (Seong et al. 2019)											
Diet FO	84 days	17.3	31.6	47.1	13.2	3.1	0.3	4.7	4.6	13.5	14.3	10.4	1.1
Diet Schizochytrium meal/RO (100%) in FM free diet		40.6	21.6	23.4	18.0	2.5	0.3	0.4	8.7	18.3	11.9	9.6	0.7
Difference (%)		+134.7	−31.6	−50.3	+36.4	−19.4	–	−91.5	+89.1	+35.6	−16.8	−7.7	−36.4
Whole body FO		20.4	30.5	43.6	12.0	2.4	0.3	3.0	4.8	12.3	12.0	19.0	1.0

(Continued)

TABLE 9.9 (CONTINUED)
Effects of Dietary FO Replacement on Fatty Acid Profile of Marine Fish Species

	Duration of Feeding Trial	SFA	OA	MUFA	LA	ALA	ARA	EPA	DHA	n-6 PUFA	n-3 PUFA	LC-PUFA	n-3/n-6 PUFA Ratio
Whole body Schizochytrium meal/RO (100%)		27.2	27.0	29.6	21.1	2.3	0.5	0.5	10.0	21.5	13.4	11.4	0.6
Difference (%)		+33.3	−13	−32.1	+75.8	−4.2	+66.7	−83.3	+108.3	+74.8	+11.7	−40	−40
Sharpsnout seabream (*Dilodus puntazzo*) (Piedecausa et al. 2007)													
Diet FO	92 days	23.6	16.3	35.8	4.3	10.0	0.6	9.3	13.0	5.8	36.5	25.3	7.1
Diet SBO (100%)		12.8	19.9	24.7	46.8	26.4	0.1	1.5	2.3	47.6	31.0	4.4	0.6
Difference (%)		−45.8	+22.1	−31	+988.4	+164	−83.3	−83.9	−82.3	+720.8	−15.1	−82.6	−91.5
Fillet FO		27.3	22.1	38.5	5.4	6.9	0.5	7.0	11.9	6.2	30.5	23.1	4.9
Whole body SBO (100%)		18.0	22.7	29.4	42.1	8.3	0.2	2.2	4.1	43.8	17.0	9.7	0.4
Difference (%)		−34	+2.7	−23.6	−220.4	20.3	−60	−68.6	−65.5	+606.5	−44.3	−58	−91.8
European seabass (*Dicentrarchus labrax*) (Campos et al. 2019)													
Diet FO	142 days	29.0	17.8	30.5	7.4	1.1	1.5	6.3	13.8	10.4	24.6	23.0	2.4
Diet PF (100%)		28.1	32.4	42.5	20.7	1.6	0.7	1.0	1.8	22.1	5.2	3.8	0.2
Difference (%)		+3.1	+82	+39.3	+179.7	+45.5	−53.3	−84.1	−87	+112.5	−78.9	−83.5	−91.7
Dorsal Fillet FO		25.8	19.2	27.7	7.3	1.2	2.2	5.4	21.3	11.2	30.4	30.2	2.7
Dorsal Fillet PF (100%)		25.0	30.2	38.5	18.8	1.7	1.5	1.9	6.5	21.8	11.4	10.7	0.5
Difference (%)		−3.1	+57.3	+39	+157.5	+41.7	−31.8	−64.8	−69.5	+94.6	−62.5	−64.6	−81.5
Red drum (*Scianops ocellatus*) (Perez-Velazquez et al. 2018)													
Diet FO	42 days	28.0	26.9	43.8	4.7	0.7	0.6	6.5	6.8	7.1	15.5	–	2.2
Diet Schyzochytrium and Arthrospira meals (50%)		48.1	10.5	17.7	6.2	0.3	0.5	3.0	9.5	12.4	14.7	–	1.2
Difference (%)		+71.8	−61	−59.6	+31.9	−57.1	−16.7	−53.8	+39.7	+74.6	−5.2	–	−45.5
Whole body FO		29.1	25.1	33.9	4.7	–	–	2.5	2.3	5.9	8.4	–	1.7
Whole body Schyzochytrium and Arthrospira meals (50%)		33.6	16.7	26.7	6.0	–	–	2.7	8.9	9.7	15.0	–	1.5
Difference (%)		+15.5	−33.5	−21.2	+27.7	–	–	+8	+287	+64.4	+78.6	–	−11.8
Cobia (*Rachycentron canadum*) (Woitel et al. 2014)													
Diet FO	56 days	31.6	9.6	21.7	12.7	1.8	1.0	11.2	11.3	14.3	30.4	27.1	2.1

(Continued)

TABLE 9.9 (CONTINUED)
Effects of Dietary FO Replacement on Fatty Acid Profile of Marine Fish Species

	Duration of Feeding Trial	SFA	OA	MUFA	LA	ALA	ARA	EPA	DHA	n-6 PUFA	n-3 PUFA	LC-PUFA	n-3/n-6 PUFA Ratio
Diet T		36.2	16.4	25.9	12.2	1.4	0.9	7.5	10.1	13.6	22.8	21.0	1.7
Difference (%)		+14.6	+70.8	+19.4	−3.9	−22.2	−10	−33	−10.6	−4.9	−25	−28.1	−19
Whole body FO		33.9	13.6	26.1	9.0	1.2	1.4	8.0	15.0	10.4	29.5	28.0	2.9
Whole body T		34.7	18.9	29.4	9.3	1.1	1.4	6.1	13.9	10.7	25.2	24.2	2.4
Difference (%)		+2.4	+39	+12.6	+3.3	−8.3	–	−23.8	−7.3	+2.9	−14.6	−13.6	−17.2
Senegalese sole (*Sole senegalensis*) (Reis et al. 2014)													
Diet FO	5 months	29.8	15.0	30.3	8.0	2.9	0.9	8.7	10.8	9.9	24.4	–	2.5
Diet RO/SBO/LO (3:2:5)(100%)		24.4	19.8	40.6	14.5	1.9	0.3	6.0	7.2	15.7	16.2	–	1.0
Difference (%)		−18.1	+32	+34	+81	−34.5	−66.7	−31	−33.3	+58.6	−33.6		−60
fillet FO after 140 days		28.0	16.6	28.5	6.8	1.5	1.4	2.8	19.1	9.8	28.2	–	2.9
fillet mix (100%) after 140 days		26.6	17.2	30.9	8.9	1.1	1.3	2.1	17.4	11.3	24.8	–	2.2
Difference (%)		−5	+3.6	+8.4	+30.9	−26.7	−7.1	−25	−8.9	+15.3	−12.1		−24.1
Fillet mix after 26 days wahout with FO diet		27.2	16.4	28.6	7.9	0.9	1.8	2.2	20.9	11.1	28.5	–	2.6
Difference (%)		−2.9	−1.2	+0.3	+16.2	−40	+28.6	−21.4	+9.4	+13.3	+1.1	–	−10.3

Abbreviations: ALS, alternative lipid source; CAO, canola oil; CO, corn oil; EO, echium oil; FXO, flaxseed oil; LO, linseed oil; OO, olive oil; PF, poultry fat; PL, pork lard; PO, palm oil; SBO, soybean oil; SFO, sunflower oil; T, tallow.

Diet difference % = (%FA diet ALS − % FA diet FO) / (%FA diet FO) ×100.

Fish difference % = (%FA fish ALS − % FA fish FO) / (%FA fish FO) ×100.

The percentage of fish oil substituted is in parentheses. The percentage difference of the fatty acid percentage between diets and fish tissues is also reported according to Turchini et al. (2009).

TABLE 9.10
Effects of Dietary FO Replacement on Fatty Acid Profile of Salmonids Species

	Duration of Feeding Trial	SFA	OA	MUFA	LA	ALA	ARA	EPA	DHA	n-6 PUFA	n-3 PUFA	LC-PUFA	n-3/n-6 PUFA Ratio
Atlantic Salmon (*Salmo salar*) (Bell et al. 2002)													
Diet FO		21.7	17.2	37.7	3.6	2.1	0.6	8.2	13.6	5.2	29.1	24.6	5.6
Diet PO (100%)		41.7	31.2	37.7	9.4	0.5	0.1	1.6	2.8	9.6	6.0	5	0.6
Difference (%)		+48	+81.4	–	+161	−76.2	−83	−80.5	−79.4	+84.6	−79.4	−79.7	−89.3
Whole body FO		25.9	18.3	37.6	3.4	1.8	0.5	5.7	16.3	4.8	29.5	26.5	6.2
Whole body PO (100%)		29.2	37.2	46.2	10.6	0.5	0.4	1.9	7.7	12.7	11.5	12.6	0.9
Difference (%)		+12.7	+103	+22.9	+211.8	−72.2	−20	−67	−52.8	+164.6	−61	−52.5	−85.5
Atlantic Salmon (*Salmo salar*) (Bell et al. 2003b)													
Diet FO	350 days	24	15.1	44.3	4.5	1.7	0.6	7.3	10.5	5.5	24.5	20.4	4.4
Diet LO (100%)		13.2	16.6	22.1	13.6	45.4	0.2	1.8	2.7	13.9	50.8	10.2	3.7
Difference (%)		−45	+9.9	−50	+202	+2570	−67	−75	−74				
Fillet FO		23.1	15.2	44	4.4	1.4	0.5	5.6	12.8	5.9	25.9	23.3	4.4
Fillet LO (100%)		14.6	16.7	23.7	11.7	37.4	0.1	1.6	4.2	12.6	49.1	11	3.9
Difference (%)		−36.8		−46.1	+165.9	+2571	−80	−71.4	−67.2	+113.6	+89.6	−52.8	−11.4
Atlantic Salmon (*Salmo salar*) (Bell et al. 2003b)													
Diet FO	350 days	24	15.1	44.3	4.5	1.7	0.6	7.3	10.5	5.5	24.5	20.4	4.4
Diet RO (100%)		11.4	48.3	55.6	17.9	8.9	0.2	1.9	2.9	18.1	14.5	5.4	0.8
Difference (%)		−52.5	+220	+25.5	+298	+423	−67	−74	−72	+229	−40.8	−73.5	−81.8
Fillet FO		23.1	15.2	44	4.4	1.4	0.5	5.6	12.8	5.9	25.9	23.3	4.4
Fillet RO (100%)		13.6	41.5	53.6	14.6	6.8	0.2	0.6	5.0	16.7	15.9	10.3	1.0
Difference (%)		−41	+173	+21.8	+231	+386	−60	−89	−61	+183	−38.6	−56	−77
Atlantic Salmon (*Salmo salar*) (Bransden et al. 2003)													
Diet FO	63 days	30.2	12.3	26.1	1.4	3.2	1	18.1	10.2	3.1	34.4	35.5	11
Diet SFO (100%)		18.3	19.1	23.3	45.1	0.7	0.6	4.7	4.4	45.9	10.9	10.9	0.2
Difference (%)		−39.4	+55.3	−10.7	+3121	−78	−40	−74	−56.8	+1380	−68	−69	−98
Whole body FO		28.9	16.3	30.8	3.5	3.0	0.9	9.8	14.0	5.4	31.3	30.3	5.9
Whole body SFO (100%)		21.0	19.8	26.7	33.2	1.1	0.7	2.8	11.5	37.6	13.5	14.6	0.4

(Continued)

TABLE 9.10 (CONTINUED)
Effects of Dietary FO Replacement on Fatty Acid Profile of Salmonids Species

	Duration of Feeding Trial	SFA	OA	MUFA	LA	ALA	ARA	EPA	DHA	n-6 PUFA	n-3 PUFA	LC-PUFA	n-3/n-6 PUFA Ratio
Difference (%)		−27	+21.5	−13.3	+848.6	−63.3	−22	−71.4	−17.9	+596	−56.8	−51.8	−93
Atlantic Salmon (*Salmo salar*) (Hixson et al. 2014a)													
Diet FO	112 days	27.7	8.5	32.5	5.6	0.8	0.7	15.5	7.9	7.2	29.2	25.9	4.1
Diet camelina oil (100%) in Solvent-extracted fish meal		12.2	20.7	40.9	19.1	22.5	0.0	0.9	1.2	21.7	24.9	2.1	1.1
Difference (%)		−56	+144	+25.8	+241	+2712	–	−94	−84.4	+201.4	−14.7	−91.9	−73
Whole body FO		25.3	18.8	33.5	5.8	1.7	1.1	6.4	16.3	8.1	29.0	26.5	3.6
Whole body camelina oil (100%)		15.9	22.6	41.9	12.9	10.6	0.4	2.5	5.3	16.3	24.6	10.8	1.5
Difference (%)		−37.2	+20.2	+25.1	+122.4	+524	−63.6	−60.9	−67	+101.2	−15.2	−59.2	−58
Atlantic Salmon (*Salmo salar*) (Sprague et al. 2015)													
Diet Northern FO		22.7	13.2	43.3	5.8	1.3	0.6	8.0	10.2	7.2	25.4	21.0	3.5
Diet RO/Schizochytrium sp. meal (6:4) (100%)		15.9	43.4	48.6	16.2	6.1	0.4	1.2	8.1	19.2	16.0	12.5	0.8
Difference (%)		−30	−229	+12.2	+179	+370	−33.3	−85	−20.6	+166.7	−37	−40.5	−77
Fillet FO		20.9	20.9	42.9	7.7	2.1	0.6	5.9	11.7	9.6	25.7	23.3	2.7
Fillet RO/Schizochytrium sp. meal (6:4) (100%)		16.4	38.0	47.3	13.4	4.6	0.6	2.6	8.9	16.7	18.9	16.5	1.1
Difference (%)		−21.5	+81.8	+10.3	+74	+119	n.d.	−55.9	−23.9	+74	−26.5	−29.2	−59
Brown trout (*Salmo trutta*) (Turchini et al. 2003)													
Diet FO	70 days	35.3	16.6	32.3	5.0	1.5	0.8	8.8	10.9	6.6	24.7	22.2	3.8
Diet PF (100%)		32.7	27.9	38.2	14.2	1.5	0.6	3.8	5.4	15.5	12.3	10.7	0.8
Difference (%)		−7.4	+68	+18.3	+184	n. d	−25	−56.8	−50.5	+134.8	−50.2	−51.8	−78.9
Fillet FO		29.4	19.4	33.7	6.2	1.3	0.8	5.0	19.0	7.9	28.8	27.5	3.7
Fillet PF (100%)		29.0	25.5	36.6	11.2	1.2	0.7	3.0	14.6	13.4	20.8	20	1.6
Difference (%)		−1.4	+31.4	+8.6	+80.6	−7.7	−12.5	−40	−23.2	+69.6	−27.8	−27.3	−56.8
Rainbow trout (*Oncorhynchus mykiss*) (Gause and Trushenski 2013)													
Diet FO	210 days	31.1	9.0	22.5	9.2	1.7	1.0	12.9	11.5	11.0	32.6	29.5	3.0
Diet T (100%)		48.0	29.0	34.1	10.6	1.0	0.3	1.9	2.3	11.2	6.2	5.2	0.6

(Continued)

TABLE 9.10 (CONTINUED)
Effects of Dietary FO Replacement on Fatty Acid Profile of Salmonids Species

Duration of Feeding Trial	SFA	OA	MUFA	LA	ALA	ARA	EPA	DHA	n-6 PUFA	n-3 PUFA	LC-PUFA	n-3/n-6 PUFA Ratio
Difference (%)	+54.3	+222	+51.6	+15.2	−41.2	−70	−85.3	−80	+1.8	−81	−82.4	−80
Fillet FO	32.1	13.1	26.0	8.6	1.3	–	7.1	17.7	10.5	31.5	30.0	3.0
Fillet T (100%) after 19 weeks of feeding	36.3	34.2	43.6	8.9	0.6	–	1.2	6.4	10.9	9.2	9.6	0.8
Difference (%)	+13.1	+161	+67.7	+3.4	−54	–	−83.1	−63.8	+3.8	−70	−68	−73
Fillet T (100%) after 4 weeks of washout with FO diet	34.5	30.6	40.8	8.9	0.7	–	2.1	9.1	11.3	13.5	14.1	1.2
Difference (%)	+7.5	+133.6	+57	+3.5	−46.2	–	−70.4	−48.6	+7.6	−57.1	−53	−60
Fillet T (100%) after 8 weeks of washout with FO diet	33.7	25.4	36.2	8.9	0.9	–	3.5	11.8	11.2	18.2	19	1.7
Difference (%)	+5	+94	39.2	–	−30	–	−50.7	−33.3	+6.7	−42.2	−36.7	−43.3
Fillet T (100%) after 12 weeks of washout with FO diet	34.0	20.1	31.4	8.2	0.9	–	4.5	15.2	10.5	24.1	24.1	2.3
Difference (%)	+5.9	+53.4	+20.8	−4.8	−30.8	–	−36.6	−14.1	–	−23.5	−19.7	−23.3

Abbreviations: ALS, alternative lipid source; CAO, canola oil; CO, corn oil; EO, echium oil; FXO, flaxseed oil; LO, linseed oil; OO, olive oil; PF, poultry fat; PL, pork lard; PO, palm oil; SBO, soybean oil; SFO, sunflower oil; T, tallow.

Diet difference % = (%FA diet ALS – % FA diet FO) / (%FA diet FO) ×100.

Fish difference % = (%FA fish ALS – % FA fish FO) / (%FA fish FO) ×100.

The percentage of fish oil substituted is in parentheses. The percentage difference of the fatty acid percentage between diets and fish tissues is also reported according to Turchini et al. (2009).

clearly visible in the fillet (Bell et al. 2002). According to the findings of many studies on different farmed aquatic species, there are two clear trends regarding the effects of replacement of FO with ALS: firstly, greater amounts of FO replacement result in important changes in FA profile and secondly, MC-PUFA-rich ALS (e.g. soybean, sunflower, and linseed oils) may exacerbate the detrimental influence of FO replacement on the nutritional value of the fillet because these FA outcompete LC-PUFA for tissue deposition compared with SFA- and MUFA-rich ALS (e.g. RAF, palm, and coconut oils). In this regard, RAF generally contain lower proportions of LNA and greater n-3/n-6 PUFA ratios than most VO, which may reduce the bioaccumulation of the LNA in the fillet and have a minor effect on tissue levels of LC-PUFA. In fact, if SFA and MUFA were provided in excess in the diet, they will be used as substrates for β-oxidation and energy production. This results in an effective sparing of LC-PUFA for deposition or other biological use associated with the so-called 'n-3 LC-PUFA sparing effect' (Trushenski et al. 2008, 2009, 2011; Turchini et al. 2011b; Bowzer et al. 2016).

The replacement of dietary FO with SFA-rich oils, such as palm oil, would also minimize the retention of LNA in fish fillets. In addition, lower levels of MC-PUFA in fillets of fish fed palm oil-based diets containing high amounts of vitamin E can minimize lipid peroxidation and prolong the shelf-life and seafood freshness. Thus, the presence of bioactive compounds, such as natural antioxidants, in some ALS (e.g. palm oil and microalgal oil extracts) by reducing lipid peroxidation may preserve the amount of n-3 LC-PUFA in the fillet. In addition, compared to ALA and LNA, stearidonic acid in echium oil could be easily elongated to LC-PUFA, namely eicosatetraenoic acid (20:4n-3) and γ-linolenic acid (GLA, 18:3n-6), in fish cells and may elevate the amount of LC-PUFA in the fillet (Ghioni et al. 2002). Regarding microbial derivates, replacement of dietary FO with *Schizochytrium sp.* oil extract (Miller et al. 2007) or biomass (Sprague et al. 2015) did not affect DHA content of fillet FA profile in Atlantic salmon, but EPA concentration was decreased due to the near absence of EPA in this heterotrophic fungoid organism.

It has been speculated that n-3 LC-PUFA retention increases when dietary provision is limited because of selective retention of LC-PUFA, as they are not considered ideal substrates for β-oxidation (Mourente et al. 2005; Turchini et al. 2009). For example, marine tropical carnivorous fish tend to deposit higher levels of n-3 LC-PUFA in their fillet compared to the concentration of n-3 LC-PUFA in respective diets, when a large amount of dietary FO is replaced by different ALS (Wu et al. 2002; Shapawi et al. 2008; Alhazzaa et al. 2011; Glencross and Rutherford 2011; Mozanzadeh et al. 2016). Furthermore, the relative resistance of DHA to β-oxidation, which requires the peroxisomal β-oxidation pathway rather than the mitochondrial pathway, and its high specificity for fatty acyl transferases, may result in its selective deposition of n-3 LC-PUFA in fish tissue (Bell et al. 2001; Sargent et al. 2002).

The LC-PUFA synthesis through desaturation and elongation pathways, and the β-oxidation of FA for energy production are the important factors that contribute to the final FA profile of the fillet in salmonids and freshwater fish species (Tocher et al. 2003). In freshwater fish species, different patterns of deposition, utilization, and the competition between ALA and LNA may result in their different influence on lipid metabolism and FA profile of the fillet (Table 9.11). For instance, by considering high levels of LNA and LA in VO, the bioconversion of ALA to DHA in salmonids and freshwater fish species might be lowered by competition with tetracosapentaenoic acid (24:5n-3) for Δ6 FAD, which would limit the bioconversion of 24:5n-3 to 24:6n-3, and subsequently reduce the synthesis of DHA (Ruyter et al. 2000b; Sprecher 2000; Tocher et al. 2001). In addition, Ruyter et al. (2000) showed that the higher dietary LNA/ALA ratio diet enhanced a higher desaturation rate of LNA to 18:3n−6 in Atlantic salmon, which indicated competition between LNA and ALA for Δ6 FAD.

There are some approaches for normalizing FA profile of the fillet in fish fed diets containing ALS during grow-out period. The outcome of this process, regardless of the strategy used, is to provide the consumer a good quality and healthy fillet with a balanced FA profile and the proper n-3 LC-PUFA levels for supporting consumers' health. The first approach is based on increasing

TABLE 9.11
Effects of Dietary FO Replacement on Fatty Acid Profile of Crustaceans and Freshwater Fish Species

	Duration of Feeding Trial	SFA	OA	MUFA	LA	ALA	ARA	EPA	DHA	n-6 PUFA	n-3 PUFA	LC-PUFA	n-3/n-6 PUFA Ratio
Giant river prawn (*Macrobrachium rosenbergii*) (Kangpanich et al. 2017)													
Tuna oil	60 days	33.8	–	28.3	17.8	4.4	1.0	3.9	9.9	19.1	18.4	–	1.0
***Schizochytrium* meal (100%)**		32.5	–	28.1	18.8	4.6	0.7	2.5	11.4	19.8	19.0	–	1.0
Difference (%)		−3.8		−0.7	+5.6	+4.5	−30	−35.9	+15.2	+3.7	+3.3		–
Fillet Tuna oil		38.4	–	28.9	13.0	2.1	2.7	7.4	6.4	15.8	16.2	–	1.0
Fillet *Schizochytrium* meal (100%)		40.4	–	28.0	14.0	2.1	2.8	5.7	6.1	17.0	13.9	–	0.8
Difference (%)		+5.2	–	−3.1	+7.7	–	+3.7	−23	−4.7	+7.6	−14.2		−20
Pacific white shrimp (*Litopenaeus vannamei*) (González-Félix et al. 2010)													
Menhaden FO	58 days	30.4	10.6	22.6	23.1	3.3	0.9	9.0	5.0	24.1	20.8	–	0.9
LO (90%)		15.1	15.9	18.1	28.7	35.4	0.1	1.1	0.6	28.8	37.8		1.3
Difference (%)		−50.3	+50	−19.9	+24.2	972.7	−88.9	−87.8	−88	+19.5	81.7		+44.4
Fillet LO (90%)		30.6	9.8	17.8	14.3	0.9	3.0	14.9	10.6	18.9	28.1		1.5
		28.4	11.3	16.2	19.0	11.5	2.3	8.4	4.2	27.7	27.3		1.0
Difference (%)		−7.2	+15.3	−9	+32.9	1177	−23.3	−43.6	−60.4	+46.6	−2.9		−33.3
Pacific white shrimp (*Litopenaeus vannamei*) (González-Félix et al. 2010)													
Menhaden FO	58 days	30.4	10.6	22.6	23.1	3.3	0.9	9.0	5.0	24.1	20.8	–	0.9
High oleic SBO (90%)		20.8	21.6	24.5	47.1	5.0	0.1	1.1	0.6	47.2	7.2		0.2
Difference (%)		−5.2	+103.8	+8.4	103.9	51.5	−88.9	−87.8	−88	+95.9	−65.4		−77.8
Menhaden FO		30.6	9.8	17.8	14.3	0.9	3.0	14.9	10.6	18.9	28.1		1.5
Fillet High oleic SBO (90%)		29.0	12.2	16.8	28.5	1.6	2.4	8.3	4.2	34.3	15.3		0.4
Difference (%)		−5.22	+24.5	−5.6	+99.3	+77.8	−20	−44.3	−60.4	+81.5	−45.6		−73.3
Red hybrid tilapia (*Oreochromis* sp.) (Ng et al. 2013)													
Diet FO	5 months	26.9	9.7	32.1	5.5	1.1	0.7	11.4	10.7	6.6	26.7	24.9	4.1
Diet PO (100%)		41.9	33.1	37.5	13.3	0.8	–	1.0	1.9	13.3	3.8	4.6	0.3
Difference (%)		−155.8	241.2	16.8	141.8	−27.3	–	−91.2	−82.2	101.5	−85.8	−81.5	−92.7
Whole body FO after 5 months		26.2	13.4	28.4	4.9	0.8	1.0	2.7	14.1	6.6	23.8	24.1	3.6

(Continued)

TABLE 9.11 (CONTINUED)
Effects of Dietary FO Replacement on Fatty Acid Profile of Crustaceans and Freshwater Fish Species

	Duration of Feeding Trial	SFA	OA	MUFA	LA	ALA	ARA	EPA	DHA	n-6 PUFA	n-3 PUFA	LC-PUFA	n-3/n-6 PUFA Ratio
Whole body PO (100%) after 5 months		28.6	33.1	40.8	11.6	0.5	1.6	0.3	5.9	14.4	7.8	10	0.5
Difference (%)		+9.2	+147	+43.7	136.7	−37.5	60	−88.9	−58.2	+118.2	−67.2	−58.5	−86.1
Largemouth bass (*Micropterus salmoides*) (Subhadra et al. 2006)													
Diet FO	84 days	33.8	24.9	37.1	11.8	1.3	1.9	8.3	4.2	14.4	14.9	–	1.0
Diet CAO (100%)		16.3	55.4	60.2	20.7	4.2	–	–	–	20.7	4.2	–	0.2
Difference (%)		−51.8	122.5	62.3	75.4	233.1	–	–	–	43.8	−71.8	–	−80
Fillet FO		32.1	26.0	36.0	9.7	0.9	2.2	3.8	7.8	10.1	14.4	–	1.1
Fillet CAO (100%)		23.3	42.1	48.2	17.8	2.5	1.8	1.1	4.4	20.0	8.1	–	0.4
Difference (%)		−27.4	61.9	33.9	83.0	177.8	−18.2	−71.1	−43.6	98.0	−43.8	–	−63.3
Hybrid striped bass (Perez-Velazquez et al. 2019)													
Diet FO	42 days	28.0	26.9	43.8	4.3	0.9	0.7	7.8	6.1	5.7	16.5		2.9
Diet Schyzochytrium and Arthrospira meals (50%)		48.1	10.5	17.7	6.2	0.3	0.5	3.0	9.5	12.4	14.7		1.2
Difference (%)		+71.8	−61	−59.6	44.2	−200	−28.6	−61.5	55.7	+117.5	−10.9		−58.6
Whole body FO		22.4	29.2	30.4	4.7	1.0	1.2	6.1	5.8	7.0	26.4	–	3.8
Whole body Schyzochytrium and Arthrospira meals (50%)		23.6	22.5	23.5	6.2	0.8	1.0	4.8	13.3	11.6	28.4	–	2.5
Difference (%)		+5.4	−22.9	−22.7	+31.9	−20	−16.7	−21.3	+129.3	+65.7	+7.6		−34.2

Abbreviations: ALS, alternative lipid source; CAO, canola oil; CO, corn oil; EO, echium oil; FXO, flaxseed oil; LO, linseed oil; OO, olive oil; PF, poultry fat; PL, pork lard; PO, palm oil; SBO, soybean oil; SFO, sunflower oil; T, tallow.

Diet difference % = (%FA diet ALS – % FA diet FO) / (%FA diet FO) ×100.

Fish difference % = (%FA fish ALS – % FA fish FO) / (%FA fish FO) ×100.

The percentage of fish oil substituted is in parentheses. The percentage difference of the fatty acid percentage between diets and fish tissues is also reported according to Turchini et al. (2009).

the levels of LC-PUFA in the fillet by using a "finishing diet" during the final weeks of farming before harvest, which is called a "washout period". During the washout period, feeds containing ALS are replaced with FO-based or LC-PUFA-enriched diets to modify the FA profile of the fillet (Turchini et al. 2009; 2011a). The efficiency of a washout period is mainly determined by different factors, such as the species, the duration of the washout period, the FA composition of the ALS used during the grow-out period, and the period of feed deprivation before feeding the finishing diet (Thanuthong et al. 2012; Reis et al. 2014; Yıldız et al. 2018). Another nutritional strategy may be the supplementation of feeds containing ALS with ingredients containing DHA, such as microalgal meals or oil extracts (Perez-Velazquez et al. 2019; Torres Rosas et al. 2019). Other strategies may include the development of GMO crops (e.g. *Camelina sativa*) to produce EPA and DHA and the application of selective breeding programs to improve the efficiency of EPA and DHA bioconversion from their precursors in ALS or even farming transgenic fish with higher capabilities in LC-PUFA biosynthesis (Glencross et al. 2020). Furthermore, nutritional programming during the early developmental stages (e.g. larvae and fry) (Geurden et al. 2013; Turkmen et al. 2017; Mellery et al. 2017) and/or through brood stock management and nutrition (Lazzarotto et al. 2015; Turkmen et al. 2017; Agh et al. 2020) has been recently investigated in order to select fish strains with better adaptation to diets containing low levels of FO. According to Hanley, "Nutritional programming is dependent on an early-life stimulus that provokes a permanent or long-term physiological effect on later life stage". With the application of nutritional programming during early developmental stages and/or parental stages, the capacity of the progeny in bioconversion of LNA and ALA to LC-PUFA increase by feeding them with LC-PUFA-deficient diets. This approach also increases the acceptance of plant-based feeds and enhances the resistance of the progeny to feeds lacking or deficient in EFA (Lazzarotto et al. 2015; Turkmen et al. 2017).

ORGANOLEPTIC AND SENSORY CHARACTERISTICS OF THE FILLET

The modification of the FA profile of the fillet in fish due to the replacement of dietary FO with ALS does not only reduce its content in particular health-promoting nutrients, particularly LC-PUFA associated with the fish consumption, but also affects its organoleptic attributes (e.g. color, taste, odor, and texture) (Table 9.12). The FA profile of dietary lipid sources can greatly impact the oxidative stability of the fillet, which consequently can affect its organoleptic properties (Ng and Bahurmiz 2009). In fact, the metabolites of lipid peroxidation, including peroxides and aldehydes, are associated with a negative impact on the flavor, color, and nutritional characteristics of the fillet, and it may also potentially produce unhealthy molecules (Trushenski and Lochmann 2009; Secci and Parisi 2016). In this sense, the peroxidation level of the fillet's lipids is dramatically related to the high PUFA concentrations in these tissues (Lopez-Bote et al. 2001; Menoyo et al. 2002). Generally, the replacement of FO with ALS extends the refrigerated shelf-life of fillets due to their low levels of LC-PUFA in ALS compared to FO. In fact, LC-PUFA-rich lipids, such as FO, are particularly susceptible to peroxidation and many studies demonstrated that FO sparing, by including ALS in feed, increases the shelf-life of fillets by reducing malondialdehyde (MDA) values (Bai and Gatlin 1993; Izquierdo et al. 2003; Menoyo et al. 2004). In this sense, high MDA concentrations and rancid odor in the fillet of fish fed on FO-based diets could be associated with remarkable lipid peroxidation due to the high amount of dietary n-3 LC-PUFA (Sargent et al. 1993b; Seno-o et al. 2008; Østbye et al. 2009).

It has been proven that the FA profile of the feed may be related to methemyoglobin (metMb) formation in the fillet. Moreover, the darkening of fish meat during their storage in the refrigerator, caused by the formation of the metMb due to the oxidation of myoglobin, has also been correlated to lipid peroxidation (Seno-o et al. 2008). In this regard, the replacement of dietary FO with olive oil in yellow tail kingfish (*Seriola lalandi*, Carangidae) prevents the darkening of the fillet by decreasing its EPA and DHA contents (Seno-o et al. 2008). In addition, VO, like olive and palm oils, contain biologically active compounds, especially antioxidants including oleuropein, hydroxytyrosol,

TABLE 9.12
Effects of Dietary FO Replacement on Organoleptic and Sensory Characteristics of Fillet

Alternative lipid	FO Replacement (%)	Effects on Sensory Attributes
		Gilthead sea bream (*Sparus aurata*) (Menoyo et al. 2004)
LO and SO	60–80	Did not affect pH and firmness of raw fillets; however, progressive inclusion of both SO and LO reduced the redness and yellowness of the fillet. Replacement of dietary FO with both SO and LO at 60% remarkably reduced the lipid peroxidation in the fillet.
		Gilthead seabream (*Sparus aurata*) and European seabass (Izquierdo et al. 2003)
SBO, RO, LO and Mix	60	Flesh quality of seabream was hardly affected by vegetal sources; however, smell and taste of fish fed the SO diet were stronger. The inclusion of vegetable oil sources slightly reduced the hardness of seabream fillets. European sea bass fed the LO diet had the lowest preferred ranking level, but the fillet texture was not affected by various vegetal oil sources.
		Gilthead sea bream (*Sparus aurata*) (Menoyo et al. 2004)
RO/SBO (1:1)	75	Significantly higher chroma values were observed in the VO group that was associated with a tendency toward more intense red and blue colors. No significant differences were found with respect to pH or water capacity between the different groups. Hardness, springiness, chewiness and gumminess of the fillets increased in fish fed on the VO diet. The sensorial attributes were not affected by dietary lipid sources.
		Red seabream, (*Pagrus auratus*) (Glencross et al. 2003)
CAO and SBO	100	Fish fed on the CAO had the whitest fillet and the weakest flavor. Fish fed on the FO diet had significantly dryer fillet than that of the SBO group. The strongest flavor being that of the FO fed fish. There were no significant differences in the level of oiliness and overall acceptability of the fish fed different dietary lipid sources.
		Brown trout (*Salmo trutta*) (Turchini et al. 2003)
OO, CO, PL, and PF	100	Fish fed vegetal oils including OO and CO had lower total and ferrous odors in fresh fish compared to fish fed FO and animal fats. There were not differences in cooked fillet sensory attributes such as color, flavor and texture among fish fed various lipid sources.
		Turbot (*Psetta maxima*) (Regost et al. 2003)
SBO and LO	100	Potatoes odor was more noticeable in dorsal fillet of fish was fed on SBO diet. Dorsal fillet of fish fed on vegetal oils diets was dryer than fish fed on FO diet. The greatest and the least exudation values were found in fish fed LO and SBO diets, respectively. Fish fed LO diet had more intensity and fat fish odor in the ventral fillet than fish fed on FO diet. Milky odor also stronger in ventral fillet of fish fed on LO diet than those fed SBO diet. Marine odor was more noticeable in the dorsal and ventral fillet of fish that fed on LO diet, but it was not statically significant.

(*Continued*)

TABLE 9.12
Effects of Dietary FO Replacement on Organoleptic and Sensory Characteristics of Fillet

Alternative lipid	FO Replacement (%)	Effects on Sensory Attributes
After washout period with FO	-	Odor and visual aspects of dorsal fillet did not affect after washout period. A fatty fish flavor was more pronounced in fish fed on only FO diet. The fillet of fish only fed on FO was less firm than that previously fed on SBO. Dorsal fillet of fish fed on SBO and then FO diets was dryer than other groups. The sweet flavor of fish that only fed on FO diet was different from those first fed on the vegetal oil diets. Fillet instrumental texture was not affected by different diets.
		Marble goby (*Oxyeleotris marmorata*) (Ti et al. 2019)
SBO and CAO	90	Fillet of fish fed on SBO diet had the highest whiteness, tenderness, juiciness, and sweetness compared to the other experimental groups.
		Rainbow trout (*Oncorhynchus mykiss*) (Turchini et al. 2013)
CAO, SFO, PF, manola oil and PO	75	In general, a high level of homogeneity was recorded amongst all treatments for the different sensorial attributes such as appearance, odor, taste and texture. . Fillets from trout fed the 90 mg g^{-1} DHA-G diet were perceived by panellists to have more fishy flavor than the fillets from fish fed the FO diet.
		Rainbow trout (*Oncorhynchus mykiss*) (Betiku et al. 2015)
DHA-algae meal		Fillet texture including springiness, stringiness, adhesiveness and cohesiveness were not altered by dietary lipid source.
		Rainbow trout (*Oncorhynchus mykiss*)(Katerina et al. 2020)
***Schizochytrium limacinum* biomass**	100	Fish that were fed *Schizochytrium limacinum* biomass supplemented diet had more red filets, stronger coloration and had lower whiteness. In addition, fish fed on FO diet had softer and juicier fillet compared to fish fed on the *Schizochytrium limacinum* biomass supplemented diet. There were no significant effects identified on the sensory quality of the filets.
		Gilthead seabream (*Sparus aurata*) (Matos et al. 2012)
SBO/RO (1:1)	30–50	Instrumental texture properties of raw fillet including hardness, cohesiveness, springiness, gumminess and chewiness were not affected by alternative oils.
		Atlantic salmon (*Salmo salar*) (Hixson et al. 2014)
Camelina oil	100	No significant differences in appearance, odor and texture between salmon fillets that were fed either FO or camelina oil.
		Red hybrid tilapia (*Oreochromis* sp.) (Ng and Bahurmiz 2009)
Crude PO, palm fatty acid distillate, refined/bleached and deodorized PO	100	Percent of drip loss of fillet after thawing was not affected by dietary PO. The texture of fillet regarding to resistance force was not affected by different lipid sources. Lipid peroxidation in the fillet of fish fed on the PO diets was significantly decreased compared to those fed on the FO diet. Fish fed on palm fatty acid distillate had the high scores in whiteness, chewiness and juiciness as well as it was sweeter, less rancid and less sour than other groups.

Abbreviations: CAO, canola oil; CO, corn oil; EO, echium oil; FXO, flaxseed oil; LO, linseed oil; OO, olive oil; PF, poultry fat; PL, pork lard; PO, palm oil; RO, rapeseed oil; SBO, soybean oil; SFO, sunflower oil; T, tallow.

tyrosol, α-tocopherol, and carotenoids that minimize the lipid peroxidation and enhance the shelf-life of the fillet (Gimeno et al. 2002; Tuck and Hayball 2002; Wang et al. 2006; Ng and Bahurmiz 2009). Furthermore, it has been confirmed that the fillet's color can be affected by dietary lipid content and FA profile. On the other hand, it has been reported that ALS may also influence the absorption and retention of astaxanthin in the fillet of Atlantic salmon (Bjerkeng et al. 1999; Turchini et al. 2013). However, other studies did not find any differences in the astaxanthin content in the fillet, when VO were used in diets for Atlantic salmon (Torstensen BE et al. 2005; Hixson et al. 2014).

The most important parameters that determine the texture quality of the fillets are their hardness, fracturability, and liquid-holding capacity. Several studies confirmed that ALS can affect fillet texture and liquid-holding capacity in different fish species (Regost et al. 2004; Izquierdo et al. 2005; Turchini et al. 2009; Hixson et al. 2014; Ti et al. 2019). In this regard, it has been suggested that SFA-enriched oils, such as RAF, palm, and coconut oils, tend to increase the firmness of the fillet and improve its textural attributes (e.g. springiness, chewiness, gumminess, and cohesiveness) compared to MUFA and PUFA-enriched oils, especially those containing high levels of LNA, ALA, and oleic acid (Fuentes et al. 2010; Stejskal et al. 2011; Xu et al. 2016). In fact, many studies have shown that oils rich in MC-PUFA and MUFA increase the fat content of the fillet due to the higher deposition of these FA, which leads to the softening of the fillet's texture (Grigorakis et al. 2003; Ginés et al. 2004; Johnston et al. 2006; Suarez et al. 2014; Xu et al. 2016). In this context, it has been reported that the inclusion of soybean oil in feed reduced fillet hardness and resulted in juicier fillets (Guillou et al. 1995; Glencross et al. 2003; Izquierdo MS et al. 2005; Ti et al. 2019). In contrast, many researchers have reported that FO replacement by the inclusion of VO in the diet did not affect fillet texture characteristics (Bell et al. 2001; Regost et al. 2003; Regost et al. 2004; Røra° et al. 2003; Montero et al. 2005; Hixson et al. 2014; Qingyuan et al. 2014). On the other hand, Ti et al. (2019) reported that dietary LNA positively influenced some fillet attributes like whiteness, tenderness, juiciness, and sweetness. In contrast, SFA-rich oils and fats resulted in lower lipid deposition in the fillet as these FA are preferable substrates for β-oxidation and energy production (Izquierdo et al. 2005; Turchini et al. 2009; Hixson et al. 2014; Xu et al. 2016). It has been suggested that fillet firmness is correlated with muscle fiber diameter and density (Hatae et al. 1990). However, some research demonstrated that the inclusion of VO in feeds has no effect on muscle fiber size (Haugen et al. 2006; Matos et al. 2012). These data indicated that whenever the evaluation of fillet textural characteristics is conducted, not only the type of lipid source and its FA profile needs to be considered but also the level of other macro- and micronutrients, as well as their interaction, must be evaluated in order to clearly understand the impact of the diet on fillet quality.

Volatile compounds in the fish fillet can be affected by dietary lipid sources (Turchini et al. 2013). In fact, lipid peroxidation is involved in the formation of volatile compounds (e.g. alkyl radicals), and these compounds are formed faster from LC-PUFA compared to SFA, MUFA, and LNA (Elmore et al. 1999). For example, it has been reported that the inclusion of soybean oil in feed could slightly increase an "earthy" flavor in the fillet of gilthead seab ream as a consequence of the activity of lipoxygenases on lipids contained in the fillet. Similar results have been observed in Atlantic salmon and turbot (*Scophthalmus maximus*, Scopthalmidae), among other species (Milo and Grosch 1993; Regost et al. 2003; Ti et al. 2019). In addition, the inclusion of sunflower and soybean oils rich in MC-PUFA could enhance a "green" or "grassy" flavor of the fillet due to the peroxidation of LNA and production of 1-hexanol by the action of 13-hydroperoxide (Belitz and Grosch 1999). In this regard, Alvarez et al. (2020) suggested that using blends of soybean oil with rapeseed oil can mask the effects of the soybean oil on the flavor of the fillet.

When comparing fish fed diets with different lipid sources, rainbow trout that were fed a diet containing FO as the main lipid source had higher levels of n-3 PUFA-derived volatile compounds (e.g. 1-hexen-3-ol, 1-penten-3-ol, 2-hexenal, 4-heptenal) in comparison to their congeners fed a diet containing n-6 PUFA-rich oils (soybean and sunflower oils) that increased the levels of n-6 PUFA-derived volatile compounds in the fillet (e.g. 1-hexanol, 1-octen-3-ol, hexanal). In addition, trout fed diets containing MUFA-rich oils (monola and canola oils) enhanced the levels of n-9-derived volatile compounds (e.g. heptanal, octanal, nonanal). These results indicate that there is a direct correlation between the flavor of fish fillets and dietary lipid sources, and consequently, the impact of the lipid source and its FA profile may directly affect fillet quality in terms of its bromatological properties, but also in terms of its flavor (Turchini et al. 2013).

EFFECTS OF DIETARY ALS ON LIPID METABOLISM

Several studies have shown that the change of dietary FA profiles can induce the modification of lipid metabolism, including lipogenesis, lipid transport by lipoproteins, lipid retention, lipid catabolism (i.e. lipolysis and FA β-oxidation), and LC-PUFA biosynthesis capacity in different fish species (Tocher 2003; Turchini et al. 2009).

The activities of lipogenic enzymes, including fatty acid synthetase (FAS), glucose-6-phosphate dehydrogenase (G6PD), and malic enzyme (ME), are not affected by the substitution of FO with VO either at the molecular or enzymatic levels (Regost et al. 2003; Torstensen et al. 2004; Richard et al. 2006; Bouraoui et al. 2011). On the contrary, an *in vitro* study conducted by Alvarez et al. (2000) showed that n-3 PUFA (EPA, DHA, and high levels of ALA) reduced the activities of FAS and G6PD in rainbow trout hepatocytes. In addition, another *in vitro* study conducted on Atlantic salmon by Vegusdal et al. (2005) confirmed that a low n-3/n-6 PUFA ratio in a hepatocyte primary cell culture promoted lipid triacylglycerol synthesis compared to a lower n-6/n-3 PUFA ratio. In agreement with the previous research, an *in vivo* study conducted by Menoyo et al. (2003) indicated that G6PD and ME activities were increased in Atlantic salmon fed a diet containing low levels of n-3 PUFA. Furthermore, a study in gilthead sea bream indicated that VO rich in n-3 PUFA, like linseed oil, reduced hepatic lipogenesis (Menoyo et al. 2004). In this regard, Vegusdal et al. (2005) demonstrated that EPA, but not DHA, and the elevation of the n-3/n-6 ratio, hampered TAG secretion in Atlantic salmon hepatocytes by stimulating mitochondrial β-oxidation; thus, reducing the availability of FA for TAG synthesis, whereas oleic acid stimulated TAG secretion. On the other hand, n-3 PUFA may decrease the synthesis of TAG by enhancing phospholipid synthesis and reducing the activity of TAG-synthesizing enzymes (Vegusdal et al. 2005).

In contrast, Qiu et al. (2017) reported that the relative gene expression levels of diacylglycerol o-acyltransferase 2 (DGAT-2), FAS, G6PD, and acyl-CoA diacylglycerol acyltransferase-2 in the large yellow croaker fed an FO-based diet were up-regulated compared to their expression in fish fed VO-based diets. However, fish fed the diet containing soybean oil showed a down-regulation of genes like DGAT-2, FAS, and G6PD, which resulted in a reduction of lipid retention in the muscle (Qiu et al. 2017). In addition, the above-mentioned study showed that the relative expression of lipoprotein lipase in fish fed the soybean oil diet was lower than fish fed the diets containing palm and rapeseed oils, suggesting lower β-oxidation capacity in this group. These studies suggest that lipogenesis is differently affected, depending on the fish species considered, and it depends on ALS and their FA profile.

Previous research proved that dietary FA profile determined the FA composition of plasma lipoproteins in fish (Lie et al. 1993; Torstensen et al. 2000, 2004; Ruyter et al. 2006; Richard et al. 2006). In this sense, it has been confirmed that diets rich in VO containing high levels of MUFA and n-6 PUFA induced liver triglyceride production and secretion due to their high levels of OA and LNA. As previously mentioned, VO with low levels of n-3/n-6 PUFA can stimulate lipogenesis and TAG synthesis that may result in an increase in very low density lipoprotein (VLDL) production by hepatocytes; thus, modulating the level of fat deposition in the hepatic parenchyma (Vegusdal et al. 2005; Ruyter et al. 2006; Jordal et al. 2007; Kjaer et al. 2008a, b). Furthermore, it has been shown that dietary VO reduces plasma cholesterol and low-density lipoproteins (LDL), which may be attributed to the low levels of cholesterol and SFA in these ALS and the presence of phytostrol, a compound with hypolipidemic effects, in VO (Richard et al. 2006; Jordal et al. 2007). Phytostrols, by the interfering with the absorption of cholesterol, and perhaps TAG in the gut, thus reduces lipogenesis. These compounds are also known for modulating the liver X receptors (LXR), which affect the expression of lipid regulatory metabolic genes (Dumolt and Rideout 2017). In this sense, blends of VO (rapeseed, palm, and linseed oils) down-regulate LDL receptor gene expression mainly in the liver and, to a lesser extent, in the adipose tissue of rainbow trout (Richard et al. 2006). Considering that high-density lipoproteins generally contain high levels of protein (48%) and phospholipids (33%) compared to other lipoproteins, they are less affected by dietary FA profiles (Ruyter et al. 2006).

Beta-oxidation of FA is a chain of processes in which FA are broken down by the activities of various proteins (e.g. FA protein transporters and cytoplasmic FA binding protein) and enzymes (e.g. fatty acyl-CoA synthase and carnitine palmitoyltransferase 1 and 2) to produce energy. Previous studies reported that, in the liver, mitochondrial and peroxisomal β-oxidation activities accounted for 30 and 50% of the total β-oxidation activity, respectively, whereas, in red muscle, β-oxidation activity in peroxisomes accounted for 20% of the total β-oxidation in several marine fish species (Crockett and Sidell 1993a,b; Frøyland et al. 2000). As previously mentioned, FO sparing remarkably changes the tissue FA profile, thus affecting the β-oxidation capacity of FA (Turchini et al. 2003; Woitel et al. 2014; Campos et al. 2019). In this sense, it has been proven that when MUFA and PUFA (LNA and ALA) levels increase in the diet, increments of FA β-oxidation occurred (Bell et al. 2003a, b; Torstensen et al. 2004; Stubhaug et al. 2007). Moreover, during the parr-smolt period in Atlantic salmon, an excess of EPA and DHA levels in the diet are β-oxidized (Stubhaug et al. 2007). In this regard, the former authors demonstrated the retention of EPA and DHA in the tissues of parr-smolt stages in Atlantic salmon fed a FO-based diet were 20% and 30%, respectively, whereas those fed a diet containing VO retained 70% EPA and 80% DHA, indicating n-3 LC-PUFA were selectively stored in these tissues. These findings suggest that FO replacement by ALS may induce a selective switch of FA substrates for their β-oxidation, and, because of the low levels of n-3 LC-PUFA in VO (blends of rapeseed, palm, and linseed oils), the β-oxidation capacity was reduced compared to FO (Stubhaug et al. 2007). In this sense, Jordal et al. (2005) demonstrated that dietary rapeseed oil lacking n-3 LC-PUFA remarkably down-regulated several mitochondrial genes (e.g. hepatic carnitine palmitoyltransferase II (CPT II)) in the liver of Atlantic salmon. Carnitine palmitoyltransferase II, together with CPT I, are mitochondrial membrane proteins whose main role is transferring FA into mitochondria for β-oxidation. Furthermore, it has been demonstrated that lipoprotein lipase gene expression was down-regulated in perivisceral adipose tissue of red seabream (Liang et al. 2002) and rainbow trout (Richard et al. 2006) that were fed an oleic acid-rich VO diet compared to an FO-based diet, indicating a reduction in β-oxidation capacity in these species.

On the other hand, *in vitro* studies with Atlantic salmon hepatocytes indicated that n-3 LC-PUFA can increase FA uptake into the mitochondria, inducing β-oxidation (Turchini et al. 2003; Torstensen and Stubhaug 2004). Furthermore, both high dietary EPA and an increase in n-3/n-6 ratio induced mitochondrial proliferation in hepatocytes, resulting in an increase of total β-oxidation capacity (Vamecq et al. 1993; Willumsen et al. 1996; Vegusdal et al. 2005; Kjaer et al. 2008a, b). In addition, Kjær et al. (2008a) reported that high levels of dietary DHA in Atlantic salmon stimulates the over-expression of some mitochondrial genes involved in lipid metabolism, like peroxisome proliferator-activated receptors (PPAR), acyl-CoA oxidase, and CPT II, compared to those fed a diet containing lower levels of n-3 LC-PUFA, changes that were linked to a higher β-oxidation capacity. In this sense, Li et al. (2015a) reported that the inclusion of soybean oil in the diet of blunt snout bream (*Megalobrama amblycephala*, Cyprinidae) juveniles resulted in an up-regulation of acyl-CoA Δ9 desaturase and down-regulation of PPARα and PPARβ, receptors that are important transcription factors in the regulation of gene expression of several enzymes (e.g. CPT I and II) involved in the β-oxidation of FA (Li et al. 2015b). However, in different studies conducted by Stubhaug et al. (2005a, b; 2007) only marginal effects of dietary FO replacement with VO on β-oxidation capacity were found in white and red muscle, as well as in the liver and heart of Atlantic salmon. These discrepancies in the effects of dietary replacement of FO with ALS on FA β-oxidation may be attributed to the species-specific capacity of LC-PUFA biosynthesis and the FA profile of ALS.

The different abundance of LC-PUFA in the fresh and marine environments has led to the different capacities in LC-PUFA biosynthesis between freshwater, salmonids, and marine fish species. However, it has been shown that the trophic level of a species may also exert a more pronounced effect on its LC-PUFA biosynthesis capacity rather than its environment (Trushenski and Rombenso 2020). Previous studies indicated that low content of dietary n-3 LC-PUFA, especially in VO-based diets, induced the activation of LC-PUFA endogenous biosynthesis in fish (Tocher et al. 2001; Fonseca-Madrigal et al. 2005; Jordal et al. 2005; Zheng et al. 2005; Jaya-Ram et al. 2008; Navarro-Guillen et al. 2014; Geay et al. 2015; Teoh and Ng 2016; Nayak et al. 2017). In fact, high levels of n-3 LC-PUFA in the diet inhibited the desaturation of both LA and LNA (Vagner and Santigosa 2011).

Regarding freshwater fish species, higher mRNA expression levels of Δ6 FAD in different tissues of silver barb (*Barbonymus gonionotus*, Cyprinidae) (Nayak et al. 2017) and common carp (Ren et al. 2012) fed VO-based diets were found when compared to their congeners fed FO-based diets. Studies with salmonids have shown that fish fed LC-PUFA-deficient diets rich in LNA and ALA showed an up-regulation of gene transcription and enzyme activity of hepatic fatty acid desaturases and elongases that promoted the biosynthesis of n-3 LC-PUFA (Bell and Dick 2004; Zheng et al. 2004; Morais et al. 2009a). However, the extent of endogenous bioconversion of LC-PUFA is restricted, and it mainly relies on several factors: i) presence of precursors (LNA and ALA) and end-products in the diet; ii) the EFA requirements of fish species that mainly depend on their trophic level (Tocher et al. 2003; Turchini et al. 2011a); and iii) the availability of other external factors in the diet, such as co-enzymes (e.g. L-carnitin), co-factor promoters (e.g. sesamin), and/or inhibitors (Pickova et al. 2010; Giri et al. 2016). Furthermore, it has been confirmed that replacement dietary FO with VO in Atlantic salmon during seawater transfer enhanced the transcription of hepatic desaturases and elongases, and their respective enzyme activities, but not during the freshwater period (Henderson et al. 1994; Ruyter et al. 2000b; Bell et al. 2001; Tocher et al. 2001; Tocher et al. 2002a,b, 2003; Stubhaug et al. 2005a, b; Zheng et al. 2005; Moya-Falcon et al. 2006). Thus, culture environmental conditions, such as temperature and salinity, may also affect LC-PUFA biosynthesis pathways, such as Δ6 desaturase activity in fish species (Vagner and Santigosa 2011). Regarding marine fish species, it has been demonstrated that Japanese seabass (*Lateolabrax japonicus*, Lateolabracidae) can biosynthesize LC-PUFA to some degree due to the activity of Δ6 FAD, while its activity can be stimulated by SFA and MUFA-rich ALS compared to FO and ALA-rich oils (Xu et al. 2014). Furthermore, it has been confirmed that groupers (Serranidae) are capable of elongating ALA into EPA but not further, to DHA. Thus, with enough FM in the diet, it seems possible to replace large amounts of FO with ALA-rich VO, such as linseed oil, for this particular group of species (Lin and Shiau 2007; Lin et al. 2007; Shapawi et al. 2008). Similar results have been observed in cobia, where molecular studies have shown a partial capacity of n-3 LC-PUFA biosynthesis, as studies dealing with partial replacement of FO with VO indicated (Zheng et al. 2009). In addition, Izquierdo et al. (2008) reported that Δ6 FAD was up-regulated in larval gilthead sea bream fed VO-based microdiets containing low levels of n-3 LC-PUFA compared to larvae fed an FO-based microdiet. In addition, some studies in marine species, such as Senegalese sole showed that the inadequacy of LC-PUFA in diet induced the up-regulation of Δ4FAD, which indicated this species has the ability to synthesize DHA from EPA through the Sprecher-independent pathway (Morais et al. 2012b). This knowledge about the ability of different species for the biosynthesis of LC-PUFA can help nutritionists to enhance the FO-sparing levels in aquafeeds without compromising fish performance, health, and welfare.

EFFECTS OF DIETARY FO REPLACEMENT WITH ALS ON FISH HEALTH

It has been proven that dietary FA composition can modulate the function of the immune system, including cellular (e.g. phagocytosis and respiratory burst activity of leucocytes) and humoral (e.g. lysozyme and complement system) responses from immune organs (e.g. head kidney, liver, and spleen). In this context, dietary FA profile may alter fish metabolism (Mourente et al. 2005;

Wassef et al. 2007), biosynthesis of eicosanoids (Montero et al. 2003; Ganga et al. 2005; Lin and Shiau 2007; Geay et al. 2010), modulate the expression of immune-related genes (Montero et al. 2008, 2010, 2015) and tissular antioxidant conditions (Castro et al. 2016), as well as modify the gut microbiome (Torrecillas et al. 2017a, 2018) (Table 9.13). These modulations of immune parameters largely depend on the levels of FO replacement by ALS, the FA profile of the ALS, and the species considered (Montero et al. 2010). In this sense, differences in the modulation of immune parameters among fish species in response to FO replacement mainly depend on their capacity of endogenous biosynthesis of LC-PUFA. In this regard, it has been confirmed that FO sparing in feed induces more severe alterations in the immune system of marine fish (Montero et al. 2003; Lin and Shiau 2007; Mourente et al. 2007) compared to freshwater fish and salmonids with high capacity for LC-PUFA biosynthesis (Seierstad et al. 2009; Kiron et al. 2011).

It should be mentioned that leucocytes and immune organs for maintaining their homeostasis can selectively incorporate and deposit FA to maintain the appropriate balance of the n-3/n-6 PUFA ratio in their bilayer phospholipid membranes (Waagbo et al. 1995; Farndale et al. 1999; Montero et al. 2005). Dietary lipids may modulate the activity of leucocytes by changing the fluidity of their membranes, and affect activities of membrane-associated enzymes and receptor sites as well as cell signaling (e.g. eicosanoids and cytokines). In this regard, Wu et al. (2003) reported that DHA increases the phagocytic function of leucocytes and T cell proliferation in Malabar grouper (*Epinephelus malabaricus*, Serranidae). Furthermore, it has been demonstrated that a higher dietary ALA/LNA ratio could improve non-specific cellular immune responses in grouper (Tan et al. 2009; Wu and Chen 2012). On the other hand, high levels of dietary n-3 PUFA have shown immune-suppressive effects in rainbow trout (Kiron et al. 1995), Atlantic salmon (Thompson et al. 1996), and yellow croaker (*Larimichthys crocea*, Scianidae) (Zuo et al. 2012a), among other species. In this context, a plethora of research has demonstrated that a well-balanced n-3/n-6 PUFA ratio in the diet can improve immune competence in fish (Fracalossi and Lovell 1994; Li et al. 1994; Bransden et al. 2003). For example, it has been reported that inappropriate dietary n-3/n-6 PUFA levels in channel catfish (*Ictalurus punctatus*, Ictaluridae; Fracalossi and Lovell 1994) and Artic char (*Salvelinus alpinus*, Salmonidae; Lodemel et al. 2001) significantly reduced their survival rate when fish were challenged against pathogenic bacteria like *Edwardsiella ictaluri* and *Aeromonas salmonicida*, respectively. In contrast, no differences in cumulative mortalities were observed in Nile tilapia (Yildirim-Aksoy et al. 2007) and Eurasian perch (*Perca fluviatilis*, Percidae; Geay et al. 2015) fed diets containing linseed oil (rich in ALA) compared to groups fed an FO-based diet, after a challenge with pathogenic bacteria – results that were attributed by the former authors to an appropriate n-3/n-6 PUFA balance in experimental diets. Furthermore, it should be mentioned that some ALS, such as coconut oil, have antibacterial, antiprotozoal, and antiviral attributes as it contains high levels of C6-C12 FA, such as lauric acid (Matsue et al. 2019). In this regard, Apraku et al. (2019) reported that complete substitution of FO with blends of coconut and corn oils (1:1) increased the expression of immune-related genes (e.g. interlukine 1β) and the survival of Nile tilapia challenged with *Aeromonas hydrophila*.

Modification of FA profile of leucocytes can modulate immune-related gene expression, such as natural killer cell enhancing factor (NEKF), which are able to detect and eliminate a broad range of pathogens (Hasegawa et al. 1998; Nuguyen et al. 2019). In this regard, Nuguyen et al. (2019) reported the down-regulation of NEKF, secreted phospholipase and prostaglandin synthase 2 in common carp (*Cyprinus carpio*, Cyprinidae) juveniles fed n-3 PUFA-rich lipid sources (i.e. linseed oil and FO) compared to fish fed a lipid source rich in n-6 PUFA (sunflower oil), due to the stimulation of synthesis of ARA-derived eicosanoids. It is well-known that changes in dietary FA profile affect eicosanoid synthesis through changing the n-3/n-6 ratio of leucocyte membranes (Farndale et al. 1999; Mourente et al. 2005). It has been demonstrated that total replacement of dietary FO with linseed oil did not affect the expression of eicosanoids-related genes (e.g. arachidonate 5-lipoxygenase, prostaglandin E synthase 2, and leukotriene α-4 hydrolase) in the liver, spleen, and head kidney of yellow perch after exposure to *Aeromonas salmonicida* due to a well-balanced n-3/n-6

TABLE 9.13
Effects of Dietary FO Replacement with ALS on General Health and Immune Competence of Aquatic Species

Alternative Lipid Source	Dietary Lipid Level (%)	% FO Replacement	Duration of Feeding Trial	Effect on Physiological Responses of Aquatic Animals
		Gilthead seabream (*Sparus aurata*) (Montero et al. 2003)		
SBO, LO, RO, and Mix	25	60	101 days	Replacement of dietary FO with SBO and LO remarkably reduced RBC count, but did not affect Hb, Hct, and serum ACH50. Replacement of FO with RO or mixture of RO, LO, and SBO did not affect hematological and ACH50. Fish fed on the LO diet had higher serum cortisol level 2 h after net chasing stress. The fatty acid profile of head kidney macrophages pronouncedly affected by dietary lipid sources.
SBO, LO, RO, and Mix	22	60–80	204 days	Replacement of dietary FO with SBO at 60 pronouncedly reduced serum ACH50. The phagocytic activity of macrophages remarkably reduced in fish fed on RO at 60% replacement level and SBO at 80% replacement level. Fish fed on the LO diet had higher serum cortisol level 2.5 h after overcrowding stress.
		Gilthead sea bream (*Sparus aurata*) (Ganga et al. 2005)		
SBO, LO, RO, and Mix	25	60	3 or 6 months	Fish fed on the SBO diet showed steatosis, with foci of swollen hepatocytes containing numerous lipid vacuoles, but fish fed on the LO, RO, or mix diets showed a similar hepatic histoarchitecture to that observed in fish fed FO.
		Gilthead sea bream (*Sparus aurata*) (Ganga et al. 2005)		
Mixture of LO, LO and PO,	22	60	281	Replacement of dietary FO with mixture of VO resulted in remarkable reduction of EPA in the blood leucocytes and drastically affected plasma fatty acid profile, but it did not affect prostaglandins concentrations including PGE_2 and PGE_3 in fish.
LO	22	100	281	Replacement of dietary FO with LO significantly reduced EPA in the blood leucocytes and profoundly affected plasma fatty acid profile as well as reduced PGE_3 concentration in the plasma.
		Gilthead sea bream (*Sparus aurata*) (Wassef et al. 2007)		
Blends of SFO, cottonseed oil, LO, and SBO	18.2	60	140	Hematological parameters were not affected by dietary lipid sources. A minor deformation in RBC shape was observed in fish fed on the VO diets. The number of both developing myelocytes and phagocytes increased in the head kidney of fish fed the VO diets. Oil droplets accumulation, peripheral shifting of organelles and nuclei were detected in the hepatocytes of fish fed on the VO diets.
		Gilthead sea bream (*Sparus aurata*) (Montero et al. 2008)		
SBO, LO, and their mixture	22	100	6 months	Phagocytic activity and serum ACH50 were significantly reduced in fish fed SBO and LO diet, but mixture of VO did not affect phagocytic index. The MX gene remarkably up-regulated in the liver of fish fed on the VO-included diets.

(*Continued*)

TABLE 9.13 (CONTINUED)
Effects of Dietary FO Replacement with ALS on General Health and Immune Competence of Aquatic Species

Alternative Lipid Source	Dietary Lipid Level (%)	% FO Replacement	Duration of Feeding Trial	Effect on Physiological Responses of Aquatic Animals
		Gilthead sea bream (*Sparus aurata*) (Fountoulaki et al. 2009)		
SBO, PO, and RO	18	69	6 months	Abnormalities were not detected in the intestinal tissue of fish fed different diet; however, fish fed on PO diet showed apparent liver steatosis, intense lipid accumulation, swelling and nuclei displacement of the hepatocytes.
		Gilthead sea bream (*Sparus aurata*) (Montero et al. 2010)		
SBO and LO	22	70–100	80 days	Serum bactericidal activity was pronouncedly reduced in fish fed on the SBO diets. Tumor necrosis factor-α gene expression remarkably up-regulated in the head kidney of fish fed on the 100% SBO diet. Interleukin-1β gene expression in the intestine and head kidney of fish was not affected by the dietary lipid sources.
		Gilthead sea bream (*Sparus aurata*) (Wassef et al. 2015a)		
LO and SFO	18	50–70	63 days	Increasing replacement of dietary FO with VO from 50% to 70% remarkably led to a lipid accumulation in the liver of fish. Fish fed on the 70% of SFO showed pyknotic nuclei and cell organelles shift toward the cell-periphery and lipid infiltration in the hepatocytes.
		European Sea bass (*Dicentrachus labrax*) (Mourente et al. 2005)		
RO, LO, and OO	23	60	238 days	Plasma concentrations of prostaglandin E2 and prostaglandin F2α were not influenced by dietary lipid sources. Total number of circulating leucocytes and macrophage respiratory burst activity pronouncedly decreased in fish fed VO diets. The fatty acid profile of RBC drastically changed by dietary VO. Fish fed on LO diet had higher lipid content in the liver. Overall the intestines of fish fed the LO diet were normal with a higher absorptive capacity than fish fed the RO diet. Cellular infiltration in the distal segment of the intestine was observed in fish fed on VO diets.
		European Sea bass (*Dicentrachus labrax*) (Torrecillas et al. 2017a)		
Blends of RO, LO, and PO in plant protein rich diets	22	60–80	153 days	Reducing FO content by VO in diets increases ex vivo and in vivo gut bacterial translocation rates. Disease resistance against *Vibrio anguillarum* remarkably reduced in fish fed on the VO blends at 80% replacement level. Reducing FO content in diets increases *ex vivo* and *in vivo* gut *Vibrio anguillarum* translocation rates in the posterior intestine.
		European Sea bass (*Dicentrachus labrax*) (Torrecillas et al. 2017b)		
Blends of RO, LO, and PO in plant protein rich diets	22	60–100	153 days	Lamina propria width in the anterior intestine remarkably increased by increasing dietary VO due to lipid deposition.

(Continued)

TABLE 9.13 (CONTINUED)
Effects of Dietary FO Replacement with ALS on General Health and Immune Competence of Aquatic Species

Alternative Lipid Source	Dietary Lipid Level (%)	% FO Replacement	Duration of Feeding Trial	Effect on Physiological Responses of Aquatic Animals
		European Sea bass (*Dicentrachus labrax*) (Monteiro et al. 2018)		
Blends of PF and mammal fat	19	50–100	114 days	The lowest glycogen deposition was found in the liver of fish fed on the 100% animal fat diet. The vacuolation degree, steatosis rate and swelling nuclei displacement were significantly higher in the liver of fish fed on the 100% animal fat diet.
		European Sea bass (*Dicentrachus labrax*) (Machado et al. 2019)		
Blends of PO, LO, and RO	18	100	73 days	After 1 h acute stress, fish fed VO diet had lower Hct, peroxidase activity, down-regulation of the head kidney glucocorticoid receptor 1 and melanocortin 2 receptor genes expression, lower neutrophil number, plasma antiprotease and bactericidal activities and down-regulation of the head kidney cyclooxygenase 2 gene expression.
		European Sea bass (*Dicentrachus labrax*) (Campos et al. 2019)		
PF	19	25–100	112 days	The liver glycogen content decreased with increasing dietary PF level, but vacuolization degree of the hepatocytes increased.
		Japanease Seabass (*Lateolabrax japonicus*) (Tan et al. 2017)		
Blend of SBO and LO	13	50–100	70 days	The serum ACH50, disease resistance and antioxidant enzymes in the liver (e.g. superoxide dismutase, catalase and glutathione peroxidase) reduced in fish fed on the VO diets. Pro-inflammatory gene expression including interlukine-1β and tumor necrosis factor-α remarkably up-regulated in fish fed on VO diets.
		Giant grouper (*Epinephelus malabaricus*) (Lin and Shiau 2007)		
CO	0.5–15.9	50	56 days	Increasing dietary lipid level enhanced WBC count, respiratory burst, lysozyme and ACH50 activities.
		Senegalese sole (*Solea senegalensis*) (Montero et al. 2015)		
LO and SBO	12	100	84 days	Fish fed on the SBO diet showed an activation of inflammatory processes in the gut due to up-regulation of immune-related genes related to complement pathway, recognizing pathogen associated to molecular patterns, defensive response against bacteria and viruses, antigen differentiation, cytokines and their receptors. An acute stress was not affect immune related genes (e.g. interleukins and interferons genes) in fish fed LO diet.
		Large yellow croaker (Tan et al. 2016)		
Blends of LO and SBO	12	50–100	70 days	The serum lysozyme, ACH50 and respiratory burst activities as well as disease resistance against *vibrio anguilarum* remarkably reduced in fish fed VO diets. The pro-inflammatory cytokines genes expression (IL-1β and TNFα) up-regulated by activation of toll like receptors-nuclear factor kappa B, but anti-inflammatory genes (e.g. IL10) down-regulated in fish fed the VO diets.

(*Continued*)

TABLE 9.13 (CONTINUED)
Effects of Dietary FO Replacement with ALS on General Health and Immune Competence of Aquatic Species

Alternative Lipid Source	Dietary Lipid Level (%)	% FO Replacement	Duration of Feeding Trial	Effect on Physiological Responses of Aquatic Animals
		Golden popano (*Trachinotus ovatus*) (You et al. 2019)		
LO and SBO	12	100	56 days	The abundance of pathogenic bacteria (*Mycoplasma* and *Vibrio*) increased, but the proportions of beneficial bacteria (*Bacillus* and *Lactococcus*) remarkably decreased in the intestine of fish fed VO diets. The activity of acid phosphatase increased in the intestine of fish fed on the VO diets, while the activity of lysozyme in the intestine and serum diamine oxide enhanced in fish fed on the SBO diet. The intestinal health impaired in fish fed on the SBO diet by reduction of the number and height of intestinal folds, muscle thickness as well as zonula occludens-1 gene expression in the intestine.
		Atlantic salmon (*Salmo salar*) (Bransden et al. 2003)		
SFO	27–30	4–100	63 days	No significant correlation was found between cumulative mortality in challenge with *Vibrio anguillarum* and inclusion of SFO in diet.
		Atlantic salmon (*Salmo salar*) (Ruyter et al. 2006)		
SBO		50–100	3 months in 12° C or 7 months in 5° C	Higher accumulation of fat in the liver of fish fed on 100% SBO diet at 5° C.
		Atlantic salmon (*Salmo salar*) (Ye et al. 2016)		
Camelina oil	-	50–100	112 days	There were not any histo-morphological changes in the dietal intestine of fish fed on the camelina oil diet, and there was not any sign of inflammation. The lamina propria of fish fed the 50% camelina oil diet was narrower than fish fed the FO diet, indicating better gut lamina propria condition.
		Rainbow trout (*Oncorhynchus mykiss*) (Caballero et al. 2002)		
Different mixture of SBO, RO, PO, OO, and lard	-	50–80	64 days	A supranuclear lipid deposition of lipid droplets was observed in the enterocytes of fish fed on the VO diets.
		Largemouth bass (*Micropterus salmoides*) (Subhadra et al. 2006)		
CAO and PF	14	50–100	84 days	Fish fed diets with FO or CAO had higher serum ACH50 than fish fed diets with PF, but hematological indices and serum lysozyme activity were not affected by different dietary lipid sources.
		Eurasian Perch (*Perca fluviatilis*) (Geay et al. 2015)		
LO	11.6	100	70 days	The use of LO in diet did not compromise hematological, humoral immune responses, leucocytes count, disease resistance, and genes related to eicosanoid synthesis in the liver of fish fed on the LO diet.

(*Continued*)

TABLE 9.13 (CONTINUED)
Effects of Dietary FO Replacement with ALS on General Health and Immune Competence of Aquatic Species

Alternative Lipid Source	Dietary Lipid Level (%)	% FO Replacement	Duration of Feeding Trial	Effect on Physiological Responses of Aquatic Animals
		Nile tilapia (*Oreochromis niloticus*) (Apraku et al. 2017)		
Coconut oil	4.3	25–100	56 days	The survival rate increased as the inclusion of coconut oil in diet was enhanced in challenge with *Streptococcus iniae*.
		Common carp (*Cyprinus carpio*) (Nuguyen et al. 2019)		
LO and SFO	10	100	63 days	The survival rate in fish was not affected after challenge with *Aeromonas hydrophyla*. The humoral immune responses and immune-related genes (e.g. interleukin-8 and lysozyme) were not affected in fish fed VO-based diets. The eicosanoids metabolism process induced in fish fed on the SFO diet.
		Nile Tilapia (*Oreochromis niloticus*) (Yildirim-Aksoy et al. 2007)		
CO, BT, FO, LO, different mixture of FO, CO, BT, and LO	7	0–100	84 days	FO-diet had higher red and white blood cell counts than other ALS. Serum protein concentration, lysozyme activity, and natural hemolytic complement activity were reduced in fish fed the BT-diet. Mortality 15 days post-challenge with *S. iniae* was lower for fish fed the BT diet compared with those fed FO or FO, CO, BT diets.
		Large yellow croaker (*Larimichthys crocea*) (Zuo et al. 2015)		
Different mixture of SFO, PO, and LO	13	70	70 days	High dietary LNA negatively impacted nonspecific immunity and antioxidant capacity. The over expression of genes associated with inflammation (cyclooxygenase-2 and interleukin-1β) were reduced by increasing dietary ALA/LNA.
		Barramundi (*Lates calcarifer*) (Alhazzaa et al. 2013)		
EO, RO, and FO	20	100	42 days	During the infection period with *Streptococcus iniae*, EO fish had reduced levels of eicosanoids, TXB2, and 6-keto-PGF1a, in their plasma compared with RO fish. Fish fed on FO and EO had a longer lasting and enduring response in their eicosanoid concentrations, following a week of bacterial infection

Abbreviations: BT, beef tallow; CAO, canola oil; CO, corn oil; EO, echium oil; FO, fish oil; FXO, flaxseed oil; LO, linseed oil; OO, olive oil; PF, poultry fat; PL, pork lard; PO, palm oil; RO, rapeseed oil; SBO, soybean oil; SFO, sunflower oil; T, tallow.

ratio in the diet (Geay et al. 2015). Eicosanoids are mainly synthetized from C20 LC-PUFA by the action of phospholipase A2, cyclooxygenases and lipooxygenases. The C20 LC-PUFA-derived eicosanoids include prostaglandins, leukotriene, thromboxanes, and lipoxins that modulate a wide range of immune processes (Knight and Rowley 1995). It has been confirmed that n-6 LC-PUFA (ARA and dihomo-gamma-linolenic acid) derived eicosanoids have pro-inflammatory functions; meanwhile n-3 LC-PUFA (EPA and DHA) derived eicosanoids have anti-inflammatory characteristics (Arts and Kohler 2008). In this context, Ganga et al. (2005) demonstrated that total replacement of dietary FO with linseed oil did not affect plasma prostaglandin E_2 (ARA-derived eicosanoid) levels but remarkably reduced plasma prostaglandin E_3 content (EPA-derived eicosanoid); changes that could be responsible for the alternations of the immune system and general health of fish. In fact, dietary n-3 PUFA can counteract with n-6 PUFA-derived eicosanoids in different ways: i) by displacement (enhancing n-3/n-6 ratio in phospholipids), ii) competing for cyclooxygenases and lipoxygenase, and iii) hindering ARA-derived eicosanoids by EPA and dihomo-gamma-linolenic acid-derived eicosanoids (Bell et al. 1994). Furthermore, FO can inhibit the synthesis of pro-inflammatory cytokines by suppressing the nuclear factor kappa beta (NF-kB) signaling pathway, whereas VO can induce the activation of NF-kB signaling pathway and consequently exacerbate inflammatory reactions as well as suppressing immune responses (Tan et al. 2017). In this regard, the former authors reported that total replacement of dietary FO with blends of VO (soybean and linseed oils) induced the transcription of pro-inflammatory cytokines (i.e. interleukin 1β and tumor necrosis factor α) in the liver of Japanese seabass, which resulted in the suppression of the humoral immune response and reduced disease resistance due to activation of the NF-kB signaling pathway and exacerbation of inflammatory responses. In mammals, it has been proven that NF-kB is responsible for the transcription of genes encoding many pro-inflammatory cytokines (Hyden and Ghosh 2011).

On the other hand, the FA profile of the diet might also regulate the immune competence by changing the redox environment and activity of antioxidant enzymes (Tan et al. 2017). In fact, FO may cause oxidative stress due to the higher susceptibility of LC-PUFA to peroxidation compared to ALS with lower levels of LC-PUFA (Jin et al. 2017b). In this sense, antioxidant enzymes such as superoxide dismutase (SOD) can elevate the defense function of macrophages by neutralizing peroxides produced during respiratory burst activity and protect them from oxidative damages (Jin et al. 2017b). Moreover, dietary appropriate substitution of FO with VO when considering optimum n-3/n-6 PUFA ratio reduced the susceptibility of lipid peroxidation in the liver, intestine, and muscle of different fish species (Stéphan et al. 1995; Menoyo et al. 2004; Lin and Shiau 2007; Peng et al. 2008; Gao et al. 2012; Ng et al. 2013; Castro et al. 2015). In contrast, imbalances in dietary n-3/n-6 PUFA ratio due to complete replacement of FO by VO have been found to compromise the anti-oxidative system in several fish species (Raso and Anderson 2003; Kiron et al. 2011; ØStbye et al. 2011; Xu et al. 2015; Peng et al. 2016; Tan et al. 2017; Yu et al. 2019; Mu et al. 2020) The mechanism in which dietary n-3 LC-PUFA can improve anti-oxidative capacity and alleviate oxidative stress in fish may be related to the activating nuclear factor erythroid-2-related factor 2 (Nrf2) signaling pathway that modulates the transcription of antioxidant enzymes (Kobayashi and Yamamoto 2005; Mu et al. 2020). In this regard, Zeng et al. (2016) reported that dietary LNA/ALA ratios can modulate antioxidant enzyme activities as well as Nrf2 gene expression in the intestine of grass carp (*Ctenopharyngodon idella*, Cyprinidae), indicating the significant influence of dietary n-3/n-6 ratio on the antioxidant status of fish.

Dietary EFA and lipotropic compound (e.g. choline) deficiencies, as well as imbalances in n-3/n-6 PUFA ratio and/or excessive level of LNA in the diet, may result in liver steatosis (an accumulation of fat in the liver) (Thompson et al. 1996; Caballero et al. 2004; Piedecausa et al. 2007; Campos et al. 2019). The impairment of hepatic function can be induced by a lack of dietary EFA and phospholipids that result in disorders in lipid metabolism due to malfunction in the production of bile salts and lipoprotein synthesis (Lall 2010). On the other hand, ALS especially RAF reduce the ADC of dietary lipids, particularly that of n-3 PUFA that result in steatosis and reduce glycogen levels in hepatocytes (Monteiro et al. 2018; Campos et al. 2019). In fact, a higher degree of FO replacement can impair lipid ADC and increase lipid deposition, particularly in the liver (Monteiro et al. 2018). The histological organization of the hepatic parenchyma and the degree of hepatic fat stores due to high levels of dietary FO replacement with ALS has been reported to be non-pathological and reversible when fish are fed with an FO-based diet during the wash-out period (Caballero et al. 2004; Mourente et al. 2005; Fountoulaki et al. 2009; Monteiro et al. 2018). It has been suggested that the main FA in VO that induce steatosis in the liver and intestine are LNA > ALA > oleic acid (Caballero et al. 2004). Thus, provision of diet with enough EFA, according to the aquatic species requirements, may avoid vacuolization of the liver and the intestine. In addition, the recovery of normal liver and intestine morphology can be achieved after 2–3 months of wash-out feeding period with a finisher diet containing FO (Wassef et al. 2007).

As an immune organ, the gut has a vital role in the immune system and its hemostasis can be affected by dietary lipid sources (Turner 2009). In this regard, it has been suggested that dietary lipids may affect gut microbiota (richness and diversity) and consequently its health (You et al. 2019). During intestinal inflammation and immune responses, the synthesis of nitric oxide (iNOS) is enhanced considerably in the gut's mucosa, *lamina propria*, and submucosa, and its production is associated with an active involvement of goblet cells (Losada et al. 2012). In this regard, it has been speculated that dietary LC-PUFA deficiency induces gut mucosal secretion due to pro-inflammatory responses and iNOS production that coincides with *ex vivo* and *in vivo* gut bacterial translocation (Fukushima et al. 1999; Schenk and Mueller 2008; Kim and Ho 2010). Thus, the replacement of dietary FO with blends of VO (rapeseed, linseed, and palm oils) could result in gut bacterial translocation due to the inflammatory responses, high mucosal secretion, and structural changes in the gut membrane that result in lower capacity of firm adhesion of certain bacterial species and their ability to physically cross the gut barrier that finally reduces gut microbiota richness (Torrecillas et al. 2017a). In this context, Torrecillas et al. (2017a) reported that pro-inflammatory cytokines responses and infiltrated submucosa leucocytes increased in the posterior intestine of European seabass fed a diet rich in VO due to translocation of bacteria from the anterior intestine. For example, it has been reported that the inclusion of linseed oil in diets for fish increased the diversity of intestinal microbiota and reduced its richness (Ringo et al. 1998; Torrecillas et al. 2017a; You et al. 2019). In fact, dietary VO enhanced the relative abundance of pathogenic bacteria, such as *Mycoplasma*, *Vibrio*, and *Brevinema*, but it reduced remarkably the relative abundance of beneficial lactic acid bacteria (LAB) such as *Bacillus*, *Streptococcus*, *Lactococcus*, *Lactobacillus*, *Globicateella*, and *Leuconnstocin* in the golden pompano (*Trachinotus ovatus*, Carangidae) (You et al. 2019). In this sense, a significantly positive correlation was found between dietary EPA and DHA with a healthy intestinal bacteria community (e.g. LAB), whereas a profound negative correlation was observed with the diversity of pathogenic bacteria in fish gut that may be related to anti-inflammatory effects of n-3 LC-PUFA (You et al. 2019). On the other hand, some studies have reported a positive correlation between *Vibrio* spp. richness and dietary LNA, as this group of bacteria prefer lipid-rich environments (Sun et al. 2018) and have the ability to use LNA (Morita et al. 1992; Jostensen and Landfald 1997). In this context, a remarkable augmentation in the population of LAB was observed in the gut of fish fed a diet containing ALA or blends of LC-PUFA, whereas a lower frequency of LAB was found in fish fed a diet containing LNA (Mills et al. 2010). However, other studies reported that the diversity indexes of gut microbiota were not significantly affected by FM and FO replacement in European sea bass (Torrecillas et al. 2017b). In general terms, there is considerable evidence that, depending on the FO replacement strategy and the ALS used, FO-sparing can lead to an imbalance between commensal bacteria with potentially pathogenic and healthy intestinal bacteria that may result into dysbiosis and the impairment of the mucosal barrier integrity (Maranduba et al. 2015; Ma et al. 2018b; Yang et al. 2018). In this regard, You et al. (2019) reported that the mRNA expression of zonula occludens-1, which is a tight junction protein in the enterocytes of golden pompano fed a soybean oil diet, significantly decreased and was associated with increased abundance of potentially pathogenic bactria (*Vibrio* and *Mycoplasma*) and reduction of intestinal probiotics (*Bacillus* and *Lactococcus*). These findings suggest that the tight junction barrier function may be impaired and the integrity of the gut and its absorption capacity may be reduced in fish that are fed a diet containing soybean oil. Therefore, dietary FO-sparing in fish may affect general health with a remarkable impact on the gut microbiome through inflammatory responses; however, species-specific studies need to be conducted since some variability exists between different studies conducted under different experimental conditions.

The replacement of FO with VO has been reported in the development of cardiac lesions in Atlantic salmon (Bell et al. 1993; Grisdale-Helland et al. 2002; Seierstad et al. 2005). These cardiac lesions coincided with extensive leucocyte infiltration at the lesion site due to the production of ARA-derived eicosanoids (e.g. thromboxane B_2 and prostaglandin E_2) by leucocytes (Bell et al. 2003). In fact, total replacement of dietary FO with sunflower oil significantly reduced the EPA and n-3/n-6 PUFA ratio, but enhanced LNA and ARA levels in the heart of fish, resulting in the production of pro-inflammatory eicosanoids (Bell et al. 1993). In contrast, Seierstad et al. (2008) reported that total replacement of dietary FO with blends of VO did not show arteriosclerosis, which was associated with an appropriate n-3/3-6 ratio in the diet. Furthermore, the FA profiling of the heart muscle demonstrated that n-3 LC-PUFA, mainly EPA and DHA, were selectively preserved and retained in the heart muscle regardless of their concentrations in the diet (Torstensen et al. 2004; Seierstad et al. 2008).

From the findings of the above-mentioned studies, it can be concluded that ALS by changing the FA composition of tissues and leucocytes can dramatically influence health and welfare of aquatic species.

RISK ASSESSMENT IN THE APPLICATION OF ALS IN AQUAFEEDS

The reliable supply and price of ALS compared to FO are two critical factors that determine the viability and sustainability of their commercial use for aquafeed production (Glencross et al. 2020). In this sense, the aquaculture industry tends to use those ALS that guarantee large supply volumes in comparison to other ALS with limited or experimental production.

On the other hand, besides cost-reduction, another positive outcome of substitution of dietary FO with ALS is the reduction of persistent organic pollutants (POPs) in aquafeeds due to the presence of these contaminants in FO (Berntssen et al. 2010). However, the use of ALS in aquafeeds also can suffer from chemical contamination. Unfavorable chemical pollutants may be carried over unintentionally through environmental pollutants, such as crop pesticides, antimicrobial residues (i.e. residues of antimicrobials in RAF), and industrial processing (van Duijn 2008; Glencross et al. 2020; Roszko et al. 2012). The most common contaminants that may be found in ALS include metals or persistent organic pollutants, heavy metals and radionuclides, natural toxins, such as mycotoxins produced by fungi, which may reduce the nutritional value of ALS and be risky for fish health and/or food safety regarding the consumer. In this sense, before the use of ALS in aquafeeds manufacture, the national and international regulatory standards along with the certain standards measures (e.g. AOAC, EU, UKAS) are required by industrial accreditation organizations. In this regard, the governments in many countries have already established the maximum residue levels (MRL) of contaminants in aquafeed raw materials, including ALS, according to evidence-based risk analyses (EC 2005; Glencross et al. 2007; Alimentarius 2016; APVMA 2018; EPA 2018). As stated before, the concentration of these contaminants below MRL in ALS is of importance for sustaining fish welfare and safe seafood production. In this sense, it is important to consider that there may be some potential sanitary processes for reducing the potential risks of pollutants from ALS into aquafeeds. The most common strategy for eliminating contaminants is the application of a "withdrawal" period before fish harvest (Burridge et al. 2010). In addition, the inclusion of binders and adsorbents to bind and chelate toxic metals and components into raw materials or feed formulation can alleviate the poisonous effects of contaminants (Binder 2007). There are many commercial products for mitigating the hazardous impact of contaminants, such as mycotoxins (Jouany 2007; Zhu et al. 2016). Furthermore, the physiological manipulation of an aquatic species can be considered to increase its metabolic turnover and excretion of contaminants (Kan and Meijer 2007).

Rendered fat from healthy non-ruminants can also be used as ALS in aquafeeds production if the rendered animal by-products are processed under rigorous sanitizing methods (USDA 2005; Council 2011a; EC 2011; APVMA 2015). However, these raw materials may have the potential of transmitting zoonotic diseases such as transmissible microbes (bacteria and viruses) and parasites from rendered animal by-products. Regarding transmissible spongiform encephalitis (TSE), it does not appear tissues of ruminants, such as the muscle and body fat, are affected by TSE. However, Salta et al. (2009) reported that gilthead sea bream fed on the neural tissue from bovine spongiform encephalopathy-infected cattle or scrapie-infected sheep did not show any clinical signs of a prion disease. But two years after feeding the TSE-infected tissue to the fish, the authors detected the signs of neurodegeneration in the brain tissue of those fish that cross-reacted with antibodies raised against endogenous seabream prion proteins. It has been suggested that extracting, purifying, and sterilizing RAF can reduce the threat of transferring ruminant infectious agents into feeds (USFDA 2001; WHO 2003; Woodgate and van der Veen 2014), and these processes also can be applied in aquafeed production (Glencross et al. 2020). In addition, a "withdrawal" period before the slaughter of poultry or ruminants can minimize the potential of inadvertent transfer of antimicrobial compounds into farmed aquatic species and, subsequently, to humans (Alimentarius 2017). Although these alternative sources are safe in terms of fish health and food safety issues, consumers may show some reluctance to consume fish that were fed diets containing rendered terrestrial farmed animals.

Regarding GMO, according to many national regulations (e.g. EC 2001; EC 2009; Council 2011b; FSANZ 2018); "fish that are fed on GMO are not themselves genetically modified and are not required to be labelled as GMO fish". Thus, the use of GMO as ALS in aquafeeds cannot be considered a potential threat to fish health and, subsequently, human consumption.

CONCLUSIONS

According to the findings of a plethora of studies, FO replacement in aquafeeds is possible with the use of various ALS when considering EFA requirements, the capacity of n-3 LC PUFA biosynthesis by the species, and the life stage and culture condition of aquatic species without compromising growth and feed utilization. In addition, the digestibility of ALS and its effects on the ADC of other nutrients in the diet also should be considered in a holistic approach toward enhancing the efficiency of aquafeed. It should be mentioned that the FA profile of fillet in farmed aquatic species extensively reflects the FA composition of the diet and the alternation of the FA profile can pronouncedly affect the sensorial attributes and the refrigerated shelf-life of the fillet. On the other hand, FO-sparing should not be evaluated only on growth and feed efficiency parameters since it can compromise the health and welfare of farmed aquatic species and is related to species-specific requirements of EFA and optimum n-3/n-6 PUFA ratio. In addition, lipid metabolism may also be influenced by FO-sparing and, consequently, affects the health, performance, and nutritional quality of aquaculture species. Thus, by applying nutritional strategies for FO-sparing in feed formulation by considering fish species, its EFA requirements, and its culture condition, the sustainability of aquafeed production will be guaranteed.

REFERENCES

Abbasi, A., A. Oujifard, M. T. Mozanzadeh, H. Habibi, and M. Nafisi-Bahabadi. 2020. Dietary simultaneous replacement of fish meal and fish oil with blends of plant proteins and vegetable oils in yellowfin seabream (*Acanthopagrus latus*) fry: Growth, digestive enzymes, antioxidant status and skin mucosal immunity. *Aquaculture Nutrition* 26:1131–1142. doi:10.1111/anu.13070

Agh, N., M. T. Mozanzadeh, F. Jafari, F. Noori, and R. Jalili. 2020. The influence of dietary fish oil replacement with mixture of vegetable oils on reproductive performance, immune responses and dynamic of fatty acids during embryogenesis in *Oncorhynchus mykiss*. *Aquaculture Research* 51:918–931. doi:10.1111/are.14437

Alhazzaa, R., A. R. Bridle, P. D. Nichols, and C. G. Carter. 2011. Up-regulated desaturase and elongase gene expression promoted accumulation of polyunsaturated fatty acid (PUFA) but not long-chain PUFA in Lates calcarifer, a tropical euryhaline fish fed a stearidonic- and c-linoleic acid enriched diet. *Journal of Agricultural and Food Chemistry* 59 (15):8423–8434.

Alhazzaa, R., A. R. Bridle, C. G. Carter, and P. D. Nichols. 2012. Sesamin modulation of lipid class and fatty acid profile in early juvenile teleost, *Lates calcarifer*, fed different dietary oils. *Food Chemistry* 134 (4):2057–2065.

Alhazzaa, R., A. R. Bridle, T. A. Mori, A. E. Barden, P. D. Nichols, and C. G. Carter. 2013. Echium oil is better than rapeseed oil in improving the response of barramundi to a disease challenge. *Food Chemistry* 141:1424–1432.

Alhazzaa, R., and K. Sase. 2016. Citrus flavonoids and hepatic n-3 LCPUFA biosynthesis. *Journal of Nutrition & Intermediary Metabolism* 4:40.

Alhazzaa, R., P. D. Nichols, and C. G. Carter. 2019. Sustainable alternatives to dietary fish oil in tropical fish aquaculture. *Reviews in Aquaculture* 11:1195–1218.

Alimentarius, C. 2016. *Codex Pesticides Residues in Food Online Database*. FAO/WHO, Rome.

Alimentarius, C. 2017. Maximum Residue Limits (MRLs) and risk management recommendations (RMRs) for residues of veterinary drugs in foods. CAC/MRL 2-2017. Updated as at the *40th Session of the Codex Alimentarius Commission* (July 2017).

Allan, G. L., S. Parkinson, M. A. Booth, D. A. J. Stone, J. R. Stuart, J. Frances, and R. Warner-Smith. 2000. Replacement of fish meal in diets for Australian silver perch, *Bidyanus bidyanus*: I. Digestibility of alternative ingredients. *Aquaculture* 186:293–310.

Almaida-Pagán, P. F., M. D. Hernández, B. García García, J. A. Madrid, J. D. Costa-Ruiz, and P. Mendiola. 2007. Effects of total replacement of fish oil by vegetable oils on n–3 and n–6 polyunsaturated fatty acid desaturation and elongation in sharpsnout seabream (*Diplodus puntazzo*) hepatocytes and enterocytes. *Aquaculture* 272:589–598.

Alvarez, M. J., A. Diez, C. Lopez-Bote, M. Gallego, and J. M. Bautista. 2000. Short-term modulation of lipogenesis by macronutrients in rainbow trout (*Onchorhynchus mykiss*) hepatocytes. *British Journal of Nutrition* 84:619–628.

Aminikhoei, Z., J. Choi, S. M. Lee, and S. H. Cho. 2014. Impacts of different dietary lipid sources on growth performance, fatty acid composition and antioxidant enzyme activity of juvenile Black Sea bream, *Acanthopagrus schlegeli*. *Iranian Journal of Fisheries Sciences* 13 (4):796–809.

Apraku, A., L. Liu, X. Leng, E. J. Rupia, and C. L. Ayisi. 2017. Evaluation of blended virgin coconut oil and fish oil on growth performance and resisteance to *Streptococcus iniae* challenge of nile tilapia (*Oreochromis niloticus*). *Egyptian Journal of Basic and Applied Science* 4:157–184.

Apraku, A., X. Huang, and C. L. Ayisi. 2019. Effects of alternative dietary oils on immune response, expression of immune-related genes and disease resistance in juvenile Nile tilapia, *Oreochromis niloticus*. *Aquaculture Nutrition* 25:597–608.

APVMA. 2015. *Agricultural and Veterinary Chemicals Legislation Amendment (Animal Feed Reform and Other Measures) Regulation 2015. Federal Register of Legislation, Rule number: 5*. Australian Pesticides and Veterinary Medicines Authority, Canberra, ACT.

APVMA. 2018. *Agricultural and Veterinary Chemicals Code Instrument No. 4 (MRL Standard) 2012. Federal Register of Legislation, Rule number: 63*. Australian Pesticides and Veterinary Medicines Authority, Canberra, ACT.

Araujo, B., R. M. Honji, A. N. Rombenso, G. Brambila, P. Mello, A. W. Hilsdorf, and R. Moreira. 2019. Influence of different arachidonic acid levels and temperature in growth performance, fatty acid profile, liver morphology and expression of lipid genes in cobia (*Rachycentron canadum*) juvenile. *Aquaculture* 511:734245.

Argyropoulou, V., N. Kalogeropoulos, and M. N. Alexis. 1992. Effect of dietary lipids on growth and tissue fatty acid composition of grey mullet (*Mugil cephalus*). *Comparative Biochemestiry Physiollogy Part A: Physiology* 101:129–135.

Arslan, M., N. Sirkecioglu, A. Bayir, H. Arslan, and M. Aras. 2012. The influence of substitution of dietary fish oil with different vegetable oils on performance and fatty acid composition of brown trout, *Salmo trutta*. *Turkish Journal of Fisheries and Aquatic Sciences* 12 (3):575–583.

Artamonova, E. Y., J. B. Svenning, T. Vasskog, E. Hansen, and H. C. Eilertsen. 2017. Analysis of phospholipids and neutral lipids in three common northern cold water diatoms: *Coscinodiscus concinnus*, *Porosira glacialis*, and *Chaetoceros socialis*, by ultrahigh performance liquid chromatography-mass spectrometry. *Journal of Applied Phycology* 29:1241–1249.

Arts, M. T., and C. C. Kohler. 2008. Health and condition in fish: The influence of lipids on membrane competency and immune response. In *Lipids in Aquatic Ecosystems*, edited by M. T. Arts, M. T. Brett and M. J. Kainz. Springer, London.

Arzel, J., F. X. M. Lopez, R. Métailler, G. Stéphan, M. Viau, G. Gandemer, and J. Guillaume. 1994. Effect of dietary lipid on growth performance and body composition of brown trout (*Salmo trutta*) reared in seawater. *Aquaculture* 123 (3–4):361–375.

Bahurmiz, O. M., and W.-K. Ng. 2007a. Effects of dietary palm oil source on growth, tissue fatty acid composition and nutrient digestibility of red hybrid tilapia, *Oreochromis* sp., raised from stocking to marketable size. *Aquaculture* 262 (2–4):382–392.

Bahurmiz, O. M., and W. K. Ng. 2007b. Effects of dietary palm oil source on growth, tissue fatty acid composition and nutrient digestibility of red hybrid tilapia, *Oreochromis* sp., raised from stocking to marketable size. *Aquaculture* 262:382–392.

Bai, S. C., and D. M. Gatlin. 1993. Dietary vitamin E concentration and duration of feeding affect tissue a-tocopherol concentrations of channel catfish (*Ictalurus punctatus*). *Aquaculture* 113:129–135.

Bandarra, N. M., P. Rema, I. Batista, P. Pousao-Ferreira, L. M. P. Valente, S. M. G. Batista, and R. O. A. Ozorio. 2011. Effects of dietary n-3/n-6 ratio on lipid metabolism of gilthead sea bream (*Sparus aurata*). *European Journal of Lipid Science and Technology* 113:1332–1341.

Baoshan, L., W. Jiying, H. Yu, H. Tiantian, W. Shixin, H. BingShan, and S. Yongzhi. 2019. Effects of replacing fish oil with wheat germ oil on growth, fat deposition, serum biochemical indices and lipid metabolic enzyme of juvenile hybrid grouper (*Epinephelus fuscoguttatus Epinephelus lanceolatus*). *Aquaculture* 505:54–62.

Bayraktar, K., and A. Bayır. 2012. The effect of the replacement of fish oil with animal fats on the growth performance, survival and fatty acid profile of rainbow trout juveniles, *Oncorhynchus mykiss*. *Turkish Journal of Fisheries and Aquatic Sciences* 12 (3):661–666.

Beheshti, F., M. C. C. Parrish, J. Wells, R. G. Taylor, M. L. Rise and F. Shahidi. 2018. Minimizing marine ingredients in diets of farmed Atlantic salmon (*Salmo salar*): Effects on growth performance and muscle lipid and fatty acid composition. *Plos One* 13 (9):e0198538. doi:10.1371/journal.pone.0198538

Belforti, M., F. Gai, C. Lussiana, M. Renna, V. Malfatto, L. Rotolo, et al. 2016. Tenebrio molitor meal in rainbow trout (*Oncorhynchus mykiss*) diets: Effects on animal performance, nutrient digestibility and chemical composition of fillets. Italian Journal of Animal. *Italian Journal of Animal Science* 14:41–70.

Belghit, I., N. S. Liland, R. Waagbø, I. Biancarosa, N. Pelusio, Y. Li, A. Krogdahl, and E.-J. Lock. 2018. Potential of insect-based diets for Atlantic salmon (*Salmo salar*). *Aquaculture* 491:72–81.

Belghit, I., R. Waagbø, E. J. Lock, and N. S. Liland. 2019. Insect-based diets high in lauric acid reduce liver lipids in freshwater Atlantic salmon. *Aquaculture Nutrition* 25:343–357.

Belitz, H. D., and W. Grosch. 1999. *Food Chemistry*. Springer-Verlag, Berlin, Germany.

Bell, J. G., J. R. Dick, A. H. McVicar, J. D. Sargent, and K. Thompson. 1993. Dietary sunflower, linseed and fish oils affect phospholipid fatty acid composition, development of cardiac lesions, phospholipase activity and eicosanoid production in Atlantic salmon (*Salmo salar*). *Prostaglandins Leukot Essent Fatty Acids* 49:665–673

Bell, J. G., D. R. Tocher, and J. R. Sargent. 1994. Effect of supplementation with 20:3(n-6), 20:4(n-6) and 20:5(n-3) on the production of prostaglandins E and F of the 1-,2-and 3-series in turbot (Scophthalmus maximus) brian astroglial cells in primary culture. *BBA-Lipid Lipid Metabolism* 1211:335–342.

Bell, J. G., I. Ashton, C. J. Secombes, B. R. Weitzel, J. R. Dick, and J. R. Sargent. 1996. Dietary lipid affects phospholipid fatty acid compositions, eicosanoid production and immune function in Atlantic salmon (*Salmo salar*). *Prostaglandins, Leukotrienes and Essential Fatty Acids* 54 (3):173–182.

Bell, J. G., D. R. Tocher, B. M. Farndale, D. I. Cox, R. W. McKinney, and J. R. Sargent. 1997. The effect of dietary lipid on polyunsaturated fatty acid metabolism in Atlantic salmon (*Salmo salar*) undergoing parr-smolt transformation. *Lipids* 32 (5):515–525.

Bell, J. G., J. McEvoy, D. R. Tocher, F. McGhee, and P. J. Campbell. 2001. Replacement of fish oil with rapeseed oil in diets of Atlantic salmon (Salmo salar) affects tissue lipid composition and hepatocyte fatty acid metabolism. *Journal of Nutrition* 131:1535–1543.

Bell, J. G., R. J. Henderson, D. R. Tocher, F. McGhee, J. R. Dick, A. Porter, et al. 2002. Substituting fish oil with crude palm oil in the diet of Atlantic salmon (*Salmo salar*) affects muscle fatty acid composition and hepatic fatty acid metabolism. *Journal of Nutrition* 132:222–230.

Bell, J. G., F. McGhee, P. J. Campbell, and J. R. Sargent. 2003a. Rapeseed oil as an alternative to marine fish oil in diets of post-smolt Atlantic salmon (*Salmo salar*): Changes in flesh fatty acid composition and effectiveness of subsequent fish oil "wash out". *Aquaculture* 218:515–528.

Bell, J. G., D. R. Tocher, R. J. Henderson, J. R. Dick, and V. O. Crampton. 2003b. Altered fatty acid compositions in Atlantic salmon (*Salmo salar*) fed diets containing linseed and rapeseed oils can be partially restored by a subsequent fish oil finishing diet. *The Journal of Nutrition* 133 (9):2793–2801.

Bell, J. G., R. J. Henderson, D. R. Tocher, and J. R. Sargent. 2004. Replacement of dietary fish oil with increasing levels of linseed oil: Modification of flesh fatty acid compositions in Atlantic salmon (*Salmo salar*) using a fish oil finishing diet. *Lipids* 39 (3):223–232.

Bell, J. G., F. McGhee, J. R. Dick, and D. R. Tocher. 2005. Dioxin and dioxin-like polychlorinated biphenyls (PCBs) in Scottish farmed salmon (*Salmo salar*): Effects of replacement of dietary marine fish oil with vegetable oils. *Aquaculture* 243 (1–4):305–314.

Bell, M. V., and J. R. Dick. 2004. Changes in capacity to synthesise 22:6n-3 during development in rainbow trout (*Oncorhynchus mykiss*). *Aquaculture* 235:393–409.

Benedito-Palos, L., J. C. Navarro, A. Sitja-Bobadilla, J. G. Bell, S. Kaushik, and J. Perez-Sanchez. 2008. High levels of vegetable oils in plant protein-rich diets fed to gilthead seabream (*Sparus aurata* L.): Growth performance, muscle fatty acid profiles and histological alterations of target tissues. *British Journal of Nutrition* 100:992–1003.

Benedito-Palos, L., J. C. Navarro, A. Bermejo-Nogales, and A. Saera-Vila. 2009. The time course of fish oil wash-out follows a simple dilution model in gilthead seabream (*Sparus aurata* L.) fed graded levels of vegetable oils. *Aquaculture* 288:98–105.

Benedito-Palos, L., G. F. Ballester-Lozano, P. Simo, V. Karalazos, A. Ortiz, J. Calduch-Giner, and J. Perez-Sanchez. 2016. Lasting effects of butyrate and low FM/FO diets on growth performance, blood haematology/biochemistry and molecular growth-related markers in gilthead sea bream (*Sparus aurata*). *Aquaculture* 454:8–18.

Berge, G. M., B. Hatlen, J. M. Odom, and B. Ruyter. 2013. Physical treatment of high EPA Yarrowia lipolytica biomass increases the availability of n–3 highly unsaturated fatty acids when fed to Atlantic salmon. *Aquaculture Nutrition* 19:11–121.

Berntssen, M. H. G., K. Julshamn, and A. K. Lundebye. 2010. Chemical contaminants in aquafeed and Atlantic salmon (*Salmo salar*) following the use of traditional versus alternative feed ingredients. *Chemosphere* 78:637–646.

Betancor, M. B., M. Sprague, O. Sayanova, S. Usher, P. J. Campbell, J. A. Napier, M. J. Caballero, and D. R. Tocher. 2015. Evaluation of a high-EPA oil from transgenic *Camelina sativa* in feeds for Atlantic salmon (*Salmo salar* L.): Effects on tissue fatty acid composition, histology and gene expression. *Aquaculture* 444:1–12.

Betancor, M. B., M. Sprague, O. Sayanova, S. Usher, C. Metochis, P. J. Campbell, J. A. Napier, and D. R. Tocher. 2016. Nutritional evaluation of an EPA-DHA oil from transgenic *Camelina sativa* in feeds for post-smolt Atlantic salmon (*Salmo salar* L.). *Plos One* 11 (7):e0159934. doi:10.1371/journal.pone.0159934

Betancor, M. B., K. Li, M. Sprague, T. Bardal, O. Sayanova, S. Usher, et al. 2017. An oil containing EPA and DHA from transgenic *Camelina sativa* to replace marine fish oil in feeds for Atlantic salmon (*Salmo salar* L.): Effects on intestinal transcriptome, histology, tissue fatty acid profiles and plasma biochemistry. *PLoS one* 12:e0175415.

Betiku, O. C., F. T. Barrows, C. Ross, and W. M. Sealey. 2016. The effect of total replacement of fish oil with DHA-Gold and plant oils on growth and fillet quality of rainbow trout (*Oncorhynchus mykiss*) fed a plant-based diet. *Aquaculture Nutrition 22*:158–169.

Bimbo, A. P. 2000. Fishmeal and oil. In *Marine & Freshwater Products Handbook*, edited by R. E. Martin, E. P. Carter, G. J. Flick and L. M. Davis. Technomic Publishing Co., Lancaster, UK, pp. 541–581.

Binder, E. M. 2007. Managing the risk of mycotoxins in modern feed production *Animal Feed Science and Technology* 133:149–166.

Bjerkeng, B., B. Hatlen and E. Wathne. 1999. Deposition of astaxanthin in fillets of Atlantic salmon (*Salmo salar*) fed diets with herring, capelin, sandeel, or Peruvian high PUFA oils. *Aquaculture* 180:307–319.

Blanchard, G., J. G. Makombu, and P. Kestemont. 2008. Influence of different dietary 18: 3n-3/18: 2n-6 ratio on growth performance, fatty acid composition and hepatic ultrastructure in Eurasian perch, *Perca fluviatilis*. *Aquaculture* 284 (1–4):144–150.

Bonacic, K., A. Martinez, E. Gisbert, A. Estevez, and S. Morais. 2017. Effect of alternative oil sources at different dietary inclusion levels on food intake and appetite regulation via enteroendocrine and central factors in juvenile *Solea sengalensis* (Kaup, 1858). *Aquaculture* 470:169–181.

Borlongan, T. G. 1992. The essential fatty acid requirement of milkfish (*Chanos chanos* Forsskal). *Fish Physiology and Biochemistry* 9 (5/6):401–407.

Bou, M., G. M. Berge, G. Baeverfjord, T. Sigholt, T. K. Østbye, O. H. Romarheim, B. Hatlen, R. Leeuwis, C. Venegas, and B. Ruyter. 2017. Requirements of n-3 very long-chain PUFA in Atlantic salmon (*Salmo salar* L): Effects of different dietary levels of EPA and DHA on fish performance and tissue composition and integrity. *British Journal of Nutrition* 117:30–47.

Bouraoui, L., J. Sánchez-Gurmaches, L. Cruz-García, J. Gutiérrez, L. Benedito-Palos, J. Pérez-Sánchez, and I. Navarro. 2011. Effect of dietary fish meal and fish oil replacement on lipogenic and lipoprotein lipase activities and plasma insulin in gilthead sea bream (*Sparus aurata* L.). *Aquaculture Nutrition* 17:54–63.

Bowyer, J., J. G. Qin, L. Adams, M. Thomson, and D. Stone. 2012a. The response of digestive enzyme activities and gut histology in yellowtail kingfish (*Seriola lalandi*) to dietary fish oil substitution at different temperatures. *Aquaculture* 396:19–28.

Bowyer, J. N., J. G. Qin, R. P. Smullen, and D. A. J. Stone. 2012b. Replacement of fish oil by poultry oil and canola oil in yellowtail kingfish (*Seriola lalandi*) at optimal and suboptimal temperature. *Aquaculture* 356:211–222.

Bowzer, J., C. Jackson, and J. Trushenski. 2016. Hybrid striped bass feeds based on fish oil, beef tallow, and EPA/DHA supplements: Insight regarding fish oil sparing and demand for n-3 long-chain polyunsaturated fatty acids. *Journal of Animal Science* 94:978–988.

Bransden, M. P., C. G. Carter, and P. D. Nichols. 2003. Replacement of fish oil with sunflower oil in feeds for Atlantic salmon (*Salmo salar* L.): Effect on growth performance, tissue fatty acid composition and disease resistance. *Comparative Biochemistry and Physiology Part B* 135:611–625.

Brufau, G., M. A. Canela, and M. Rafecas. 2008. Phytosterols: Physiologic and metabolic aspects related to cholesterol-lowering properties. *Nutrition Research* 28:217–225.

Bureau, D. P. 1997. *Nutritive value of megalac, seal oil and beef tallow for rainbow trout (Oncorhynchus mykiss).* Report submitted to Church and Dwight Co., Princeton, NJ.

Bureau, D. P., J. Gibson, and A. El-Mowafi. 2002. Use of animal fats in aquaculture feeds. In *Avances en nutricion acuicola VI. Memorias del VI simposium internacional de nutricion acuicola 3*, 6th September 2002, Cancún, Quintana Roo, México.

Bureau, D. P., K. Hua, and A. M. Harris. 2008. The effect of dietary lipid and long-chain n-3 PUFA levels on growth, energy utilization, carcass quality, and immune function of rainbow trout, *Oncorhynchus mykiss*. *Journal of the World Aquaculture Society* 39 (1):1–21.

Bureau, D. P., and D. L. Meeker. 2011. *Terrestrial Animal Fats. Fish Oil Replacement and Alternative Lipid Sources in Aquaculture Feeds.* CRC Press, Boca Raton, FL, 245–266.

Burja, A. M., H. Radianingtyas, A. Windust, and C. J. Barrow. 2006. Isolation and characterization of polyunsaturated fatty acid producing *Thraustochytrium* species: Screening of strains and optimization of omega-3 production. *Applied Microbiology and Biotechnology* 72:1161–1169.

Burridge, L., J. S. Weis, F. Cabello, J. Pizarro, and K. Bostick. 2010. Chemical use in salmon aquaculture: A review of current practices and possible environmental effects. *Aquaculture* 306:7–23.

Buzzi, M., R. J. Henderson, and J. R. Sargent. 1996. The desaturation and elongation of linolenic acid and eicosapentaenoic acid by hepatocytes and liver microsomes from rainbow trout (*Oncorhynchus mykiss*) fed diets containing fish oil or olive oil. *Biochim Biophys Acta* 1299:235–244.

Caballero, M. J., A. Obach, G. Roselund, D. Montero, M. Gisvold, and M. S. Izquierdo. 2002. Impact of different dietary lipid sources on growth, lipid digestibility, tissue fatty acid composition and histology of rainbow trout, *Oncorhynchus mykiss*. *Aquaculture* 214:253–271.

Caballero, M. J., M. S. Izquierdo, E. Kjørsvik, A. J. Fernndez, and G. Rosenlund. 2004. Histological alterations in the liver of sea bream, *Sparus aurata* L., caused by short- or long-term feeding with vegetable oils. Recovery of normal morphology after feeding fish oil as the sole lipid source. *Journal of Fish Diseases* 27:531–541.

Campos, I., E. Matos, and L. Valente. 2016. Fatty acids apparent digestibility of differente animal fat sources in European sea bass (*Dicentrarchus labrax*). In *Paper presented at the Aquaculture Europe 16.* Edinburgh, Scotland.

Campos, I., E. Matos, M. R. G. Maia, A. Marques, and L. M. P. Valente. 2019. Partial and total replacement of fish oil by poultry fat in diets for European sea bass (*Dicentrarchus labrax*) juveniles: Effects on nutrient utilization, growth performance, tissue composition and lipid metablism. *Aquaculture* 502:107–120.

Caro, M., A. Sansone, J. Amézaga, V. Navarro, C. Ferreri, and I. Tueros. 2017. Wine lees modulate lipid metabolism and induce fatty acid remodelling in Zebrafish. *Food and Function* 8:1652–1659.

Carvalho, M., H. Peres, R. Saleh, R. Fontanillas, G. Rosenlund, A. Oliva-Teles, and M. Izquierdo. 2018. Dietary requirement for n-3 long-chain polyunsaturated fatty acids for fast growth of meagre (*Argyrosomus regius*, Asso 1801) fingerlings. *Aquaculture* 488:105–113.

Castell, J., J. G. Bell, D. R. Tocher, and J. R. Sargent. 1994. Effects of purified diets containig different combinations of arachidonic and docosahexaenoic acid on survival, growth and fatty acid composition of juvenile turbot (*Scophthalmus maximus*). *Aquaculture* 128:315–333.

Castro, C., G. Corraze, S. Panserat, and A. O. Teles. 2014. Effects of fish oil replacement by a vegetable oil blend on digestibility, postprandial serum metabolite profile, lipid and glucose metabolism of European sea bass (*Dicentrarchus labrax*) juveniles. *Aquaculture Nutrition* 21:592–603.

Castro, C., A. F. Diógenes, F. Coutinho, S. Panserat, G. Corraze, A. Pérez-Jiménez, H. Peres, and A. Oliva-Teles. 2016. Liver and intestine oxidative status of gilthead sea bream fed vegetable oil and carbohydrate rich diets. *Aquaculture* 464:665–672.

Castro, C., A. Peréz-Jiménez, F. Coutinho, P. Díaz-Rosales, C. A. dos Reis Serra, S. Panserat, H. Peres, and A. Oliva-Teles. 2015. Dietary carbohydrate and lipid sources affect differently the oxidative status of European sea bass (*Dicentrarchus labrax*) juveniles. *British Journal of Nutrition* 114:1584–1593.

Chen, C. Y., B. L. Sun, X. X. Li, P. Y. Li, W. T. Guan, Y. Z. Bi, and Q. Pan. 2013. N-3 essential fatty acids in Nile tilapia, *Oreochromis niloticus*: Quantification of optimum requirement of dietary linolenic acid in juvenile fish. *Aquaculture* 416–417:99–104.

Chen, Y., Z. Sun, Z. Liang, Y. Xie, J. Su, Q. Luo, J. Zhu, Q. Liu, T. Han, and A. Wang. 2020. Effects of dietary fish oil replacement by soybean oil and L-carnitine supplementation on growth performance, fatty acid composition, lipid metabolism and liver health of juvenile largemouth bass, *Micropterus salmoides. Aquaculture* 516:734596.

Cho, C. Y., and S. J. Kaushik. 1990. Nutritional energetics in fish: Energy and protein utilization in rainbow trout (*Salmo gairdneri*). *World Review of Nutrition and Dietetics* 61:132–172.

Chou, B.-S., and S.-Y. Shiau. 1999. Both n-6 and n-3 fatty acids are required for maximal growth of juvenile hybrid tilapia. *North American Journal of Aquaculture* 61 (1):13–20.

Choubert, G., M. M. Mendes-Pinto, and R. Morais. 2006. Pigmenting efficacy of astaxanthin fed to rainbow trout *Oncorhynchus mykiss*: Effect of dietary astaxanthin and lipid sources. *Aquaculture* 257 (1–4):429–436.

Christie, W. W. 2003. *Lipid Analysis*. Oily Press, Bridgwater, pp 1–289.

Corl, B. A., S. A. M. Oliver, X. Lin, W. T. Oliver, Y. Ma, R. J. Harrell, et al. 2008. Conjugated linoleic acid reduces body fat accretion and lipogenic gene expression in, neonatal pigs fed low- or high-fat formulas. *Journal of Nutrition* 138:449–454.

Council, C. S. 2011a. *Regulation on the Administration of Feed and Feed Additives* 《饲料和饲料添加剂管理条例》. *State Council Gazette, Issue Number 33, Serial No. 1392, Rule Number: 609*. People's Republic of China State Council, Beijing.

Council, C. S. 2011b. *Regulations on Administration of Agricultural Genetically Modified Organisms Safety* (2011 Revision) 《农业转基因生物安全管理条例 (2011修订)》. State Council Gazette Supplement, *Rule number: 304*. People's Republic of China State Council, Beijing.

Craig, S. R., and I. D. M. Gatlin. 1995. Coconut and Beef tallow, but not tricaprylin, can replace menhaden oil in the diet of red drum (*Sicaenops ocellatus*) without Adversely affecting growth or fatty acid composition. *Journal of Nutrition*:3041–3048.

Cravedi, J. P., G. Choubert, and G. Delous. 1987. Digestibility of chloraphenicol, oxolinic acid and oxytetracycline in rainbow trout and influence of these antibiotics on lipid digestibility. *Aquaculture* 60:133–141.

Crockett, E. L., and B. D. Sidell. 1993a. Peroxisomal b-oxidation is a significant pathway for catabolism of fatty-acids in a marine teleost. *American Journal of Physiology* 264:R1004–R1009.

Crockett, E. L., and B. D. Sidell. 1993b. Substrate selectivities differ for hepatic mitochondrial and peroxisomal b-oxidation in an Antarctic fish, Notothenia gibberifrons. *Biochemical Journal* 289:427–433.

D'Abramo, L. R., and S. S. Sheen. 1993. Polyunsaturated fatty acid nutrition in juvenile freshwater prawn *Macrobrachium rosenbergii*. *Aquaculture* 115:63–86.

Dalbir, S. P., G. Roopma, K. Ritu, G. Vaini, and R. Shivalika. 2015. Effect of fsh oil substitution with sunflower oil in diet of juvenile *Catla catla* (Ham) on growth performance and feed utilization. *Journal of Fisheries Livestock Production* 3:144. doi:110.4172/2332-2608.1000144

De Silva, S. S., and T. A. Anderson. 1995. *Fish Nutrition in Aquaculture*. Chapman and Hall, London, 319 pp.

Deering, M. J., D. R. Fielder, and D. R. Hewitt. 1997. Growth and fatty acid composition of juvenile leader prawns, *Penaeus monodon*, fed different lipids. *Aquaculture* 151 (1–4):131–141.

Dernekbaşi, D., I. Karayücel, and A. Öksüz. 2011. Effect of dietary canola oil level on fatty acid composition of rainbow trout (*Oncorhynchus mykiss* L.). *Israeli Journal of Aquaculture*. 63:535–546.

Deshimaru, O., K. Kuroki, and Y. Yone. 1982. Nutritive value of various oils for yellowtail. *Bulletin of the Japanese Society of Scientific Fisheries* 48:1155–1157.

Diaz-Lopéz, M., M. J. Perez, N. G. Acosta, et al. 2009. Effect of dietary substitution of fish oil by Echium oil on growth, plasma parameters and body lipid composition in gilthead seabream (*Sparus aurata* L.) *Aquaculture Nutrition* 15:500–512.

Diez, A., D. Menoyo, S. Perez-Benavente, J. A. Calduch-Giner, S. V. R. de Celis, A. Obach, et al. 2007. Conjugated linoleic acid affects lipid composition, metabolism, and gene expression in Gilthead sea bream (*Sparus aurata* L). *Journal of Nutrition* 137:1363–1369.

Dineshbabu, G., G. Goswami, R. Kumar, A. Sinha, and D. Das. 2019. Microalgae–nutritious, sustainable aqua- and animal feed source. *Journal of Functional Foods* 62:103545.

Dosanjh, B. S., D. A. Higgs, M. D. Plotnikoff, J. R. Markert, and J. T. Buckley. 1988. Preliminary evaluation of canola oil, pork lard and marine lipid singly and in combination as supplemental dietary lipid sources for juvenile fall chinook salmon (*Oncorhynchus tshawytscha*). *Aquaculture* 68 (4):325–343.

Dosanjh, B. S., D. A. Higgs, D. J. McKenzie, D. J. Randall, J. G. Eales, N. Rowshandeli, M. Rowshandeli, and G. Deacon. 1998. Influence of dietary blends of menhaden oil and canola oil on growth, muscle lipid composition, and thyroidal status of Atlantic salmon (*Salmo salar*) in sea water. *Fish Physiology and Biochemistry* 19 (2):123–134.

Du, Z. Y., P. Clouet, L. M. Huang, P. Degrace, W. H. Zheng, J. G. He, L. X. Tian, and Y. J. Liu. 2008. Utilization of different dietary lipid sources at high level in herbivorous grass carp (*Ctenopharyngodon idella*): Mechanism related to hepatic fatty acid oxidation. *Aquaculture Nutrition* 14 (1):77–92.

Duan, Q., K. Mai, J. Shentu, Q. Ai, H. Zhong, Y. Jiang, L. Zhang, C. Zhang, and S. Guo. 2014. Replacement of dietary fish oil with vegetable oils improves the growth and flesh quality of large yellow croaker (*Larmichthys crocea*). *Journal of Ocean University of China (Oceanic and Coastal Sea Research)* 13 (3):445–452.

Dumolt, J. H., and T. C. Rideout. 2017. The lipid-lowering effects and associated mechanisms of dietary phytosterol supplementation. *Current Pharmaceutical Design* 23 (34):5077–5085.

(EC), E. C. 2001. Directive 2001/18/EC of the European Parliament and of the Council of 12 March 2001 on the deliberate release into the environment of genetically modified organisms and repealing Council Directive 90/220/EEC. *Official Journal of the European Union*, Rule number: Directive 2001/18/EC. European Commission, Brussels.

(EC), E. C. 2005. Commission Regulation (EC) 396/2005 of February 2005 on maximum residue levels of pesticides in or on food and feed of plant and animal origin and amending Council Directive 91/414/EEC. Available from: https://eur-lex.europa.eu/legal-content/EN/TXT/PDF/?uri=CELEX:02005R0396-20160513.

EC. 2009. Directive 2009/41/EC of the European Parliament and of the Council of 6 May 2009 on the contained use of genetically modifi ed micro-organisms. *Official Journal of the European Union*, Rule number: 2009/41/EC. European Commission, Brussels.

EC. 2011. Commission Regulation (EU) No 142/2011 of 25 February 2011 implementing Regulation (EC) No 1069/2009 of the European Parliament and of the Council laying down health rules as regards animal by-products and derived products not intended for human consumption and implementing Council Directive 97/78/EC as regards certain samples and items exempt from veterinary checks at the border under that Directive Text with EEA relevance. *Official Journal of the European Union*, Rule: 142/2011. European Commission, Brussels.

El-Kerdawy, A., and M. Salama. 1997. Effect of dietary lipid sources on the growth and fatty acid composition of gilthead seabream (*Sparus aurata*). In *Proc. Workshop "Feeding Tomorrow's Fish"*, edited by A. Tacon and B. Basurco. Cahier Options Méditerranéennes, Mazarron, 235–242.

Elmore, J. S., D. S. Mottram, M. Enser, and J. D. Wood. 1999. Effect of the polyunsaturated fatty acid composition of beef muscle on the profile of aroma volatiles. *Journal of Agriculture and Food Chemistry* 47:1619–1625.

Emery, J. A., R. P. Smullen, and G. M. Turchini. 2014. Tallow in Atlantic salmon feed. *Aquaculture* 422–423:98–108.

Emery, J. A., J. Smullen, R. S. J. Keast, and G. M. Turchini. 2016. Viability of tallow inclusion in Atlantic salmon diet, as assessed by an on-farm grow out trial. *Aquaculture* 451:289–297.

EPA. 2018. *Part 40, Section 180 – Tolerances and Exemptions for Pesticide Chemical Residues in Food.* United States Environmental Protection Agency, Washington, DC.

Eroldoğan, T. O., A. H. Yilmaz, G. M. Turchini, M. Arslan, A. N. Sirkecioğlu, K. Engin, I. Ozsahinoglu, and P. Mumogullarinda. 2013. Fatty acid metabolismin in European sea bass (*Dicentrarchus labrax*): Effects of n-6 PUFA and MUFA in fish oil replaced diets. *Fish Physiology and Biochemistry* 39:941–955.

FAO. 2018. *The state of world fisheries and aquaculture 2018.* Food and Agricultural Organization of the United Nations, Rome.

FAO. 2020. *Fishery and Aquaculture Statistics. Global Aquaculture Production 1950–2018 (FishstatJ).* FAO Fisheries and Aquaculture Department, Rome. Updated 2020. www.fao.org/fishery/statistics/software/fishstatj/en

Farndale, B. M., J. G. Bell, M. P. Bruce, N. R. Bromage, F. Oyen, S. Zanuy, et al. 1999. Dietary lipid composition affects blood leucocyte fatty acid compositions and plasma eicosanoid concentrations in European sea bass (*Dicentrarchus labrax*). *Aquaculture* 179:335–350.

Ferraz, R. B., N. Kabeya, M. Lopes-Marques, A. M. Machado, R. A. Ribeiro, A. L. Salaro, R. Ozorio, L. F. C. Castro, and O. Monroig. 2019. A complete enzymatic capacity for long-chain polyunsaturated fatty acid biosynthesis is present in the Amazonia teleost tambaqui, *Colossoma macropomum. Comparative Biochemistry and Physiology, Part B* 227:90–97.

Finco, A. M. O., L. D. G. Mamani, J. C. Carvalho, G. V. de Melo Pereira, V. Thomaz-Soccol, and C. R. Soccol. 2017. Technological trends and market perspectives for production of microbial oils rich in omega-3. *Critical Reviews in Biotechnology* 37:656–671.

Fonseca-Madrigal, J., V. Karalazos, P. Campbell, J. G. Bell, and D. R. Tocher. 2005. Influence of dietary palm oil on growth, tissue fatty acid compositions, and fatty acid metabolism in liver and intestine in rainbow trout (*Oncorhynchus mykiss*). *Aquaculture Nutrition* 11:241–250.

Fountoulaki, E., A. Vasilaki, A. Hurtado, K. Girigorakis, I. Karacostas, I. Nengas, G. Rigos, Y. Kotzamanis, B. Venou, and M. N. Alexis. 2009. Fish oil substitution by vegetable oils in commercial diets for gilthead sea bream (*Sparus aurata* L.): Effects on growth performance, flesh quality and fillet fatty acid profile recovery of fatty acid profiles by a fish oil finishing diet under flucctuating water temperatures. *Aquaculture* 289:317–326.

Fox, J. M., and P. V. Zimba. 2018. Minerals and trace elements in microalgae. In Levine, I. A., and Fleurence, J. (Eds.), *Microalgae in Health and Disease Prevention* (1st ed.). Academic Press, Cambridge, MA, pp. 177–193.

Fracalossi, D. M., and R. T. Lovell. 1994. Dietary lipid sources influence responses of channel catfish (*Ictalurus punctatus*) to challenge with the pathogen *Edwardsiella ictaluri*. *Aqucaulture* 119:287–298.

Fracalossi, D. M., and R. T. Lovell. 1995. Growth and liver polar fatty acid composition of year-1 channel catfish fed various lipid sources at two water temperatures. *Aquaculture* 57:107–113.

Francis, D. S., G. M. Turchini, P. L. Jones, and S. S. De Silva. 2007b. Growth performance, feed efficiency and fatty acid composition of juvenile Murray cod, *Maccullochella peelii peelii*, fed graded levels of canola and linseed oil. *Aquaculture Nutrition* 13:335–350.

Francis, D. S., G. M. Turchini, P. L. Jones, and S. S. DeSilva. 2007a. Effects of fish oil substitution with a mix blend vegetable oil on nutrient digestibility in Murray cod, *Maccullochella peelii peelii*. *Aquaculture* 269:447–455.

Francis, D. S., T. Thanuthong, S. P. S. D. Senadheera, M. Paolucci, E. Coccia, S. S. De Silva, and G. M. Turchini. 2014. n-3 LC-PUFA deposition efficiency and appetite-regulating hormones are modulated by the dietary lipid source during rainbow trout grow-out and finishing periods. *Fish Physiology and Biochemistry* 40 (2):577–593.

Froese, R., Winker, H., Gascuel, D., Somaila, U. R. and Pauly. 2016. Minimizing the impact of fishing. *Fish and Fisheries* 17:785–802.

Froese, R., and D. Pauly. 2019. Fishbase. Website. Available from: www.fishbase.org (7 September 2019).

Frøyland, L., Ø. Lie, and R. K. Berge. 2000. Mitochondrial and peroxisomal b-oxidation capacities in various tissues from Atlantic salmon *Salmo salar*. *Aquaculture Nutrition* 6:85–89.

FSANZ. 2018. *Standard 1.5.2 – Food Produced Using Gene Technology. Federal Register of Legislation, Rule number*: Standard 1.5.2. Food Standards Australia New Zealand, Canberra, ACT.

Fuentes, A., I. Fernandez-Segovia, J. A. Serra, and J. A. Barat. 2010. Comparison of wild and cultured sea bass (*Dicentrarchus labrax*) quality. *Food Chemistry* 119 (4):1514–1518.

Fukushima, K., I. Sasaki, H. Ogawa, H. Naito, Y. Funayama, and S. Matsuno. 1999. Colonization of microflora in mice: Mucosal defense against luminal bacreia. *Journal of Gastroenterology* 34:54–60.

Ganga, R., J. G. Bell, D. Montero, L. Robaina, M. J. Caballero, and M. S. Izquierdo. 2005. Effect of dietary lipids on plasma fatty acid profiles and prostaglandin and leptin production in gilthead seabream (*Sparus aurata*). *Comparative Biochemistry and Physiology, Part B* 142:410–418.

Gao, J., S. Koshio, M. Ishikawa, S. Yokoyama, T. Ren, C. F. Komilus, and Y. Han. 2012. Effects of dietary palm oil supplements with oxidized and non-oxidized fish oil on growth performances and fatty acid compositions of juvenile Japanese sea bass, *Lateolabrax japonicus*. *Aquaculture* 324–325:97–103.

Garcia-Ortega, A., T. J. Trushenski, and K. Kissinger. 2016. Evaluation of fish meal and fish oil replacement by soybean protein and algal meal from *Schizochytrium limacinum* in diets for giant grouper *Epinephelus lanceolatus*. *Aquaculture* 452:1–8.

Garrido, D., N. Kabeya, M. B.Betancor, J. A. Pérez, N. G. Acosta, D. R. Tocher, C. Rodríguez, O. Monroig. 2019. Functional diversification of teleost Fads2 fattyacyl desaturases occurs independently of the trophic level. *Scientific Reports* 9, 11199.

Gasco, L., M. Henry, G. Piccolo, S. Marono, F. Gai, M. Renna, et al. 2016. Tenebrio molitor meal in diets for European sea bass (*Dicentrarchus labrax* L.) juveniles: Growth performance, whole body composition and in vivo apparent digestibility. *Animal Feed Science and Technology* 220:34–45.

Gasco, L., F. Gai, G. Maricchiolo, L. Genovese, S. Ragonese, T. Bottari, G. Caruso. 2018. Fishmeal alternative protein sources for aquaculture feeds. In Gasco, L., F. Gai, G. Maricchiolo, L. Genovese, S. Ragonese, T. Bottari, and G. Caruso (Eds.), *Feeds for the Aquaculture Sector*. Springer, Berlin/Heidelberg, Germany, pp. 1–28.

Gause, B. R., and J. T. Trushenski. 2013. Sparing fish oil with beef tallow in feeds for rainbow trout: Effects of inclusion rates and finishing on production performance and tissue fatty acid composition. *North American Journal of Aquaculture* 75 (4):495–511.

Gbadamosi, O. K., and I. Lupatsch. 2018. Effects of dietary *Nannochloropsis salina* on the nutritional performance and fatty acid profile of Nile tilapia, *Oreochromis niloticus*. *Algal Research* 33:48–54.

Geay, F., J. Mellery, E. Tinti, J. Douxfils, Y. Larondelle, S. N. M. Mandiki, and P. S. Kestemont. 2015. Effects of dietary linseed oil on innate immune system of Eurasian perch and disease resistance after exposure to *Aeromonas salmonicida achromogen*. *Fish and Shellfish Immunology* 47:782–796.

Geay, F., I. Santigosa, E. Culi, C. Corporeau, P. Boudry, Y. Dreano, L. Corcos, et al. 2010. Regulation of FADS2 expression and activity in European sea bass (*Dicentrarchus labrax* L.) fed a vegetable diet. *Comparative Biochemestiry Physiollogy Part B: Biochemistry of Molecular and Biology* 156:237–243.

Geurden, I., F. Jutfelt, R.-E. Olsen, and K. S. Sundell. 2009. A vegetable oil feeding history affects digestibility and intestinal fatty acid uptake in juvenile rainbow trout *Oncorhynchus mykiss*. *Comparative Biochemistry and Physiology Part A: Molecular & Integrative Physiology* 152 (4):552–559.

Geurden, I., P. Borchert, M. N. Balasubramanian, J. W. Schrama, M. Dupont-Nivet, E. Quillet, S. J. Kaushik, S. Panserat, and F. Medale. 2013. The positive impact of the early-feeding of a plant-based diet on its future acceptance and utilization in rainbow trout. *Plos One* 8:e83162.

Ghioni, C., A. E. Porter, G. W. Taylor, and D. R. Tocher. 2002. Metabolism of 18:4n-3 (stearidonic acid) and 20:4n-3 in salmonid cells in culture and inhibition of the production of prostaglandin F2a (PGF2a) from20:4n-6 (arachidonic acid). *Fish Physiology and Biochemistry* 27:81–96.

Gimeno, E., A. I. Castellote, M. C. Lamuela-Raventós De la Torre, and M. C. López-Sabater. 2002. The effects of harvest and extraction methods on the antioxidant content (phenolics, a-tocopherol, and b-carotene) in virgin olive oil. *Food Chemistry* 78:207–211.

Ginés, R., T. Valdimarsdottir, K. Sveinsdottir, and H. Thorarensen. 2004. Effects of rearing temperature and strain on sensory characteristics, texture, color and fat of Arctic charr (*Salvelinus alpinus*). *Food Quality and Preference* 15:177–185.

Giri, S. S., J. Graham, N. K. A. Hamid, J. A. Donald, and G. M. Turchini. 2016. Dietary micronutrients and in vivo n–3 LC-PUFA biosynthesis in Atlantic salmon. *Aquaculture* 452:416–425.

Glencross, B. D., and D. M. Smith. 1999. The linoleic and linolenic acids requirements of the prawn, *Penaeus monodon*. *Aquaculture Nutrition* 5:53–64.

Glencross, B. D., and D. M. Smith. 2001. Optimising the essential fatty acids, eicosapentaenoic and docosahexaenoic acid in the diet of the prawn, *Penaeus monodon*. *Aquaculture Nutrition* 7:101–112.

Glencross, B., W. Hawkins, and J. Curnow. 2003. Evaluation of canola oils as alternative lipid resources in diets for juvenile red seabream, *Pagrus auratus*. *Aquaculture Nutrition* 9:305–315.

Glencross, B. D., M. Booth, and G. L. Allan. 2007. A feed is only as good as its ingredients – A review of ingredient evaluation for aquaculture feeds. *Aquaculture Nutrition* 13:17–34.

Glencross, B. D. 2009. Exploring the nutritional demand for essential fatty acids by aquaculture species. *Aquaculture* 1:71–124.

Glencross, B. D., and N. R. Rutherford. 2011a. A determination of the quantitative requirements for docosahexaenoic acid (DHA) requirements for juvenile arramundi (*Lates calcarifer*). *Aquaculture Nutrition* 17:536–548.

Glencross, B. D., and G. M. Turchini, eds. 2011b. *Fish Oil Replacement in Starter, Grow-Out and Finishing Feeds for Farmed Aquatic Animals*. Edited by G. M. Turchini, W. K. Ng and D. R. Tocher. CRC Press, Taylor & Francis group, Boca Raton, FL.

Glencross, B., D. Blyth, S. Irvin, N. Bourne, M. Campet, P. Boisot, et al. 2016. An evaluation of the complete replacement of both fishmeal and fish oil in diets for juvenile Asian seabass, *Lates calcarifer*. *Aquaculture* 451:298–309.

Glencross, B. D., J. Baily, M. H. G. Berntssen, R. Hardy, S. MacKenzie, and D. R. Tocher. 2020. Risk assessment of the use of alternative animal and plant raw material resources in aquaculture feeds. *Reviews in Aquaculture* 12:703–758.

Gonzalez-Felix, M. L., D. M. Gatlin, A. L. Lawrence, and M. Perez-Velazquez. 2002. Effect of dietary phospholipid on essential fatty acid requirements and tissue lipid composition of *Litopenaeus vannamei* juveniles. *Aquaculture* 207:151–167.

González-Félix, M. L., F. S. D. da Silva, D. A. Davis, T. M. Samocha, T. C. Morris, J. S. Wilkenfeld, and M. Perez-Velazquez. 2010. Replacement of fish oil in plant based diets for Pacific white shrimp (*Litopenaeus vannamei*). *Aquaculture* 309 (1–4):152–158.

Grant, A. A. M., D. Baker, D. A. Higgs, C. J. Brauner, J. G. Richards, S. K. Balfry, and P. M. Schulte. 2008. Effects of dietary canola oil level on growth, fatty acid composition and osmoregulatory ability of juvenile fall chinook salmon (*Oncorhynchus tshawytscha*). *Aquaculture* 277 (3–4):303–312.

Grigorakis, K., K. D. A. Taylor, and M. N. Alexis. 2003. organoleptic and volatile aroma compounds comparison of wild and cultured gilthead sea bream (*Sparus aurata*): Sensory differences and possible chemical basis. *Aquaculture* 225 (1–4):109–119.

Grisdale-Helland, B., B. Ruyter, G. Rosenlund, A. Obach, S. J. Helland, M. G. Sandberg, H. Standal, and C. Røsjø. 2002. Influence of high contents of dietary soybean oil on growth, feed utilization, tissue fatty acid composition, heart histology and standard oxygen consumption of Atlantic salmon (*Salmo salar*) raised at two temperatures. *Aquaculture* 207 (3–4):311–329.

Guillou, A., P. Soucy, M. Khalil, and L. Adambounou. 1995. Effects of dietary vegetable and marine lipid on growth, muscle fatty acid composition and organoleptic quality of flesh of brrok charr (*Salvelinus fontinalis*). *Aquaculture* 136:351–362.

Hardy, R. W., T. M. Scott, and L. W. Harrell. 1987. Replacement of herring oil with menhaden oil, soybean oil, or tallow in the diets of Atlantic salmon raised in marine net-pens. *Aquaculture* 65 (3–4):267–277.

Hardy, R. W. 2010. Utilization of plant proteins in fish diets: Effects of global demand and supplies of fishmeal. *Aquaculture Research* 41:770–776.

Harwood, J. L. 2019. Algae: Critical sources of very long-chain polyunsaturated fatty acids. *Biomolecules* 9:1–14. doi:10.3390/biom9110708

Hasegawa, S., C. Nakayasu, T. Yoshitomi, T. Nakanishi, and N. Okamoto. 1998. Specific cellmediated cytotoxicity against an allogenic target cell line in isogeneic ginbuna crucian carp. *Fish and Shellfish Immunology* 8:303–313.

Haslam, R. P., S. Usher, O. Sayanova, J. A. Napier, M. B. Betancor, and D. R. Tocher. 2015. The supply of fish oil to aquaculture: A role for transgenic oilseed crops. *World Agriculture* 5:15–23.

Hatae, K., F. Yoshimatsu, and J. Matsumoto. 1990. Role of muscle fibres in contributing to firmness of cooked fish. *Journal of Food Science* 55:693–696.

Hatlen, B., G. M. Berge, J. M. Odom, H. Mundheim, and B. Ruyter. 2012. Growth performance, feed utilisation and fatty acid deposition in Atlantic salmon, *Salmo salar* L., fed graded levels of high-lipid/high EPA *Yarrowia lipolytica* biomass. *Aquaculture* 364–365:39–47.

Haugen, T., A. Kiessling, R. E. Olsen, M. B. Rora, E. Slinde, and R. Nortvedt. 2006. Seasonal variations in muscle growth dynamics and selected quality attributes in Atlantic halibut (*Hippoglossus hippoglossus* L.) fed dietary lipids containing soybean and/or herring oil under different rearing regimes. *Aquaculture* 261:565–579.

Henderson, R. J., M. T. Park, and J. R. Sargent. 1994. The desaturation and elongation of 14C-labelled polyunsaturated fatty acids by pike (*Esox lucius* L.) in vivo. *Fish Physiology and Biochemistry* 14:223–235.

Hernández-Vergara, M. P., D. B. Rouse, M. A. Olvera-Novoa, and D. A. Davis. 2003. Effects of dietary lipid level and source on growth and proximate composition of juvenile redclaw (*Cherax quadricarinatus*) reared under semi-intensive culture conditions. *Aquaculture* 223 (1–4):107–115.

Hixson, S. M., C. C. Parrish, and D. M. Anderson. 2014a. Full substitution of fish oil with camelina (*Camelina sativa*) oil, with partial substitution of fish meal with camelina meal, in diets for farmed Atlantic salmon (*Salmo salar*) and its effect on tissue lipids and sensory quality. *Food Chemistry* 157:51–61.

Hixson, S. M., C. C. Parrish, and D. M. Anderson. 2014b. Use of camelina oil to replace fish oil in diets for farmed salmonids and Atlantic cod. *Aquaculture* 431:44–52.

Hua, K., and D. P. Bureau. 2009. Development of a model to estimate digestible lipid content of salmonid fish feeds. *Aquaculture* 286:271–276.

Huang, C.-H., M.-C. Huang, and P.-C. Hou. 1998. Effect of dietary lipids on fatty acid composition and lipid peroxidation in sarcoplasmic reticulum of hybrid tilapia, *Oreochromis niloticus*× *O. aureus*. *Comparative Biochemistry and Physiology Part B: Biochemistry and Molecular Biology* 120 (2):331–336.

Huang, C. H., W. J. Shyong, and W. Y. Lin. 2001. Dietary lipid supplementation affects the body fatty acid composition but not the growth of juvenile river chub, *Zacco barbata* (Regan). *Aquaculture Research* 32 (12):1005–1010.

Huang, S. S. Y., A. N. Oo, D. A. Higgs, C. J. Brauner, and S. Satoh. 2007. Effect of dietary canola oil level on the growth performance and fatty acid composition of juvenile red seabream, *Pagrus major*. *Aquaculture* 271:420–431.

Huang, S. S. Y., C. H. L. Fu, D. A. Higgs, S. K. Balfry, P. M. Schulte, and C. J. Brauner. 2008. Effects of dietary canola oil level on growth performance, fatty acid composition and ionoregulatory development of spring chinook salmon parr, *Oncorhynchus tshawytscha*. *Aquaculture* 274 (1):109–117.

Huguet, C. T., F. Norambuena, J. A. Emery, K. Hermon, and G. M. Turchini. 2015. Dietary n-6/n-3 LC-PUFA ratio, temperature and time interactions on nutrient and fatty acids digestibility in Atlantic salmon. *Aquaculture* 436:160–166.

Hyden, M. S., and S. Ghosh. 2011. NF-kB in immunobiology. *Cell Research* 21:223–244.

Ibeas, C., J. Cejas, T. Gómez, S. Jerez, and A. Lorenzo. 1996. Influence of dietary n –3 highly unsaturated fatty acids levels on juvenile gilthead seabream (*Sparus aurata*) growth and tissue fatty acid composition. *Aquaculture* 142 (3):221–235.

Ibeas, C., M. S. Izquierdo, and A. Lorenzo. 1994. Effect of different levels of n–3 highly unsaturated fatty acids on growth and fatty acid composition of juvenile gilthead seabream (*Sparus aurata*). *Aquaculture* 127 (2):177–188.

IFFO. 2016. *IFFO RS Standard – Introduction*. Available at http://www.iffo.net/default.asp?contentID=636.

Izquierdo, M. S., J. Socorro, L. Arantzamedi, and C. M. Hernandez-Cruz. 2000. Recebt advances in lipid nutrition in fish larvae. *Fish Physiology and Biochemistry* 22:97–107.

Izquierdo, M. S., A. Obach, L. Arantzamedi, D. Montero, L. Robaina, and G. Rosenlund. 2003. Dietary lipid sources for seabream and seabass: Growth performance, tissue composition and flesh quality. *Aquaculture Nutrition* 9:397–407.

Izquierdo, M. S., D. Montero, L. Robina, M. J. Caballero, G. Rosenlund, and R. Gines. 2005a. Alternations in fillet fatty acid profile and flesh quality in gilthead seabream (*Sparus aurata*) fed vegetable oils for a long term period. Recovery of fatty acid profiles by fish oil feeding. *Aquaculture* 250:431–444.

Izquierdo, M. S., L. Robaina, E. Juarez-Carrillo, V. Oliva, C. M. Hernandez-Cruz, et al. 2008. Regulation of growth, fatty acid composition and delta 6 desaturase expression by dietary lipids in gilthead seabream larvae (*Sparus aurata*). *Fish Physiology and Biochemistry* 34:117–127.

Jackson, and Newton. 2016. Project to model the use of fisheries byproduct in the production of marine ingredients, with special reference to the omega 3 fatty acids EPA and DHA. Institute of Aquaculture, University of Stirling, UK and IFFO, The Marine Ingredients Organisation.

Jaya-Ram, A., M.-K. Kuah, P.-S. Lim, S. Kolkovski, and A. C. Shu-Chien. 2008. Influence of dietary HUFA and levels on reproductive performance, tissue fatty acid profile and desaturase and elongase mRNAs expression in female zebrafish *Danio rerio*. *Aquaculture* 277:275–281.

Ji, H., J. Li, and P. Liu. 2011. Regulation of growth performance and lipid metabolism by dietary n-3 highly unsaturated fatty acids in juvenile grass carp, *Ctenpharyngodon idelilus*. *Comparative Biochemestiry Physiollogy Part B* 159:49–56.

Jin, M., O. Monroing, Y. Lu, Y. Yuan, Y. Li, L. Ding, D. R. Tocher, and Q. Zhou. 2017a. Dietary DHA/EPA ratio affected tissue fatty acid profiles, antioxidant capacity, hematological characteristics and expression of lipid-related genes but not growth in juvenile black seabream (*Acanthopagrus schlegelii*). *Plos One* 12 (4):e0176216.

Jin, M., Y. Lu, Y. Yuan, Y. Li, H. Qiu, P. Sun, H.-N. Ma, L.-Y. Ding, and Q.-C. Zhou. 2017b. Regulation of growth, antioxidant capacity, fatty acid profiles, hematological characteristics and expression of lipid related genes by different dietary n-3 highly unsaturated fatty acids in juvenile black seabream (*Acanthopagrus schlegelii*). *Aquaculture* 471:55–65.

Jobling, M. 2003a. Do changes in Atlantic salmon, *Salmo salar* L., fillet fatty acids following a dietary switch represent wash-out or dilution? Test of a dilution model and its application *Aquaculture Research* 34:1215–1221.

Jobling, M., and E. Å. Bendiksen. 2003b. Dietary lipids and temperature interact to influence tissue fatty acid compositions of Atlantic salmon, *Salmo salar* L., parr. *Aquaculture Research* 34 (15):1423–1441.

Johnson, I. A., Alderson, R., Sandham, C., Dingwall, A., et al. 2000. Muscle fibre density in relation to the colour and texture of smoked Atlantic salmon (*Salmo salar* L.). *Aquaculture* 189:335–349.

Johnston, I. A., X. Li, V. L. A. Vieira, D. Nickell, A. Dingwall, R. Alderson, P. Campbell, and R. Bickerdike. 2006. Muscle and flesh quality traits in wild and farmed Atlantic salmon. *Aquaculture* 256:323–336.

Jordal, A. E., B. E. Torstensen, S. Tsoi, D. R. Tocher, S. P. Lall, and S. E. Douglas. 2005. Dietary rapeseed oil affects the expression of genes involved in hepatic lipid metabolism in Atlantic salmon (*Salmo salar* L.). *Journal of Nutrition* 135:2355–2361.

Jordal, A.-E. O., Ø. Lie, and B. E. Torstensen. 2007. Complete replacement of dietary fish oil with a vegetable oil blend affect liver lipid and plasma lipoprotein levels in Atlantic salmon (*Salmo salar* L.). *Aquaculture Nutrition* 13:114–130.

Jostensen, J.-P., and B. Landfald. 1997. High prevalence of polyunsaturated fatty acid producing bacteria in arctic invertebrates. *FEMS (Fed. Eur. Microbiol. Soc.) Microbiol. Lett.* 151 (1):95–101.

Jouany, J. P. 2007. Methods for preventing, decontaminating and minimizing the toxicity of mycotoxins in feeds. *Animal Feed Science and Technology* 137:342–362.

Kabeya, N., S. Yevzelman, A. Oboh, D. R. Tocher, and O. Monroig. 2018. Essential fatty acid metabolism and requirements of the cleaner fish, ballan wrasse *Labrus bergylta*: Defining pathways of long-chain polyunsaturated fatty acid biosynthesis. *Aquaculture* 488:199–206.

Kalogeropoulos, N., M. N. Alexis, and R. J. Henderson. 1992. Effects of dietary soybean and cod-liver oil levels on growth and body composition of gilthead sea bream (*Sparus aurata*). *Aquaculture* 104:293–308.

Kan, C. A., and G. A. L. Meijer. 2007. The risk of contamination of food with toxic substances present in animal feed. *Animal Feed Science and Technology* 133:84–108.

Kanazawa, A., S. Teshima, and K. Ono. 1979. Relationship between essential fatty acid requirements of aquatic animals and the capacity for bioconversion of linolenic acid to highly unsaturated fatty acids. *Comparative Biochemistry and Physiology* 63B:295–298.

Kangpanich, C., J. Pratoomyot, and W. Senanan. 2017. Effects of alternative oil sources in feed on growth and fatty acid composition of juvenile giant river prown (*Macrobrachium rosenbergii*). *Agriculture and Natural Resources* 51:103–108.

Karalazos, V., E. Å. Bendiksen, J. R. Dick, and J. G. Bell. 2007. Effects of dietary protein, and fat level and rapeseed oil on growth and tissue fatty acid composition and metabolism in Atlantic salmon (*Salmo salar* L.) reared at low water temperatures. *Aquaculture Nutrition* 13 (4):256–265.

Karalazos, V., E. Å. Bendiksen, and J. G. Bell. 2011. Interactive effects of dietary protein/lipid and oil source on growth, feed utilization and nutrient and fatty acid digestibility of Atlantic salmon. *Aquaculture* 311:193–200.

Karapanagiotidis, I. T., M. V. Bell, D. C. Little, and A. Yakupitiyage. 2007. Replacement of dietary fish oils by alpha-linolenic acid-rich oils lowers omega 3 content in tilapia flesh. *Lipids* 42 (6):547–559.

Kawaguchi, S., and S. Nicol. 2007. Learning about *Antarctic krill* from the fishery. *Antarctic Science* 19:219–230.

Kenari, A. A., M. T. Mozanzadeh, and R. Pourgholam. 2011. Effects of total fish oil replacement to vegetable oils at two dietary lipid levels on the growth, body composition, haemato-immunological and serum biochemical parameters in caspian brown trout (*Salmo trutta caspius* Kessler, 1877). *Aquaculture Research* 42 (8):1131–1144.

Kim, D.-K., K.-D. Kim, J.-Y. Seo, and S.-M. Lee. 2012. Effects of dietary lipid source and level on growth performance, blood parameters and flesh quality of sub-adult olive flounder (*Paralichthys olivaceus*). *Asian-Australasian Journal of Animal Science* 25 (6):869–879.

Kim, K.-D., S.-M. Lee, H. G. Park, S. C. Bai, and Y.-H. Lee. 2002. Essentiality of dietary n-3 highly unsaturated fatty acids in juvenile Japanese flounder *Paralichthys olivaceus. Journal of World Aquaculture Society* 33:432–440.

Kim, K.-D., and S.-M. Lee. 2004. Requirement of dietary n-3 highly unsaturated fatty acids for juvenile flounder (*Paralichthys olivaceus*). *Aquaculture* 229 (1–4):315–323.

Kim, Y. S., and S. B. Ho. 2010. Intestinal goblet cells and mucins in health and disease: Recent insights and progress. *Current Gastroenterology Reports* 12 (5):319–330.

Kiron, V., H. Fukuda, T. Takeuchi, and T. Watanabe. 1995. Essential fatty acid nutrition and defence mechanisms in rainbow trout *Oncorhynchus mykiss. Comparative Biochemestiry Physiollogy Part A* 111:361–367.

Kiron, V., J. Thawonsuwan, A. Panigrahi, J. Scharsack, and S. Satoh. 2011. Antioxidant and immune defences of rainbow trout (*Oncorhynchus mykiss*) offered plant oils differing in fatty acid profiles from early stages. *Aquaculture Nutrition* 17:130–140.

Kissinger, K. R., A. G. Ortega, and J. T. Turshenski. 2016. Partial fish meal replacement by soy protein concentrate, squid and algal meals in low fish-oil diets containing *Schizochytrium limacinum* for longfin yellotail *Seriola rivoliana. Aquaculture* 452:37–44.

Kjær, M. A., M. Todorcˇevic´, B. E. Torstensen, A. Vegusdal, and B. Ruyter. 2008a. Dietary n-3 HUFA affects mitochondrial fatty acid b-oxidation capacity and susceptibility to oxidative stress in Atlantic salmon. *Lipids* 43:813–827.

Kjaer, M. A., A. Vegusdal, T. Gjoen, A. C. Rustan, M. Todorcevic, and B. Ruyter. 2008b. Effect of rapeseed oil and dietary n-3 fatty acids on triacylglycerol synthesis and secretion in Atlantic salmon hepatocytes. *Biochimica et Biophysica Acta* 1781:112–122.

Knight, J., and A. Rowley. 1995. Immunoregulatory activities of eicosanoids on rainbow trout, *Oncorhynchus mykiss*, leucocyte proliferation. *Vetenary Immunology Immunopathology* 42:367–378.

Kobayashi, M., and M. Yamamoto. 2005. Molecular mechanisms activating the Nrf2-Keapl pathway of antioxidant gene regulation. *Antioxidants and Redox Signaling* 7:385–394.

Kousoulaki, K., T. K. Ostbye, A. Krasnov, J. S. Torgersen, T. Morkore, and J. Sweetman. 2015. Metabolism, health and fillet nutritional quality in Atlantic salmon (*Salmo salar*) fed diets containing n-3-rich microalgae. *Journal of Nutritional Science* 4:1–13.

Koven, W. M., R. J. Henderson, and J. R. Sargent. 1994. Lipid digestion in turbot (*Scophtalmus maximus*): I. Lipid class and fatty acid composition of digesta from different segments of the digestive tract. *Fish Physiology and Biochemistry* 13:69–79.

Lall, S. P. 2010. The health benefits of farmed salmon: Fish oil decontamination processing removes persistent organic pollutants. *British Journal of Nutrition* 103:1931–1932.

Lane, R. L., and C. C. Kohler. 2006. Effects of dietary lipid and fatty acids on white bass reproductive performance, egg hatchability, and overall quality of progeny. *North American Journal of Aquaculture* 68 (2):141–150.

Lane, R. L., J. T. Trushenski, and C. C. Kohler. 2006. Modification of fillet composition and evidence of differential fatty acid turnover in sunshine bass *Morone chrysops* × *M. saxatilis* following change in dietary lipid source. *Lipids* 41:1029–1038.

Laporte, J., and J. Trushenski. 2011. Growth performance and tissue fatty acid composition of Largemouth Bass fed diets containing fish oil or blends of fish oil and soy-derived lipids. *North American Journal of Aquaculture* 73 (4):435–444.

Lazzarotto, V., G. Corraze, A. Leprevost, E. Quillet, M. Dupont-Nivet, and F. Medale. 2015. Three-year breeding cycle of rainbow trout (Oncorhynchus mykiss) fed a plant-based diet, totally free of marine resources: Consequences for reproduction, fatty acid composition and progeny survival. *Plos One* 10:e0117609.

Leaver, M. J., L. A. N. Villeneuve, A. Obach, L. Jensen, J. E. Bron, D. R. Tocher, and J. B. Taggart. 2008. Functional genomics reveals increases in cholesterol biosynthetic genes and highly unsaturated fatty acid biosynthesis after dietary substitution of fish oil with vegetable oils in Atlantic salmon (*Salmo salar*). *Bmc Genomics* 9 (1):299.

Lee, S.-M., J. Y. Lee, Y. J. Kang, H.-D. Yoon, and S. B. Hur. 1993. n-3 highly unsaturated fatty acid requirement of the Korean rockfish *Sebastes schlegeli*. *Korean Journal of Fisheries and Aquatic Sciences* 26 (5):477–492.

Lee, S. M., J. H. Lee, and K. D. Kim. 2003. Effect of dietary essential fatty acids on growth, body composition and blood chemistry of juvenile starry flounder (*Platichthys stellatus*). *Aquaculture* 225:269–281.

Lee, S. M., and S. H. Cho. 2009. Influences of dietary fatty acid profile on growth, body composition and blood chemistry in juvenile fat cod (*Hexagrammos otakii* Jordan et Starks). *Aquaculture Nutrition* 15:19–28.

Lee, S., F. A. Aya, S. Won, A. Hamidoghli, and S. C. Bai. 2020. Effects of replacing dietary fish oil with beef tallow on growth performance, serological parameters, and fatty acid composition in juvenile olive flounder, *Paralichthys olivaceus.Journal of World Aquaculture Society* 51:393–406.

Legendre, M., N. Kerdchuen, G. Corraze, and P. Bergot. 1995. Larval rearing of an African catfish *Heterobranchus longifilis* (Teleostei, Clariidae): Effect of dietary lipids on growth, survival and fatty acid composition of fry. *Aquatic Living Resources* 8 (4):355–363.

Leu, M.-Y., S.-D. Yang, and C.-H. Liou. 1994. Effect of dietary n-3 highly unsaturated fatty acids on growth, feed efficiency and fatty acid composition of juvenile silver bream *Rhabdosargus sarba* (Sparidae). *Asian Fisheries Science* 7:233–240.

Lewis, T. E., Nichols, P. D. and T. A. McMeekin. 1999. The biological potential of thraustochytrids. *Marine Biotechnolology* 1:580–587.

Li, M., M. Zhang, Y. Ma, R. Ye, M. Wang, H. Chen, D. Xie, Y. Dong, L. Ning, C. You, S. Wang, and Y. Li. 2020. Dietaty supplementation with n-3 high unsaturated fatty acids decreases serum lipid levels and improves flesh quality in the marine teleost golden pompano *Trachinotus ovatus*. *Aquaculture* 516:734632.

Li, M. H., D. J. Wise, M. R. Johnson, and E. H. Robinson. 1994. Dietary mehaden oil reduced resistance of channel catfish (*Ictalurrus punctatus*) to *Edwardsiella ictaluri*. *Aquaculture* 128:335–334.

Li, Y. Y., C. B. Hu, Y. J. Zheng, X. A. Xia, W. J. Xu, W. Z. Chen, Z. W. Sun, and J. H. Huang. 2008a. The effects of dietary fatty acids on liver fatty acid composition and Δ6-desaturase expression differ with ambient salinities in *Siganus canaliculatus*. *Comparative Biochemistry and Physiology, Part B* 151:183–190.

Li, Y. Y., X. J. Zheng, C. B. Hu, X. Xia, w. J. Xu, W. Z. Chen, Z. w. Sun, and J. H. Huang. 2008b. Comparison of capability in utilizing linoleic and alpha-linolenic acids in euryhaline rabbitfish *Siganus oramin* reared in different ambient salinity. *Comp Biochem Physiol Part C* 148:456–457.

Li, Y., O. Monroig, L. Zhang, S. Wang, X. Zheng, J. R. Dick, C. You, and D. R. Tocher. 2010. Vertebrate fatty acyl desaturase with A4 activity. *Proc. Natl. Acad. Sci. U. S. A.* 107:16840–16845.

Li, Y., Y. Zhao, Y. Zhang, X. Liang, Y. Zhang, and J. Gao. 2015b. Growth performance, fatty acid composition, peroxisome proliferator-activated receptors gene expressions, and antioxidant abilities of blunt snout bream, *Megalobrama amblycephala*, fingerlings fed different dietary oil sources. *Journal of World Aquaculture Society* 46:395–406.

Li, Y., X. Liang, Y. Zhang, and J. Gao. 2016. Effects of different dietary soybean oil levels on growth, lipid deposition, tissues fatty acid composition and hepatic lipid metabolism related gene expressions in blunt snout bream (*Megalobrama amblycephala*) juvenile. *Aquaculture* 451:16–23.

Liang, X., H. Y. Ogata, and H. Oku. 2002. Effect of dietary fatty acids on lipoprotein lipase gene expression in the liver and visceral adipose tissue of fed and starved red sea bream *Pagrus major*. *Comparative Biochemistry and Physiology Part A* 132 (4):913–919.

Lie, O., A. Sandvin, and R. Waagbo. 1993. Influence of dietary fatty acids on the lipid composition of lipoproteins in farmed Atlantic salmon (*Salmo salar*). *Fish Physiology and Biochemistry* 12:249–260.

Liland, N. S., I. Biancarosa, P. Araujo, D. Biemans, C. G. Bruckner, R. Waagbø, B. Torstensen, and E.-J. Lock. 2017. Modulation of nutrient composition of black soldier fly (*Hermetia illucens*) larvae by feeding seaweed-enriched media. *PLoS ONE* 12 (8):e0183188.

Lim, C., H. Ako, C. L. Brown, and K. Hahn. 1997. Growth response and fatty acid composition of juvenile *Penaeus vannamei* fed different sources of dietary lipid. *Aquaculture* 151 (1–4):143–153.

Lim, P. K., P. L. Boey, and W. K. Ng. 2001. Dietary palm oil level affects growth performance, protein retention and tissue vitamin E concentration of African catfish, *Clarias gariepinus*. *Aquaculture* 202:101–112.

Lin, H. Z., Y. J. Liu, J. G. He, W. H. Zheng, and L. Z. Tian. 2007. Alternative vegetable lipid sources in diets for grouper, *Epinephelus coioides* (Hamilton): Effects on growth, and muscle and liver fatty acid composition. *Aquaculture Research* 38:1605–1611.

Lin, Y. H., and S. Y. Shiau. 2007. Effects of dietary blend of fish oil with corn oil on growth and non-specific immune responses of grouper, *Epinephelus malabaricus*. *Aquaculture Nutrition* 13:137–144.

Liu, K. K. M., F. T. Barrows, R. W. Hardy, and F. M. Dong. 2004. Body composition, growth performance, and product quality of rainbow trout (*Oncorhynchus mykiss*) fed diets containing poultry fat, soybean/corn lecithin, or menhaden oil. *Aquaculture* 238 (1–4):309–328.

Lo´pez-Bote, C. J., A. Diez, G. Corraze, J. Arzel, M. Alvarez, J. Dias, S. J. Kaushik, and J. M. Bautista. 2001. Dietary protein source affects the susceptibility to lipid peroxidation of rainbow trout (*Oncorhynchus mykiss*) and sea bass (*Dicentrarchus labrax*) muscle. *Animal Science* 73:433–449.

Lochmann, R. T., and I. D. M. Gatlin. 1993. Essential fatty acid requirement of juvenile red drum (*Sciaenops ocellatus*). *Fish Physioloigy and Biochemistry* 12:221–235.

Lodemel, J. B., T. M. Mayhew, R. Myklebust, R. E. Olsen, S. Espelid, and E. Ringo. 2001. Effect of three dietary oils on disease susceptibility in Arctic charr (*Salvelinus alpinus* L.) during cohabitant challenge with *Aeromonas salmonicida ssp. salmonicida*. *Aquaculture Research* 32:935–945.

Losada, A. P., R. Bermudez, L. D. Failde, and M. I. Quiroga. 2012. Quantitative and qualitative evaluation of iNOS expression in turbot (*Psetta maxima*) infected with *Enteromyxum scophthalmi*. *Fish and Shellfish Immunology* 32:243–248.

Lund, I., C. Rodriguez, M. S. Izquierdo, N. El Kertaoui, P. Kestemont, B. D. Reis, D. Dominguez, and J. A. Perez. 2019. Influence of salinity and linoleic or -linolenic acid based diets on ontogenetic development and metabolism of unsaturated fatty acids in pike perch larvae (*Sander lucioperca*). *Aquaculture* 500:550–561.

Luo, L., M. Xue, C. Vachot, I. Geurden, and S. Kaushik. 2014. Dietary medium chain fatty acids from coconut oil have little effect on profiles in rainbow trout (*Oncorhynchus mykiss*). *Aquaculture* 420–421:24–31.

Ma, J. J., Q. J. Shao, Z. Xu, and F. Zhou. 2013. Effect of dietary n-3 highly unsaturated fatty acids on growth, body composition and fatty acid profiles of juvenile black seabream, *Acanthopagrus schlegeli* (Bleeker). *Journal of World Aquaculture Society* 44 (3):311–325.

Ma, H. N., M. Jin, T. T. Zh, C. C. Li, Y. Lu, Y. Yuan, J. Xiong, and Q. C. Zhou. 2018a. Effect of dietary arachidonic acid levels on growth performance, fatty acid profiles and lipid metabolism of juvenile yellow catfish (*Pelteobagrus fulvidraco*). *Aquaculture* 486:31–41.

Ma, N., P. Guo, J. Zhang, T. He, S. W. Kim, G. Zhang, and X. Ma. 2018b. Nutrients mediate intestinal bacteria-mucosal immune crosstalk. *Frontiers in Immunology* 9:5.

Machado, M., C. Castro, A. O. Teles, and B. Costas. 2019. Interactive effects of dietary vegetable oil and carbohydrate incorporation on the immune response of European sea bass (*Dicentrarchus labrax*) juveniles subjected to acute stress. *Aquaculture* 498:171–180.

Maranduba, G. M., S. B. De Castro, G. T. de Souza, G. Rossato, G. F. Da, M. A. Valente, J. V. M. Rettore, C. P. Maranduba, C. M. de Souza, and A. M. do Carmo. 2015. Intestinal microbiota as modulators of the immune system and neuroimmune system: Impact on the host health and homeostasis. *Journal of Immunological Research* 1–14. doi:10.1155/2015/931574

Martinez-Llorens, S., A. Toma´s Vidal, A. V. Mon˜ino, and M. Pla Torres. 2007. Effects of dietary soybean oil concentration on growth nutrient utilization and muscle fatty acid composition of gilthead sea bream (*Sparus aurata* L.). *Aquaculture Research* 38:76–81.

Martino, R. C., J. E. P. Cyrino, L. Portz, and L. C. Trugo. 2002. Performance and fatty acid composition of surubim (*Pseudoplatystoma coruscans*) fed diets with animal and plant lipids. *Aquaculture* 209 (1–4):233–246.

Martins, D. L., E. Gomes, P. Rema, J. Dias, R. O. A. Ozorio, and L. M. P. Valente. 2005. Growth, digestibility and nutrient utilization of rainbow trout (*Oncorhynchus mykiss*) and European sea bass (*Dicentrarchus labrax*) juveniles fed different dietary soybean oil levels. *Aquaculture International* 14:285–295.

Martins, M. A., L. M. P. Valente, and S. P. Lall. 2009. Apparent digestibility of lipid and fatty acids in fish oil, poultry fat and vegetable oil diets by Atlantic halibut, *Hippoglossus hippoglossus* L. *Aquaculture* 294:132–137.

Martins, D. A., E. Gomes, P. Rema, J. Dias, R. O. A. Ozorio, and L. M. P. Valente. 2006. Growth, digestibility and nutrient utilization of rainbow trout (*Oncorhynchus mykiss*) and European sea bass (*Dicentrarchus labrax*) juveniles fed different dietary soybean oil levels. *Aquaculture International* 14:285–295.

Martins, A. D., L. Valente, and S. P. Lall. 2011. Partial replacement of fish oil by flaxseed oil in Atlantic halibut (*Hippoglossus hippoglossus* L.) diets: Effects on growth, nutritional and sensory quality. *Aquaculture Nutrition* 17:671–684.

Mata-Sotres, J. A., A. Tinajero-Chavez, F. Barreto-Curiel, G. Pares-Serra, O. B. D. Rio-Zaragoza, M. T. Viana, and A. N. Rombenso. 2018. DHA (22:6n3) supplementation is valuable in *Totoaba mcdonald* fish oil-free diets containing poultry by-product meal and beef tallow. *Aquaculture* 497:440–451.

Matos, E., A. Goncalves, N. Bandarra, R. Colen, M. L. Nunes, L. M. P. Valente, M. T. Dinis, and J. Dias. 2012. Plant proteins and vegetable oil do not have detrimental effects on post-mortem muscle instrumental texture, sensory properties and nutritional value of gilthead seabream. *Aquaculture* 358–359:205–212.

Matsue, M., Y. Mori, S. Nagase, Y. Sugiyama, R. Hirano, K. Ogai, K. Ogura, S. Kurihara, and S. Okamoto. 2019. Measuring the antimicrobial activity of lauric acid against various bacteria in human gut microbiota using a new method. *Cell Transplantation* 28 (12):1528–1541.

Mellery, J., J. Brel, J. Dort, F. Geay, P. Kestemont, D. S. Francis, Y. Larondelle, and X. Rollin. 2017. A n-3 PUFA depletion applied to rainbow trout fry (*Oncorhynchus mykiss*) does not modulate its subsequent lipid bioconversion capacity. *British Journal of Nutrition* 117:187–199.

Menoyo, D., C. J. Lo´pez-Bote, J. M. Bautista, and A. Obach. 2002. Herring vs. anchovy fish oils in salmon feeding. *Aquatic Living Resource* 15:217–223.

Menoyo, D., M. S. Izquierdo, L. Robaina, R. Ginés, C. J. López, and J. M. Bautista. 2004. Adaptation of lipid metabolism, tissue composition and flesh quality in gilthead sea bream (*Sparus aurata*) to the fish oil replacement by linseed and soybean oils. *British Journal of Nutrition* 92:41–52.

Menoyo, D., C. J. Lopez-Bote, J. M. Bautista, and A. Obach. 2003. Growth, digestibility and fatty acid utilization in large Atlantic salmon (*Salmo salar*) fed varying levels of n-3 and saturated fatty acids. *Aquaculture* 225:295–307.

Menoyo, D., C. J. López-Bote, A. Obach, and J. M. Bautista. 2005. Effect of dietary fish oil substitution with linseed oil on the performance, tissue fatty acid profile, metabolism, and oxidative stability of Atlantic salmon. *Journal of Animal Science* 83 (12):2853–2862.

Menoyo, D., C. J. Lopez-Bote, A. Diez, A. Obach, and J. M. Bautista. 2007. Impact of n-3 fatty acid chain length and n-3/n-6 ratio in Atlantic salmon (*Salmo salar*) diets. *Aquaculture* 267 (1–4):248–259.

Miller, M. R., P. D. Nichols, and C. G. Carter. 2007. Replacement of fish oil with thraustochytrid *Schizochytrium* sp. L. oil in Atlantic salmon parr (*Salmo salar* L) diets. *Comparative Biochemistry Physiology Part A* 148:382–392.

Miller, M. R., P. D. Nichols, and C. G. Carter. 2008a. The digestibility and accumulation of dietary phytosterols in Atlantic salmon (*Salmo salar* L.) smolt fed diets with replacement plant oils. *Lipids* 43 (6):549–557.

Miller, M. R., P. D. Nichols, and C. G. Carter. 2008b. n-3 Oil sources for use in aquaculture–alternatives to the unsustainable harvest of wild fish. *Nutrition Research Reviews* 21 (2):85–96.

Miller, M. R., P. D. Nichols, and C. G. Carter, eds. 2011. *New Alternative n–3 Long Chain Polyunsaturated Facid-Rich Oil Sources*. Edited by G. M. Turchini, W.-K. Ng and D. R. Tocher. Taylor & Francis, CRC Press, Boca Raton, FL.

Mills, S. C., A. G. Windsor, and S. C. Knight. 2010. The potential interactions between polyunsaturated fatty acids and colonic inflammatory processes. *Clinical Experiment Immunology* 142 (2):216–228.

Milo, C., and W. Grosch. 1993. Changes in the odorants of broiled trout (*Salmo salar*) as affected by the storage of the raw material. *Journal of Agriculture and Food Chemistry* 41:2076–2081.

Monge-Ortiz, R., A. Tomas-Vidal, D. Rodriguez-Barreto, S. Martınez-Llorens, J. A. Perez, M. Jover-Cerda, and A. Lorenzo. 2018. Replacement of fish oil with vegetable oil blends in feeds for greater amberjack (*Seriola dumerili*) juveniles: Effect on growth performance, feed efficiency, tissue fatty acid composition and flesh nutritional value. *Aquaculture Nutrition* 24:605–615.

Monroig, Ó., Y. Li, and D. R. Tocher. 2011. Delta-8 desaturation activity varies among fatty acyl desaturases of teleost fish: High activity in delta-6 desaturases of marine species. *Comparative Biochemistry Physiology, Part B* 159:206–213.

Monroig, Ó., D. R. Tocher, and J. C. Navarro. 2013. Biosynthesis of polyunsaturated fatty acids in marine invertebrates: Molecular studies in cephalopods. *Marine Drugs* 11:3998–4018.

Monteiro, M., E. Matos, R. Ramos, I. Campos, and L. M. P. Valente. 2018. A blend of land animal fats can replace up to 75% fish oil without affecting growth and nutrient utilization of European seabass. *Aquaculture* 487:22–31.

Montero, D., T. Kalinowski, A. Obach, R. L. L. Tort, M. J. Caballero, and M. S. Izquierdo. 2003. Vegetable lipid sources for gilthead seabream (*Sparus aurata*): Effects on fish health. *Aquaculture* 225:353–370.

Montero, D., L. Robaina, M. J. Caballero, R. Ginés, and M. Izquierdo. 2005a. Growth, feed utilization and flesh quality of European sea bass (*Dicentrarchus labrax*) fed diets containing vegetable oils: A time-course study on the effect of a re-feeding period with a 100% fish oil diet. *Aquaculture* 248:121–134.

Montero, D., T. Kalinowski, M. J. Caballero, A. Obach, L. Tort, L. Robaina, et al. 2005b. Effect of dietary vegetable lipid sources in gilthead sea bream (*Sparus aurata*) immnue status and stress resistance. In *Mediterranean fish Nutrition*, edited by B. Basurco, M. Izquierdo, D. Montero, I. Nengas and M. Alexis. CIHEAM, Zaragoza, pp. 103–112.

Montero, D., V. Grasso, M. S. Izquierdo, R. Ganga, F. Real, L. Tort, et al. 2008. Total substitution of fish oil by vegetable oils in gilthead sea bream (*Sparus aurata*) diets: Effects on hepatic Mx expression and some immune parameters. *Fish and Shellfish Immunology* 24:147–155.

Montero, D., F. Mathlouthi, L. Tort, J. M. Afonso, S. Torrecillas, A. F. Vaquero, D. Negrin, and M. S. Izquierdo. 2010. Replacement of dietary fish oil by vegetable oils effects humoral immnunity and experssion of pro-inflammatory cytokines genes in gilthead sea bream *Sparus aurata. Fish and Shellfish Immunology* 29:1073–1081.

Montero, D., V. B. Dorta, M. J. Caballero, M. Ponce, S. Torrecillas, M. Izquierdo, M. J. Zamorano, and M. Manchado. 2015. Dietary vegetable oils: Effects on the expression of immnue- related genes in Senegles sole (*Sole senegalensis*) intestine. *Fish and Shellfish Immunology* 44:100–108.

Morais, S., O. Monroig, X. Zheng, M. J. Leaver, and D. R. Tocher. 2009a. Highly unsaturated fatty acid synthesis in Atlantic salmon: Characterization of ELOVL5-and ELOVL2-like elongases. *Marine Biotechnology* 11:627–639.

Morais, S., O. Monroig, X. Zheng, M. J. Leaver, and D. R. Tocher. 2009b. Highly unsaturated fatty acid synthesis in Atlantic salmon: Isolation of genes of fatty acyl elongases and characterisation of ELOVL5- and ELOVL2-like elongase cDNAs. *Marine Biotechnology* 11:627–639.

Morais, S., F. Castanheira, L. Martínez-Rubio, L. E. C. Conceição, and D. R. Tocher. 2012. Longchain polyunsaturated fatty acid synthesis in a marine vertebrate: Ontogenetic and nutritional regulation of a fatty acyl desaturase with Δ4 activity. *Biochimica et Biophysica Acta* 1821, 660–671.

Morita, N., N. Okajima, M. Gotoh, H. Hayashi, H. Okuyama, and S. Sasaki. 1992. Synthesis in vitro of very long chain fatty acids in *Vibrio* sp. *strain ABE-1. Archives of Microbiology* 157 (3):223.

Morton, K. M., D. Blyth, N. Bourne, S. Irvin, and B. D. Glencross. 2014. Effect of ration level and dietary docosahexaenoic acid content on the requirements for long-chain poly-unsaturated fatty acids by juvenile barramundi (*Lates calcarifer*). *Aquaculture* 433:164–172.

Mourente, G., J. E. Good, and J. G. Bell. 2005. Partial substitution of fish oil with rapeseed, linseed and olive oils in diets for European sea bass (*Dicentrarchus labrax* L.): Effects on flesh fatty acid composition, plasma prostaglandins E2 and F2a, immune function and effectiveness of a fish oil finishing diet. *Aquaculture Nutrition* 11:25–40.

Mourente, G., J. E. Good, K. D. Thompson, and J. G. Bell. 2007. Effects of partial substitution of dietary fish oil with blends of vegetable oils, on blood leucocyte fatty acid compositions, immune function and histology in European sea bass (*Dicentrarchus labrax* L.). *British Journal of Nutrition* 98:770–779.

Mourente, G., and D. R. Tocher. 2009. Tuna nutrition and feeds: Current status and future perspectives. *Reviews in Fisheries Science* 17:373–390.

Moya-Falco´n, C., E. Hvattum, T. N. Tran, M. S. Thomassen, M. S. Skorve, and B. Ruyter. 2006. Phospholipid molecular species, B-oxidation, desaturation and elongation of fatty acids in Atlantic salmon hepatocytes: Effects of temperature and 3-thia fatty acids. *Comparative Biochemistry and Physiology Part B* 145:68–80.

Moya-Falcón, C., M. S. Thomassen, J. V. Jakobsen, and B. Ruyter. 2005. Effects of dietary supplementation of rapeseed oil on metabolism of [1-14C] 18: 1n− 9,[1-14C] 20: 3n− 6, and [1-14C] 20: 4n− 3 in atlantic salmon heaptocytes. *Lipids* 40 (7):709–717.

Mozanzadeh, M. T., N. Agh, V. Yavari, J. G. Marammazi, T. Mohammadian, and E. Gisbert. 2016. Partial or total replacement of dietary fish oil with alternative lipid sources in silvery-black porgy (*Sparidentex hasta*). *Aquaculture* 451:232–240.

Mozanzadeh, M. T., J. G. Marammazi, V. Yavari, N. Agh, T. Mohammadian, and E. Gisbert. 2015. Dietary n-3 LC-PUFA requirements in silvery-black porgy juveniles (*Sparidentex hasta*). *Aquaculture* 448:151–161.

Mu, H., C. wei, W. Xu, W. Gao, W. Zhang, and K. Mai. 2020. Effects of replacement of dietary fish oil by rapeseed oil on growth performance, anti-oxidative capacity and inflammatory response in large yellow croaker *Larimichthys crocea*. *Aquaculture Reports* 16:100251.

Mukhopadhayay, P. K., and S. K. Rout. 1996. Effects of different dietary lipids on growth and tissue fatty acid changes in fry of carp *Catla catla* (Hamilton). *Aquaculture Research* 27:623–630.

Mukhopadhyay, P. K., and S. Mishra. 1998. Effect of feeding different lipid sources on growth, feed efficiency and tissue fatty acid composition of *Clarias batrachus* fry and fingerlings. *Journal of Applied Ichthyology* 14 (1–2):105–107.

Muralisankar, T., P. S. Bhavan, S. Radhakrishnan, C. Seenivasan, N. Manickam, and R. Shanthi. 2014. Effects of dietary supplementation of fish and vegetable oils on the growth performance and muscle compositions of the freshwater prawn *Macrobrachium rosenbergii*. *The Journal of Basic and Applied Zoology* 67:34–39.

Murray, D. S., H. Hager, D. R. Tocher, and M. J. Kainz. 2014. Effect of partial replacement of dietary fish meal and oil by pumpkin kernel cake and rapeseed oil on fatty acid composition and metabolism in Arctic charr (*Salvelinus alpinus*). *Aquaculture* 431:85–91.

Napier, J. A., R. E. Olsen, and D. R. Tocher. 2019. Update on GM canola crops as novel sources of omega-3 fish oils. *Plant Biotechnology Journal* 17:703–705.

Navarro-Guillen, C., S. Engrola, F. Castanheira, N. Bandarra, I. Hachero-Cruzado, D. R. Tocher, et al. 2014. Effect of varying dietary levels of LC-PUFA and vegetable oil sources on performance and fatty acids of snegalese sole post larvae: Puzzling results suggest complete biosynthesis pathway from C18 PUFA to DHA. *Comparative Biochemestiry Physiollogy Part B* 167:51–58.

Nayak, M., A. Saha, A. Pradhan, M. Samanta, and S. S. Giri. 2017. Dietary fish oil replacement by linseed oil: Effect on growth, nutrient utilization, tissue fatty acid composition and desaturase gene expression in silver barb (*Puntius gonionotus*) fingerlings. *Comparative Biochemestiry Physiollogy Part B* 205:1–12.

Ng, W. K., M. C. Tee, and P. L. Boey. 2000. Evaluation of crude palm oil and refined palm olein as dietary lipids in pelleted feeds for a tropical bagrid catfish *Mystus nemurus* (Cuvier & Valenciennes). *Aquaculture Research* 31 (4):337–347.

Ng, W.-K., P.-K. Lim, and H. Sidek. 2001. The influence of a dietary lipid source on growth, muscle fatty acid composition and erythrocyte osmotic fragility of hybrid tilapia. *Fish Physiology and Biochemistry* 25 (4):301–310.

Ng, W.-K., P.-K. Lim, and P.-L. Boey. 2003. Dietary lipid and palm oil source affects growth, fatty acid composition and muscle a-tocopherol concentration of African catfish, *Clarias gariepinus*. *Aquaculture* 215:229–243.

Ng, W. K., T. Sigholt, and J. G. Bell. 2004. The influence of environmental temperature on the apparent nutrient and fatty acid digestibility in Atlantic salmon (*Salmo salar* L.) fed finishing diets containing different blends of fish oil, rapeseed oil and palm oil. *Aquaculture Research* 35:1228–1237.

Ng, W. K., Y. E. Chang, and P. G. Lee. 2006. Effects of incorporating spent bleaching clay from palm oil refining in the diets of African catfish, *Clarias gariepinus*. *Asian Fisheries Science* 19:157–164.

Ng, W. K., and O. M. Bahurmiz. 2009. The impact of dietary oil source and frozen storage on the physical, chemical and sensorial quality of fillets from market-size red hybrid tilapia, *Oreochromis* sp. *Food Chemistry* 113:1041–1048.

Ng, W. K., C. Y. Chong, and N. Romano. 2013. Effects of dietary fish and vegetable oils on the growth, tissue fatty acid composition, oxidative stability and vitamin E content of red hybrid tilapia and the efficacy of using fish oil finishing diets. *Aquaculture* 372–375:97–110.

Nichols, L. A., D. E. Jackson, J. A. Manthey, S. D. Shukla, and L. J. Holland. 2011. Citrus flavonoids repress the mRNA for stearoyl-CoA desaturase, a key enzyme in lipid synthesis and obesity control, in rat primary hepatocytes. *Lipids in Health and Disease* 10:36.

Nobrega, R. O., C. F. Correa, B. Mattioni, and D. M. Fracalossi. 2017. Dietary α-linolenic for juvenile Nile tilapia at cold suboptimal temperature. *Aquaculture* 471:66–71.

Nogales-Mérida, S., A. Toma´s-Vidal, M. J. Cerda´, and S. Martı´nez-Llorens. 2011. Growth performance, histological alterations and fatty acid profile in muscle and liver of sharp snout sea bream (*Diplodus puntazzo*) with partial replacement of fish oil by pork fat. *Aquacult International* 19:917–929.

Nogales-Mérida, S., S. Martínez-Llorens, A. V. Moñino, M. J. Cerdá, and A. Tomás-Vidal. 2017. Fish oil substitution by soybean oil in Sharpsnout seabream *Diplodus puntazzo*: Performance, fatty acid profile, and liver histology. *Journal of Applied Aquaculture* 29:1–16. doi:10.1080/10454438.2016.1274933

Norambuena, F., A. Rombenso, and G. M. Turchini. 2016. Towards the optimization of Atlantic salmon reared at different water temperatures via the manipulation of dietary ARA/EPA ratio. *Aquaculture* 450:48–57.

NRC. 2011. *Nutrient Requirements of Fish and Shrimp*, 1st ed. The National Academy Press, Washington, DC.

Nuguyen, T. M., S. N. M. Mandiki, T. N. T. Tran, Y. Larondelle, J. Mellery, E. Mignolet, V. Cornet, E. Flamion, and P. Kestemont. 2019. Growth performance and immnue status in common carp *Cyprinus carpio* as affected by plant oil-based diets complemented with B-glucan. *Fish and Shellfish Immunology* 92:288–299.

Oboh, A., M. B. Betancor, D. R. Tocher, and O. Monroig. 2016. Biosynthesis of long-chain polyunsaturated fatty acids in the African catfish *Clarias gariepinus*: Molecular cloning and functional characterisation of fatty acyl desaturase (fads2) and elongase (elovl2) cDNAs7. *Aquaculture* 462:70–79.

OGTR. 2018. *Risk Assessment and Risk Management Plan for DIR 155 Commercial Release of Canola Genetically Modified for Omega-3 Oil Content (DHA canola NS B5ØØ27-4).* Office of the Gene Technology Regulator, Canberra, ACT.

Oliva-Teles, A., P. Enes, and H. Peres. 2015. Replacing fishmeal and fish oil in industrial aquafeeds for carnivorous fish. In *Feed and Feeding Practices in Aquaculture*, edited by A. D. Davis. Elsevier, Cambridge, UK, pp. 203–233.

Olsen, R. E., and R. J. Henderson. 1997. Muscle fatty acid composition and oxidative stress indices of Arctic charr, *Salvelinus alpinus* (L.), in relation to dietary polyunsaturated fatty acid levels and temperature. *Aquaculture Nutrition* 3 (4):227–238.

Olsen, R. E., R. J. Henderson, and E. Ringø. 1998. The digestion and selective absorption of dietary fatty acids in Arctic charr, Salvelinus alpinus. *Aquaculture Nutrition* 4:13–21.

Olsen, R. E., R. J. Henderson, J. Sountama, G. Hemre, E. Ringo, W. Melle, and D. R. Tocher. 2004. Atlantic salmon, *Salmo salar*, utilizes wax ester-rich oil from *Calanus finmarchicus* effectively. *Aquaculture* 240:433–449.

Olsen, R. E., R. Myklebust, T. Kaino, and E. Ringø. 1999. Lipid digestibility and ultrastructural changes in the enterocytes of Arctic charr (*Salvelinus alpinus* L.) fed linseed oil and soybean lecithin. *Fish Physiology and Biochemistry* 21:35–44.

Olsen, R. E., R. Myklebust, E. Ringø, and T. M. Mayhew. 2000. The influences of dietary linseed oil and saturated fatty acids on caecal enterocytes in Arctic charr (*Salvelinus alpinus* L.): A quantitative ultrastructural study. *Fish Physiology and Biochemistry* 22:207–216.

Olsen, R. E., and E. Ringø. 1997. Lipid digestibility in fish: A review. *Recent Research Developments in Lipid Research* 1:199–264.

Olsen, R. E., J. Suontama, E. Langmyhr, H. Mundheim, E. Ringø, W. Melle, M. K. Malde, and G. I. Hemre. 2006. The replacement of fish meal with Antarctic krill, *Euphausia superba* in diets for Atlantic salmon, *Salmo salar Aquaculture Nutrition* 12:280–290.

Olsen, R. E., R. Waagbø, W. Melle, E. Ringø, and S. P. Lall. 2011. Alternative marine sources. In *Fish Oil Replacement and Alternative Lipid Sources in Aquaculture Feeds*, edited by G. M. Turchini, W.-K. Ng and D. R. Tocher. CRC Press, Florida, pp. 267–324.

Olsen, R. L., J. Toppe, and I. Karunasagar. 2014. Challenges and realistic opportunities in the use of byproduct from processing of fish and shellfish. *Trends in Food Science and Technology* 36:144–151.

Oo, A. N., S. Satoh, and N. Tsuchida. 2007. Effect of replacements of fishmeal and fish oil on growth and dioxin contents of rainbow trout. *Fisheries Science* 73 (4):750–759.

Orozco Colonia, B. S., G. V. de Melo Pereira, and C. R. Soccol. 2020. Omega-3 microbial oils from marine thraustochytrids as a sustainable and technological solution: A review and patent landscape. *Trends in Food Science and Technology* 99:244–256. doi:10.1016/j.tifs.2020.03.007

Østbye, T. K., M. A. Kjaer, A. M. B. Rora, B. Torstensen, and B. Ruyter. 2009. High n-3 HUFA levels in the diet of Atlantic Salmon (*Salmo salar*) affect muscle and mitochondrial membrane lipids and their susceptibility to oxidative stress. *Aquaculture Nutrition* 17:177–190.

ØStbye, T. K., M. A. Kjaer, A. M. B. Rørå, B. Torstensen, and B. Ruyter. 2011. High n-3 HUFA levels in the diet of Atlantic salmon affect muscle and mitochondrial membrane lipids and their susceptibility to oxidative stress. *Aquaculture Nutrition* 17:177–190.

Oswald, A., M. Ishikawa, S. Koshio, S. Yokoyama, A. Moss, and S. Dossou. 2019. Nutritional evaluation of Nannochloropsis powder and lipid as alternative to fish oil for kuruma shrimp, *Marsupenaeus japonicus*. *Aquaculture* 504:427–436. doi:10.1016/j.aquaculture.2019.02.028

Özşahinoğlu, I., T. Eroldoğan, P. Mumoğullarında, S. Dikel, K. Engin, A. H. Yılmaz, M. Arslan, and A. N. Sirkecioğlu. 2013. Partial replacement of fish oil with vegetable oils in diets for European seabass (*Dicentrarchus labrax*): Effects on growth performance and fatty acids profile. *Turkish Journal of Fisheries and Aquatic Sciences* 13:819–825.

Papiol, G. G., and A. Estevez. 2019. Effects of dietary arachidonic acid and eicosapentaenoic acids on common dentex (*Dentex dentex* Linnaeus 1758) larval performance. *Journal of World Aquaculture Society* 50:908–921. doi:10.1111/jwas.12599

Parpoura, A. C. R., and M. N. Alexis. 2001. Effects of different dietary oils in sea bass (*Dicentrarchus labrax*) nutrition. *Aquaculture International* 9:463–476.

Paulino, R. R., R. T. Pereira, T. V. Fontes, A. Oliva-Teles, H. Peres, D. J. Carneiro, and P. V. Rosa. 2018. Optimal dietary linoleic acid to linolenic acid ratio improved fatty acid profile of the juvenile tambaqui (*Colossoma macropomum*). *Aquaculture* 488:9–16.

Peng, S., L. Chen, J. G. Qin, J. Hou, N. Yu, Z. Long, J. Ye, and X. Sun. 2008. Effects of replacement of dietary fish oil by soybean oil on growth performance and liver biochemical composition in juvenile black seabream, *Acanthopagrus schlegeli*. *Aquaculture* 276:154–161.

Peng, X., F. Li, S. Lin, and Y. Chen. 2016. Effects of total replacement of fish oil on growth performance, lipid metabolism and antioxidant capacity in tilapia (*Oreochromis niloticus*). *Aquacult International* 24:145–156.

Perez-Velazquez, M., D. M. Gatlin III, M. L. Gonzalez-Felix, and A. Garcia-Ortega. 2018. Partial replacement of fishmeal and fish oil by algal meals in diets of red drum *Sciaenops ocellatus*. *Aquaculture* 487:41–50.

Perez-Velazquez, M., D. M. Gatlin III, M. L. Gonzalez-Felix, A. Garcia-Ortega, C. R. de Cruz, M. L. Juarez-Gomez, and k. Chen. 2019. Effect of fishmeal and fish oil replacement by algal meals in biological performance and fatty acid profile of hybrid striped bass (*Morone crhysops M. saxatolis*). *Aquaculture* 507:83–90.

Pérez, J. A., C. Rodríguez, A. Bolaños, J. R. Cejas, and A. Lorenzo. 2014. Beef tallow as an alternative to fish oil in diets for gilthead sea bream (*Sparus aurata*) juveniles: Effects on fish performance, tissue fatty acid composition, health and flesh nutritional value. *European Journal of Lipid Science and Technology* 116:571–583.

Petropoulos, I. K., K. D. Thompson, A. Morgan, J. R. Dick, D. R. Tocher, and J. G. Bell. 2009. Effects of substitution of dietary fish oil with a blend of vegetable oils on liver and peripheral blood leucocyte fatty acid composition, plasma prostaglandin E2 and immune parameters in three strains of Atlantic salmon (*Salmo salar*). *Aquaculture Nutrition* 15 (6):596–607.

Pettersson, A., L. Johnsson, E. Brännäs, and J. Pickova. 2009. Effects of rapeseed oil replacement in fish feed on lipid composition and self-selection by rainbow trout (*Oncorhynchus mykiss*). *Aquaculture Nutrition* 15 (6):577–586.

Pickova, J., S. Sampels, and M. H. Berntssen. 2010. 11 Minor components in fish oil and alternative oils with potential physiological effect. In Turchini, G.M., Ng, W-K., and Tocher, D.R (Eds.), *Fish Oil Replacement and Alternative Lipid Sources in Aquaculture Feeds*. CRC Press, Taylor and Francis, Boca Raton, FL, pp. 351.

Piedecausa, M. A., M. J. Mazón, B. GarcíaGarcía, and M. D. Hernández. 2007. Effects of total replacement of fish oil by vegetable oils in sharpsnout seabream (*Diplodus puntazzo*) diets. *Aquaculture* 263:211–219.

Pratoomyot, J., E. Å. Bendiksen, J. G. Bell, and D. R. Tocher. 2008. Comparison of effects of vegetable oils blended with southern hemisphere fish oil and decontaminated northern hemisphere fish oil on growth performance, composition and gene expression in Atlantic salmon (*Salmo salar* L.). *Aquaculture* 280 (1–4):170–178.

Qiao, H., H. Wang, Z. Song, J. Ma, B. Li, X. Liu, S. Zhang, J. Wang, and L. Zhang. 2014. Effects of dietary fish oil replacement by microalgae raw materials on growth performance, body composition and fatty acid profile of juvenile olive flounder, *Paralichthys olivaceous*. *Aquaculture Nutrition* 20:646–653.

Qingyuan, D., M. Kangsen, S. Jikang, A. Qinghui, Z. Huiying, J. Yujian, Z. Lu, Z. Chunxiao, and G. Sitong. 2014. Replacement of dietary fish oil with vegetable oils improves the growth and flesh quality of large yellow croaker (*Larmichthys crocea*). *Journal of Ocean University of China* 13:445–452.

Qiu, H., M. Jin, Y. Li, Y. Lu, Y. Hou, and Q. Zhou. 2017. Dietary lipid sources influence fatty acid composition in tissue of large yellow croaker (*Larmichthys crocea*) by regulating triacylglycerol synthesis and catabolism at the transcriptional level. *Plos One* 12 (1):e0169985. doi:0169910.0161371/journal.pone.0169985

Raso, S., and T. A. Anderson. 2003. Effects of dietary fish oil replacement on growth and carcass proximate composition of juvenile barramundi (*Lates calcarifer*). *Aquaculture Research* 34:813–819.

Regost, C., J. Arzel, M. Cardinal, G. Rosenlund, and S. J. Kaushik. 2003. Total replacement of fish oil by soybean or linseed oil with a return to fish oil in Turbot (*Psetta maxima*) 2. Flesh quality properties. *Aquaculture* 220:737–747.

Regost, C., J. Jakobsen, and A. Rora. 2004. Flesh quality of raw and smoked fillets of Atlantic salmon as influenced by dietary oil sources and frozen storage. *Food Research International* 37:259–271.

Reis, B., E. M. Cabral, T. J. R. Fernandes, M. Castro-Cunha, M. B. P. P. Oliveira, L. M. Cunha, and L. M. P. Valente. 2014. Long-term feeding of vegetable oils to *Senegalese sole* until market size: Effects on growth and flesh quality. Recovery of fatty acid profiles by a fish oil finishing diet. *Aquaculture* 434:425–433.

Reis, D. B., J. A. Perez, I. Lund, N. G. Acosta, B. Abdul-Jalbar, A. Bolanos, and C. Rodtiguez. 2020. Esterification and modification of [1–14C] n-3 and n-6 polyunsaturated fattyacids in pikeperch (*Sander lucioperca*) larvae reared under linoleic orα-linolenic acid-based diets and variable environmental salinities. *Comparative Biochemistry and Physiology B* 246–247:110449.

Ren, H.-t., J.-h. Yu, P. Xu, and Y.-k. Tang. 2012. Influence of dietary fatty acids on muscle fatty acid composition and expression levels of Δ6 desaturase-like and Elovl5-like elongase in common carp (*Cyprinus carpio* var. Jian). *Comparative Biochemistry and Physiology. Part B, Biochemistry and Molecular Biology* 163:184–192.

Richard, N., S. Kaushik, L. Larroquet, S. Panserat, and G. Corraze. 2006. Replacing dietary fish oil by vegetable oils has little effect on lipogenesis, lipid transport and tissue lipid uptake in rainbow trout (*Oncorhynchus mykiss*). *British Journal of Nutrition* 96:299–309.

Ringo, E., H. R. Bendiksen, S. J. Gausen, A. Sundsfjord, and R. E. Olsen. 1998. The effect of dietary fatty acids on lactic acid bacteria associated with the epithelial mucosa and from faecalia of Arctic charr, *Salvelinus alpinus* (L.). *Journal of Applied Micobiology* 85 (5):855–864.

Robin, J. H., C. Regost, J. Arzel, and S. J. Kaushik. 2003. Fatty acid profile of fish following a change in dietary fatty acid source: Model of fatty acid composition with a dilution hypothesis. *Aquaculture* 225 (1–4):283–293.

Rombenso, A. N., J. T. Trushenski, D. Jirsa, and M. Drawbridge. 2015. Successful fish oil sparing in white seabass feeds using saturated fatty acid-rich soybean oil and 22:6n-3 (DHA) supplementation. *Aquaculture* 448:176–185.

Rombenso, A. N., J. T. Trushenski, D. Jirsa, and M. Drawbridge. 2016. Docosahexaenoic acid (DHA) and arachidonic acid (ARA) are essential to meet LC-PUFA requirements of juvenile California yellowtail (*Seriola dorsalis*). *Aquaculture* 463:123–134.

Rombenso, A. N., J. T. Trushenski, and M. H. Schwarz. 2017. Beef tallow is suitable as a primary lipid source in juvenile Florida Pompano feeds. *Aquaculture Nutrition* 23:1274–1286. doi:10.1111/anu.12502

Røra, A. M. B., C. Regost, and J. Lampe. 2003. Liquid holding capacity, texture and fatty acid profile of smoked fillets of Atlantic salmon fed diets containing fish oil or soybean oil. *Food Research International* 36:231–239.

Rosenlund, G., A. Obach, M. G. Sandberg, H. Standal, and K. Tveit. 2001. Effect of alternative lipid sources on long-term growth performance and quality of Atlantic salmon (*Salmo salar* L.). *Aquaculture Research* 32:323–328.

Roszko, M., A. Szterk, K. Szymczyk, and B. Waszkiewicz-Robak. 2012. PAHs, PCBs, PBDEs and pesticides in cold-pressed vegetable oils. *Journal of the American Oil Chemists Society* 89:389–400.

Ruiz-Lopez, N., R. P. Haslam, J. A. Napier, and O. Sayanova. 2014. Successful high-level accumulation of fish oil omega-3 long-chain polyunsaturated fatty acids in a transgenic oilseed crop. *Plant Journal* 77:198–208.

Ruyter, B., C. Røsjø, O. Einen, et al. 2000a. Essential fatty acids in Atlantic salmon: Effects of increasing dietary doses of n-6 and n-3 fatty acids on growth, survival and fatty acid composition of liver, blood and carcass. *Aquaculture Nutrition* 6:119–127.

Ruyter, B., C. Røsjø, K. Måsøval, O. Einen, and M. S. Thomassen. 2000b. Influence of dietary n–3 fatty acids on the desaturation and elongation of [1–14C] 18:2n–6 and [1-14C] 18:3n–3 in Atlantic salmon hepatocytes. *Fish Physiology and Biochemistry* 23:151–158.

Ruyter, B., C. M. Falcon, G. Roselund, and A. Vegusdal. 2006. Fat content and morphology of liver and intestine of Atlantic salmon (*Salmo salar*): Effects of temprature and dietary soybean oil. *Aquaculture* 252:441–452.

Sales, J., and B. Glencross. 2011. A meta-analysis of the effects of dietary marine oil replacement with vegetable oils on growth, feed conversion and muscle fatty acid composition of fish species. *Aquaculture Nutrition* 17:271–287.

Salini, M. J., S. J. Irvin, N. Bourne, D. Blyth, S. Cheers, N. Habilay, and B. D. Glencross. 2015a. Marginal efficiencies of long chain-polyunsaturated fatty acid use by barramundi (*Lates calcarifer*) when fed diets with varying blends of fish oil and poultry fat. *Aquaculture* 449:48–57.

Salini, M. J., N. M. Wade, B. C. Arauji, G. M. Turchini, and B. D. Glencross. 2016. Eicosapentaenoic acid, arachidonic acid and eicosanoid metabolism in juvenile barramundi *Lates calcarifer*. *Lipids* 51:973–988.

Salini, M. J., G. M. Turchini, and B. D. Glencross. 2017. Effect of dietary saturated and monounsaturated fatty acids in juvenile barramundi *Lates calcarifer. Aquaculture Nutrition* 23:264–275.

Salta, E., C. Panagiotidis, K. Teliousis, S. Petrakis, E. Eleftheriadis, et al. 2009. Evaluation of the possible transmission of BSE and scrapie to gilthead Sea bream (*Sparus aurata*). *Plos One* 4:e6175. doi:10.1371/journal.pone.0006175

Sanders, T. H., J. A. Lansden, R. L. Greene, J. S. Drexler, and E. J. Williams. 1992. Oil characteristics of peanut fruit separated by a nondestructive maturity classification method. *Peanut Science* 9:20–23.

Sargent, J., R. J. Henderson, and D. R. Tocher, eds. 1993a. *The Lipids*. Edited by J. E. Halver. Academic Press; San Diego.

Sargent, J. R., J. G. Bell, M. V. Bell, R. J. Henderson, and D. R. Tocher, eds. 1993b. *The Metabolism of Phospholipids and Polyunsaturated Fatty Acids in Fish*. Edited by B. Callou and P. Vittelo. American Geophysical Union, Washington, DC, pp. 103–124.

Sargent, J. R., D. R. Tocher, and J. G. Bell, eds. 2002. *The Lipids*. In Halver, J. E., and Hardy, R. W. (Eds.), *Fish Nutrition*. 3rd ed. Elsevier (Academic Press), San Diego, CA, pp. 181–257. http://www.elsevierdirect.com/imprint.jsp?iid=5

Sarker, P. K., A. R. Kapuscinski, A. J. Lanois, D. Livesey, K. P. Bernhard, and M. L. Coley. 2016a. Towards sustainable aquafeeds: Complete substitution of fish oil with marine microalga *Schizochytrium* sp. improves growth and fatty acid deposition in jevenile Nile tilapia (*Oreochromis niloticus*). *Plos One* 11 (6):e0156684.

Sarker P.K, M. Gamble S. Kelson and A. Kapuscinski. 2016b. Nile tilapia (*Oreochromis niloticus*) show high digestibility of lipid and fatty acids from marine Schizochytrium sp. and of protein and essential amino acids from freshwater Spirulina sp. feed ingredients. *Aquaculture Nutrition* 22:109–119.

Sarker, P. K., A. R. Kapuscinski, G. W. Vandenberg, E. Proulx, and A. J. Sitek. 2020. Towards sustainable and ocean-friendly aquafeeds: Evaluating a fish-free feed for rainbow trout (*Oncorhynchus mykiss*) using three marine microalgae species. *Elementa Science of the Anthroposence* 8:5. doi:10.1525/elementa.404

Satoh, S., W. E. Poe, and R. P. Wilson. 1989. Effect of dietary n-3 fatty acids on weight gain and liver polar lipid fatty acid composition of fingerling channel catfish. *The Journal of Nutrition* 119 (1):23–28.

Schenk, M., and C. Mueller. 2008. The mucosal immune system at the gastrointestinal barrier. *Best Practice and Research Clinical Gatroenterology* 22:391–409.

Schiller Vestergren, A. L., S. Trattner, J. Mráz, B. Ruyter, and J. Pickova. 2011. Fatty acids and gene expression responses to bioactive compounds in Atlantic salmon (*Salmo salar* L.) hepatocytes. *Neuroendocrinology Letter* 32 Supplement 2:41–50.

Schulz, C., U. Knaus, M. Wirth, and B. Rennert. 2005. Effects of varying dietary fatty acid profile on growth performance, fatty acid, body and tissue composition of juvenile pike perch (*Sander lucioperca*). *Aquaculture Nutrition* 11 (6):403–413.

Secci, G., and G. Parisi. 2016. From farm to fork: Lipid oxidation in fish products. A review. *Italian Journal of Animal Science* 15:124–136.

Seierstad, S. L., O. Haugland, S. Larsen, R. Waagbo, and O. Evensen. 2009. Pro-inflammatory cytokine expression and respiratiry burst activity following replacement of fish oil with rapeseed oil in the feed for Alantic salmon (*Salmo salar* L.). *Aquaculture* 289:212–218.

Seierstad, S. L., T. T. Poppe, E. O. Koppang, A. Svindland, G. Rosenlund, L. Froyland, and S. Larsen. 2005. Influence of dietary lipid composition on cardiac patholog in farmed Atlantic salmon, *Salmo salar. Journal of Fish Diseases* 11:677–690.

Seierstad, S. L., A. Svindland, S. Larsen, G. Rosenlund, B. E. Torstensen, and O. Evensen. 2008. Development of intimal thickening of coronary arteries over the lifetime of Atlantic salmon, *Salmo salar* L., fed different lipid sources. *Journal of Fish Diseases* 31:401–413.

Seno-o, A., F. Takakuwa, T. Hashiguchi, K. Morioka, T. Masumoto, and H. Fukada. 2008. Replacement of dietary fish oil with olive oil in young yellowtail *Seriola quinqueradiata*: Effects on growth, muscular fatty acid composition and prevention of dark muscle discoloration during refrigerated storage. *Fisheries Science* 74:1297–1306.

Seong, T., H. Matsutani, Y. Haga, R. Kitagima, and S. Satoh. 2019. First step of non-fish meal, non-fish oil diet development for red seabream, (*Pagrus major*), with plant protein sources and microalgae *Schizochytrium* sp. *Aquaculture Research* 50:2460–2468.

Shapawi, R., S. Mustafa, and W. K. Ng. 2008. Effects of dietary fish oil replacement with vegetable oils on growth and tissue fatty acid composition of humpback grouper, *Cromileptes altivelis* (Valenciennes). *Aquaculture Research* 39:315–323.

Sigurgisladottir, S., S. P. Lall, C. C. Parrish, and R. G. Ackman. 1992. Cholestane as a digestibility marker in the absorption of polyunsaturated fatty acid ethyl esters in Atlantic salmon. *Lipids* 27:418–424.

Silva-Brito, F., F. Timoteo, A. Esteves, M. J. Peixoto, R. Ozorio, and L. Magnoni. 2019. Impact of the replacement of dietary fish oil by animal fats and environmental salinity on the metabolic response of European seabass (*Dicentrarchus labrax*). *Comparative Biochemestiry Physiollogy Part B* 233:46–59.

Šimat, V., J. Vlahovic, B. Soldo, D. Skroza, I. Ljubenkov, and I. Generalic Mekinic. 2019. Production and refinement of omega-3 rich oils from processing by-products of farmed fish species *Foods* 8 (125). doi:10.3390/foods8040125

Simopoulos, A. P. 2008. The importance of the omega-6/omega-3 fatty acid ratio in cardiovascular disease and other chronic diseases. *Bulletin of Experimental Biology and Medicine* 233:674–688.

Skalli, A., and J. Robin. 2004. Requirement of n-3 long chain polyunsaturated fatty acids for European sae bass (*Dicentrarchus labrax*) juveniles: Growth and fatty acid composition. *Aquaculture* 240:399–415.

Smith, D., B. Hunter, G. Allan, D. Roberts, M. Booth, and B. Glencross. 2004. Essential fatty acids in the diet of silver perch (*Bidyanus bidyanus*): Effect of linolenic and linoleic acid on growth and survival. *Aquaculture* 236 (1–4):377–390.

Soller, F., M. A. Rhodes, and D. A. Davis. 2017. Replacement of fish oil with alternative lipid sources in plant-based practical feed formulations for marine shrimp (*Litopenaeus vannamei*) reared in outdoor ponds and tanks. *Aquaculture Nutrition* 23 (1):63–75.

Sprague, M., J. Walton, P. J. Campbell, F. Strachan, J. R. Dick, and J. G. Bell. 2015. Replacement of fish oil with a DHA-rich algal meal derived from *Schizochytrium* sp. on the fatty acid and persistent organic pollutant levels in diets and flesh of Atlantic salmon (*Salmo salar* L.) post-smolts. *Food Chemistry* 185:413–4210.

Sprague, M., M. B. Betancor, and D. R. Tocher. 2017. Microbial and genetically engineered oils as replacements for fish oil in aquaculture feeds. *Biotechnology Letters* 39:1599–1609.

Sprecher, H. 2000. Metabolism of highly unsaturated n–3 and n–6 fatty acids. *Biochimica et Biophysica Acta* 1486:219–231.

St. John, M. A., A. Borja, G. Chust, M. Heath, I. Grigorov, P. Mariani, A. P. Martin, and R. S. Santos. 2016. A dark hole in our understanding of marine ecosystems and their services: Perspectives from the mesopelagic community. *Frontiesrs in Marine Science* 3:31. doi:10.3389/fmars.2016.00031

Stejskal, V., P. Vejsada, M. Cepak, J. Spicka, F. Vacha, J. Kouril, and T. Polcar. 2011. Sensory and textural attributes and fatty acid profiles of fillets of extensively and intensively farmed Eurasian perch (*Perca fluviatilis* L.). *Food Chemistry* 129 (3):1054–1059.

Stéphan, G., J. Guillaume, and F. Lamour. 1995. Lipid peroxidation in turbot (*Scophalmus maximus*) tissue: Effect of dietary vitamin E and dietary n-6 or n-3 polyunsaturated fatty acids. *Aquaculture* 130:251–268.

Stevens, J. R., R. W. Newton, M. Tlusty, and D. C. Little. 2018. The rise of aquaculture by-products: Increasing food production, value, and sustainability through strategic utilisation. *Marine Policy* 90:115–124.

Stubhaug, I., L. Froyland, and B. E. Torstensen. 2005a. Beta-oxidation capacity of red and white muscle and liver in Atlantic salmon (*Salmo salar* L.)–effects of increasing dietary rapeseed oil and olive oil to replace capelin oil. *Lipids* 40:39–47.

Stubhaug, I., D. R. Tocher, J. G. Bell, J. R. Dick, and B. E. Torstensen. 2005b. Fatty acid metabolism in Atlantic salmon (*Salmo salar* L.) hepatocytes and influence of dietary vegetable oil. *Biochimica et Biophysica Acta* 1734:277–288.

Stubhaug, I., Ø. Lie, and B. E. Torestensen. 2007. Fatty acid productive value and b-oxidation capacity in Atlantic salmon (*Salmo salar* L.) fed on different lipid sources along the whole growth period. *Aquaculture Nutrition* 13:145–155.

Suarez, M. D., M. Garcia-Gallego, C. E. Trenzado, J. L. Guil-Guerrero, M. Furne, A. Domezain, I. Albae, and A. Sanz. 2014. Influence of dietary lipids and culture dnsity on rainbow trout (Oncorhynchus mykiss) flesh composition and quality parameter. *Aquaculture Engeneering* 63:16–24.

Subhadra, B., R. Lochmann, S. Rawles, and R. Chen. 2006. Effect of dietary lipid source on the growth, tissue composition and hematological parameters of largemouth bass (Micropterus salmoides). *Aquaculture* 255:210–222.

Sun, P., M. Jin, L. Ding, Y. Lu, H. Ma, Y. Yuan, and Q. Zhou. 2018. Dietary lipid levels could improve growth and intestinal microbiota of juvenile swimming crab, *Portunus trituberculatus. Aquculture* 490:208–216.

Sun, C., B. Liu, Q. Zhou, Z. Xiong, F. Shan, and H. Zhang. 2020. Response of *Macrobrachium rosenbergii* to vegetable oils replacing dietary fish oil: Insights from antioxidant defense. *Frontiers in Physiology*. doi:10.3389/fphys.2020.00218

Tacon, A. G., and M. Metian. 2015. Feed matters: Satisfying the feed demand of aquaculture. *Review of Fish Science and Aquaculture* 23:1–10.

Takeuchi, T., and T. Watanabe. 1977. Requirement of carp for essential fatty acids. *Japanies Society Science of Fish* 43:541–551.

Takeuchi, T., Y. Shiina, and T. Watanabe. 1992. Suitable levels of n-3 highly unsaturated fatty acids in diet for fingerlings of red sea bream. *Nippon Suisan Gakkaishi* 58:509–514.

Takeuchi, T. 1997. Essential fatty acid requirements of aquatic animals with emphasis on fish larvae and fingerlings. *Reviews in Fisheries Science* 5 (1):1–25.

Tan, X. Y., Z. Luo, P. Xie, and X. J. Liu. 2009. Effect of dietary linolenic acid/linoleic acid ratio on growth performance, hepatic fatty acid profiles and intermediary metabolism of juvenile yellow catfish *Pelteobagrus fulvidraco. Aquaculture* 296:96–101.

Tan, P., X. Dong, K. Mai, W. Xu, and Q. Ai. 2016. Vegetable oil induced inflammatory response by altering TLR-NF-kB signalling, macrophages infiltration and polarization in adipose tissue of large yellow croaker (*Larimichthys crocea*). *Fish and Shellfish Immunology* 59:398–405.

Tan, P., X. Dong, H. Xu, K. Mai, and Q. Ai. 2017. Dietary vegetable oil suppressed non-specific immunity and liver antioxidant capacity but induced inflammatory response in Japanese sea bass (*Lateolabrax japonicus*). *Fish and Shellfish Immunology* 63:139–146.

Teoh, C.-Y., and W.-K. Ng. 2016. The implications of substituting dietary fish oil with vegetable oils on the growth performance, fillet fatty acid profile, and modulation of the fatty acid elongase, desaturase and oxidation activities of red hybrid tilapia, *Oreochromis* sp. *Aquaculture* 465:311–322.

Teshima, S. I., A. Kanazawa, and S. Koshio. 1992. Ability for bioconversion of n-3 fatty acids in fish and crustaceans. *Oceanis* 17:67–75.

Thanuthong, T., D. S. Francis, S. P. S. D. Senadheera, P. L. Jones, and G. M. Turchini. 2012. Short-term food deprivation before a fish oil finishing strategy improves the deposition of n–3 LC-PUFA, but not the washing-out of C18 PUFA in rainbow trout. *Aquaculture Nutrition* 18:441–456.

Thompson, K. D., M. F. Tatner, and R. J. Henderson. 1996. Effects of dietary (n-3) and (n-6) polyunsaturated fatty acid ratio on the imune response of Atlantic salmon, *Salmo salar* L. *Aquaculture Nutrition* 2:21–31.

Ti, W. M., M. K. Ong, and C. Y. Teoh. 2019. Assessment of the effects of dietary fatty acids on growth performance, body compositions, plasma lysozyme activity and sensorial quality of juvenile marble goby, *Oxyeleotris marmorata. Aquaculture Reports* 14:1–9.

Tibbetts, S. M., J. E. Milley, and S. P. Lall. 2006. Apparent protein and energy digestibility of common and alternative feed ingredients by Atlantic cod, *Gadus morhua* (Linnaeus, 1758). *Aquaculture* 261:1314–1327.

Tibbetts, S. M., J. Mann, and A. Dumas. 2017. Apparent digestibility of nutrients, energy, essential amino acids and fatty acids of juvenile Atlantic salmon (*Salmo salar* L.) diets containing whole-cell or cell-ruptured *Chlorella vulgaris* meals at five dietary inclusion levels. *Aquaculture* 481:25–39.

Tibbetts, S. M., M. A. Scaife, R. E. Armenta. 2020. Apparent digestibility of proximate nutrients, energy and fatty acids in nutritionally-balanced diets with partial or complete replacement of dietary fish oil with microbial oil from a novel *Schizochytrium* sp. (T18) by juvenile Atlantic salmon (*Salmo salar* L.). *Aquaculture* 520:735003.

Tidwell, J. H., S. Coyle, and L. A. Bright. 2007. Effects of different types of dietary lipids on growth and fatty acid composition of largemouth bass. *North American Journal of Aquaculture* 69 (3):257–264.

Tocher, D. R., J. G. Bell, J. R. Dick, R. J. Henderson, F. McGhee, D. Michell, and P. C. Morris. 2000. Polyunsaturated fatty acid metabolism in Atlantic salmon (*Salmo salar*) undergoing parr-smolt transformation and the effects of dietary linseed and rapeseed oils. *Fish Physiology and Biochemistry* 23 (1):59–73.

Tocher, D. R., J. G. Bell, P. MacGlaughlin, F. McGhee, and J. R. Dick. 2001. Hepatocyte fatty acid desaturation and polyunsaturated fatty acid composition of liver in salmonids: Effects of dietary vegetable oil. *Comparative Biochemistry and Physiology Part B* 130:257–270.

Tocher, D., J. Fonseca-Madrigal, J. G. Bell, J. Dick, R. J. Henderson, and J. Sargent. 2002a. Effects of diets containing linseed oil on fatty acid desaturation and oxidation in hepatocytes and intestinal enterocytes in Atlantic salmon (*Salmo salar*). *Fish Physiology and Biochemistry* 26 (2):157–170.

Tocher, D. R., M. Agaba, N. Hastings, J. G. Bell, J. R. Dick, and A. J. Teale. 2002b. Nutritional regulation of hepatocyte fatty acid desaturation and polyunsaturated fatty acid composition in zebrafish (*Danio rerio*) and tilapia (*Oreochromis niloticus*). *Fish Physiology and Biochemistry* 24:309–320.

Tocher, D. R. 2003a. Metabolism and functions of lipids and fatty acids in teleost fish. *Reviews in Fisheries Science* 11:107–184.

Tocher, D. R., J. G. Bell, F. McGhee, J. R. Dick, and J. Fonseca-Madrigal. 2003b. Effects of dietary lipid level and vegetable oil on fatty acid metabolism in Atlantic salmon (*Salmo salar* L.) over the whole production cycle. *Fish Physiology and Biochemistry* 29:193–209.

Tocher, D. R., J. R. Dick, P. MacGlaughlin, and J. G. Bell. 2006. Effect of diets enriched in Δ6 desaturated fatty acids (18: 3n– 6 and 18: 4n– 3), on growth, fatty acid composition and highly unsaturated fatty acid synthesis in two populations of Arctic charr (*Salvelinus alpinus* L.). *Comparative Biochemistry and Physiology Part B: Biochemistry and Molecular Biology* 144 (2):245–253.

Tocher, D. R., E. Å. Bendiksen, C. P. J., and J. G. Bell. 2008. The role of phospholipids in nutrition and metabolism of teleost fish. *Aquaculture* 280:21–34.

Tocher, D. R. 2010. Fatty acid requirements in ontogeny of marine and freshwater fish. *Aquaculture Research* 41:717–732.

Torno, C., S. Staats, S. C. Michl, S. de Pascual-Teresa, M. Izquierdo, G. Rimbach, et al. 2018. Fatty acid composition and fatty acid associated gene-expression in gilthead sea bream (*Sparus aurata*) are affected by low-fish oil diets, dietary resveratrol, and holding temperature. *Marine Drugs* 16:1–22.

Torrecillas, S., M. J. Caballero, D. Mompel, D. Montero, M. J. Zamorano, L. Robaina, F. R. Ramirez, V. Karalazos, S. Kaushik, and M. Izquierdo. 2017a. Disease resistance and response against *Vibrio anguillarum* intestinal infection in European sea bass (*Dicentrarchus labrax*) fed low fish meal and fish oil diets. *Fish and Shellfish Immunology* 67:302–311.

Torrecillas, S., D. Mompel, M. J. Caballero, D. Montero, D. Merrifield, A. Rodiles, L. Robaina, M. J. Zamorano, V. Karalazos, S. Kaushik, and M. Izquierdo. 2017b. Effect of fishmeal and fish oil replacement by vegetable meals and oils on gut health of European sea bass (*Dicentrarchus labrax*). *Aquaculture* 468:386–398.

Torrecillas, S., F. Rivero-Ramirez, M. S. Izquierdo, M. J. Caballero, A. Makol, P. Suarez-Bregue, A. Fernandez-Montero, J. Rotllant, and D. Montero. 2018. Feeding European sea bass (*Dicentrarchus labrax*) juveniles with a functional synbiotic additive (mannan oilgosaccharides and *Pediococcus acidilactici*): An effective tool to reduce low fish meal and fish oli gut health effects? *Fish and Shellfish Immunology* 81:10–20.

Torres Rosas, V., J. M. Monserrat, M. Bessonart, L. Magnone, L. A. Romano, and M. B. Tesser. 2019. Fish oil and meal replacement in mullet (*Mugil liza*) diet with Spirulina (*Arthrospira platensis*) and linseed oil. *Comparative Biochemestiry and Physiollogy Part C* 218:46–54.

Torstensen, B. E., O. Lie, and L. Froyland. 2000. Lipid metabolism and tissue composition in Atlantic salmon (*Salmo salar* L.) effects of capelin oil, palm oil, and oleic acid-enriched sunflower oil as dietary lipid sources. *Lipids* 35:653–664.

Torstensen, B. E., L. Frøyland, and O. Lie. 2004a. Replacing dietary fish oil with increasing levels of rapeseed oil and olive oil – effects on Atlantic salmon (*Salmo salar* L.) tissue and lipoprotein lipid composition and lipogenic enzyme activities. *Aquaculture Nutrition* 10 (3):175–192.

Torstensen, B. E., and I. Stubhaug. 2004b. Beta-oxidation of 18:3n-3 in Atlantic salmon (*Salmo salar* L.) hepatocytes treated with different fatty acids. *Lipids* 39:1–8.

Torstensen, B. E., J. G. Bell, G. Rosenlund, R. J. Henderson, I. E. Graff, et al. 2005. Tailoring of Atlantic salmon (*Salmo salar*) flesh lipid composition and sensory quality by replacing fish oil with a vegetable oil blend. *Journal of Agriculture and Food Chemistry* 53:10166–10178.

Torstensen, B. E., M. Espe, M. Sanden, I. Stubhaug, R. Waagbø, G.-I. Hemre, R. Fontanillas, U. Nordgarden, E. M. Hevrøy, and P. Olsvik. 2008. Novel production of Atlantic salmon (*Salmo salar*) protein based on combined replacement of fish meal and fish oil with plant meal and vegetable oil blends. *Aquaculture* 285 (1–4):193–200.

Trattner, S., B. Ruyter, T. K. Østbye, T. Gjøen, V. Zlabek, A. Kamal-Eldin, and J. Pickova. 2008. Sesamin increases alpha-linolenic acid conversion to docosahexaenoic acid in Atlantic salmon (*Salmo salar* L.) hepatocytes: Role of altered gene expression. *Lipids* 43:999–1008.

Trushenski, J. T., H. A. Lewis, and C. C. Kohler. 2008. Fatty acid profile of sunshine bass: I. Profile change is affected by initial composition and differs among tissues. *Lipids* 43:629–641.

Trushenski, J. T., J. Boesenberg, and C. C. Kohler. 2009a. Influence of growout feed fatty acid composition on finishing success in Nile tilapia. *North American Journal of Aquaculture* 71:242–251.

Trushenski, J. T., and R. T. Lochmann. 2009b. Potential, implications and solutions regarding the use of rendered animal fats in aquafeeds. *American Journal of Animal and Veterinary Sciences* 4 (4):108–128.

Trushenski, J. T., P. Blaufuss, B. Mulligan, and J. Laporte. 2011. Growth performance and tissue fatty acid composition of rainbow trout reared on feeds containing fish oil or equal blends of fish oil and traditional or novel alternative lipids. *North American Journal of Aquacultures* 73:194–203.

Trushenski, J., M. Schwarz, A. Bergman, A. Rombenso, and B. Delbos. 2012. DHA is essential, EPA appears largely expendable, in meeting the n-3 long-chain polyunsaturated fatty acid requirements of juvenile cobia *Rachycentron canadum*. *Aquaculture* 326–329:81–89.

Trushenski, J. T., and A. N. Rombenso. 2020. Trophic levels predict the nutritional essentiality of polyunsaturated fatty acids in fish—introduction to a special section and a brief synthesis. *North American Journal of Aquaculture* 82:241–250.

Tuck, K. L., and P. J. Hayball. 2002. Major phenolic compounds in olive oil: Metabolism and health effects. *Journal of Nutrition Biochemical* 13:636–644.

Tucker, J., J. W. Lellis, G. K. Vermeer, D. E. Roberts Jr., and P. N. Woodward. 1997. The effects of experimental starter diets with different levels of soybean or menhaden oil on red drum (*Sciaenops ocellatus*). *Aquaculture* 149:323–339.

Turchini, G. M., T. Mentasti, L. Frøyland, E. Orban, F. Caprino, V. M. Moretti, et al. 2003. Effects of alternative dietary lipid sources on performance, tissue chemical composition, mitochondrial fatty acid oxidation capabilities and sensory characteristics in brown trout (*Salmo trutta* L.). *Aquaculture* 225:251–267.

Turchini, G., B. Torstensen, and W. Ng. 2009. Fish oil replacement in finfish nutrition. *Aquaculture* 1:10–57.

Turchini, G. M., T. Mentasti, F. Caprino, I. Giani, S. Panseri, et al. 2005. The relative absorption of fatty acids in brown trout (*Salmo trutta*) fed a commercial extruded pellet coated with different lipid sources. *Italian Journal of Animal Science* 4:241–252.

Turchini, G. M., D. S. Francis, and S. S. De Silva. 2006. Modification of tissue fatty acid composition in Murray cod (*Maccullochella peelii peelii*, Mitchell) resulting from a shift from vegetable oil diets to a fish oil diet. *Aquaculture Research* 37:570–585.

Turchini, G. M., D. S. Francis, R. S. J. Keast, and A. J. Sinclair. 2011a. Transforming salmonid aquaculture from a consumer to a producer of long chain omega-3 fatty acids. *Food Chemistry* 124:609–614.

Turchini, G. M., D. S. Francis, S. P. S. D. Senadheera, T. Thanuthong, and S. S. De Silva. 2011b. Fish oil replacement with different vegetable oils in Murray cod: Evidence of an "omega-3 sparing effect" by other dietary fatty acids. *Aquaculture* 315:250–259.

Turchini, G. M., W. K. Ng, and D. R. Tocher. 2011c. *Fish Oil Replacement and Alternative Lipid Soruces in Aquaculture Feeds*: CRC Press, Taylor & Francis group, Boca Raton, FL.

Turchini, G. M., K. Hermon, V. M. Moretti, F. Caprino, M. L. Busetto, F. Bellagamba, T. Rankin, and D. S. Francis. 2013. Seven fish oil substitutes over a rainbow trout growout cycle: II) Effects on final eating quality and a tentative estimation of feed related production costs. *Aquaculture Nutrition* 19:95–109.

Turchini, G. M., J. T. Trushenski, and B. D. Glencross. 2018. Thoughts for the future of aquaculture nutrition: Realigning perspectives to reflect contemporary issues related to judicious use of marine resources in aquafeeds. *North American Journal of Aquaculture* 81:13–39.

Turkmen, S., P. I. Castro, M. J. Caballero, C. M. Hernandez-Cruz, R. Saleh, M. J. Zamorano, J. Regidor, and M. Izquierdo. 2017. Nutritional stimuli of gilthead seabream (*Sparus aurata*) larvae by dietary fatty acids: Effects on larval performance, gene expression and neurogenesis. *Aquaculture Research* 48:202–213.

Turner, J. R. 2009. Intestinal mucosal barrier function in health and disease. *Nature Reviews Immunology* 9 (11):799–809.

Tzompa-Sosa, D. A., L. Yi, H. J. F. van Valenberg, M. A. J. S. van Boekel, and C. M. M. Lakemond. 2014. Insect lipid profile: Aqueous versus organic solvent-based extraction methods. *Food Research International* 62:1087–1094.

Ulven, S. M., and K. B. Holven. 2015. Comparison of bioavailability of krill oil versus fish oil and health effect. *Vascular Health Risk Management* 11:511–524.

USFDA. 2001. *Cattle Materials Prohibited in Animal Food or Feed to Prevent the Transmission of Bovine Spongiform Encephalopathy. Code of Federal Regulations, Rule Number: Sec. 589.2001.* U.S. Food and Drug Administration, Silver Spring, MD.

USDA. 2005. *Bovine Spongiform Encephalopathy; Minimal-Risk Regions and Importation of Commodities. Federal Register, Rule Number: 70 FR 459.* Animal and Plant Health Inspection Service, US Department of Agriculture, Riverdale, MD.

USDA. 2019. *Oilseeds: World Market and Trades.* Foreign Agriculture Service, Office of Global Analysis, December, 2019.

Usher, S., L. Han, R. P. Haslam, L. V. Michaelson, D. Sturtevant, M. Aziz, K. D. Chapman, O. Sayanova, and J. A. Napier. 2017. Tailoring seed oil composition in the real world: Optimising omega-3 long chain polyunsaturated fatty acid accumulation in transgenic *Camelina sativa*. *Scientific Reports.* doi:10.1038/s41598-017-06838-0, 7:6570.

Vagner, M., and E. Santigosa. 2011. Characterization and modulation of gene expression and enzymatic activity of delta-6 desaturase in teleosts: A review. *Aquaculture* 315:131–143.

Vamecq, J., L. Vallee, P. L. de la Porte, M. Fontaine, D. de Craemer, C. van den Branden, H. Lafont, R. Grataroli, and G. Nalbone. 1993. Effect of various n-3/n-6 fatty acid ratio contents of high fat diets on rat liver and heart peroxisomal and mitochondrial [beta]-oxidation. *Biochimica et Biophysica Acta* 1170:151–156.

van Duijn, G. 2008. Industrial experiences with pesticide removal during edible oil refining. *European Journal of Lipid Science and Technology* 110:982–989.

Vang, B., A. M. Pedersen, and R. L. Olsen. 2013. Oil extraction from the copepod *Calanus finmarchicus* using proteolytic enzymes. *Journal of Aquatic Food Product Technology* 22:619–628.

Vargas, R. J., S. M. Guimarães de Souza, A. M. Kessler, and S. R. Baggio. 2008. Replacement of fish oil with vegetable oils in diets for jundiá (*Rhamdia quelen* Quoy and Gaimard 1824): Effects on performance and whole body fatty acid composition. *Aquaculture Research* 39 (6):657–665.

Varghese, S., and O. V. Oommen. 2000. Long-term feeding of dietary oils alters lipid metabolism, lipid peroxidation, and antioxidant enzyme activities in a teleost (*Anabas testudineus bloch*). *Lipids* 35 (7):757–762.

Vegusdal, A., T. Gjøen, R. K. Berge, M. S. Thomassen, and B. Ruyter. 2005. Effect of 18:1n-9, 20:5n-3 and 22:6n-3 on lipid accumulation and secretion by Atlantic salmon hepatocytes. *Lipids* 40:477–486.

Waagbo, R., G. Hemre, and J. Holm. 1995. Tissue fatty acid composition, haematology and immunity in adult cod, *Gadus morhua* L., fed three dietary lipid sources. *Journal of Fish Diseases* 18:615–622.

Wang, Y., K. H. Yuen, and W. K. Ng. 2006. Deposition of tocotrinols and tocopherols in the tissues of red hybrid tilapia, *Oreochromis* sp., fed a tocotrienol-rich fraction extracted from crude palm oil and its effect on lipid peroxidation. *Aquaculture* 253:583–591.

Wang, F., Y. Wu, Z. Chen, G. Zhang, J. Zhang, S. Shan, and G. Kattner. 2019. Trophic interactions of mesopelagic fishes in the South China Sea illustrated by stable Isotopes and fatty acids. *Frontiers in Marine Science*. doi:10.3389/fmars.2018.00522

Ward, O. P., and A. Singh. 2005. Omega-3/6 fatty acids: Alternative sources of production. *Process Biochemistry* 40:3627–3652.

Wassef, E. A., O. M. Wahby, and E. M. Sakr. 2007. Effect of dietary vegetable oils on health and liver histology of gilthead seabream (*Sparus aurata*) growers. *Aquaculture Research* 38:852–861.

Wassef, E. A., N. E. Saleh, and H. A. El-Abd El-Hady. 2009. Vegetable oil blend as alternative lipid resource in diets for gilthead seabream, *Sparus aurata Aquaculture International* 17:421–435.

Wassef, E. A., S. H. Shalaby, and E. S. Norhan. 2015a. Comparative evaluation of sunflower oil and linseed oil as dietary ingredient for gilthead seabream (*Sparus aurata*) fingerlings. *Oil Seeds and Fat Crops and Lipids* 22 (2): A201.

Wassef, E. A., S. H. Shalaby, and N. E. Saleh. 2015b. Cottonseed oil as a complementary lipid source in diets for gilthead seabream *Sparus aurata* juveniles. *Aquaculture Research* 46 (10):2469–2480.

Watanabe, T., and T. Takeuchi. 1976. Evaluation of pollock liver oil as a supplement to diets for rainbow trout *Bulletin of the Japanese Society of Scientific Fisheries* 42:893–906.

Whalen, K. S. 1999. *Lipid Utilization and Feeding of Juvenile Yellowtail Flounder (Pleuronectes ferrugineus)*. Masters thesis, Memorial University of Newfoundland, Newfoundland, Canada.

WHO. 2003. *WHO Guidelines on Transmissible Spongiform Encephalopathies in Relation to Biological and Pharmaceutical Products*. WHO, Geneva.

Williams, K. C., C. G. Barlow, L. J. Rodgers, and C. Agcopra. 2006. Dietary composition manipulation to enhance the performance of juvenile barramundi (*Lates calcarifer* Bloch) reared in cool water. *Aquaculture Research* 37:914–927.

Willumsen, N., H. Vaagenes, O. Lie, A. C. Rustan, and R. K. Berge. 1996. Eicosapentaenoic acid, but not docosahexaenoic acid, increases mitochondrial fatty acid oxidation and upregulates 2,4-dienoyl-CoA reductase gene expression in rats. *Lipids* 31:579–592.

Woitel, F. R., J. T. Trushenski, M. H. Schwarz, and M. L. Jahncke. 2014. More judicious use of fish oil in cobia feeds: II. effects of graded fish oil sparing and finishing. *North American Journal of Aquaculture* 76:232–241.

Wonnacott, E. J., R. L. Lane, and C. C. Kohler. 2004. Influence of dietary replacement of menhaden oil with canola oil on fatty acid composition of sunshine bass. *North American Journal of Aquaculture* 66 (4):243–250.

Woodgate, S. L., and J. T. van der Veen. 2014. Fats and oils – animal fats and oils – animal based. In *Food Processing: Principles and Applications*, edited by S. Clark, S. Jung and B. Lamsal. John Wiley & Sons, Ltd., Chichester, West Sussex, pp. 481–499.

Wu, F. C., and H. Y. Chen. 2012. Effects of dietary linolenic acid to linoleic acid ratio on growth, tissue fatty acid profile and immune response of the juvenile grouper *Epinephelus malabaricus*. *Aquaculture* 324–325:111–117.

Wu, F. C., Y. Y. Ting, and H. Y. Chen. 2002. Docosahexaenoic acid is superior to eicosapentaenoic acid as the essential fatty acid for growth of grouper, *Epinephelus malabaricus*. *Journal of Nutrition* 132:72–79.

Xie, D., F. Chen, S. Lin, C. You, S. Wang, Q. Zhang, O. Monroig, D. R. Tocher, and Y. Li. 2016. Longchain polyunsaturated fatty acid biosynthesis in the euryhaline herbivorous teleost *Scatophagus argus*: Functional characterization, tissue expression and nutritional regulation of two fatty acyl elongases. *Comparative Biochemistry and Physiology, Part B* 198:37–45.

Xie, D., M. Gong, W. Wei, J. Jin, X. Wang, X. Wang, and Q. Jin. 2019. Antarctic krill (*Euphausia superba*) oil: A comprehensive review of chemical composition,extraction technologies, health benefits, and current applications. *Comprehensive Review in Food Sience and Food Safety* 18:514–534.

Xu, H., L. Cao, Y. Wei, Y. Zhang, and M. Liang. 2018. Lipid contents in farmed fish are influenced by dietary DHA/EPA ratio: A study with the marine flatfish, tongue sole (*Cynoglossus semilaevis*). *Aquaculture* 485:183–190.

Xu, H., X. Dong, R. Zuo, K. Mai, and Q. Ai. 2016. Response of juvenile Japanese seabass (*Lateolabrax japonicus*) to different dietary fatty acid profiles: Growth performance, tissue lipid accumulation, liver histology and flesh texture. *Aquaculture* 461:40–47.

Xu, H., Y. Zhang, J. Wang, R. Zùo, K. Mai, and Q. Ai. 2015. Replacement of fish oil with linseed oil or soybean oil in feeds for Japanese seabass, *Lateolabrax japonicus*: Effects on growth performance, immune response, and tissue fatty acid composition. *Journal of the World Aquaculture Society* 46:349–362.

Xu, X. L., W. J. Ji, J. D. Castell, and R. K. O'Dor. 1994. Essential fatty acid requirements of the Chinese prawn, *Penaeus chinensis*. *Aquaculture* 127:29–40.

Xu, X., and P. Kestemont. 2002. Lipid metabolism and FA composition in tissues of Eurasian perch *Perca fluviatilis* as influenced by dietary fats. *Lipids* 37 (3):297–304.

Xu, H. G., Q. H. Ai, K. S. Mai, W. Xu, J. Wang, H. M. Ma, W. B. Zhang, X. J. Wang, and Z. G. Liufu. 2010. Effects of dietary arachidonic acid on growth performance, survival, immune response and tissue fatty acid composition of juvenile Japanese seabass, *Lateolabrax japonicus*. *Aquaculture* 307:75–82.

Xu, H. G., X. J. Dong, Q. H. Ai, K. S. Mai, W. Xu, Y. J. Zhang, and R. T. Zuo. 2014. Regulation of tissue LC-PUFA contents, Δ6 fatty acyl desaturase (FADS2) gene expression and the methylation of the putative FADS2 gene promoter by different dietary fatty acid profiles in Japanese seabass (*Lateolabrax japonicus*). *Plos One* 9 (1):e87726. doi:10.81371/journal.pone.0087726

Xue, M., L. Luo, X. Wu, Z. Ren, P. Gao, Y. Yu, and G. Pearl. 2006. Effects of six alternative lipid sources on growth and tissue fatty acid composition in Japanese sea bass (*Lateolabrax japonicus*). *Aquaculture* 260:206–214.

Xue, Z., P. L. Sharpe, S. P. Hong, N. S. Yadav, D. Xie, D. R. Short, H. G. Damude, R. A. Rupert, J. E. Seip, J. Wang, D. W. Pollak, M. W. Bostick, M. D. Bosak, D. J. Macool, and D. H. Hollerbach. 2013. Production of omega-3 eicosapentaenoic acid by metabolic engineering of *Yarrowia lipolytica*. *Nature Biotechnology* 31:734–740.

Yaakob, Z., E. Ali, A. Zainal, M. Mohamad and M. S. Takriff. 2014. An overview: Biomolecules from microalgae for animal feed and aquaculture. *Journal of Biological Research-Thessaloniki* 21:6.

Yang, P., H. Hu, Y. Liu, Y. Li, Q. Ai, W. Xu, W. Zhang, Y. Zhang, Y. Zhang, and K. Mai. 2018. Dietary stachyose altered the intesinal microbiota profile and improved the intestinal mucosal barrier function of juvenile turbot, *Scophthalmus maximus* L. *Aquaculture* 486:98–106.

Ye, C. L., D. M. Anderson, and S. P. Lall. 2016. The effects of camelina oil and solvent extracted camelina meal on the growth, carcass composition and hindgut histology of Atlantic salmon (*Salmo salar*) parr in freshwater. *Aquaculture* 450:397–404.

Yildirim-Aksoy, M., C. lim, D. A. Davis, R. Shelby, and P. H. Klesius. 2007. Influence of dietary lipid sources on the growth performance, immune response and resistance of Nile tilapia, *Oreochromis niloticus*, to *Streptococcus iniae* challenge. *Journal of Applied Aquaculture* 19:29–49.

Yıldız, M., T. O. Eroldoğan, S. Ofori-Mensah, K. Engin, and M. A. Baltacı. 2018. The effects of fish oil replacement by vegetable oils on growth performance and fatty acid profile of rainbow trout: Re-feeding with fish oil finishing diet improved the fatty acid composition. *Aquaculture* 488:123–133.

You, C., B. Chen, M. wang, S. Wang, M. Zhang, Z. Sun, A. J. Juventus, H. Ma, and Y. Li. 2019. Effects of dietary lipid sources on the intestinal mocrobiome and health of golden pompano (*Trachintus ovatus*). *Fish and Shellfish Immunology* 89:187–197.

Yu, J., S. Li, H. Niu, J. Chang, Z. Hu, and Y. Han. 2019. Influence of dietary linseed oil as substitution of fish oil on whole fish fatty acid composition, lipid metabolism and oxidative status of juvenile Manchurian trout, *Brachymystax lenok*. *Scientific Reports* 9:13846. doi:10.11038/s41598-13019-50243-13848

Zeng, Y. Y., W. D. Jiang, Y. Liu, P. Wu, J. Zhao, J. Jiang, et al. 2016. Dietary alpha-linolenic acid/linoleic acid ratios modulate intestinal immunity, tight junctions, anti-oxidant status and mRNA levels of NF-κB p65, MLCK and Nrf2 in juvenile grass carp (*Ctenopharyngodon idella*). *Fish and Shellfish Immunology* 51:351–364.

Zhang, M., C. Chen, C. You, B. Chen, S. Wang, and Y. Li. 2019. Effects of dietary dietary ratios of docosahexaenoic to eicosapentaenoic acid (DHA/EPA) on the growth, non-specific immune indices, tissue faty acid compositions and expression of genes related to LC-PUFA biosynthesis in juvenile golden pompano *Trachinotus ovatus*. *Aquaculture* 505:488–495.

Zheng, X., D. R. Tocher, C. A. Dickson, J. G. Bell, and A. J. Teale. 2004. Effects of diets containing vegetable oil on expression of genes involved in highly unsaturated fatty acid biosynthesis in liver of Atlantic salmon (*Salmo salar*). *Aquaculture* 236:467–483.

Zheng, X., B. E. Torestensen, D. R. Tocher, J. R. Dick, R. J. Henderson, and J. G. Bell. 2005. Environmental and dietary influences on highly unsaturated fatty acid biosynthesis and expression of fatty acyl desaturase and elongase genes in liver of Atlantic salmon (*Salmo salar*). *Biochemistry and Biophysical Acta* 1734:13–24.

Zheng, X., Z. Ding, Y. Xu, O. Monroig, S. Morais, and D. R. Tocher. 2009. Physiological roles of fatty acyl desaturases and elongases inmarine fish: Characterisation of cDNAs of fatty acyl Δ6 desaturase and elovl5 elongase of cobia (*Rachycentron canadum*). *Aquaculture* 290:122–131.

Zhou, Q. C., C. C. Li, S. Y. Liu, S. Y. Chi, and Q. H. Yang. 2007. Effects of dietary lipid sources on growth and fatty acid composition of juvenile shrimp, *Litopenaeus vannamei*. *Aquaculture Nutrition* 13:222–229.

Zhou, J. C., D. Han, J. Y. Jin, S. Q. Xie, Y. X. Yang, and X. M. Zhu. 2014. Compared to fish oil alone, a corn and fish oil mixture decreases the lipid requirement of a freshwater fish species, *Carassius auratus gibelio*. *Aquaculture* 428:272–279.

Zhu, Y., Y. I. Hassan, C. Watts, and T. Zhou. 2016. Innovative technologies for the mitigation of mycotoxins in animal feed and ingredients—A review of recent patents. *Animal Feed Science and Technology* 216:19–29.

Zuo, R., Q. Ai, K. Mai, W. Xu, J. Wang, H. Xu, Z. Liufu, and Y. Zhang. 2012a. Effects of dietary docosahexaenoic to eicosapentaenoic acid ratio (DHA/EPA) on growth, nonspecific immunity, expression of some immune related genes and disease resistance of large yellow croaker (*Larmichthys crocea*) following natural infestation of parasites (*Cryptocaryon irritans*). *Aquaculture* 334:101–109.

Zuo, R., Q. Ai, K. Mai, W. Xu, J. Wang, H. Xu, Z. Liufu, and Y. Zhang. 2012b. Effects of dietary n-3 highly unsaturated fatty acids on growth, nonspecific immunity, expression of some immune related genes and disease resistance of large yellow croaker (*Larmichthys crocea*) following natural infestation of parasites (*Cryptocaryon irritans*). *Fish and Shellfish Immunology* 32 (2):249–258.

Zuo, R., K. Mai, W. Xu, G. M. Turchini, and Q. Ai. 2015. Dietary ALA, but not LNA, increase growth, reduce inflammatory processes, and increase anti-oxidant capacity in the marine finfish *Larimichthys crocea*. *Lipids* 50:149–163.

10 Enhancing Feed Utilization in Cultured Fish
A Multilevel Task

Jurij Wacyk, Jose Manuel Yañez M.V., and Rodrigo Pulgar

CONTENTS

INTRODUCTION

Considering factors like human population growth and consumption habits, FAO projects that animal protein consumption will double worldwide by 2050. Different players in the animal protein industry will contribute to this increase in demand, and aquaculture being one of the fastest-growing industries in the animal protein sector, can play a central role in global food security (FAO, 2020)

Currently, aquaculture provides 17% of global animal protein. It is projected to reach 204 million tons of production by 2030 (FAO, 2020); however, aquaculture's contribution to the global food system is not guaranteed, especially with the current use of non-renewable energy and natural resources (Troell et al., 2014). For this increase to be sustainable, aquaculture as a part of the global food system depends on the balance between the amount of product harvested and the sustainable management of natural resources used to produce it (Troell et al., 2014; Ytrestøyl et al., 2015).

DOI: 10.1201/9780429331664-10

FAO defines sustainability for aquaculture as "long-term production of safe products concerning natural resources and in such a way as to deliver socioeconomic development not only for local fishery communities but also for other resource users and globally" (FAO, 2007).

The most in-demand natural resources by any animal production system, including aquaculture, are related to the production and use of ingredients in animal diets (Hua et al., 2019a; Naylor et al., 2009). If we consider that aquafeed production between 1995 and 2015 increased by almost 40 million tons and that aquaculture is projected to have a central role in satisfying the global demand for animal protein, this will further increase the pressure over natural resources associated with fish diets and demands on natural resources (FAO, 2020, 2011).

A WIDE VIEW OF THE EFFICIENCY OF NUTRIENT USE

Historically, feeds for intensive aquaculture relied on marine resources (i.e., fish meal and oil), but the stagnation and sustainability issues of fisheries' productivity, due to overexploitation, stimulated research to reduce the inclusion of marine ingredients. To quantify the efficiency of use of marine resources and how this translates into net production of marine protein, indices like fish in/fish out (FIFO), Forage fish dependency ratio, and Marine protein and oil dependency ratio have been proposed (Crampton et al., 2010; Naylor et al., 2009; Ytrestøyl et al., 2015). Even though from 1990 to 2013, the ratio of wild fish used to produce farmed fish has steadily decreased from 1.04 to 0.63 kg/kg (Hua et al., 2019b; Naylor et al., 2009), with a reduction in inclusion rates for fish meal and oil in Atlantic salmon diets from 65 to 24 and 19 to 11 percent, respectively, (Ytrestøyl et al., 2015), the total use of marine resources has increased due to the fast growth in aquaculture productivity and demand for omega-3 fatty acids (SOFIA, 2018).

The decrease in fish meal and oil inclusion has been possible due to extensive research on terrestrial sources of proteins and oil. Several different alternatives have been studied, among them unicellular microorganism sources (algae and microorganisms), terrestrial animal by-products, krill, fisheries sub-products, and terrestrial plant protein and oil (i.e., soybean meal, barley, lupin, pea, raps, corn, wheat, and sub-products) (Gatlin et al., 2007; Tacon et al., 2009). Worldwide, the use of different plant-derived ingredients has been the strategy of choice to minimize marine ingredients and optimize the cost-benefit ratio for diets in the aquaculture industry. This strategy has been possible due to the oil and protein content of some of the plants used to produce them, adequate amino acid profiles, and acceptable palatability by the fish (Halver and Hardy, 2002). Data from the last decade indicate that plant origin ingredients will continue to be used in increasing proportions, even in carnivorous fish diets (Naylor et al., 2009; FAO, 2018). However, while replacing half of the marine ingredients can be achieved without significant changes in productive fish performance (Naylor et al., 2009), higher levels of replacement are still challenging, without unbalancing (i.e., deficiencies or excesses) the nutrient content of the diets, compromising nutrient use efficiency and fish productivity (Oliva-Teles et al., 2015).

The use of land-based ingredients to formulate diets for cultivated fish has undoubtedly helped reduce pressure over the oceans, but despite this advance toward aquaculture sustainability, we have to keep in mind that worldwide the generation of food is continuously increasing the pressure on natural resources. Current assessments indicate that the global food system is responsible for approximately 30% of global energy consumption more than 20% of global greenhouse gas emissions (FAO, 2011; Gerber et al., 2013; Poore and Nemecek, 2018). If aquaculture is turning to land-based ingredients to formulate diets, this may only shift the burdens from ocean to land (Pahlow et al., 2015).

Using standardized methodologies, like life cycle assessment (LCA), several studies have evaluated aquaculture impacts on ecosystems throughout the entire production cycle, considering variables from the extraction of natural resources, processing, production, waste management, and final disposal of systems outputs (Abdou et al., 2018). These studies use different frameworks defined by sets of boundaries to direct the evaluation of aquaculture impacts in terms of farming systems

(Aubin et al., 2009; d'Orbcastel et al., 2009), management practices (Henriksson et al., 2017; Pelletier et al., 2009a) or complete life cycles including all the processes up to the "end of life" of the product (Bohnes et al., 2019).

Despite differences in scope, most LCA studies suggest that improving the way fish use dietary nutrients (Abdou et al., 2018, 2017; Aubin et al., 2009; Bohnes et al., 2019; Boissy et al., 2011; Cao et al., 2013; d'Orbcastel et al., 2009; Mungkung et al., 2013; Pelletier et al., 2009b; Philis et al., 2019) and using genetic selection of fish (Henriksson et al., 2018, 2017) are the most effective ways to reduce aquaculture environmental impacts. Improving dietary nutrient use is relevant hence from an ecosystem perspective because it has the potential to reduce the use of natural resources and rates of material being released to the environment (d'Orbcastel et al., 2009) and from an aquaculture perspective because feed represents 30 – 70% of total production costs (FAO, 2020).

ENHANCING FEED UTILIZATION IN FISH: BASIC CONCEPTS

Increasing pressure over natural resources to cover projected human nutrient demands makes feed utilization efficiency central for a sustainable increase in aquaculture production. Several parameters are used to monitor aquaculture productivity, but the most frequently used to measure the efficiency with which fish transform feed into tissue is the feed conversion ratio (FCR) or its reciprocal, feed efficiency (Table 10.1).

FCR is defined as the amount of feed needed to generate a unit of growth (Table 10.1) and is widely used as one of the variables used to evaluate cultured fish's productive performance (Table 10.2). The conversion of feed in fish growth can be influenced by environmental and culture conditions (Ytrestøyl et al., 2020) but directly defined as a function of variables associated with the feed and fish growth (Halver and Hardy, 2003). Despite the wide use of FCR, some aspects of its determination need to be kept in mind, particularly when looking for room to make improvements.

FCR assessment implies the determination of feed intake which is generally assessed at the available fish study unit (i.e., tank, cage, raceway), so group-level data is generally collected, assuming similar intake among fish in the experimental unit (Halver and Hardy, 2002; Houlihan et al., 2001). Hence, trying to address this issue and improve feed intake data generation, different studies have worked with individual fish, but the effect of social interactions over the representative on this data is not clear, as if it represents feed intake in real productive conditions (Nicieza and Metcalfe, 1999; Rodde et al., 2020).

TABLE 10.1
Most Commonly Used Parameters to Monitor Productive Fish Performance

Mean Weight	(Total Biomass) / (Number of Fish)
Weight gain	Wt – W0
Percentage of weight gain	((Wt – W0)) / W0 × 100
Specific growth rate	(ln Wt – ln W0) × 100 / (t (days))
Thermal growth coefficient	$((wt^{\wedge(1/3)} - W0^{\wedge(1/3)})) / ((T \times D) \times 100)$
Feed conversion ratio	(total feed intake) / (body weight gain)
Feed efficiency	(body weight gain) / (total feed fed) × 100
Nutrient retention	(Wt × fnc – W0 × inc) × 100 / (n intake)
Protein efficiency ratio	(Wt – W0) / (Ni × 6.25)

Wt: final body weight; W0: initial body weight; ln: natural logarithm; T: water temperature; D: number of days; fnc: final nutrient or energy content; inc: initial nutrient or energy content; n: nutrient or energy; ni: nitrogen intake.

TABLE 10.2
FCR Values Reported in the Literature (Adapted from FAO, 2020; Tacon et al., 2009)

Fish Groups / Species	Average FCR Range Reported
Carps (*Ctenopharyngodon idellus, Cyprinus carpio, Carassius carassius, Parabramis pekinensis, Mylopharyngodon piceus*)	1.3–2.5
Tilapia (*Oreochromis niloticus, O. mossambicus, O. aureus, O. andersonii, O. spilurus*)	1.3–2.5
Catfish (*Pangasius spp, Ictalurus punctatus, Silurus asotus, C. gariepinus, C. macrocephalus, Pelteobagrus fulvidraco, Clarias gariepinus, P. hypophthalmus, Leiocassis longirostris, C. anguillaris, P. pangasius*)	0.9–2.9
Salmon (*Salmo salar, Oncorhynchus kisutch, O. tshawytscha*)	1.2–1.5
Trout (*Oncorhynchus mykiss, Salvelinus fontinalis, Salmo trutta*)	1.1–1.6
Eel (*Anguilla japonica, A. Anguilla, A. australis*)	1.4–1.7
European sea bass (*Dicentrarchus labrax*)	1.6–2.6
Gilthead sea bream (*Sparus aurata*)	1.5–2.2
Yellowtail (*Seriola lalandi*)	1.2–2.3
Barramundi (*Lates calcarifer*)	1.3–1.8
Average world aquaculture (2012)	1.75

Another issue to consider is that feed intake changes depending on fish size, and FCR is generally assessed using defined periods that are restricted to the duration of the study, which do not necessarily represent nutrient use over more extended periods (Jobling, 1993a; Silverstein, 2006). This is particularly relevant considering FCR have been described to be modulated by temperature (Handeland et al., 2008), fish size and body composition (Cook et al., 2000), diseases (Roberts, 2012), and culture conditions that may generate different hierarchies or stress (Ytrestøyl et al., 2020). Moreover, feeding strategies between studies can differ, and feeding the fish to apparent satiation or using restrictive feeding makes comparisons difficult.

We also need to keep in mind that the FCR calculation considers the intake of feed, which means that it does not consider the differential contribution of nutrients contained in that feed to different biological needs, like requirements associated with maintenance or voluntary activity and the nutrient composition of growth (Azevedo et al., 2002; Jobling, 1993b; Silverstein, 2006). This concept is important from a practical point of view because in fish, like rainbow trout, along with weight gain, basal metabolic rate increases daily (Azevedo et al., 1998), implying slow but steady changes in resources investment away from new growth.

New growth implies the consumption, digestion, absorption, and deposition of nutrients like protein, lipids, minerals, water, and other compounds, retained in different tissue compartments (i.e., muscle, liver, bones), with rates of deposition per unit of mass that differ between tissues, fish weight, and nutrient availability, among other factors (i.e., physiological status) implying a dynamic change of needs (Bureau et al., 2006; Halver and Hardy, 2002; Houlihan et al., 2001).

Hence, given the general nature of the FCR calculation, the use of nutrient retention, and other variables (Table 10.1) are necessary to obtain a more accurate determination on how efficiently fish are using nutrients. In this regard, Fry et al. (2018), using available data in terms of FCR, nutrient retention, as well as edible portion and its nutritional content, highlights the need to use a combination of variables at the same time and focus on nutrients in order to assess more precisely the utilization of feed. Improving the way dietary nutrients are used for growth or another productive variable of interest is a multifactorial process influenced by different aspects of diet formulation and the fish and its potential to obtain and use them (i.e., genetic improvement, further ahead in the text).

DIET NUTRIENT CONTENT: GENERAL CONSIDERATIONS AND SOME EFFECTS OVER FCR AND NUTRIENT RETENTION

Diet Space

Diet formulation defines both ingredient and nutrient contents of aquafeeds in a process that balances different criteria. Fish nutrient requirements and ingredient cost, not necessarily in that order, open the way to the formulation and manufacture constraints, like not altering pigmentation targets in fish fillet or producing pellets with unsuitable hardness or water stability (Hardy and Barrows, 2003; Saez et al., 2016; Zettl et al., 2019). To this, we can add other features: storage capacity without losing nutritional value, particularly important for remote production locations away from the feed plant; and reduced nutrient leaching to the environment, to mention some and to underscore the importance of pellet structural characteristics along with its nutrient content (Gasco et al., 2018; Tacon and Forster, 2003). All these and other considerations, depending on the objective of the diet (i.e., functional diets), make the process of formulating and manufacturing a sequence that, as we move forward towards the final steps, reduces the space available to make changes and include other ingredients without compromising the initial objective (Cornell, 2002). It is within this available dietary space where we can make changes to improve FCR and nutrient retention via studying how fish use ingredients and nutrients.

Nutrient Content

The nutrient content of the pellet is defined by the mixture of ingredients used in its production (currently via extrusion), and, since we are matching fish needs with the amount of nutrients we can accommodate in the diet, other constraints appear that can impact nutrient use. Fish feeding habits (i.e., carnivores vs. herbivores), for example, have determined not only the digestive capacities of fish to digest and absorb nutrients from the intestinal lumen, but also fish differ in the way that nutrients are used to cover requirements, hence defining limits for available diet space (Hardy and Barrows, 2003; Steinberg, 2018).

For most aquaculture species, levels of dietary protein to optimize growth is in the range of 25 to 50% of crude protein, usually achieved by mixing different protein sources. In general, carnivorous fish require higher levels of protein in their diets than herbivorous fish; this is also observed for juvenile states of different species (NRC, 2011). This higher demand restricts the range of ingredients used to formulate and manufacture diets (i.e., diet space), even when high protein contents can be found in some ingredients like animal by-products (i.e., 50–90%). This is associated with fish not presenting a requirement for protein but for amino acids, and those that cannot be synthesized by the fish are considered essential and must be included in the diet. Hence, the low content of lysine, methionine, and tryptophan, in plants and some animal by-products limits the inclusion of these ingredients in fish with high protein demands (Teles et al., 2019). Amino acid deficiencies in aquafeeds can cause significant reductions in FCR and nutrient retention, depending on factors like which amino acid is deficient, the magnitude of the deficiency, and how long the fish have been fed the deficient diet (Li et al., 2008). The strategy of choice to avoid these adverse effects has been to formulate in line with fish requirements and to use synthetic amino acids to supplement formulations as needed (Nunes et al., 2014).

On the contrary, fish do not require carbohydrates (CHOs) because simple sugars, like glucose, used by the fish for different purposes can be synthesized from other nutrients like amino acids. However, this does not mean that CHOs should be avoided from diet formulation because several benefits have been reported for its inclusion in fish diets, particularly in the form of starch (NRC, 2011). Digestible starch is a low-cost energy source for the fish, and when included adequate levels, have been reported to improve the use and deposit of dietary protein, an effect associated with sparing carbohydrate skeletons derived from amino acids to be used as an energy source (Krogdahl et al.,

2005; Stone, 2003). This effect over the retention of protein changes depending on factors like the type of the starch, gelatinization level, and the feeding habit of the fish, with a more substantial sparing effect described for omnivores than carnivores, in response to associated differences in metabolic capabilities for CHO use (Kamalam et al., 2017). The influence of dietary CHOs over FCR and nutrient retention is also associated with its use during the diet production process to adjust the pellet structure. This aspect includes features like porosity, critical for oil absorption capacity, and adjusting sinking rates according to the needs of the fish species of interest; salmonids, for example, feed more efficiently with slow-sinking pellets. Current recommended levels of digestible CHO for salmonids and marine species are < 20% and 20–50% for omnivorous fish (NRC, 2011).

Regarding dietary lipids, this group of nutrients represents one of the most important sources of energy and essential fatty acids (EFA), critical for efficient growth, normal health, reproduction, and fillet quality (NRC, 2011). Fish require alpha-linolenic (18:3 n3) and linoleic (18:2 n6) acids since, like other vertebrates, they lack the necessary enzymes to synthesize these FAs. On the other hand, for eicosapentaenoic (EPA – 20:5 n3) and Docosahexaenoic (DHA – 22:6 n3) acids, fish seem to present relative essentiality because the rate at which they are required cannot be covered by the fish metabolism. Hence, these FAs need to be provided by the diet of fresh and marine stages in salmonids or fish with different feeding habits (i.e., carnivorous vs. herbivorous) (NRC, 2011; Tocher et al., 2019). Deficiency of EFA is associated with poor reproductive performance, increases in mortality, and reduced growth rate and nutrient conversion; available data indicates that while EFA requirements are covered, reduction in fish growth performance can be minimized (Rosenlund et al., 2016). The recommendation for EFA inclusion has been described in the range of 0.5 – 1% of the diet for salmonids; however, recently, values closer to 1.6% have been suggested in order to adapt to current use of marine ingredients. Variation for this value has been reported for different species and the specific FA under consideration (NRC, 2011; Ruyter et al., 2019; Sissener, 2018; Tocher et al., 2019; Tocher, 2010).

As described above, adequate nutrients in diet formulation covering fish requirements and defining some of the pellets' structural features are the first steps to avoid adverse impacts over growth and nutrient use. Nevertheless, for the dietary levels of nutrients to accomplish their many functions, such as building blocks of macromolecules (i.e., muscle tissue for growth), regulatory signals in different pathways (i.e., feed intake regulation), keeping gradients across membranes (i.e., metabolism integration), or sources of energy (i.e., fuel for metabolic work), fish need to consume, digest, and absorb nutrients to make them available for use.

Nutrient Intake

To correctly evaluate the effect of a diet over any parameter of interest, including of course growth and nutrient use, it is necessary to have an accurate measurement of the amount of diet consumed by the fish (Jobling et al., 1995; Nikki et al., 2004). Several considerations have already been presented for this point concerning FCR, and many aspects of the regulation of feed intake are reviewed elsewhere (Houlihan et al., 2001), but taking good care of proper acclimation periods every time a diet is changed and feeding to apparent satiety are necessary management practices to reduce variability associated with feeding intake. This aspect is relevant considering its impact on nutrient use; even when many different variables can influence feed intake, estimates of losses due to uneaten pellets are in the range of 10–20% of what is offered in different aquaculture systems (Boyd et al., 2007).

Digestibility

Since nutrients need to be digested and absorbed to be biologically used, information on digestibility coefficients is a practical and essential diet formulation tool. Nutrient digestibility of ingredients and diets can be determined via different methods, with variations in terms of experimental diet formulation, markers to be used, and fecal sample collection methods, which can influence the final value obtained. Despite these differences and since not all consumed nutrients are absorved after digestion, all share the same purpose. Digestibility coefficients indicate available nutrients in fish diets or those not lost in feces (Shomorin et al., 2019; Storebakken et al., 1998), and are assumed to be additive. This means that the sum of the proportional contribution of each ingredient's digestibility should be the same as the digestibility of the whole diet. Since interactions between nutrients can be present for some mixtures of ingredients, for most ingredient mixes, additivity holds correct, allowing to formulate diets based on digestibility values (Halver and Hardy, 2003). Feeding frequency, temperature, salinity, and fish species, even closely related ones as salmon and trout, have been described as factors influencing digestibility coefficients and are considered when changing or adjusting a diet's formulation.

Digestible Protein/Digestible Energy – High Energy

Along with nutrient content and digestibility coefficients, the proportion between dietary nutrients are an essential factor when considering nutrient conversion and use. Nutrient interactions are described at different levels, between amino acids, vitamins and minerals, and macronutrients, all influencing dietary nutrient use in different degrees. Nutrients as building blocks and sources of energy need to be adequately provided to the fish to stimulate fast and efficient muscle growth (NRC, 2011). In practical terms, nutrients are grouped into six categories – water, carbohydrates, lipids, proteins, minerals, and vitamins – all needed for efficient growth. However, given their chemical nature, only proteins, carbohydrates, and lipids, can provide nutrients needed by fish (i.e., amino acids, fatty acids) along with carbon skeletons for energy production (Nelson et al., 2017). Considering the caloric content of each macronutrient and the use of extrusion technology to fabricate aquafeeds, several studies have shown the benefits of high energy diets (> 20 MJ/kg), which increase fish growth and reduce environmental impacts with lower levels of nitrogen-containing compounds released into the culture water (Moc̆k et al., 2019). The increase in growth responds to more efficient use of dietary protein for muscle deposition due to a sparing protein effect of lipids, more substantial than that observed for carbohydrates. Several studies have reported positive effects of these diets on the growth of salmon (Einen and Roem, 1997; Hillestad and Johnsen, 1994; Martinez-Rubio et al., 2013) and other fish species (Kabir et al., 2020, 2019). However, to maximize the positive effects of high energy diets, several aspects of their use need to be addressed, such as the increase in visceral fat accumulation or negative impacts over feed intake, which is in line with a lipostatic regulation of the process (Jobling and Johansen, 1999). Some of these have already started to be addressed, adjusting the ratio of digestible protein and energy of high energy salmon diets (Dessen et al., 2017; Weihe et al., 2018). Today, however, the high inclusion of marine ingredients used in some of the studies directly addressing the effect of high energy diets are no longer used. Hence, the balance of dietary proteins and energy needs attention, but the content and balance of other nutrients like vitamins, minerals, balance between PUFA, or even pigments need to be addressed in this new context.

Marine Ingredient Replacement/Plant Ingredient Use

One primary driver in studying how ingredients can influence fish growth and how the fish use nutrients has been the low sustainability of fish meal and fish oil production. Notwithstanding the massive efforts to reduce the inclusion of marine ingredients in aquafeeds (roughly a 17% reduction in the use of fishmeal and fish oil have been possible in the last 20 years), the fast growth of the aquaculture industry and the need for more feed makes it an ongoing challenge (FAO, 2020; Fry et al., 2016; Pahlow et al., 2015).

Due to cost and generally favorable nutrient composition, plant protein and oil sources are currently the most widely used alternatives to replace marine ingredients. This situation has implied the study of many different aspects like the determination of digestibility coefficients, the effect of the presence of antinutritional factors, nutrient content and balance as well as ingredient processing (Aksnes et al., 1996; Barrows et al., 2008; Gatlin et al., 2007; Glencross et al., 2005; Gomes et al., 1995; Hardy and Barrows, 2003; Iwashita et al., 2008; Wacyk et al., 2012).

The list of plant-derived ingredients currently used in aquafeeds is ample and has high variability in dietary inclusion, depending on the fish species under consideration. The main species from which protein meals, concentrates, and other ingredients are produced are soybeans, corn, raps, peas, and lupins for carnivorous species; rice needs to be added to this list if we consider Nile tilapia. For fish oil replacement, canola and soybean have been the main alternatives (Pahlow et al., 2015). For a more detailed consideration of inclusion levels depending on species, the reader may consult (Molina-Poveda, 2016; Oliva-Teles et al., 2015).

Based on current literature, both from individual ingredient studies and meta-analytic reviews, the inclusion of plant proteins up to 30–50% seems not to have a significant effect on carnivore, fish growth, and FCR, particularly salmonids, where more information is available. However, at higher inclusion levels, the evidence consistently indicates that the growth rate and the FCR are negatively impacted with variations in the magnitude of this effect, depending on the protein source being used. The causes of this phenomenon seem to be nutrient imbalances (such as amino acid deficiencies), lower energy digestibility, presence of antinutritional factors, and low palatability (Choi et al., 2020; Collins et al., 2013; Egerton et al., 2020; Hardy, 2010; Turchini et al., 2018).

In terms of dietary oil, depending on plant source and diet formulation covering essential fatty acid requirements, replacement of 60–75% of the diet fish oil does not substantially reduce growth, FCR, and feed intake (Sales and Glencross, 2011; Turchini et al., 2009). On the other hand, replacing 100% of fish oil with plant oil has been reported to significantly reduce FCR, growth, and lower deposition of unsaturated fatty acids (i.e., EPA-DHA), with the effect depending on the fatty acid profile and the proportion of the different saturation categories of the plant oil being used (Glencross and Turchini, 2010; Lazzarotto et al., 2018; Torstensen and Tocher, 2010).

The replacement levels described are mainly based on carnivorous fish studies and have been further advanced, working with ingredient blends to replace fish meal and fish oil. In this sense, the use of mixtures of vegetable oils, like linseed, rapeseed/canola, soybean, sunflower, and olive oils, can be used to meet LC-PUFA requirements having in mind a minimum of 5–15% FO inclusion in order to cover EFA requirements (Foroutani et al., 2018; Glencross and Turchini, 2010). Similarly, the use of protein blends and specific amino acid supplementation, mainly methionine, lysine, threonine, and taurine, has allowed for the reduction of fish meal inclusion in carnivorous fish diets to levels of 5 to 10% with minor or no effects over fish growth performance (Altan et al., 2010; Cabral et al., 2011; Hansen et al., 2007; Kaushik et al., 2004; Kissil and Lupatsch, 2004; Salze et al., 2010; Zhang et al., 2012). Moreover, with the same approach, and using more refined forms of plant protein, like protein concentrates or isolates, the use of plant protein blends, farm animal by-products (i.e., poultry by-product meal), and specific amino acids supplementation (i.e., varies with the mix), have been used to formulate fish meal-free diets that have been reported not to compromise growth performance in sea bream, rainbow trout, and Atlantic salmon (Belghit et al., 2019; Burr et al., 2012; Davidson et al., 2016; Kader et al., 2012; Slawski et al., 2013, 2012).

Despite these advances in fish meal and fish oil replacement, having in mind current rates projected for aquaculture growth and the use of marine ingredients, in particular carnivorous fish (Hua et al., 2019b), improving the fish (i.e., genetically), specifically in terms of feed conversion is highly relevant to keep improving the use of natural resources to feed cultured fish.

TOWARD THE GENETIC IMPROVEMENT OF NUTRIENT UTILIZATION IN FISH

The genetic improvement of fish is one of the critical components to increasing the sustainability of aquaculture. Different strategies can be used to improve cultivated fish's productive response, including tolerance to more stringent culture conditions (i.e., temperature, stress, diseases) as well as the way fish are able to use nutrients. Moreover, today with the rapid development and application of molecular techniques and the availability of genomic resources (i.e., genome sequencing) we now can look at specific physiological processes in more detail in order to understand them better and improve the use of natural resources used to feed the fish.

When it comes to selection, the most important breeding objective included into genetic improvement programs in fish is the rapid growth rate (Lhorente et al., 2019; Yáñez et al., 2020). However, the final increase in profitability also depends on feed efficiency; therefore, FCR should also be included in the breeding goal. Technical and biological aspects of selection for FCR in fish are presented as follows.

GENETIC IMPROVEMENT: GROWTH AND FCR

Genetic improvement programs aim to enhance both production efficiency and sustainability in terrestrial and aquatic animal systems. In this regard, genetic selection aimed at generating animal populations with a better feed conversion rate (FCR) is essential to reach these goals for the optimization of feed use, which in turn will have positive economic and environmental impacts. However, selective breeding for FCR in fish has been limited due to the evident difficulty in obtaining phenotypic records of individual feed intake in the aquatic environment. It has been shown that fast-growing animals are more efficient in the utilization of feed resources than slow-growing animals. In this regard, selective breeding for FCR has been indirectly carried out by selection for fast-growing individuals in livestock species. This is due to a favorable genetic association between feed intake and growth-related traits (Knap and Kause, 2018). For instance, a correlated selection response for rapid growth and FCR under restricted feeding has been observed in terrestrial animals (i.e., pigs and rabbits), suggesting that individuals who grow more rapidly under a restricted feeding regime have better FCR (Drouilhet et al., 2016; Nguyen and McPhee, 2005).

Moreover, heritability values for FCR under both an *ad libitum* and a restricted feeding regime are similar in rabbits and pigs (Drouilhet et al., 2013; Hermesch, 2004). Even the correlated selection response has shown similarities between the two different feeding regimes (Drouilhet et al., 2016). Similar results were found in a breeding population of sea bass, in which FCR and growth rate under restricted feeding were strongly correlated ($r_p = -0.78$ and $r_g = -0.98$) (Besson et al., 2019). This implies that improving animals for faster growth rate indirectly improve FCR as a correlated response. In aquaculture species, some studies identified the same negative correlation between FCR and growth-related traits, while other works have identified an absence of association between both traits (de Verdal et al., 2017; Knap and Kause, 2018). Therefore, an essential proportion of the genetic variance of FCR might be due to other sources rather than accelerated growth in fish.

RECORDING FCR FOR GENETIC IMPROVEMENT

Given the difficulties of measuring individual feed intake in fish, various approaches have been tested to approximate an estimation of the genetic component of FCR. For instance, family-based approaches have been used to measure feed intake, which is based on using tanks comprised of animals from only one family and quantifying the difference between the amount of feed provided and the feed waste collected (Kolstad et al., 2004). This approach exploits between-family information only to estimate the genetic variance and predict the genetic merit of feed efficiency for different animals and, therefore, will result in limited genetic progress if applied to actual selection (Doupé and Lymbery, 2003; Sonesson and Meuwissen, 2009).

Another approach to individually measure feed intake is the use of pellets, including radio-opaque material inside (e.g., glass beads). Using x-rays, it is possible to count the number of eaten pellets and thus record feed intake on an individual basis (Kause et al., 2016, 2006). The drawbacks of this approach are related to the management of fish, which need to be sedated and x-rayed to be individually phenotyped, and the need for multiple records for the same animal in order to account for the variance of feed intake across different feeding times.

Feed intake can also be measured on an individual basis in tanks to calculate FCR. For instance, phenotypic variation for feed efficiency and a high negative correlation ($r_p = -0.57$) between growth and residual feed intake under a restricted feeding regime was found by using individually isolated rainbow trout in a small-scale experiment (n=55 fish) (Silverstein, 2006). Moreover, genetic parameters for FCR have been estimated by rearing fish isolated in individual aquariums and recording intake of a limited amount of feed in 588 animals from a breeding population of sea bass, showing a moderate heritability value (0.25) (Besson et al., 2019). The evident drawback of recording FCR on an individual basis using aquariums is that it is a high resource- and time-consuming activity to be carried out on all the selection candidates. A recent study has shown the utility of stable isotope profile as an individual indicator trait to record feed efficiency in Atlantic salmon. Thus, the inclusion of 1–2% of ^{15}N in the diet showed reliable results of protein absorption and retention as a way of quantifying FCR in this species (Dvergedal et al., 2019b). Although this approach is technically feasible and available, it is still expensive for phenotyping FCR, considering that thousands of individuals have to be measured in the routine operation of a breeding program.

SELECTION FOR FCR USING INDIRECT TRAITS

Other indirect measures of FCR can be related to weight loss and fat deposition. For instance, weight loss during fasting is related with the levels of energy required for maintaining the metabolic rate of an individual, and there should be a correlated response on improving FCR when selecting for animals with lower weight loss during fasting, due to the fact that fish with lower maintenance requirements are converting food more efficiently. In fact, it has been observed that weight loss during fasting is correlated to residual feed intake in sea bass (Grima et al., 2010) and FCR in rainbow trout (Grima et al., 2008). Fish use their stored energy to cover maintenance costs. Regarding fat deposition, it has been suggested that selection for leaner animals will lead to the generation of more efficient individuals, due to the fact that fat retention requires more energy per unit of weight gain than protein retention. This has been demonstrated given the positive correlations found between fat deposition (measured as back fat depth) and feed efficiency in pigs (Knap and Wang, 2012).

THE USE OF GENOMICS TO IMPROVE FCR

The use of genomic information, through marker-assisted selection (MAS) or genomic prediction (GP) could enhance selection by improving the accuracy of selection in traits which are difficult to measure in selection candidates (e.g., FCR). These approaches are now feasible in a number of fish species by the development and implementation of dense single nucleotide polymorphism (SNP) panels. The association between a high number of SNP markers and the phenotype can be identified through genome-wide association analyses (GWAS). MAS schemes can be implemented if the SNPs associated explain a relatively high proportion of the genetic variance for the trait. If this is not the case for any particular trait, GP is the method of choice to accelerate the genetic progress for the trait, by using all the SNPs (i.e., thousands) available to predict the genetic merit of the breeders (Meuwissen et al., 2001). GP has been shown to be more accurate than conventional pedigree-based genetic evaluations for the improvement of several economically important traits in fish, including disease resistance (Bangera et al., 2017; Correa et al., 2017; Yoshida et al., 2018) and carcass quality traits (Horn et al., 2020; Yoshida et al., 2019). There is only one study in which FCR has been evaluated using genomic information. Genomic heritability of 0.47 was calculated for FCR in sea bass, and in terms of selection accuracy, genomic-based methods outperformed pedigree-based methods in the genetic evaluation for the trait (Besson et al., 2019). More recently, the use of stable isotope profiling as a trait indicator for feed efficiency has been used to identify genomic regions and signaling pathways associated with the trait in Atlantic salmon

(Dvergedal et al., 2020). In addition, the same approach has been demonstrated to represent reliable selection criteria to improve FCR in this species (Dvergedal et al., 2019b, 2019a). However, the high cost of this approach has to be taken into account when implementing it as a practical phenotyping measurement in real breeding programs.

GLOBAL GENE EXPRESSION TO UNDERSTAND THE RESPONSE TO DIETARY FISH REPLACEMENT WITH DIETARY VEGETABLE IN TELEOST FISH

The replacement of fish meal and fish oil as the main protein and lipid source in aquafeeds is a complex task considering it is not only high-quality protein and oil that is being removed (i.e., cholesterol and minerals). Also, when we use alternative protein and lipid sources, we add other components (i.e., carbohydrate, antinutritional factors) into the mix.

Considering that nutrients can act as metabolic signals (Panserat and Kaushik, 2010; Tacchi et al., 2011; Wacyk et al., 2012), molecular tools, particularly those evaluating transcriptional changes, have shed light on how metabolic pathways change in different fish tissues.

One of the primary omics tools for fish nutrition studies has been transcriptomics. Microarrays and RNA-seq data reveal the physiological effects of diets and ingredient replacements on the fish. As stated above, due to the increasing demand for oil and protein in aquafeeds, more sustainable alternatives, such as plant-derived ingredients, are needed. However, the consequences surrounding fish growth and health after such substitution are not yet fully understood. Hence the use of transcriptomic approaches to study the effects of FM and FO replacement in aquafeeds on different fish tissues is helping to move forward diet design and nutrient use. Most of these studies have been developed mainly in the juvenile stage of fish species like Salmo salar, Gadus morhua, Oncorhynchus mykiss, Paralichthys olivaceus Sparus aurata – Salmo salar being the most studied.

The main focus has been the effect of dietary replacement on the transcriptional profile of organs, like the fish intestine and liver.

Regarding the study of fish oil replacement, most of the studies used plant oil extracted from single species, such as Camelina seed (*Camelina sativa*), linseed (*Linum usitatissimum*), rapeseed (*Brassica napus*), soybean (*Glycine max*), or olive (*Olea europaea*). However, blends of rapeseed/Camelina seed, and palm or rapeseed/linseed and palm were also used in the range of 40–100% fish oil replacement.

A study by Jordal et al. in 2005 (Jordal et al., 2005) compared the liver gene expression profiling of 2 groups of Atlantic salmon fed diets containing 100% FO or 75% rapeseed oil (RO) for 42 weeks. Using a small-scale cDNA microarray of only 73 features associated with lipid metabolism, these authors indicated that mainly lipoprotein metabolism and cholesterol gene-associated pathways were affected by the diet in the liver. Later studies were done using a medium-scale cDNA microarray of 17K features (Leaver et al., 2008; Morais et al., 2011a, 2011b; Taggart et al., 2008), and then Betancor et al. used a high-scale oligonucleotides microarray of 44K features (Betancor et al., 2016, 2015), covering the complete genome of Atlantic salmon. Despite the diversity of variables and the fragmentation of information, these studies have allowed us to identify relevant dietary lipid replacement aspects. For instance, in Atlantic salmon, under 40% of dietary replacement had no significant effect on both growth and mortality of fish, but the (n-3):(n-6) ratio in the VO diet is less than the control (FO) in the liver, which is explained by a significant decrease of n-3 and an increase of n-6 fatty acid composition in tissues (Jordal et al., 2005; Leaver et al., 2008; Taggart et al., 2008; Torstensen et al., 2005). This aspect is relevant since Atlantic salmon flesh has a naturally high content of very-long-chain n–3 polyunsaturated fatty acids such as EPA and DHA. Interestingly, the use of wild type and metabolically engineered Camelina sativa oilseed in the diets had no detrimental effects on fish performance, feed efficiency, metabolic responses, neither decrease of n-3 fatty acid in the muscle tissue of Atlantic salmon compared with fish fed with the FO diet (Betancor et al., 2016, 2015). Moreover, the flesh of fish fed the metabolically engineered Camelina sativa diet had a higher EPA level than fish fed the FO diet, emphasizing the relevant role that this type of VO can play in replacing FO in salmon diets.

In this way, current evidence indicates that overrepresented gene ontology (GO) categories disturbed by dietary replacement included sterol biosynthesis, isoprenoid biosynthesis, fatty acid desaturation, mRNA processing, translational elongation, metabolism of carbohydrates, and immune response, among others. Notably among those genes that consistently were up-regulated among the above-mentioned studies were the Δ5/6 fatty acyl desaturases, which function in HUFA biosynthesis. However, in some cases, this up-regulation was not sufficient to reach the concentrations of long-chain n–3 polyunsaturated fatty acids as in fish fed with FO. On the contrary, even though the dietary cholesterol concentrations were always lower in the VO diets, the plasma concentrations in the fish fed both diets were similar, indicating that the Mevalonate pathway's activation was enough to reach the concentration of plasma cholesterol through its endogenous synthesis. This information observed in several fish species is also supported by the consistent up-regulation of the transcriptional regulator SREBP2 in response to different VO sources (Leaver et al., 2008; Morais et al., 2011b).

Morais and coworkers reported that processes, such as metabolism of lipoprotein, cholesterol, and carbohydrates, were disturbed in both Atlantic salmon family genogroups defined as "Lean" or "Fat", despite the genetic background of the fish in response to a dietary replacement, while other processes were particularly perturbed in each group (Morais et al., 2011a, 2011b). This fact emphasizes that, although dietary replacement with plant lipids impacts common biological processes among the fish species studied, there is a wide variability of response even within the same species.

Similarly, molecular tools to study the effect of fish meal replacement have shown us several physiological processes and metabolic pathways that change when using alternative protein sources.

One of the most widely used proteins to replace fish meal is produced from soy, and the use of soybean meal (SBM) has been associated with the generation of inflammation in the distal portion of the fish intestine (Baeverfjord and Krogdahl, 1996; Booman et al., 2018). This condition's development has been described to depend on ANF content, particularly saponins (Krogdahl et al., 2015), and that carnivorous fish, like salmonids, seem to be more susceptible than omnivorous or herbivorous fish (Urán et al., 2008). Using microarray to study the early response of Atlantic salmon intestine to the inclusion of SBM (i.e., 20%), Sahlmann et al. (2013) reported that, along with the first histological changes in the intestine, most changes in gene transcription seem to occur after five days of being fed the experimental diets. Different gene transcription patterns were observed in this period, with genes associated with an immune response like NF-KB related genes and regulators of T and B cell function being up-regulated, while downregulation of genes was associated with transport processes like endocytosis, exocytosis, transporters, and pathways associated with metabolism. These gene expression changes indicate tissue dysfunction and are associated with a negative impact on the productive performance of the fish. Also looking at the effect of replacing fish meal with plant protein, Król et al. (2016) evaluated the effect of the inclusion of soy and faba bean protein concentrates (more refined products) and compared the transcriptome changes in the Salmo salar intestine with fish fed SBM- and FM-based diets. In this study, the authors report lower levels of transcriptional alterations when protein blends were used, compared to when the fish were fed diets with only one plant protein source, likely due to a reduced ANF content. Recently, Kiron et al. (2020), using RNA-seq to describe the effects of the inclusion of SBM in Salmo salar diets over intestinal transcriptome, report that among disturbed pathways were those associated with taurine metabolism, cytochrome p450 biosynthesis, and transport mechanisms in this tissue, along with several other markers associated with the development of inflammation. Working with Seriola lalandi, Dam et al. (2020) used several plant protein sources and farm animal protein meals to evaluate intestinal transcriptome changes. In this study, they identified changes in gene expression related to nutrient metabolism and protein digestion in favor of poultry by-product meal and faba bean meal compared to blood meal and corn gluten meal in practical aquafeeds. Changes in microbiota have also been reported in response to the inclusion of SBM in salmonid diets (Catalán et al., 2018), digestive enzyme activity, and gene expression (Lilleeng et al., 2007; Perera and Yúfera, 2017).

Efficient nutrient absorption is an important component to keep in mind, but to improve feed conversion and nutrient retention, we have to consider the central role the liver is playing. Working with Atlantic cod Lie et al. (2011), studied the effects of different levels of a plant protein blend (soy protein concentrate, soybean meal, and wheat gluten) on growth performance and gene expression in the livers of fish. In this study, the replacement of 75% of the fish meal by plant protein did not reduce fish growth performance but feeding the fish diets with 100% replacement significantly reduced feed intake, growth, and feed conversion. In these last fish, genes associated with growth regulation (i.e., IGFIIR) and with protein turnover (i.e., CatD) were down-regulated in the liver. Working with rainbow trout, Panserat et al. (2008, 2009) completely replaced fish meal with plant protein and reported significant alternations in different metabolic pathways in the liver of fish, particularly with depression in protein metabolism response to the plant protein. Caballero-Solares et al. (2018), working with Salmo salar that were fed diets with only 5% of FM reported significant changes in the transcription of genes involved in carbohydrate and lipid metabolism, cell growth, apoptosis, and immune response. Similar to the intestine, proper liver function is central for efficient use of dietary nutrients, particularly in terms of the metabolic use fish are given to absorbed nutrients (Halver and Hardy, 2003; Wacyk et al., 2012),

Available data from replacement studies using molecular tools suggest that adaptation to dietary replacement is a complex process. Hence, to optimize the efficient use of nutrients, such as plant-derived ingredients or others currently being tested, it is necessary to integrate disciplines and expand the number of studies to other species of plants and fish, other tissues, and other stages of fish development. Studying the effects of the inclusion of new protein sources on the expression of genes in fish intestines or organs like the liver and muscle helps improve the use of plant ingredients further and pave the road for new ingredient alternatives.

CONCLUDING REMARKS

Intensive aquaculture relies on the use of formulated feeds to cover the energy and nutrient requirements of fish. The requirements need to be known and used along with technical and commercial criteria to formulate and fabricate diets. With the increasing awareness regarding the environmental impact of our food, current projection of aquaculture growth, and the likely scenarios in which this will occur (Gephart et al., 2020), in order for this growth to be a sustainable, efficient use of natural resources to feed the fish, improving feed conversion and nutrient retention is crucial.

Improving the use of dietary nutrients will move forward, taking advantage of the knowledge generated by studying the use of alternatives to marine ingredients, particularly protein and oil derived from terrestrial plants. Among current challenges associated with the diet, the refinement of nutrient requirements and metabolic function of nutrients in different fish species and growth stages is needed to frame further advances in reducing fish meal and fish oil use. Given the complexity and nutrient-rich nature of marine ingredients, studying the role of amino acids like taurine, minerals, vitamins, and energy-yielding macronutrients regarding their dietary proportions will continue to receive attention from carnivorous fish.

Feed efficiency is a highly desirable trait to be included into the breeding goal of genetic improvement programs for many aquaculture species, from both profitability and sustainability perspectives. However, this trait is hard and expensive to measure at an individual level and different approaches have been tested to directly or indirectly phenotype FCR in a breeding context. Some of these methods requires that the animals are sacrificed, and others can be obtained from live candidates. For the former, phenotypic records on the full-sibs can be used to predict the genetic merit on the non-tested selection candidates. Genomic predictions, which use phenotype and genotype information on reference animals for selection of the genotyped selection candidates will be much more accurate than conventional pedigree-based genetic evaluations. Thus, genomic selection is expected to increase the rate of genetic progress for feed efficiency, especially if full-sib testing is used to directly or indirectly measure FCR.

Over the past decade, the aquaculture industry has incorporated several research tools and technologies to improve feed utilization. Among the tools are genetics and omics sciences, particularly genomics. This strategy has allowed the selection of fish families with greater efficiency in using dietary ingredients and identifying critical biological processes to determine the effects of dietary replacements on fish.

Considering the trend observed in other research areas, such as biomedicine and terrestrial animal production, other strategies derived from biotechnology, such as the generation of transgenic plants and animals, use new tools CRISPR/ Cas9 that could be put to use in aquaculture production. Similarly, other omics sciences, such as proteomics, metabolic, and interactomics, should be incorporated into the industry in a more standardized way. On the other hand, the broad availability of existing data (big data) and new informatic analysis technologies, such as machine learning, will allow meta-analyzing the information to integrate it and optimize food use processes in the aquaculture industry. For this to happen, it will also be necessary to incorporate other elements inherent to the production, such as the efficient use of water and energy and environmentally sustainable production, all of which must be accompanied by a close relationship between industry and academia.

REFERENCES

Abdou, K., Aubin, J., Romdhane, M.S., Le Loc'h, F., Lasram, F.B.R., 2017. Environmental assessment of seabass (*Dicentrarchus labrax*) and seabream (*Sparus aurata*) farming from a life cycle perspective: A case study of a Tunisian aquaculture farm. *Aquaculture* 471, 204–212. doi:10.1016/j.aquaculture.2017.01.019

Abdou, K., Ben Rais Lasram, F., Romdhane, M.S., Le Loc'h, F., Aubin, J., 2018. Rearing performances and environmental assessment of sea cage farming in Tunisia using life cycle assessment (LCA) combined with PCA and HCPC. *Int. J. Life Cycle Assess.* 23, 1049–1062. doi:10.1007/s11367-017-1339-2

Aksnes, A., Hjertnes, T., Opstvedt, J., 1996. Comparison of two assay methods for determination of nutrient and energy digestibility in fish. *Aquaculture* 140, 343–359. doi:10.1016/0044-8486(95)01200-1

Altan, O., Gamsiz, K., Korkut, A.Y., 2010. Soybean meal and rendered animal protein ingredients replace fishmeal in practical diets for sea bass. Isr. *J. Aquac. – Bamidgeh* 62, 56–62.

Aubin, J., Papatryphon, E., van der Werf, H.M.G., Chatzifotis, S., 2009. Assessment of the environmental impact of carnivorous finfish production systems using life cycle assessment. *J. Clean. Prod.* 17, 354–361. doi:10.1016/j.jclepro.2008.08.008

Azevedo, P.A., Bureau, D.P., Leeson, S., Cho, Y., 2002. Growth and efficiency of feed usage by Atlantic salmon (*Salmo salar*) fed diets with different dietary protein: Energy ratios at two feeding levels. *Fish. Sci.* 68, 878–888. doi:10.1046/j.1444-2906.2002.00506.x

Azevedo, P.A., Cho, C.Y., Leeson, S., Bureau, D.P., 1998. Effects of feeding level and water temperature on growth, nutrient and energy utilization and waste outputs of rainbow trout (*Oncorhynchus mykiss*). *Aquatic Living Resources* 11(4), 227–238.

Baeverfjord, G., Krogdahl, A., 1996. Development and regression of soybean meal induced enteritis in Atlantic salmon, *Salmo salar* L., distal intestine: A comparison with the intestines of fasted fish. *J. Fish Dis.* 19, 375–387. doi:10.1046/j.1365-2761.1996.d01-92.x

Bangera, R., Correa, K., Lhorente, J.P., Figueroa, R., Yáñez, J.M., 2017. Genomic predictions can accelerate selection for resistance against *Piscirickettsia salmonis* in Atlantic salmon (*Salmo salar*). *BMC Genomics* 18, 121. doi:10.1186/s12864-017-3487-y

Barrows, F.T., Gaylord, T.G., Sealey, W.M., Porter, L., Smith, C.E., 2008. The effect of vitamin premix in extruded plant-based and fish meal based diets on growth efficiency and health of rainbow trout, *Oncorhynchus mykiss*. *Aquaculture* 283, 148–155. doi:10.1016/j.aquaculture.2008.07.014

Belghit, I., Liland, N.S., Gjesdal, P., Biancarosa, I., Menchetti, E., Li, Y., Waagbø, R., Krogdahl, Å., Lock, E.J., 2019. Black soldier fly larvae meal can replace fish meal in diets of sea-water phase Atlantic salmon (*Salmo salar*). *Aquaculture* 503, 609–619. doi:10.1016/j.aquaculture.2018.12.032

Besson, M., Allal, F., Chatain, B., Vergnet, A., Clota, F., Vandeputte, M., 2019. Combining individual phenotypes of feed intake with genomic data to improve feed efficiency in sea Bass. *Front. Genet.* 10, 219. doi:10.3389/fgene.2019.00219

Betancor, M.B., Sprague, M., Sayanova, O., Usher, S., Campbell, P.J., Napier, J.A., Caballero, M.J., Tocher, D.R., 2015. Evaluation of a high-EPA oil from transgenic *Camelina sativa* in feeds for Atlantic salmon (*Salmo salar* L.): Effects on tissue fatty acid composition, histology and gene expression. *Aquaculture* 444, 1–12. doi:10.1016/j.aquaculture.2015.03.020

Betancor, M.B., Sprague, M., Sayanova, O., Usher, S., Metochis, C., Campbell, P.J., Napier, J.A., Tocher, D.R., 2016. Nutritional evaluation of an EPA-DHA oil from Transgenic *Camelina sativa* in feeds for post-smolt Atlantic salmon (*Salmo salar* L.). *PLoS One* 11, e0159934. doi:10.1371/journal.pone.0159934

Bohnes, F.A., Hauschild, M.Z., Schlundt, J., Laurent, A., 2019. Life cycle assessments of aquaculture systems: A critical review of reported findings with recommendations for policy and system development. *Rev. Aquac.* 11, 1061–1079. doi:10.1111/raq.12280

Boissy, J., Aubin, J., Drissi, A., van der Werf, H.M.G., Bell, G.J., Kaushik, S.J., 2011. Environmental impacts of plant-based salmonid diets at feed and farm scales. *Aquaculture* 321, 61–70. doi:10.1016/j.aquaculture.2011.08.033

Booman, M., Forster, I., Vederas, J.C., Groman, D.B., Jones, S.R.M., 2018. Soybean meal-induced enteritis in Atlantic salmon (*Salmo salar*) and Chinook salmon (*Oncorhynchus tshawytscha*) but not in pink salmon (*O. gorbuscha*). *Aquaculture* 483, 238–243. doi:10.1016/j.aquaculture.2017.10.025

Boyd, C.E., Tucker, C., McNevin, A., Bostick, K., Clay, J., 2007. Indicators of resource use efficiency and environmental performance in fish and crustacean aquaculture. *Rev. Fish. Sci.* 15, 327–360. doi:10.1080/10641260701624177

Bureau, D.P., Hua, K., Cho, C.Y., 2006. Effect of feeding level on growth and nutrient deposition in rainbow trout (*Oncorhynchus mykiss* Walbaum) growing from 150 to 600 g. *Aquac. Res.* 37, 1090–1098. doi:10.1111/j.1365-2109.2006.01532.x

Burr, G.S., Wolters, W.R., Barrows, F.T., Hardy, R.W., 2012. Replacing fishmeal with blends of alternative proteins on growth performance of rainbow trout (*Oncorhynchus mykiss*), and early or late stage juvenile Atlantic salmon (*Salmo salar*). *Aquaculture* 334–337, 110–116. doi:10.1016/j.aquaculture.2011.12.044

Caballero-Solares, A., Xue, X., Parrish, C.C., Foroutani, M.B., Taylor, R.G., Rise, M.L., 2018. Changes in the liver transcriptome of farmed Atlantic salmon (*Salmo salar*) fed experimental diets based on terrestrial alternatives to fish meal and fish oil. *BMC Genomics* 19, 1–26. doi:10.1186/s12864-018-5188-6

Cabral, E.M., Bacelar, M., Batista, S., Castro-Cunha, M., Ozório, R.O.A., Valente, L.M.P., 2011. Replacement of fishmeal by increasing levels of plant protein blends in diets for Senegalese sole (*Solea senegalensis*) juveniles. *Aquaculture* 322–323, 74–81. doi:10.1016/j.aquaculture.2011.09.023

Cao, L., Diana, J.S., Keoleian, G.A., 2013. Role of life cycle assessment in sustainable aquaculture. *Rev. Aquac.* 5, 61–71. doi:10.1111/j.1753-5131.2012.01080.x

Catalán, N., Villasante, A., Wacyk, J., Ramírez, C., Romero, J., 2018. Fermented soybean meal increases lactic acid bacteria in gut microbiota of Atlantic salmon (*Salmo salar*). *Probiotics Antimicrob. Proteins* 10, 566–576. doi:10.1007/s12602-017-9366-7

Choi, D.G., He, M., Fang, H., Wang, X.L., Li, X.Q., Leng, X.J., 2020. Replacement of fish meal with two fermented soybean meals in diets for rainbow trout (*Oncorhynchus mykiss*). *Aquac. Nutr.* 26, 37–46. doi:10.1111/anu.12965

Collins, S.A., Øverland, M., Skrede, A., Drew, M.D., 2013. Effect of plant protein sources on growth rate in salmonids: Meta-analysis of dietary inclusion of soybean, pea and canola/rapeseed meals and protein concentrates. *Aquaculture*, 400–401, 85–100. doi:10.1016/j.aquaculture.2013.03.006

Cook, J.T., McNiven, M.A., Richardson, G.F., Sutterlin, A.M., 2000. Growth rate, body composition and feed digestibility/conversion of growth-enhanced transgenic Atlantic salmon (*Salmo salar*). *Aquaculture* 188, 15–32. doi:10.1016/S0044-8486(00)00331-8

Cornell, J.A., 2002. *Experiments with Mixtures. The Journal of Ecology*, Wiley Series in Probability and Statistics. Wiley. doi:10.1002/9781118204221

Correa, K., Bangera, R., Figueroa, R., Lhorente, J.P., Yáñez, J.M., 2017. The use of genomic information increases the accuracy of breeding value predictions for sea louse (*Caligus rogercresseyi*) resistance in Atlantic salmon (*Salmo salar*). *Genet. Sel. Evol.* 49, 1–5. doi:10.1186/s12711-017-0291-8

Crampton, V.O., Nanton, D.A., Ruohonen, K., Skjervold, P.-O., El-Mowafi, A., 2010. Demonstration of salmon farming as a net producer of fish protein and oil. *Aquac. Nutr.* 16, 437–446. doi:10.1111/j.1365-2095.2010.00780.x

d'Orbcastel, E.R., Blancheton, J.P., Aubin, J., 2009. Towards environmentally sustainable aquaculture: Comparison between two trout farming systems using life cycle assessment. *Aquac. Eng.* 40, 113–119. doi:10.1016/j.aquaeng.2008.12.002

Dam, C.T.M., Ventura, T., Booth, M., Pirozzi, I., Salini, M., Smullen, R., Elizur, A., 2020. Intestinal transcriptome analysis highlights key differentially expressed genes involved in nutrient metabolism and digestion in yellowtail kingfish (*Seriola lalandi*) fed terrestrial animal and plant proteins. *Genes (Basel)* 11, 1–17. doi:10.3390/genes11060621

Davidson, J., Barrows, F.T., Kenney, P.B., Good, C., Schroyer, K., Summerfelt, S.T., 2016. Effects of feeding a fishmeal-free versus a fishmeal-based diet on post-smolt Atlantic salmon *Salmo salar* performance, water quality, and waste production in recirculation aquaculture systems. *Aquac. Eng.* 74, 38–51. doi:10.1016/j.aquaeng.2016.05.004

de Verdal, H., Mekkawy, W., Lind, C.E., Vandeputte, M., Chatain, B., Benzie, J.A.H., 2017. Measuring individual feed efficiency and its correlations with performance traits in Nile tilapia, *Oreochromis niloticus*. *Aquaculture* 468, 489–495. doi:10.1016/j.aquaculture.2016.11.015

Dessen, J.E., Weihe, R., Hatlen, B., Thomassen, M.S., Rørvik, K.A., 2017. Different growth performance, lipid deposition, and nutrient utilization in in-season (S1) Atlantic salmon post-smolt fed isoenergetic diets differing in protein-to-lipid ratio. *Aquaculture* 473, 345–354. doi:10.1016/j.aquaculture.2017.02.006

Doupé, R.G., Lymbery, A.J., 2003. Toward the genetic improvement of feed conversion efficiency in Fish. *J. World Aquac. Soc.* 34, 245–254. doi:10.1111/j.1749-7345.2003.tb00063.x

Drouilhet, L., Achard, C.S., Zemb, O., Molette, C., Gidenne, T., Larzul, C., Ruesche, J., Tircazes, A., Segura, M., Bouchez, T., Theau-Clément, M., Joly, T., Balmisse, E., Garreau, H., Gilbert, H., 2016. Direct and correlated responses to selection in two lines of rabbits selected for feed efficiency under ad libitum and restricted feeding: I. Production traits and gut microbiota characteristics. *J. Anim. Sci.* 94, 38–48. doi:10.2527/jas.2015-9402

Drouilhet, L., Gilbert, H., Balmisse, E., Ruesche, J., Tircazes, A., Larzul, C., Garreau, H., 2013. Genetic parameters for two selection criteria for feed efficiency in rabbits. *J. Anim. Sci.* 91, 3121–3128. doi:10.2527/jas.2012-6176

Dvergedal, H., Harvey, T.N., Jin, Y., Ødegård, J., Grønvold, L., Sandve, S.R., Våge, D.I., Moen, T., Klemetsdal, G., 2020. Genomic regions and signaling pathways associated with indicator traits for feed efficiency in juvenile Atlantic salmon (*Salmo salar*). *Genet. Sel. Evol.* 52, 66. doi:10.1186/s12711-020-00587-x

Dvergedal, H., Ødegård, J., Mydland, L.T., Øverland, M., Hansen, J.Ø., Ånestad, R.M., Klemetsdal, G., 2019a. Stable isotope profiling for large-scale evaluation of feed efficiency in Atlantic salmon (*Salmo salar*). *Aquac. Res.* 50, 1153–1161. doi:10.1111/are.13990

Dvergedal, H., Ødegård, J., Øverland, M., Mydland, L.T., Klemetsdal, G., 2019b. Selection for feed efficiency in Atlantic salmon using individual indicator traits based on stable isotope profiling. *Genet. Sel. Evol.* 51, 1–14. doi:10.1186/s12711-019-0455-9

Egerton, S., Wan, A., Murphy, K., Collins, F., Ahern, G., Sugrue, I., Busca, K., Egan, F., Muller, N., Whooley, J., McGinnity, P., Culloty, S., Ross, R.P., Stanton, C., 2020. Replacing fishmeal with plant protein in Atlantic salmon (*Salmo salar*) diets by supplementation with fish protein hydrolysate. *Sci. Rep.* 10, 1–16. doi:10.1038/s41598-020-60325-7

Einen, O., Roem, A.J., 1997. Dietary protein/energy ratios for Atlantic salmon in relation to fish size: Growth, feed utilization and slaughter quality. *Aquac. Nutr.* 3, 115–126. doi:10.1046/j.1365-2095.1997.00084.x

FAO, 2007. *A Qualitative Assessment of Standards and Certification Schemes Applicable to Aquaculture in the Asia–Pacific Region*, Bangkok, Thailand. FAO [WWW Document].

FAO, 2011. "ENERGY-SMART" *FOOD FOR PEOPLE AND CLIMATE*, FAO Rome, Italy.

FAO, 2018. *El estado mundial de la pesca y la acuicultura*, FAO Rome, Italy.

FAO, 2020. *The State of World Fisheries and Aquaculture 2020, The State of World Fisheries and Aquaculture 2020*, FAO. Rome, Italy. doi:10.4060/ca9229en

Foroutani, M.B., Parrish Id, C.C., Wells, J., Taylor, R.G., Rise, M.L., Shahidi, F., 2018. Minimizing marine ingredients in diets of farmed Atlantic salmon (*Salmo salar*): Effects on growth performance and muscle lipid and fatty acid composition. *PLoS One* 13(9), e0198538. doi:10.1371/journal.pone.0198538

Fry, J.P., Love, D.C., MacDonald, G.K., West, P.C., Engstrom, P.M., Nachman, K.E., Lawrence, R.S., 2016. Environmental health impacts of feeding crops to farmed fish. *Environ. Int.* 91, 201–214. doi:10.1016/j.envint.2016.02.022

Fry, J.P., Mailloux, N.A., Love, D.C., Milli, M.C., Cao, L., 2018. Feed conversion efficiency in aquaculture: Do we measure it correctly? *Environ. Res. Lett.* 13, 024017. doi:10.1088/1748-9326/aaa273

Gasco, L., Gai, F., Maricchiolo, G., Genovese, L., Ragonese, S., Bottari, T., Caruso, G., 2018. Supplementation of vitamins, minerals, enzymes and antioxidants in fish feeds. . In: *Feeds for the Aquaculture Sector. Springer Briefs in Molecular Science.* Springer, Cham, pp. 63–103. doi:10.1007/978-3-319-77941-6_4

Gatlin, D.M., Barrows, F.T., Brown, P., Dabrowski, K., Gaylord, T.G., Hardy, R.W., Herman, E., Hu, G., Krogdahl, Å., Nelson, R., Overturf, K., Rust, M., Sealey, W., Skonberg, D., Souza, E.J., Stone, D.,

Wilson, R., Wurtele, E., 2007. Expanding the utilization of sustainable plant products in aquafeeds: A review. *Aquac. Res.* 38, 551–579. doi:10.1111/j.1365-2109.2007.01704.x

Gephart, J.A., Golden, C.D., Asche, F., Belton, B., Brugere, C., Froehlich, H.E., Fry, J.P., Halpern, B.S., Hicks, C.C., Jones, R.C., Klinger, D.H., Little, D.C., McCauley, D.J., Thilsted, S.H., Troell, M., Allison, E.H., 2020. Scenarios for global aquaculture and its role in human nutrition. *Rev. Fish. Sci. Aquac.* 29(1), 1–17. doi:10.1080/23308249.2020.1782342

Gerber, P.J., Steinfeld, H., Henderson, B., Mottet, A., Opio, C., Dijkman, J., Falcucci, A., Tempio, G., 2013. *Tackling Climate Change through Livestock – A Global Assessment of Emissions and Mitigation Opportunities* FAO, Rome, Italy.

Glencross, B., Evans, D., Dods, K., McCafferty, P., Hawkins, W., Maas, R., Sipsas, S., 2005. Evaluation of the digestible value of lupin and soybean protein concentrates and isolates when fed to rainbow trout, *Oncorhynchus mykiss*, using either stripping or settlement faecal collection methods. *Aquaculture* 245, 211–220. doi:10.1016/j.aquaculture.2004.11.033

Glencross, B., Turchini, G., 2010. Fish oil replacement in starter, grow-out, and finishing feeds for farmed aquatic animals, in: Turchini GM, Ng W-K, Tocher DR (Eds.) *Fish Oil Replacement and Alternative Lipid Sources in Aquaculture Feeds.* CRC Press, pp. 373–404. doi:10.1201/9781439808634-c12

Gomes, E.F., Rema, P., Kaushik, S.J., 1995. Replacement of fish meal by plant proteins in the diet of rainbow trout (*Oncorhynchus mykiss*): Digestibility and growth performance. *Aquaculture* 130, 177–186. doi:10.1016/0044-8486(94)00211-6

Grima, L., Quillet, E., Boujard, T., Robert-Granié, C., Chatain, B., Mambrini, M., 2008. Genetic variability in residual feed intake in rainbow trout clones and testing of indirect selection criteria (Open Access publication). *Genet. Sel. Evol.* 40, 607. doi:10.1186/1297-9686-40-6-607

Grima, L., Vandeputte, M., Ruelle, F., Vergnet, A., Mambrini, M., Chatain, B., 2010. In search for indirect criteria to improve residual feed intake in sea bass (*Dicentrarchus labrax*). Part I: Phenotypic relationship between residual feed intake and body weight variations during feed deprivation and re-feeding periods. *Aquaculture* 300, 50–58. doi:10.1016/j.aquaculture.2010.01.003

Halver, J.E., Hardy, R.W., 2002. *Fish Nutrition*. Academic Press, San Diego, CA.

Halver, J.E., Hardy, R.W., 2003. 14 – Nutrient flow and retention, in: Halver, J.E., Hardy, R.W. (Eds.), *Fish Nutrition* (3rd ed.). Academic Press, San Diego, CA, pp. 755–770. doi:10.1016/B978-012319652-1/50015-X

Handeland, S.O., Imsland, A.K., Stefansson, S.O., 2008. The effect of temperature and fish size on growth, feed intake, food conversion efficiency and stomach evacuation rate of Atlantic salmon post-smolts. *Aquaculture* 283, 36–42. doi:10.1016/j.aquaculture.2008.06.042

Hansen, A.C., Rosenlund, G., Karlsen, Ø., Koppe, W., Hemre, G.I., 2007. Total replacement of fish meal with plant proteins in diets for Atlantic cod (*Gadus morhua* L.) I – Effects on growth and protein retention. *Aquaculture* 272, 599–611. doi:10.1016/j.aquaculture.2007.08.034

Hardy, R.W., 2010. Utilization of plant proteins in fish diets: Effects of global demand and supplies of fishmeal. *Aquac. Res.* 41(5), 770–776. doi:10.1111/j.1365-2109.2009.02349.x

Hardy, R.W., Barrows, F.T., 2003. 9 – Diet formulation and manufacture, in: Halver, J.E., Hardy, R.W. (Eds.), *Fish Nutrition* (3rd ed.). Academic Press, San Diego, CA, pp. 505–600. doi:10.1016/B978-012319652-1/50010-0

Henriksson, P.J.G., Belton, B., Murshed-E-Jahan, K., Rico, A., 2018. Measuring the potential for sustainable intensification of aquaculture in Bangladesh using life cycle assessment. *Proc. Natl. Acad. Sci. U. S. A.* 115, 2958–2963. doi:10.1073/pnas.1716530115

Henriksson, P.J.G., Dickson, M., Allah, A.N., Al-Kenawy, D., Phillips, M., 2017. Benchmarking the environmental performance of best management practice and genetic improvements in Egyptian aquaculture using life cycle assessment. *Aquaculture* 468, 53–59. doi:10.1016/j.aquaculture.2016.09.051

Hermesch, S., 2004. Genetic improvement of lean meat growth and feed efficiency in pigs. *Austr. J. Exp. Agric.* 44, 383–391. doi:10.1071/ea04017

Hillestad, M., Johnsen, F., 1994. High-energy/low-protein diets for Atlantic salmon: Effects on growth, nutrient retention and slaughter quality. *Aquaculture* 124, 109–116. doi:10.1016/0044-8486(94)90366-2

Horn, S.S., Meuwissen, T.H.E., Moghadam, H., Hillestad, B., Sonesson, A.K., 2020. Accuracy of selection for omega-3 fatty acid content in Atlantic salmon fillets. *Aquaculture* 519, 734767. doi:10.1016/j.aquaculture.2019.734767

Houlihan, D., Boujard, T., Jobling, M., 2001. *Food Intake in Fish, Food Intake in Fish.* Blackwell Science Ltd, Oxford, UK. doi:10.1002/9780470999516

Hua, K., Cobcroft, J.M., Cole, A., Condon, K., Jerry, D.R., Mangott, A., Praeger, C., Vucko, M.J., Zeng, C., Zenger, K., Strugnell, J.M., 2019a. One earth the future of aquatic protein: Implications for protein sources in aquaculture diets. *One Earth* 1, 316–329. doi:10.1016/j.oneear.2019.10.018

Hua, K., Cobcroft, J.M., Cole, A., Condon, K., Jerry, D.R., Mangott, A., Praeger, C., Vucko, M.J., Zeng, C., Zenger, K., Strugnell, J.M., 2019b. One earth the future of aquatic protein: Implications for protein sources in aquaculture diets. *One Earth*, 1(3), 316–329. doi:10.1016/j.oneear.2019.10.018

Iwashita, Y., Yamamoto, T., Furuita, H., Sugita, T., Suzuki, N., 2008. Influence of certain soybean antinutritional factors supplemented to a casein-based semipurified diet on intestinal and liver morphology in fingerling rainbow trout *Oncorhynchus mykiss*. *Fish. Sci.* 74, 1075–1082. doi:10.1111/j.1444-2906.2008.01627.x

Jobling M. 1993. Bioenergetics: feed intake and energy partitioning. In: Rankin J.C., Jensen F.B. (eds) *Fish Ecophysiology*. Chapman & Hall Fish and Fisheries Series, vol 9. Springer, Dordrecht. https://doi.org/10.1007/978-94-011-2304-4_1.

Jobling, M., Arnesen, A.M., Baardvik, B.M., Christiansen, J.S., Jørgensen, E.H., 1995. Monitoring feeding behaviour and food intake: Methods and applications. *Aquac. Nutr.* 1, 131–143. doi:10.1111/j.1365-2095.1995.tb00037.x

Jobling, M., Johansen, S.J.S., 1999. The lipostat, hyperphagia and catch-up growth. *Aquac. Res.* 30, 473–478. doi:10.1046/j.1365-2109.1999.00358.x

Jordal, A.E.O., Torstensen, B.E., Tsoi, S., Tocher, D.R., Lall, S.P., Douglas, S.E., 2005. Dietary rapeseed oil affects the expression of genes involved in hepatic lipid metabolism in Atlantic salmon (*Salmo salar* L.). *J. Nutr.* 135, 2355–2361. doi:10.1093/jn/135.10.2355

Kabir, K.A., Verdegem, M.C.J., Verreth, J.A.J., Phillips, M.J., Schrama, J.W., 2019. Effect of dietary protein to energy ratio, stocking density and feeding level on performance of Nile tilapia in pond aquaculture. *Aquaculture* 511, 634200. doi:10.1016/j.aquaculture.2019.06.014

Kabir, K.A., Verdegem, M.C.J., Verreth, J.A.J., Phillips, M.J., Schrama, J.W., 2020. Effect of dietary carbohydrate to lipid ratio on performance of Nile tilapia and enhancement of natural food in pond aquaculture. *Aquac. Res.* 51, 1942–1954. doi:10.1111/are.14546

Kader, M.A., Bulbul, M., Koshio, S., Ishikawa, M., Yokoyama, S., Nguyen, B.T., Komilus, C.F., 2012. Effect of complete replacement of fishmeal by dehulled soybean meal with crude attractants supplementation in diets for red sea bream, Pagrus major. *Aquaculture* 350–353, 109–116. doi:10.1016/j.aquaculture.2012.04.009

Kamalam, B.S., Medale, F., Panserat, S., 2017. Utilisation of dietary carbohydrates in farmed fishes: New insights on influencing factors, biological limitations and future strategies. *Aquaculture*. doi:10.1016/j.aquaculture.2016.02.007

Kause, A., Kiessling, A., Martin, S.A.M., Houlihan, D., Ruohonen, K., 2016. Genetic improvement of feed conversion ratio via indirect selection against lipid deposition in farmed rainbow trout (*Oncorhynchus mykiss* Walbaum). *Br. J. Nutr.* 116, 1656–1665. doi:10.1017/S0007114516003603

Kause, A., Tobin, D., Houlihan, D.F., Martin, S.A.M., Mäntysaari, E.A., Ritola, O., Ruohonen, K., 2006. Feed efficiency of rainbow trout can be improved through selection: Different genetic potential on alternative diets. *J. Anim. Sci.* 84, 807–817. doi:10.2527/2006.844807x

Kaushik, S.J., Covès, D., Dutto, G., Blanc, D., 2004. Almost total replacement of fish meal by plant protein sources in the diet of a marine teleost, the European seabass, *Dicentrarchus labrax*. *Aquaculture* 230, 391–404. doi:10.1016/S0044-8486(03)00422-8

Kiron, V., Park, Y., Siriyappagouder, P., Dahle, D., Vasanth, G.K., Dias, J., Fernandes, J.M.O., Sørensen, M., Trichet, V.V., 2020. Intestinal transcriptome analysis reveals soy derivative-linked changes in Atlantic salmon. *Front. Immunol.* 11, 3013. doi:10.3389/fimmu.2020.596514

Kissil, G.W., Lupatsch, I., 2004. Successful replacement of fishmeal by plant proteins in diets for the gilthead seabream, *Sparus aurata* L. Isr. *J. Aquac. – Bamidgeh* 56, 188–199.

Knap, P.W., Kause, A., 2018. Phenotyping for genetic improvement of feed efficiency in fish: Lessons from pig breeding. *Front. Genet.* 9, 24. doi:10.3389/fgene.2018.00184

Knap, P.W., Wang, L., 2012. Pig breeding for improved feed efficiency, in: John F. Patience (Ed.), *Feed Efficiency in Swine*. Wageningen, Wageningen Academic Publishers, pp. 167–181. doi:10.3920/978-90-8686-756-1_8

Kolstad, K., Grisdale-Helland, B., Gjerde, B., 2004. Family differences in feed efficiency in Atlantic salmon (*Salmo salar*). *Aquaculture* 241, 169–177. doi:10.1016/j.aquaculture.2004.09.001

Krogdahl, Å., Gajardo, K., Kortner, T.M., Penn, M., Gu, M., Berge, G.M., Bakke, A.M., 2015. Soya saponins induce enteritis in Atlantic salmon (*Salmo salar* L.). *J. Agric. Food Chem.* 63, 3887–3902. doi:10.1021/jf506242t

Krogdahl, Å., Hemre, G.I., Mommsen, T.P., 2005. Carbohydrates in fish nutrition: Digestion and absorption in postlarval stages. *Aquac. Nutr.* 11(2), 103–122. doi:10.1111/j.1365-2095.2004.00327.x

Król, E., Douglas, A., Tocher, D.R., Crampton, V.O., Speakman, J.R., Secombes, C.J., Martin, S.A.M., 2016. Differential responses of the gut transcriptome to plant protein diets in farmed Atlantic salmon. *BMC Genomics* 17, 156–171. doi:10.1186/s12864-016-2473-0

Lazzarotto, V., Médale, F., Larroquet, L., Corraze, G., 2018. Long-term dietary replacement of fishmeal and fish oil in diets for rainbow trout (*Oncorhynchus mykiss*): Effects on growth, whole body fatty acids and intestinal and hepatic gene expression. *PLoS One* 13, 1–25. doi:10.1371/journal.pone.0190730

Leaver, M.J., Villeneuve, L.A.N., Obach, A., Jensen, L., Bron, J.E., Tocher, D.R., Taggart, J.B., 2008. Functional genomics reveals increases in cholesterol biosynthetic genes and highly unsaturated fatty acid biosynthesis after dietary substitution of fish oil with vegetable oils in Atlantic salmon (*Salmo salar*). *BMC Genomics* 9, 1–15. doi:10.1186/1471-2164-9-299

Lhorente, J.P., Araneda, M., Neira, R., Yáñez, J.M., 2019. Advances in genetic improvement for salmon and trout aquaculture: The Chilean situation and prospects. *Rev. Aquac.* 11, 340–353. doi:10.1111/raq.12335

Li, P., Kangsen, A.E., Ae, M., Ae, J.T., Wu, G., 2008. New developments in fish amino acid nutrition: Towards functional and environmentally oriented aquafeeds. doi:10.1007/s00726-008-0171-1

Lie, K.K., Hansen, A.C., Eroldogan, O.T., Olsvik, P.A., Rosenlund, G., Hemre, G.I., 2011. Expression of genes regulating protein metabolism in Atlantic cod (*Gadus morhua* L.) was altered when including high diet levels of plant proteins. *Aquac. Nutr.* 17, 33–43. doi:10.1111/j.1365-2095.2009.00704.x

Lilleeng, E., Frøystad, M.K., Vekterud, K., Valen, E.C., Krogdahl, Å., 2007. Comparison of intestinal gene expression in Atlantic cod (*Gadus morhua*) fed standard fish meal or soybean meal by means of suppression subtractive hybridization and real-time PCR. *Aquaculture* 267, 269–283. doi:10.1016/j.aquaculture.2007.01.048

Martinez-Rubio, L., Wadsworth, S., González Vecino, J.L., Bell, J.G., Tocher, D.R., 2013. Effect of dietary digestible energy content on expression of genes of lipid metabolism and LC-PUFA biosynthesis in liver of Atlantic salmon (*Salmo salar* L.). *Aquaculture* 384–387, 94–103. doi:10.1016/j.aquaculture.2012.12.010

Meuwissen, T.H.E., Hayes, B.J., Goddard, M.E., 2001. Prediction of total genetic value using genome-wide dense marker maps. *Genetics* 157, 1819–1829.

Mock, T.S., Francis, D.S., Jago, M.K., Glencross, B.D., Smullen, R.P., Keast, R.S.J., Turchini, G.M., 2019. The impact of dietary protein: Lipid ratio on growth performance, fatty acid metabolism, product quality and waste output in Atlantic salmon (*Salmo salar*). *Aquaculture* 501, 191–201. doi:10.1016/j.aquaculture.2018.11.012

Molina-Poveda, C., 2016. Nutrient requirements, in: Sergio F. Nates (Ed.), *Aquafeed Formulation*. Elsevier Inc., pp. 75–216. Oxford, UK. doi:10.1016/B978-0-12-800873-7.00004-X

Morais, S., Pratoomyot, J., Taggart, J.B., Bron, J.E., Guy, D.R., Bell, J.G., Tocher, D.R., 2011a. Genotype-specific responses in Atlantic salmon (*Salmo salar*) subject to dietary fish oil replacement by vegetable oil: A liver transcriptomic analysis. *BMC Genomics* 12, 255. doi:10.1186/1471-2164-12-255

Morais, S., Pratoomyot, J., Torstensen, B.E., Taggart, J.B., Guy, D.R., Gordon Bell, J., Tocher, D.R., 2011b. Diet × genotype interactions in hepatic cholesterol and lipoprotein metabolism in Atlantic salmon (*Salmo salar*) in response to replacement of dietary fish oil with vegetable oil. *Br. J. Nutr.* 106, 1457–1469. doi:10.1017/S0007114511001954

Mungkung, R., Aubin, J., Prihadi, T.H., Slembrouck, J., Van Der Werf, H.M.G., Legendre, M., 2013. Life cycle assessment for environmentally sustainable aquaculture management: A case study of combined aquaculture systems for carp and tilapia. *J. Clean. Prod.* 57, 249–256. doi:10.1016/j.jclepro.2013.05.029

Naylor, R.L., Hardy, R.W., Bureau, D.P., Chiu, A., Elliott, M., Farrell, A.P., Forster, I., Gatlin, D.M., Goldburg, R.J., Hua, K., Nichols, P.D., 2009. Feeding aquaculture in an era of finite resources. *Proc. Natl. Acad. Sci.* 106, 15103–15110. doi:10.1073/pnas.0905235106

Nelson, D.L., Cox, M.M., Lehninger, A.L., 2017. *Lehninger Principles of Biochemistry*. New York: Freeman and Company.

Nguyen, N., McPhee, C.P., 2005. Genetic parameters and responses of performance and body composition traits in pigs selected for high and low growth rate on a fixed ration over a set time. *Genet. Sel. Evol.* 37, 199. doi:10.1186/1297-9686-37-3-199

Nicieza, A.G., Metcalfe, N.B., 1999. Costs of rapid growth: The risk of aggression is higher for fast-growing salmon. *Funct. Ecol.* 13, 793–800. doi:10.1046/j.1365-2435.1999.00371.x

Nikki, J., Pirhonen, J., Jobling, M., Karjalainen, J., 2004. Compensatory growth in juvenile rainbow trout, *Oncorhynchus mykiss* (Walbaum), held individually. *Aquaculture* 235, 285–296. doi:10.1016/j.aquaculture.2003.10.017

NRC, 2011. *Nutrient Requirements of Fish and Shrimp*. The National Academies Press, Washington, DC. doi:10.17226/13039

Nunes, A.J.P., Sá, M.V.C., Browdy, C.L., Vazquez-Anon, M., 2014. Practical supplementation of shrimp and fish feeds with crystalline amino acids. *Aquaculture*. 431, 20–27. doi:10.1016/j.aquaculture.2014.04.003

Oliva-Teles, A., Enes, P., Peres, H., 2015. Replacing fishmeal and fish oil in industrial aquafeeds for carnivorous fish. *Feed Feed. Pract. Aquac.* 203–233. doi:10.1016/B978-0-08-100506-4.00008-8

Pahlow, M., van Oel, P.R., Mekonnen, M.M., Hoekstra, A.Y., 2015. Increasing pressure on freshwater resources due to terrestrial feed ingredients for aquaculture production. *Sci. Total Environ.* 536, 847–857. doi:10.1016/j.scitotenv.2015.07.124

Panserat, S., Hortopan, G.A., Plagnes-Juan, E., Kolditz, C., Lansard, M., Skiba-Cassy, S., Esquerré, D., Geurden, I., Médale, F., Kaushik, S., Corraze, G., 2009. Differential gene expression after total replacement of dietary fish meal and fish oil by plant products in rainbow trout (*Oncorhynchus mykiss*) liver. *Aquaculture* 294, 123–131. doi:10.1016/j.aquaculture.2009.05.013

Panserat, S., Kaushik, S.J., 2010. Regulation of gene expression by nutritional factors in fish. *Aquac. Res.* 41, 751–762. doi:10.1111/j.1365-2109.2009.02173.x

Panserat, S., Kolditz, C., Richard, N., Plagnes-Juan, E., Piumi, F., Esquerré, D., Médale, F., Corraze, G., Kaushik, S., 2008. Hepatic gene expression profiles in juvenile rainbow trout (*Oncorhynchus mykiss*) fed fishmeal or fish oil-free diets. *Br. J. Nutr.* 100, 953–967. doi:10.1017/S0007114508981411

Pelletier, N., Tyedmers, P., Sonesson, U., Scholz, A., Ziegler, F., Flysjo, A., Kruse, S., Cancino, B., Silverman, H., 2009a. Supporting information: Not all salmon are created equal: Life cycle assessment (LCA) of global salmon farming systems. *Environ. Sci. Technol.* 43, 8730–6. doi:10.1021/es9010114

Pelletier, N., Tyedmers, P., Sonesson, U., Scholz, A., Ziegler, F., Flysjo, A., Kruse, S., Cancino, B., Silverman, H., 2009b. Not all salmon are created equal: Life Cycle Assessment (LCA) of global salmon farming systems. *Environ. Sci. Technol.* 43, 8730–8736. doi:10.1021/es9010114

Perera, E., Yúfera, M., 2017. Effects of soybean meal on digestive enzymes activity, expression of inflammation-related genes, and chromatin modifications in marine fish (*Sparus aurata* L.) larvae. *Fish Physiol. Biochem.* 43, 563–578. doi:10.1007/s10695-016-0310-7

Philis, G., Ziegler, F., Gansel, L.C., Jansen, M.D., Gracey, E.O., Stene, A., 2019. Comparing life cycle assessment (LCA) of salmonid aquaculture production systems: Status and perspectives. *Sustainability* 11, 2517–2544. doi:10.3390/su11092517

Poore, J., Nemecek, T., 2018. Reducing food's environmental impacts through producers and consumers. *Science* 360, 987–992. doi:10.1126/science.aaq0216

Roberts, R.J., 2012. *Fish Pathology, Fish Pathology.* 4th ed. Wiley-Blackwell, Oxford, UK. doi:10.1002/9781118222942

Rodde, C., Chatain, B., Vandeputte, M., Trinh, T.Q., Benzie, J.A.H., de Verdal, H., 2020. Can individual feed conversion ratio at commercial size be predicted from juvenile performance in individually reared Nile tilapia *Oreochromis niloticus*? *Aquac. Reports* 17, 100349. doi:10.1016/j.aqrep.2020.100349

Rosenlund, G., Torstensen, B.E., Stubhaug, I., Usman, N., Sissener, N.H., 2016. Atlantic salmon require long-chain n-3 fatty acids for optimal growth throughout the seawater period. *J. Nutr. Sci.* 5, 1–13. doi:10.1017/jns.2016.10

Ruyter, B., Sissener, N.H., Ostbye, T.K., Simon, C.J., Krasnov, A., Bou, M., Sanden, M., Nichols, P.D., Lutfi, E., Berge, G.M., 2019. N-3 Canola oil effectively replaces fish oil as a new safe dietary source of DHA in feed for juvenile Atlantic salmon. *Br. J. Nutr.* 122, 1329–1345. doi:10.1017/S0007114519002356

Saez, P.J., Abdel-Aal, E.-S.M., Bureau, D.P., 2016. Feeding increasing levels of corn gluten meal induces suboptimal muscle pigmentation of rainbow trout (*Oncorhynchus mykiss*). *Aquac. Res.* 47, 1972–1983. doi:10.1111/are.12653

Sahlmann, C., Sutherland, B.J.G., Kortner, T.M., Koop, B.F., Krogdahl, Å., Bakke, A.M., 2013. Early response of gene expression in the distal intestine of Atlantic salmon (*Salmo salar* L.) during the development of soybean meal induced enteritis. *Fish Shellfish Immunol.* 34, 599–609. doi:10.1016/j.fsi.2012.11.031

Sales, J., Glencross, B., 2011. A meta-analysis of the effects of dietary marine oil replacement with vegetable oils on growth, feed conversion and muscle fatty acid composition of fish species. *Aquac. Nutr.* 17, e271–e287. doi:10.1111/j.1365-2095.2010.00761.x

Salze, G., McLean, E., Battle, P.R., Schwarz, M.H., Craig, S.R., 2010. Use of soy protein concentrate and novel ingredients in the total elimination of fish meal and fish oil in diets for juvenile cobia, *Rachycentron canadum*. *Aquaculture* 298, 294–299. doi:10.1016/j.aquaculture.2009.11.003

Shomorin, G.O., Storebakken, T., Kraugerud, O.F., Øverland, M., Hansen, B.R., Hansen, J.Ø., 2019. Evaluation of wedge wire screen as a new tool for faeces collection in digestibility assessment in fish: The impact of nutrient leaching on apparent digestibility of nitrogen, carbon and sulphur from fishmeal, soybean meal and rapeseed meal-based diets in rainbow trout (*Oncorhynchus mykiss*). *Aquaculture* 504, 81–87. doi:10.1016/j.aquaculture.2019.01.051

Silverstein, J.T., 2006. Relationships among feed intake, feed efficiency, and growth in juvenile rainbow trout. *N. Am. J. Aquac.* 68, 168–175. doi:10.1577/a05-010.1

Sissener, N.H., 2018. Are we what we eat? Changes to the feed fatty acid composition of farmed salmon and its effects through the food chain. *Journal of Experimental Biology*, 221, 1–11. doi:10.1242/jeb.161521

Slawski, H., Adem, H., Tressel, R.P., Wysujack, K., Koops, U., Kotzamanis, Y., Wuertz, S., Schulz, C., 2012. Total fish meal replacement with rapeseed protein concentrate in diets fed to rainbow trout (*Oncorhynchus mykiss* Walbaum). *Aquac. Int.* 20, 443–453. doi:10.1007/s10499-011-9476-2

Slawski, H., Nagel, F., Wysujack, K., Balke, D.T., Franz, P., Schulz, C., 2013. Total fish meal replacement with canola protein isolate in diets fed to rainbow trout (*Oncorhynchus mykiss* W.). *Aquac. Nutr.* 19, 535–542. doi:10.1111/anu.12005

SOFIA, 2018. *World Fisheries and Aquaculture.* doi:*issn* 10

Sonesson, A.K., Meuwissen, T.H., 2009. Testing strategies for genomic selection in aquaculture breeding programs. *Genet. Sel. Evol.* 41, 37. doi:10.1186/1297-9686-41-37

Steinberg, C.E.W., 2018. Trophic diversification and speciation – 'Your eating fuels evolution', in: Christian E. W. Steinberg (Ed.), *Aquatic Animal Nutrition.* Cham, Switzerland, Springer International Publishing, pp. 431–474. doi:10.1007/978-3-319-91767-2_7

Stone, D.A.J., 2003. Dietary carbohydrate utilization by fish. *Rev. Fish. Sci.* 11(4), 337–369. doi:10.1080/10641260390260884

Storebakken, T., Kvien, I.S., Shearer, K.D., Grisdale-Helland, B., Helland, S.J., Berge, G.M., 1998. The apparent digestibility of diets containing fish meal, soybean meal or bacterial meal fed to Atlantic salmon (*Salmo salar*): Evaluation of different faecal collection methods. *Aquaculture* 169, 195–210. doi:10.1016/S0044-8486(98)00379-2

Tacchi, L., Bickerdike, R., Douglas, A., Secombes, C.J., Martin, S.A.M., 2011. Transcriptomic responses to functional feeds in Atlantic salmon (*Salmo salar*). *Fish Shellfish Immunol.* 31, 704–715. doi:10.1016/j.fsi.2011.02.023

Tacon, A.G.J., Forster, I.P., 2003. Aquafeeds and the environment: Policy implications. *Aquaculture*, 226(1–4), 181–189. Elsevier. doi:10.1016/S0044-8486(03)00476-9

Tacon, A.G.J., Metian, M., Turchini, G.M., De Silva, S.S., 2009. Responsible aquaculture and trophic level implications to global fish supply. *Rev. Fish. Sci.* 18, 94–105. doi:10.1080/10641260903325680

Taggart, J.B., Bron, J.E., Martin, S.A.M., Seear, P.J., Høyheim, B., Talbot, R., Carmichael, S.N., Villeneuve, L.A.N., Sweeney, G.E., Houlihan, D.F., Secombes, C.J., Tocher, D.R., Teale, A.J., 2008. A description of the origins, design and performance of the TRAITS-SGP Atlantic salmon *Salmo salar* L. cDNA microarray. *J. Fish Biol.* 72, 2071–2094. doi:10.1111/j.1095-8649.2008.01876.x

Teles, A.O., Couto, A., Enes, P., Peres, H., 2019. Dietary protein requirements of fish – a meta-analysis. *Rev. Aquac.* 12, raq.12391. doi:10.1111/raq.12391

Tocher, D., Betancor, M., Sprague, M., Olsen, R., Napier, J., 2019. Omega-3 long-chain polyunsaturated fatty acids, EPA and DHA: Bridging the gap between supply and demand. *Nutrients* 11, 89. doi:10.3390/nu11010089

Tocher, D.R., 2010. Fatty acid requirements in ontogeny of marine and freshwater fish. *Aquac. Res.* 41(5), 717–732. doi:10.1111/j.1365-2109.2008.02150.x

Tocher, D.R., Betancor, M.B., Sprague, M., Olsen, R.E., Napier, J.A., 2019. Omega-3 long-chain polyunsaturated fatty acids, EPA and DHA: Bridging the gap between supply and demand. *Nutrients.* doi:10.3390/nu11010089

Torstensen, B.E., Bell, J.G., Rosenlund, G., Henderson, R.J., Graff, I.E., Tocher, D.R., Lie, Ø., Sargent, J.R., 2005. Tailoring of atlantic salmon (*Salmo salar* L.) flesh lipid composition and sensory quality by replacing fish oil with a vegetable oil blend. *J. Agric. Food Chem.* 53, 10166–10178. doi:10.1021/jf051308i

Torstensen, B.E., Tocher, D.R., 2010. The effects of fish oil replacement on lipid metabolism of fish. In: Turchini GM, Ng W- & Tocher DR (Eds.), *Fish Oil Replacement and Alternative Lipid Sources in Aquaculture Feeds.* Boca Raton, Florida: Taylor & Francis (CRC Press), pp. 405–437. http://www.crcpress.com/product/isbn/9781439808627

Troell, M., Naylor, R.L., Metian, M., Beveridge, M., Tyedmers, P.H., Folke, C., Arrow, K.J., Barrett, S., Crépin, A.S., Ehrlich, P.R., Gren, Å., Kautsky, N., Levin, S.A., Nyborg, K., Österblom, H., Polasky, S., Scheffer, M., Walker, B.H., Xepapadeas, T., De Zeeuw, A., 2014. Does aquaculture add resilience to the global food system? *Proc. Natl. Acad. Sci. U. S. A.* 111(37), 13257–13263.doi:10.1073/pnas.1404067111

Turchini, G.M., Hermon, K.M., Francis, D.S., 2018. Fatty acids and beyond: Fillet nutritional characterisation of rainbow trout (*Oncorhynchus mykiss*) fed different dietary oil sources. *Aquaculture* 491, 391–397. doi:10.1016/j.aquaculture.2017.11.056

Turchini, G.M., Torstensen, B.E., Ng, W.K., 2009. Fish oil replacement in finfish nutrition. *Rev. Aquac.* 1(1), 10–57. doi:10.1111/j.1753-5131.2008.01001.x

Urán, P.A., Gonçalves, A.A., Taverne-Thiele, J.J., Schrama, J.W., Verreth, J.A.J., Rombout, J.H.W.M., 2008. Soybean meal induces intestinal inflammation in common carp (*Cyprinus carpio* L.). *Fish Shellfish Immunol.* 25, 751–760. doi:10.1016/j.fsi.2008.02.013

Wacyk, J., Powell, M., Rodnick, K., Overturf, K., Hill, R.A., Hardy, R., 2012. Dietary protein source significantly alters growth performance, plasma variables and hepatic gene expression in rainbow trout (*Oncorhynchus mykiss*) fed amino acid balanced diets. *Aquaculture* 356–357, 223–234. doi:10.1016/j.aquaculture.2012.05.013

Weihe, R., Dessen, J.E., Arge, R., Thomassen, M.S., Hatlen, B., Rørvik, K.A., 2018. Improving production efficiency of farmed Atlantic salmon (*Salmo salar* L.) by isoenergetic diets with increased dietary protein-to-lipid ratio. *Aquac. Res.* 49, 1441–1453. doi:10.1111/are.13598

Yáñez, J.M., Joshi, R., Yoshida, G.M., 2020. Genomics to accelerate genetic improvement in tilapia. *Anim. Genet.* 51, 658–674. doi:10.1111/age.12989

Yoshida, G.M., Bangera, R., Carvalheiro, R., Correa, K., Figueroa, R., Lhorente, J.P., Yáñez, J.M., 2018. Genomic prediction accuracy for resistance against *Piscirickettsia salmonis* in farmed rainbow trout. *G3 Genes, Genomes, Genet.* 8, 719–726. doi:10.1534/g3.117.300499

Yoshida, G.M., Carvalheiro, R., Rodríguez, F.H., Lhorente, J.P., Yáñez, J.M., 2019. Single-step genomic evaluation improves accuracy of breeding value predictions for resistance to infectious pancreatic necrosis virus in rainbow trout. *Genomics* 111, 127–132. doi:10.1016/j.ygeno.2018.01.008

Ytrestøyl, T., Aas, T.S., Åsgård, T., 2015. Utilisation of feed resources in production of Atlantic salmon (*Salmo salar*) in Norway. *Aquaculture* 448, 365–374. doi:10.1016/j.aquaculture.2015.06.023

Ytrestøyl, T., Takle, H., Kolarevic, J., Calabrese, S., Timmerhaus, G., Rosseland, B.O., Teien, H.C., Nilsen, T.O., Handeland, S.O., Stefansson, S.O., Ebbesson, L.O.E., Terjesen, B.F., 2020. Performance and welfare of Atlantic salmon, <scp> *Salmo salar* </scp> L. post-smolts in recirculating aquaculture systems: Importance of salinity and water velocity. *J. World Aquac. Soc.* 51, 373–392. doi:10.1111/jwas.12682

Zettl, S., Cree, D., Soleimani, M., Tabil, L., Yildiz, F., 2019. Mechanical properties of aquaculture feed pellets using plant-based proteins. *Cogent Food Agric.* 5, 1656917. doi:10.1080/23311932.2019.1656917

Zhang, Y., Øverland, M., Xie, S., Dong, Z., Lv, Z., Xu, J., Storebakken, T., 2012. Mixtures of lupin and pea protein concentrates can efficiently replace high-quality fish meal in extruded diets for juvenile black sea bream (*Acanthopagrus schlegeli*). *Aquaculture* 354–355, 68–74. doi:10.1016/j.aquaculture.2012.03.038

11 Feed Industry Initiatives
Probiotics, Prebiotics, and Synbiotics

Vanesa Robles, Marta F. Riesco, and David G. Valcarce

CONTENTS

INTRODUCTION

The aquaculture industry has great potential to give response to animal protein demand. In order to be competitive, this industry must achieve adequate production of the domesticated species and avoid diseases that could hinder such cultures. Prebiotics, probiotics, and synbiotics offer possibilities for approaching both targets.

WHAT ARE PREBIOTICS, PROBIOTICS, AND SYNBIOTICS?

Prebiotics have been described as non-digestible ingredients that improve health by the stimulation of beneficial bacteria in the gut (Guerreiro et al. 2016). Some carbohydrates, peptides, and lipids could act as prebiotics. According to Guerreiro et al. (2016) fructooligosaccharides (FOS), mannanoligosaccharides (MOS), Galactooligosaccharides (GOS), arabinoxylan-oligosaccharides, and polyoligosaccharides, like inulin, are considered among the most effective prebiotics in aquaculture. Some authors have reported these compounds are able to increase immune responses (Shokryazdan et al. 2017).

Probiotics has been defined by Fuller (1989) as a "live microbial feed supplement which beneficially affect the host animal by improving its intestinal microbial balance". Several bacteria have been used as probiotic species in aquaculture, such as *Bacillus* sp., *Vibrio* sp., *Lactobacillus acidophilus*, *Enterococcus faecium*, *Pseudomonas fluorescents*, *Lactococcus lactis*, *L. casei*, *L. acidophilus*, *Pseudomonas* sp., *Streptococcus thermopiles*, *Clostridium butyricum*, among others (Das et al. 2017). The most obvious mode of action of probiotics is the competitive exclusion of possible pathogens in the digestive tract. The following antibacterial activities have been attributed to

DOI: 10.1201/9780429331664-11

the aforementioned bacteria: inhibition of pathogen epithelia adhesion, modulation of the immune system, and production of bacteriostatic and bactericidal substances due to nutrient competition. The use of probiotics in aquaculture is generally related either to the improvement of growth (Gobi et al. 2018), due to their potential effect on nutrient digestibility, or to infection prevention, due to their antibacterial, antiviral, and antifungal activities. However, recent studies have demonstrated that these bacteria could also have a positive impact at other unexpected levels. As an example, benefits in gamete quality have been reported (Valcarce et al. 2019a, 2019b), and even a possible fish behavior modification has been demonstrated (Valcarce et al. 2020). Considering these unexpected results, it is not only important to improve our knowledge about how pre and probiotics perform their action, but also it would be really interesting to analyze in detail their effect at the molecular level. These studies could be particularly important when probiotics are able to modify germ cell quality, considering that progeny will depend on these cells.

Synbiotics could be considered those supplements that combine probiotics and prebiotics. Several combinations have been used in different species; as an example, *Bacillus* sp. has been combined with MOS; FOS, and Chitosan (Ai et al. 2011; Ye et al. 2011; Geng et al. 2011) and *Enterococcus* with MOS and polyhydroxybutyrate acid (PHB) (Rodriguez -Estrada et al. 2009).

Due to the interesting and diverse range of potential beneficial effects, these ingredients are widely accepted in the market. There are some probiotic supplements commercially available for aquaculture practices like Bactocell®, Primalac®, and AquaStar®.

Administration Methods

The main methods of probiotic administration in aquaculture are oral administration and addition to the water. The first method aims to improve microbial flora in the digestive tract and prevent diseases; the second method usually aims to change the organic matter in, thereby improving the quality of the water by adding these microorganisms. Some papers have reported an improvement in water quality, particularly with gram-positive bacteria that they claim are more efficient in organic matter transformation (Schwartz 1979).

Probiotics could be administered either singly or in combination (Hai 2015). It has been published that multi-strain probiotics enhance protection against pathogens (Kesarcodi-Watson et al. 2012). Although debatable, these bacteria can be administered live, dead, or inactivated (Hai 2015). Dosages and periods of administration are also key factors that have yet to be determined. Depending on the probiotic, the target species, and the desired effect, probiotic diet supplementation could vary from 10^8 to 10^{11} CFU/g (Hai 2015). The administration period could also range from a few days to several months (Hai 2015). The moment of administration could be also a crucial factor, and it could significantly differ in juveniles and adults.

In this chapter we will outline the uses of prebiotics, probiotics, and symbiotics in the field of aquaculture, providing detailed tables showing the studies using probiotics, prebiotics, and synbiotics with different fish species of commercial interest during recent years. This chapter pays special attention to those recent papers which report the novel and unexpected effects of these compounds on fish.

BENEFITS OF PREBIOTICS, PROBIOTICS, AND SYNBIOTICS FOR THE AQUACULTURE INDUSTRY

Improvement of Nutrient Digestibility and Growth

The growth of the individuals is an essential element in any process related to animal production. Two relevant objectives in aquaculture sector are the search for rapid growth and the optimization in fish biomass. The use of functional feeds (probiotics, mainly, but also of prebiotics and synbiotics) has been a strategy used in a myriad of fish species in order to achieve higher growth rates.

The improvement of nutrient digestibility is measured by indirect parameters. The percent weight gain (PWG), specific growth rate (SGR), feed conversion ratio (FCR), protein efficiency ratio (PER), and protein productive value (PPV) are common parameters evaluated in the experiments regarding functional feeds in aquaculture (Table 11.1). However, there exist other parameters linked to animal growth that have been explored, such as the expression of genes involved in muscle growth restraining or muscle growth promotion. For example, recently, Gong et al. (2019) reported a downregulation of myostatin genes (MSTN-1 and MSTN-2) and an overexpression of insulin-like growth factor (IGF-1 and IGF-2) in grass carp (*Ctenopharyngodon idellus*) fed with *Pediococcus pentosaceus* for 30 days.

The positive effect of probiotic consumption on growth parameters has been shown in animals from different ages. Taking Nile tilapia (*Oreochromis niloticus*) as an example, during recent years, the positive effects of probiotic consumption have been demonstrated in many different types of experimental animals in terms of initial body weight from fries or juveniles (Zhang et al. 2019; Won et al. 2020; Xia et al. 2020) to older fish (Li et al. 2019).

The genera *Bacillus*, *Lactobacillus* or *Lactococcus*, are among the most used in terms of growth promotion in aquaculture since they belong to the most popular probiotics group (Hoseinifar et al. 2018) in the fish farms. However, new genera and species have been recently explored with this purpose. Tilapias fed with *Psychrobacter maritimus* (Makled et al. 2019) or *Rummeliibacillus stabekisii* (Tan et al. 2019) reported better growth rates compared to controls.

This probiotic-derived improvement to fish growth may be correlated to an enhancement of digestibility of feed via intestinal microvilli density and morphology alteration. Probiotics may increase the intestinal absorptive surface area by improving the microvilli density and microvilli length (Won et al. 2020; Xia et al. 2020).

Growth performance improvements may also be attributed to the stimulation of digestive enzyme activities and inhibition of other harmful flora in the host intestine (Adel et al. 2017). The functional feeds may stimulate digestive enzymes, which promote feed digestibility (De Schrijver & Ollevier 2000) and therefore, the energetic benefits affect growth rate. Makled and colleagues reported a significant activation of intestinal enzymatic protease, amylase, and lipase activities in Nile tilapia fed with *Psychrobacter maritimus* during 50 days, which was concomitant to growth improvement (Makled et al. 2019).

Not only probiotics have a positive effect on growth performance. Prebiotics and symbiotics have been described as functional feeds promoting fish growth. The supplementation of the diet with the prebiotic β-(1,3) (1,6)-D-glucan, improved WGR, SGR, and FCR in common carp (*Cyprinus carpio*) (Kühlwein et al. 2014). Similarly, a 12-week diet supplemented with corncob-derived xylooligosaccharides (CDXOS) also reported better WGR, SCR, and FCR in *Oreochromis niloticus* juveniles (Van Doan et al. 2019). Interestingly, and as an example of symbiotics affecting fish growth, in the same experimental design this prebiotic was tested together with *Lactobacillus plantarum*. The symbiotic not only showed a better growth performance compared to the control group it also provided the highest values compared to the probiotic and prebiotic individually (Van Doan et al. 2019).

Prevention of Infections and Effects on Fish Immune Response

Animal health improvement is an integral part of the culture practices, and the main drivers for the application of probiotics, prebiotics, and/or synbiotics are, in fact, the prevention of infection and health promotion in fish cultures. The irresponsible use of antibiotics as prophylactics for decades has produced a wide range of undesirable effects, such as the production of resistant bacteria in aquaculture facilities, the enlargement of antibiotic residues, the eradication of gastrointestinal beneficial microbes in the fish gut, and the promotion of disarrangements in the environmental microbiota (Chauhan & Singh 2019). Legislation enacted by the European Union in 2006 (EU Regulation No. 1831/2003), which forbids the use of antibiotics, opened a new scenario for functional feeds

TABLE 11.1
Compilation of Several *in vivo* Experiments in Aquaculture Evaluating the Effect of Probiotics on Growth Parameters. The Table Includes the Works Done on the Most Cultured Species Worldwide

Functional Feed	Animals	Supplementation	Time	Immunity Improvement Data	Ref.
***Ctenopharyngodon idellus* (grass carp)**					
PROB ***Pediococcus pentosaceus* SL001**	32.1 ± 9 g	Included in the diet pellets	30 d	• Improvement of WGR, SGR, FI, and FCR • Downregulation of MSTN-1 and MSTN-2 (muscle growth restraining) • Overexpression of IGF-1 and IGF-2 (muscle growth promoters)	(Gong et al. 2019)
***Cyprinus carpio* (common carp)**					
PROB ***Lactococcus lactis* Q-8** ***Lactococcus lactis* Q-9** ***Lactococcus lactis* Z-2**	33.07 ± 0.55 g	Sprayed on the diet pellets	8 wks	• Improvement of WGR, SGR	(Feng et al. 2019)
PROB ***Pediococcus acidilactici*MA18/5M**	10.0 ± 2.5 g	Included in the diet pellets	60 d	• Improvement of FCR	(Hoseinifar et al. 2019)
PROB ***Bacillus coagulans* MTCC 9872** ***Bacillus licheniformis* MTCC 6824** ***Paenibacillus polymyxa* MTCC 122**	0.329 ± 0.01 g	Included in the diet pellets	80 d	• Improvement of WGR, SGR, FI, and FCR	(Gupta et al. 2014)
PREB **β-(1,3)(1,6)-D-glucan**	11.1 ± 0.0 g	Included in the diet pellets	8 wks	• Improvement of WGR, SGR, and FCR	(Kühlwein et al. 2014)
***Oreochromis niloticus* (Nile tilapia)**					
PROB ***Bacillus cereus* NY5** ***Bacillus cereus* NY5+** ***Bacillus subtilis***	0.20 ± 0.05 g	Included in the diet pellets	6wks	• Improvement of WGR and FCR	(Xia et al. 2020)
PROB ***Bacillus subtilis* WB60** ***Lactococcus lactis***	2.83 ± 0.05 g	Included in the diet pellets	8 wks	• Improvement of WGR, SGR, FCR, and PER	(Won et al. 2020)
PROB ***Clostridium butyricum*M2014537**	56.21 ± 0.81 g	Sprayed on the diet pellets	56 d	• Improvement of SGR, FI	(Li et al. 2019)

(Continued)

TABLE 11.1 (CONTINUED)
Compilation of Several *in vivo* Experiments in Aquaculture Evaluating the Effect of Probiotics on Growth Parameters. The Table Includes the Works Done on the Most Cultured Species Worldwide

Functional Feed	Animals	Supplementation	Time	Immunity Improvement Data	Ref.
PROB ***Bacillus velezensis* LF01**	4.0 g aprox	Included in the diet	9 wks	• Improvement of WGR, SGR and spleen index	(Zhang et al. 2019)
PROB ***Psychrobactermaritimus***	5.4 ± 0.1 g	Sprayed on the diet pellets	50 d	• Improvement of WGR, SGR, FCR, and PER	(Makled et al. 2019)
PROB ***Rummeliibacillus stabekisii***	4.1 ± 0.34 g	Included in the diet pellets	8 wks	• Improvement of WGR, FCR, and FER	(Tan et al. 2019)
PROB ***Lactobacillus plantarum* CR1T5**	4.97 ± 0.04 g	Included in the diet pellets	12 wks	• Improvement of WG, WGR, SCR, and FCR	(Van Doan et al. 2019)
PREB **Corncob-derived xylooligosaccharides (CDXOS)**	4.97 ± 0.04 g	Included in the diet pellets	12 wks	• Improvement of WG, WGR, SCR, and FCR	(Van Doan et al. 2019)
SYMB ***Lactobacillus plantarum* CR1T5 +** **Corncob-derived xylooligosaccharides (CDXOS)**	4.97 ± 0.04 g	Included in the diet pellets	12 wks	• Improvement of WG, WGR, SCR, and FCR	(Van Doan et al. 2019)
Carassius spp.					
PROB| *Exiguobacterium acetylicum* S01	58 g aprox.	Included in the diet (vehicle: vegetable oil)	4 wks	• Improvement of final weight, WGR, SGR, and FCR	(Jinendiran et al. 2019)
PROB| *Bacillus coagulans*	14.33±0.15g	Included in the diet pellets	8 wks	• Improvement of final weight, WGR, SGR, and FCR, PER	(Yu et al. 2018)
***Salmo salar* (Atlantic salmon)**					
PROB ***Aliivibrio* sp.,VI1, NCIMB 42593;** ***Aliivibrio* sp., VI2, NCIMB 42592** ***Aliivibrio* sp., VI3, NCIMB 42594**	84.33 ± 22.60 g	1 single "probiotic bath"	4-6 m	• Improvement of FCR	(Klakegg et al. 2020)
***Labeo rohita* (Roho labeo)**					
PROB ***G. candidum* QAUGC01**	20 ± 2.34 g	Encapsulated in sodium alginate	11 wk	• Improvement of WGR, SGR, and survival	(Amir et al. 2019)
***Megalobrama amblycephala* (Wuchang bream)**					
PROB ***Streptococcus faecalis***	46.32 ± 0.09 g	Included in the diet pellets	10 wks	• Improvement of FCR	(Zhong et al. 2019)

(Continued)

TABLE 11.1 (CONTINUED)
Compilation of Several *in vivo* Experiments in Aquaculture Evaluating the Effect of Probiotics on Growth Parameters. The Table Includes the Works Done on the Most Cultured Species Worldwide

Functional Feed	Animals	Supplementation	Time	Immunity Improvement Data	Ref.
***Oncorhynchus mykiss* (rainbow trout)**					
PROB *Lactobacillus delbrukei* **subsp.***bulgaricus* *Lactobacillus acidophilus* *Citrobacter farmeri*	19.08 - 32.9 g	Included in the diet	60 d	• Improvement of WGR, SGR, RGR, FER, DWG, PER	(Mohammadian et al. 2019)
PROB *Lactobacillus rhamnosus* **ATCC 7469**	18.41 ± 0.32 g	Microencapsulated with alginate and hi-maize starch and coated with chitosan	60 d	• Improvement of WGR and FCR	(Hooshyar et al. 2020)
PROB *Saccharomyces cerevisiae*	250 ± 50 g	Included in the diet	130 d	• Increase of FCR	(Vazirzadeh et al. 2020)
PROB **PrimaLac® (***Lactobacillus acidophilus***,** *Lactobacillus casei Enterococcus faecium* **and** *Bifidobacterium bifidium***)**	25 ± 1.8 g	Included in the diet	8 wks	• Improvement of WGR and SGR	(Naderi Farsani et al. 2020)
PROB| *Lactobacillus fermentum***1744 (ATCC 14931)**	114.2 ± 9.4 g	Encapsulated in sodium alginate	56 days	• Improvement of FCR	(Madreseh et al. 2019)
PROB| *Enterococcus casseliflavus*	38.3±1 g	Included in the diet	8 wks	• Improvement of WGR, SGR, and FCR	(Safari et al. 2016)
PROB | Bifidobacterium *animalis* **PTCC-1631 and** *Bifidobacterium lactis* **PTCC-1736**	0.58 ± 0.19 g	Sprayed on the diet pellets	8 wks	• Increase of WG, SGR, MGR, RFI, higher growth, nutrient utilization, digestibility, and lower FCR • Higher protein gain, lipid gain, PPV and lipid productive value	(Sahandi et al. 2019)
PROB| *Lactobacillus fermentum***1744 (ATCC 14931)**					
PROB *Lactococcus lactis* **L19;** *Enterococcus faecalis* **W24** *Lactococcus lactis* **L19 +** *Enterococcus faecalis* **W24**	9.50 ± 0.03 g	Included in the diet	8 wks	• Improvement of FBW, WG,FER, SGR, and PER	(Kong et al. 2020)

since they present innumerable advantages to overcome the limitations and side-effects of antibiotics. Therefore, many trajectories of research focus on the study of the beneficial effects of functional feeds for the prevention of infections and promotion of fish immune response at many levels.

Antibacterial Activity

Inhibition of Pathogen Epithelial Adhesion

The most feared disease-causing bacterial pathogens among fish farmers are gram-negative genera like *Aeromonas*, *Flavobacterium*, *Pseudomonas*, *Vibrio*, and *Yersinia*. Usually, when a functional feed is tested for animal production purposes, the studies include a pathogen challenge experiment. Survival and health parameters are evaluated after pathogen infection in fish in order to assess the protective activity of the feeds (Zhang et al. 2019; Cao et al. 2019). This ability is one of the key screening steps for the selection of isolated bacteria as ideal probiotics (Chauhan & Singh 2019).

Probiotics stop pathogens from multiplying in the gut tract on the superficial surfaces and in the culture environment of the farm animals. Probiotics can block pathogen attachment to the gastrointestinal epithelium by blocking epithelial surface receptors or stimulating mucins, carbohydrates which generate a barrier along the epithelial monolayer (Corr et al. 2009). This mechanism of action is known as competitive inhibition or competitive exclusion (Zorriehzahra et al. 2016). The biological basis of this process is bacterial antagonism, a common phenomenon in nature. The competition for space between pathogenic bacteria and probiotics is a proposed mode of action of beneficial microbes in aquaculture (Zorriehzahra et al. 2016) when used in husbandry practices and included in the fish environment.

One of the most used probiotic genus in aquaculture is *Bacillus*, and it has been proven to better combat a wide range of fish pathogens compared to other probiotic strains (Kuebutornye, Abarike, et al. 2020). Bacillus has been proven to fight against fish pathogens by many different mechanisms, as recently reviewed by Kuebutornye et al. (2020). The genus can produce of bacteriocins, suppress virulence gene expression, compete for adhesion sites, produce lytic enzymes, produce antibiotics, promote host's immune stimulation, compete for nutrients and energy, and produce organic acids.

Modulation of the Immune System and Production of Bacteriostatic and Bactericidal Substances

Functional feeds significantly influence the host immune system. The innate immune system of fish is considered to be the first line of defense against a broad spectrum of pathogens, and it plays a more important role for fish as compared with mammals (Saurabh & Sahoo 2008). The effect of probiotics in stimulating the systemic immune responses are now well documented in several fish species. There exist vast bibliography reporting the use of probiotic strains promoting a number of enzymatic activities such as lysozyme (Ahmadifar et al. 2019), phagocytic (Makled et al. 2019), alternative complement pathway (Modanloo et al. 2017), respiratory burst (Noor-Ul et al. 2020), and superoxide dismutase activities (Mohammadi et al. 2020) (Table 11.2). Lysozyme is a vital bactericidal enzyme of innate immunity, and shows a crucial defense mechanism against various pathogens in fish (Nayak 2010). Aside from serum lysozyme content in different species (Panigrahi et al. 2004; Kim & Austin 2006; Balcázar et al. 2007), probiotics can also enhance the lysozyme level in the skin mucosa of fish (Taoka et al. 2006; Hoseinifar et al. 2019). It is well documented that lysozyme activates the complement system (Saurabh & Sahoo 2008). In the innate humoral response, complement activation plays a central role in phagocytosis through chemotaxis and the opsonization of phagocytic cells (Saurabh & Sahoo 2008). The higher levels of lysozyme or serum complement activity are regularly evaluated in experiments testing the efficiency of functional feedings (Vidal et al. 2016; Zaineldin et al. 2018) as well as the phagocytosis activity and killing ability of head kidney leukocytes (Salinas et al. 2006). These immune responses correlate with the modulation of expression of pro-inflammatory cytokines like interleukins in the probiotic-supplemented fish. Usually, publications evaluating probiotics include gene expression analysis (Panigrahi et al. 2011; Tang

TABLE 11.2
Compilation of Several *in vivo* Experiments in Aquaculture Evaluating the Effect of Probiotics on Immunity Parameters. The Table Includes the Works Done on the Most Cultured Species Worldwide

Functional Feed	Animals	Supplementation	Time	Immunity Improvement Data	Ref.
***Ctenopharyngodon idellus* (grass carp)**					
PROB ***Bacillus velezensis* BvL03**	2-year-old	Intraperitoneal injection		• Decrease in fish mortality when challenged with *A. hydrophila*	(Cao et al. 2019)
PROB ***Pediococcus pentosaceus* SL001**	32.1 ± 9 g	Included in the diet pellets	30 d	• Decrease in fish mortality when challenged with *A. hydrophila* • Increase of the expression levels of IgM and C3 • 舦┌┌┌Downregulation of IL-8	(Gong et al. 2019)
	71.42 ± 4.36 g	Included in the diet pellets	42 d	• Lower oxidative stress (MDA) and an increase in antioxidant activities (Increased levels on T-AOC, SOD, CAT, GSH) when challenged with *A. hydrophila* • Upregulation of antioxidant enzymes expression (SOD, CAT, GPX); Improvement of the immune response (upregulation of IL-10; dowregulation of TNF-α, IL-1β, and IL-8) when challenged with *A. hydrophila*	(Tang et al. 2019)
PROB ***Bacillus subtilis* WB600 spores expressing cysteine protease of *Clonorchis sinensis***	25 ± 2 g	Included in the diet pellets	8 wks	• Higher levels of IgM in samples of serum, bile, mucus of surface • Longer intestinal microvilli and higher number of intraepithelial lymphocytes • Effective protection under cercariae challenge	(Tang et al. 2017)
***Cyprinus carpio* (common carp)**					
PROB ***Lactobacillus fermentum* PTCC 1638**	3.90 ± 0.2 g	Included in the diet pellets	4 wks	• Higher levels of white blood cells, serum total protein and albumin levels • Enhacement of serum respiratory burst and lysozyme activity	(Ahmadifar et al. 2019)
PROB ***Lactococcus lactis* Q-8** ***Lactococcus lactis* Q-9** ***Lactococcus lactis* Z-2**	33.07 ± 0.55 g	Sprayed on the diet pellets	8 wks	• Better survival rate • Up-regulated protein levels of pro-inflammatory cytokines (TNF-α, IL-1β, IL-6, IL-12) after ingestion • Reduction of pro-inflammatory cytokines (TNF-α, IL-1β, IL-6, IL-12) when challenged with *A. hydrophila* • Increase of IL-10, TGF-β under infection or not	(Feng et al. 2019)

(Continued)

TABLE 11.2 (CONTINUED)
Compilation of Several *in vivo* Experiments in Aquaculture Evaluating the Effect of Probiotics on Immunity Parameters. The Table Includes the Works Done on the Most Cultured Species Worldwide

Functional Feed	Animals	Supplementation	Time	Immunity Improvement Data	Ref.
PROB ***Pediococcus acidilactici* MA18/5M**	10.0 ± 2.5 g	Included in the diet pellets	60 d	• Increase of skin mucus, total immunoglobulin (Ig), and protein • Increase of skin mucus protease activity • Increase of lysozime gene expression in skin	(Hoseinifar et al. 2019)
PROB ***Pediococcus acidilactici* MA18/5M**	Juvenile	Included in the diet by commercial supplement	8 wks	• Higher serum alternative haemolytic complement (ACH50) activity • serum and skin mucus total Ig levels • Increase in serum and skin mucus total Ig levels • Downregulation of TNF-alpha	(Modanloo et al. 2017)
PROB ***Bacillus coagulans* MTCC 9872** ***Bacillus licheniformis* MTCC 6824** ***Paenibacilluspolymyxa* MTCC 122**	0.329 ± 0.01 g	Included in the diet pellets	80 d	• Higher values of lysozyme activity, respiratory burst assay, and myeloperoxidase content • Enhanced resistance of fish fry against *A. hydrophila* challenge	(Gupta et al. 2014)
PROB ***Lactobacillus plantarum* 225/1, 155/1, 211/1B, 226/1, and 274/1 (Mixture)**	45 g	Included in the diet pellets	14 d	• Increase in the proliferative activity of LPS-stimulated B lymphocytes, γ-globulins levels, and total protein • Better survival rates under *A. hydrophila* challenge	(Kazuń et al. 2018)
PREB **Raffinose**	10.0 ± 2.5 g	Included in the diet pellets	60 d	• Increase of skin mucus protease activity • Increase of lysozime gene expression in skin	(Hoseinifar et al. 2019)
PREB **Galactooligosaccharide**	Juvenile	Included in the diet by commercial supplement	8 wks	• Higher serum alternative haemolytic complement (ACH50) activity • serum and skin mucus total Ig levels. • Increase in serum and skin mucus total Ig levels • Downregulation of TNF-alpha	(Modanloo et al. 2017)
PREB **β-(1,3)(1,6)-D-glucan**	11.1 ± 0.0 g	Included in the diet pellets	8 wks	• Higher infiltration of leucocytes into the epithelial layer in the anterior intestine • Elevated hematocrit value, higher monocyte fraction	(Kühlwein et al. 2014)

(*Continued*)

TABLE 11.2 (CONTINUED)
Compilation of Several *in vivo* Experiments in Aquaculture Evaluating the Effect of Probiotics on Immunity Parameters. The Table Includes the Works Done on the Most Cultured Species Worldwide

Functional Feed	Animals	Supplementation	Time	Immunity Improvement Data	Ref.
SYMB ***Pediococcus acidilactici* MA18/5M + Raffinose**	10.0 ± 2.5 g	Included in the diet pellets	60 d	• Increase of skin mucus protein • Increase of skin mucus protease activity • Increase of lysozyme gene expression in skin	(Hoseinifar et al. 2019)
SYMB ***Pediococcus acidilactici* MA18/5M + Galactooligosaccharide**	Juvenile	Included in the diet by commercial supplement	8 wks	• Higher serum alternative haemolytic complement (ACH50) activity • serum and skin mucus total Ig levels • Increase on serum and skin mucus total Ig levels • Downregulation of TNF-alpha	(Modanloo et al. 2017)
***Oreochromis niloticus* (Nile tilapia)**					
PROB ***Bacillus cereus* NY5** ***Bacillus cereus* NY5+ *Bacillus subtilis***	0.20 ± 0.05 g	Included in the diet pellets	6wks	• Longer gut microvilli length and density • Higher c-type lysozyme (lyzc) gene expression • Better survival rates under*Streptococcus agalactiae* challenge	(Xia et al. 2020)
PROB ***Lactobacillus plantarum* KC426951**	8 g approx.	Sprayed on the diet pellets	90 d	• Enhancement of skin mucosal total protein (TP), lysozyme (LYZ), alkaline phosphatase (ALP), and protease (PRO) activities • Higher levels of serum superoxide dismutase (SOD) and alternative complement (ACH50)	(Mohammadi et al. 2020)
PROB ***Bacillus subtilis* WB60** ***Lactococcus lactis***	2.83 ± 0.05 g	Included in the diet pellets	8 wks	• Higher lysozyme (LYZ), Superoxide dismutase (SOD) and myeloperoxidase activity (MOP) • Longer gut microvilli length and thicker muscular layer • Overexpression of HSP70, IL-1β, IFN-γ, and TNF-α • Better survival rates under*A. hydrophila* challenge	(Won et al. 2020)
PROB ***Bacillus velezensis* TPS3N** ***Bacillus subtilis* TPS4** ***Bacillus amyloliquefaciens* TPS17** ***B.velezensis* TPS3N+*B. subtilis* TPS4+*B.amyloliquefaciens* TPS17**	46.24 ± 0.48 g	Included in the diet	4 wks	• Higher intestinal superoxide dismutase (SOD) and lipase activity • Thicker muscular layer	(Kuebutornye, Wang, et al. 2020)

(Continued)

TABLE 11.2 (CONTINUED)

Compilation of Several *in vivo* Experiments in Aquaculture Evaluating the Effect of Probiotics on Immunity Parameters. The Table Includes the Works Done on the Most Cultured Species Worldwide

Functional Feed	Animals	Supplementation	Time	Immunity Improvement Data	Ref.
PROB ***Clostridium butyricum*** **M2014537**	56.21 ± 0.81 g	Sprayed on the diet pellets	56 d	• Better survival rates under*Streptococcus agalactiae* challenge • Increase of serum total antioxidant capacity (T-AOC) • Decrease of MDA serum contents • Higher serum complement 3 (C3) and complement 4 (C4) concentrations • Intestinal overexpression of t TNF-α, IL-8, IL-10, TLR2, MyD88, IRAK-4	(Li et al. 2019)
PROB ***Bacillus velezensis*** **LF01**	4.0 g approx	Included in the diet	9 wks	• Increased lysozyme (LZY) and superoxide dismutase (SOD) activities • Overexpression of C3, lyzc, and MHC-IIβ • Improved survival rates after *S. agalactiae* challenge	(Zhang et al. 2019)
PROB ***Psychrobacter maritimus***	5.4 ± 0.1 g	Sprayed on the diet pellets	50 d	• Increase in digestive enzyme (protease, amylase, lipase) activity • Phagocytic activity, lysozyme activity, alternative complement hemolysis, and hematological parameters were also significantly increased • Upregulation of IL-4 and IL-12 and downregulation of HSP70	(Makled et al. 2019)
PROB ***Rummeliibacillus stabekisii***	4.1 ± 0.34 g	Included in the diet pellets	8wks	• Increase of intestinal digestive enzymes (protease, cellulase, and xylanase) • Increase of phagocytic activity, respiratory bursts, and superoxide dismutase (SOD) of head kidney leukocytes and improvement of serum lysozyme activity • Overexpression of IL- 1β, TNF-α, TGF-β and Hsp70 • Better survival rates under*A. hydrophila*and *S. iniae* challenges	(Tan et al. 2019)
PROB ***Lactobacillus plantarum*** **CR1T5**	4.97 ± 0.04 g	Included in the diet pellets	12 wks	• Higher skin lysozyme and peroxidase activities • Higher serum lysozyme, peroxidase, alternative. complement, phagocytosis, and respiratory burst activities • Better survival rates under*A. hydrophila* challenge.	(Van Doan et al. 2019)

(Continued)

TABLE 11.2 (CONTINUED)
Compilation of Several *in vivo* Experiments in Aquaculture Evaluating the Effect of Probiotics on Immunity Parameters. The Table Includes the Works Done on the Most Cultured Species Worldwide

Functional Feed	Animals	Supplementation	Time	Immunity Improvement Data	Ref.
PROB ***Bacillus cereus* QSI-1**	15.2 ± 0.53 g	Included in the diet pellets	15 d	• An increase in alternative complement activity and lysozyme activity in skin mucus • Decrease in superoxide dismutase activity in skin mucus • Better survival rate when challenge with *A. hydrophila*	(Jiang et al. 2019)
PROB ***Exiguobacterium acetylicum* S01**	58 g approx.	Included in the diet (vehicle: vegetable oil)	4 wks	• Better survival rate when challenge with *A. hydorphila* • Enhancement of hematological profile, respiratory burst, phagocytic activities and antimicrobial enzymes • Improvement of total immunoglobulin levels • Upregulation of lysozymes and pro- and anti-inflammatory cytokines (IL-1β, IL-10, and TGFβ)	(Jinendiran et al. 2019)
PROB ***Lactococcus lactis 16-7***	18 ± 3 g	Included in the diet pellets	42 d	• Enhancement serum superoxide dismutase activity, phagocytic activities of innate immune cells • Over-expression levels of immune-related genes (INF-γ, IL-1β, IL-11, TNF-α, MyD88). • Reduction of intestinal mucosal barrier damage and inflammation when challenged with *A. hydrophila* • Antagonize the colonization of *A. hydrophila* in the intestine	(Dong et al. 2018)
PROB ***Bacillus coagulans***	14.33±0.15g	Included in the diet pellets	8 wks	• Higher blood respiratory burst (RB), myeloperoxidase (MPO), and anti-superoxide anion free radical (AFASER) activities • Increment on White blood cell count (WBC) and glutathione (GSH) content • Overexpression of NOX2, Nrf2, Keap1, Bach1, Prx2 • Lower MDA content in supplemented groups	(Yu et al. 2018)
PROB ***Bacillus amyloliquefaciens* FPTB16**	25.98 ± 2.57 g	Included in the diet	4 wks	• Enhancement of oxygen radical production, serum lysozyme activity, total serum protein, myeloperoxidase activity, phosphatase activity • Overexpression of IL-1β, TNF-α, C3, and iNOS. • Downregulation of IFN-γ	(Singh et al. 2017)

(*Continued*)

TABLE 11.2 (CONTINUED)

Compilation of Several *in vivo* Experiments in Aquaculture Evaluating the Effect of Probiotics on Immunity Parameters. The Table Includes the Works Done on the Most Cultured Species Worldwide

Functional Feed	Animals	Supplementation	Time	Immunity Improvement Data	Ref.
PREB **Corncob-derived xylooligosaccharides (CDXOS)**	4.97 ± 0.04 g	Included in the diet pellets	12 wks	• Higher skin lysozyme and peroxidase activities • Higher serum lysozyme, peroxidase, alternative complement, phagocytosis, and respiratory burst activities • Better survival rates under*A. hydrophila* challenge	(Van Doan et al. 2019)
SYMB ***Lactobacillus plantarum* CR1T5** + **Corncob-derived xylooligosaccharides (CDXOS)**	4.97 ± 0.04 g	Included in the diet pellets	12 wks	• Higher skin lysozyme and peroxidase activities • Higher serum lysozyme, peroxidase, alternative complement, phagocytosis, and respiratory burst activities • Better survival rates under*A. hydrophila* challenge (higher than probiotic or prebiotic alone)	(Van Doan et al. 2019)
***Catla catla* (catla)**					
PROB *Bacillus subtilis* FPTB13	40.0±1.9g	Included in the diet	2 wks	• Enhancement of oxygen radical production, myeloperoxidase activity, lysozyme activity, total protein content, and alkaline phosphatase activity • Better survival rate when challenged with *E. tarda*	(Sangma & Kamilya 2015)
***Salmo salar* (Atlantic salmon)**					
PROB ***Aliivibrio* sp.,Vl1, NCIMB 42593;** ***Aliivibrio* sp., Vl2, NCIMB 42592** ***Aliivibrio* sp., Vl3, NCIMB 42594**	84.33±22.60 g	1 single "probiotic bath"	4-6 m	• Better survival rate • Lower ulcer prevalence	(Klakegg et al. 2020)
PROB ***Pediococcus acidilactici* MA18/5M**	35±3.4 g	Included in the diet pellets	12 wk	• Higher antiviral response. Overexpression of *mx*-1 and *tlr3*	(Jaramillo-Torres et al. 2019)
***Labeo rohita* (Roho labeo)**					

(Continued)

TABLE 11.2 (CONTINUED)
Compilation of Several *in vivo* Experiments in Aquaculture Evaluating the Effect of Probiotics on Immunity Parameters. The Table Includes the Works Done on the Most Cultured Species Worldwide

Functional Feed	Animals	Supplementation	Time	Immunity Improvement Data	Ref.
PROB *Bacillus subtilis*	26.4 ± 1.3 g	Included in the diet pellets	8 wks	• Higher white blood cell (WBC) count, globulin (GB) and total protein (TP) levels • Increase of serum phagocytic activity (PA), respiratory burst activity (RBA), complement C3 (CC3) level, alternative complement pathway (ACP),lysozyme activity (LA), and immunoglobulin M (IgM) levels in head kidney (HK) leucocytes • Increase of superoxide dismutase (SOD) and catalase (CAT) and glutathione peroxidase (GPx) activities • Upregulation of L-1β, IL-8, TNF-α, and NF-κB • Better survival rate when challenge with *A. hydrophila*	(Devi et al. 2019b)
PROB ***Bacillus aerophilus* KADR3**	35–40 g	Included in the diet pellets	6 wks	• Enhancement of serum lysozyme, phagocytosis, serum total protein, respiratory bursts, serum IgM levels, superoxide dismutase, and alternative complement pathway activities • Better survival rate when challenged with *A. hydrophila*	(Ramesh et al. 2017)
PROB ***Bacillus amyloliquefaciens* CCF7**	20.23 g	Included in the diet	70 d	• After *A. hydrophila* challenge test, lower serum aspartate transaminase (AST), serum alanine transaminase (ALT) activity, and liver malondialdehyde levels and higher liver catalase and superoxide dismutase activities, serum globulin concentration and IgM levels	(Nandi et al. 2018)
PROB ***G. candidum* QAUGC01**	20 ± 2.34 g	Encapsulated in sodium alginate	11 wk	• Higher intestinal enzyme activities (protease, amylase, and cellulase) and hemato-immunological indices (RBCs, Hb, HCT, WBCs, MCHC, respiratory bursts, and phagocytic activity, total protein, lysozyme, IgM) • Upregulation of heat shock protein HSP 70 gene in muscle, intestine, and liver • Reduction of serum AST and ALT activities, total cholesterol, and triglyceride	(Amir et al. 2019)
PREB **Galactooligosaccharides**	26.4 ± 1.3 g	Included in the diet pellets	8 wks	• Increase of superoxide dismutase (SOD) and catalase (CAT) and glutathione peroxidase (GPx) activities • Upregulation of L-1β, IL-8, TNF-α, and NF-κB • Better survival rate when challenged with *A. hydrophila*	(Devi et al. 2019b)

(Continued)

TABLE 11.2 (CONTINUED)
Compilation of Several *in vivo* Experiments in Aquaculture Evaluating the Effect of Probiotics on Immunity Parameters. The Table Includes the Works Done on the Most Cultured Species Worldwide

Functional Feed	Animals	Supplementation	Time	Immunity Improvement Data	Ref.
SYMB ***Bacillus subtilis* +** **Galactooligosaccharides**	26.4 ± 1.3 g	Included in the diet pellets(Ramesh et al. 2015)	8 wks	• Higher white blood cell (WBC) count, globulin (GB), and total protein (TP) levels • Increase of serum phagocytic activity (PA), respiratory burst activity (RBA), complement C3 (CC3) level, alternative complement pathway (ACP), lysozyme activity (LA), and immunoglobulin M (IgM) levels in head kidney (HK) leucocytes • Increase of superoxide dismutase (SOD) and catalase (CAT) and glutathione peroxidase (GPx) activities • Upregulation of L-1β, IL-8, TNF-α, and NF-κB • Better survival rate when challenged with *A. hydrophila*	(Devi et al. 2019b)
***Megalobrama amblycephala* (Wuchang bream)**					
PROB ***Streptococcus faecalis***	46.32 ± 0.09 g	Included in the diet pellets	10 wks	• Increase of SOD, CAT, and GPx hepatic activities • Increase of LZM, MPO, ACP, and AKP plasma activities • Increase of hemato-immunological C3, C4, and IgM activities • Lower MDA and NO levels • Overexpression of*Leap-I*,*Leap-II*,*muc2*,and *muc5b* • Better survival rate when challenged with *A. hydorphila*	(Zhong et al. 2019)
***Oncorhynchus mykiss* (rainbow trout)**					
PROB ***Lactobacillus delbrukei*subsp.*bulgaricus*** ***Lactobacillus acidophilus*** ***Citrobacter farmeri***	19.08–32.9 g	Included in the diet	60 d	• Increase of hemoglobin and hematocrit • Increase of serum lysozyme and complement activities • Lower NBT levels • Overexpression of I*GF-1*, *FATP*, *γ-GTP*, *IL-1β*, *IL-8*, *and IL-10* • Better survival rate (*L. bulgaricus*and *L. acidophilus*) when challenged with *L. garvieae*	(Mohammadian et al. 2019)
PROB ***Lactobacillus rhamnosus* ATCC 7469**	18.41 ± 0.32 g	Microencapsulated with alginate and hi-maize starch and coated with chitosan	60 d	• Increase of serum lysozyme and complement activities • Increase of SOD and CAT. • Overexpression of *IL-1 andTNF-1α* • Better survival rate when challenged with*Y. Ruckeri*	(Hooshyar et al. 2020)

(Continued)

TABLE 11.2 (CONTINUED)
Compilation of Several *in vivo* Experiments in Aquaculture Evaluating the Effect of Probiotics on Immunity Parameters. The Table Includes the Works Done on the Most Cultured Species Worldwide

Functional Feed	Animals	Supplementation	Time	Immunity Improvement Data	Ref.
PROB ***Saccharomyces cerevisiae***	250 ± 50 g	Included in the diet	130 d	• Increase of white blood cells • Increase of respiratory burst • Overexpression of *IL-1β* and *TNF-1α*. • Lower serum cholesterol	(Vazirzadeh et al. 2020)
PROB **PrimaLac® (*Lactobacillus acidophilus*, *Lactobacillus casei*, *Enterococcus faecium*, and *Bifidobacterium bifidium*)**	25 ± 1.8 g	Included in the diet	8 wks	• Increase of lipase, protease, and amylase activities • Lower of ALP, ALT, and AST activities	(Naderi Farsani et al. 2020)
PROB ***Enterococcus casseliflavus***	38.3±1 g	Included in the diet	8 wks	• Increase of lipase, trypsin, and amylase activities • Increase of total serum protein, and albumin • Higher serum IgM • Increase of respiratory burst activity of blood leukocytes • Better survival rate when challenged with *Streptococcus iniae*	(Safari et al. 2016)
***Channa argus* (snakehead)**					
PROB ***Lactococcus lactis* L19;** ***Enterococcus faecalis* W24** ***Lactococcus lactis* L19 + *Enterococcus faecalis* W24**	9.50 ± 0.03 g	Included in the diet	8 wks	• Increased levels of IgM, ACP, AKP, LZ, C3, and C4 activities in serum • Upregulation of L-1β, IL-6, IL-10, TNF-α, IFN-γ, HSP70, HSP90, and TGF-β • Better survival rate when challenged with *A. veronii*	(Kong et al. 2020)

et al. 2019; Feng et al. 2019). Other common protocols performed in publications and focused on probiotics, prebiotics and symbiotics are phagocyte respiratory burst (RB) evaluation (Ahmadifar et al. 2019) and antioxidant enzyme quantification (Devi et al. 2019a; Tang et al. 2019). RB has been widely recognized since decades ago as a key indicator of cellular immunity mechanisms in teleost fish when evaluating the innate immune defense against pathogens (Miyazaki 1998).

The previously described modulation of the innate immune system is not only bound to probiotics. Prebiotics such as raffinose (Hoseinifar et al. 2019), Galactooligosaccharide (Modanloo et al. 2017), β-(1,3)(1,6)-D-glucan (Kühlwein et al. 2014), Corncob-derived xylooligosaccharides (CDXOS) (Van Doan et al. 2019), and their combinations with probiotic strains, as symbiotics as supplements *Pediococcus acidilactici* MA18/5M + Raffinose, *Pediococcus acidilactici* MA18/5M + Galactooligosaccharide, or *Lactobacillus plantarum* CR1T5 + Corncob-derived xylooligosaccharides (CDXOS), have been described to modulate enzymatic activities, gene expression, or hematocrit values in species with high commercial interest, like common carp and Nile tilapia.

The probiotic bacteria genus, like *Lactobacillus* and *Bifidobacterium*, can generate inhibitory substances with bactericidal or bacteriostatic tendencies which may have an effect on pathogenic microbes (Servin 2004). Within this group of molecules are included siderophores, lysozymes, hydrogen peroxide, bacteriocins, or proteases (Panigrahi & Azad 2007; Tinh et al. 2008). Some strains produce volatile fatty acids such as acetic, butyric, lactic, and propionic acids. The production of this acid environment results in a pH reduction of the gastrointestinal tract (Chauhan & Singh 2019). Consequently, the proliferation of opportunistic pathogens is withdrawn (Tinh et al. 2008; Chauhan & Singh 2019). Thus, the search for probiotic microorganisms with bactericidal or bacteriostatic activities does not stop. For example, a description of the fatty acid profile of *Kluyveromyces lactis* M3 for use as probiotics in *Sparus aurata* has recently been published (Guluarte et al. 2019). Studies focused on these activities are usually carried out in experiments directly designed for species of commercial interest or on model organisms, such as zebrafish (*Danio rerio*). For example, a study was recently published in which the bacteriocin-like activity of the potential probiotic *Chromobacterium aquaticum* was evaluated in the model species (Yi et al. 2019).

Nutrient Competition

One possible method of probiotic action against pathogens is nutrient competition (Ringø et al. 2016). Probiotics can consume the nutrients that are essential for pathogens, and, in this way, they can restrict their target's presence in the intestinal tract. For example, iron is necessary for bacterial growth; therefore bacteria that are able to sequester ferric iron by siderophore production can be used as probiotics because they significantly decrease available iron for pathogens (Tinh et al. 2008; Chauhan & Singh 2019).

Antiviral and Antifungal Activity

Some studies suggest that probiotics could also engage in antiviral activity. However, the precise mechanism by which this occurs is still under study. One hypothesis is that some enzymes produced and secreted by probiotic bacteria could have an inhibitory effect on virus. Botić et al. (2007) used eukaryotic cell cultures to study the antiviral activity potential of probiotics. They concluded that mechanisms of action could be related to a direct antiviral activity of probiotic metabolites or to a probiotic effect hindering the adsorption and cell internalization of the virus. It has been described that feeding supplementation of shrimp with the probiotic *Bacillus megaterium* increases resistance against white-spot syndrome virus (Li et al. 2009). Probiotic antiviral activity has been also studied in other species such as olive flounder (*Paralichthys olivaceus*) (Harikrishnan et al. 2010). There are also a few studies that reported probiotic antifungal activity. As an example, in catfish, *Lactobacillus plantarum* had an effect against *Saprolegnia parasitica* (Nurhajati et al. 2012).

Stress Tolerance and Reproduction Improvement

Apart from pathogen pressure, aquaculture species are subjected to different environmental perturbations related to the physicochemical parameters provoking different stress responses that severely affect their physiological state (Mohapatra et al. 2013). It has been reported that stress can induce suppression of reproduction in different species (Chabbi & Ganesh 2015; Chabbi & Ganesh 2016; Marcon et al. 2018). Reproduction and brood stock management are two of the main bottlenecks in the development of the aquaculture industry which exists to enable a captive group of fish to undergo reproductive maturation, spawning, and production of fertilized eggs. Although many cultured fish species achieve successful reproductive performance under cultured conditions, there remains a large number of important species which still exhibit reproductive dysfunction or present diminished reproductive performance under captivity conditions (Zohar & Mylonas 2001). This reduced reproduction success is often due to the combination of captivity-induced stress and the lack of a natural spawning environment (Zohar & Mylonas 2001). Considerable efforts have been made in order to find new compounds to mitigate stress related to captivity conditions in aquaculture fish species. Regarding this, the present section summarizes the main articles in the study of dietary probiotic supplementation as a promising tool to improve fish reproduction performance.

In the first study on this topic, the administration of the probiotic *Bacillus subtilis* improved the reproductive performance of four ornamental fish species: guppy, Mexican molly, green swordtail and southern platy fish (Ghosh et al. 2007). Probiotic administration in this study provoked the improvement of reproductive parameters, such as the gonadosomatic index (GSI: a well-used indicator of ovary development and fecundity) and the fecundity of spawning females in all fish species (Ghosh et al. 2007). Moreover, the authors described that the fry length and the weight were significantly higher after probiotic supplementation in all four species. The same beneficial effects concerning the increase in GSI and fecundity were observed in a later study in platy fish after dietary incorporation of a commercial probiotic mixture of *L. acidophilus*, *L. casei*, *E. faecium*, and *B. thermophilum* (Abasali & Mohamad 2010). Nevertheless, in this case weight and deformation rate were not affected by probiotic supplementation. These studies opened a new field on the use of probiotics to improve fish reproductive performance. However, the mechanisms behind these benefits were lately studied.

Reproduction and fertility processes are orchestrated by the hypothalamic-pituitary-gonadal (HPG) axis (Acevedo-Rodriguez et al. 2018). Stress could affect these HPG-involved tissues at different levels, triggering consequences on the reproduction process. Interestingly, it has been demonstrated that microbial colonization can affect the HPG axis, resulting in modulation of stress-related responses (Sudo 2006; Palermo et al. 2011). To expand our knowledge on the effects of these types of compounds, it is necessary to understand the action mechanism in an integrative way, at different levels of this axis. To that end, Dr Carnevali and his group has widely employed the zebrafish model in order to characterize the probiotic supplementation effects on females at different levels of the HPG axis. In their initial analyses, this group reported that after supplementing zebrafish with experimental diets containing *L. rhamnosus* for a short period of time, several beneficial responses were observed such as: an increase in the number of ovulated eggs, higher hatching rates, and faster embryonic development (Gioacchini et al. 2010). More importantly, the authors found higher levels of *L. rhamnosus* by qPCR in the gut microbiota of treated fish females compared to the controls (Gioacchini et al. 2011), underlining the importance of diet and gut microbes in the reproductive processes and supporting the hypothesis that feed additives could affect reproductive performance and fertility capacity. This probiotic supplementation was able to not only colonize the gastrointestinal tract but also to modulate the microbial communities, causing a microbial population remodeling and enhancing the presence of another *Lactobacillus*, *Streptococcus thermophilus* (Gioacchini et al. 2012).

The Dr Carnevali group also studied the effects of *L. rhamnosus* administration on zebrafish ovary development using approaches from the cellular to the molecular level. Several HPG pathways were described to be affected by this probiotic supplementation, provoking altered processes related to vitellogenin uptake and testosterone conversion, among others, triggering gonadal differentiation and producing beneficial influences on the zebrafish physiological performance (Gioacchini et al. 2011; Carnevali et al. 2013). Moreover, the *L. rhamnosus* administration effect on the mechanism for the acquisition of oocyte maturational competence was demonstrated by several lines of evidence, including qPCR analysis and secondary protein structure studies (Gioacchini et al. 2012). In their analyses, the authors also characterized some hormonal signaling related to ovary development that was affected by probiotic administration. Specifically, they demonstrated that the probiotic may act indirectly by activating the gene expression of neuropeptide hormones and metabolic signals, such as kisspeptin and leptin – both at the CNS level and at the peripheral leptin level (Gioacchini et al. 2010). Both hormones are widely known for their role in stimulation of the reproductive axis activating GnRH as evidenced by several authors (Barb et al. 2005). As a consequence of this, the Dr Carnevali group observed improvements in zebrafish follicle maturation, fecundity, and egg quality (Gioacchini et al. 2010; Carnevali et al. 2013). These findings allowed the authors to conclude that these hormones (either alone or with others) could represent the bond between the metabolic and reproductive systems through which the probiotic may act (Gioacchini et al. 2012). Specifically, dietary administration of *L. rhamnosus* probiotic upregulates *lep* expression both at peripheral (intestinal) and central (brain) levels. In addition, its increase was accomplished through the upregulation of kisspeptins (Gioacchini et al. 2010), suggesting the positive role of probiotics with regard to reproduction (Gioacchini et al. 2014b).

Concerning probiotic consequences in male reproduction, first studies were performed by our group, providing evidence of a positive role of the probiotic *P. acidilactici* on zebrafish testicular cells (Valcarce et al. 2015). Probiotic supplementation improved the levels of some sperm transcripts previously published as sperm quality markers, due to their correlation with sperm motility and their predictive fertility value (Guerra et al. 2013). A significant increase of *leptin*, *bdnf*, and *dmrt1* gene expression after a ten-day probiotic administration was reported (Valcarce et al. 2015). In a second study performed by our group, we tested on zebrafish male two strains of probiotics, *Lactobacillus rhamnosus* and *Bifidobacterium longum*, which has been successfully used in human sperm to improve sperm motility, DNA integrity, and reduce oxidative stress (Valcarce et al. 2017). This mixture of probiotics was used as feed supplementation in zebrafish model for one spermatogenic cycle. The antioxidant and anti-inflammatory properties of this probiotic preparation triggered an increase in sperm concentration, total motility, progressive motility, and fast spermatozoa subpopulations (Valcarce et al. 2019a). Moreover, in this study, some additional effects on fish behavior were observed. Interestingly, the animals fed with the supplement showed different behavior patterns compared to control groups, suggesting a diet-related modulation on the lower stress-like conduct (Valcarce et al. 2019a). Surprisingly, the administration of this probiotic combination for only 4.5 months had an effect on progeny, improving survival at 24 hours post-fertilization (hpf) (Valcarce et al. 2019b).

The effect of probiotics on male reproduction has also been studied in other marine and fresh water fish species relevant to aquaculture. One of them is the European eel. Males were intraperitoneally treated with recombinant hCG and exposed to *L. rhamnosus* probiotic for two weeks (Vílchez et al. 2015). The authors observed that probiotic supplementation triggered the enhancement of spermatogenesis through a significant higher sperm volume and improved motility parameters. These results were accompanied with the increasing levels of spermatogenic genes, such as *activin*, *arα*, *prl*, and *fshr* (Vílchez et al. 2015).

In goldfish (*Carassius auratus*), Mehdinejad et al. demonstrated that combined probiotic dietary supplementation of *Pediococcus acidilactici* and nucleotide significantly improved reproductive performance. These probiotic mixtures contributed to the improved sperm motility and fertilization success in treated males (Mehdinejad et al. 2019).

Taking into account all these findings, probiotics possess a broad spectrum of action that includes a combined effect on reproduction, metabolism, and behavior in fish. In this sense, different strains and probiotic mixtures could be incorporated in aquaculture diets as potential beneficial compounds to mitigate the stress-related effects on captive fish reproduction performance.

Water Quality Improvement

Some studies suggest that the Gram + bacteria used as probiotics can improve the water quality. On the one hand, they can efficiently convert organic matter into CO_2 (Chauhan & Singh 2019), they have significant algicidal tendencies (Fukami et al. 1997), and the nitrifying probiotic bacteria can eliminate ammonia and nitrates (Zorriehzahra et al. 2016). On the other hand, they also modify pH, temperature, and dissolved oxygen, among other parameters.

FUTURE PERSPECTIVES AND CONCLUDING REMARKS

In this chapter, we have revisited several studies that demonstrated the positive effects of prebiotics, probiotics, and synbiotics on several relevant species for aquaculture. The main benefits are related to animal growth and water quality improvement as well as protection against some diseases. But probiotics can also have effects at other levels, improving gamete quality or modifying fish behavior. However, probiotic mechanisms of action still need to be studied at the cellular and molecular levels for a better understanding of these unexpected effects. Improving and standardizing probiotic administration procedures, understanding the durability of probiotic effects in the host, and learning the best way of administration in each particular case, are important aspects to be determined. These studies will undoubtedly improve the efficiency in the use of probiotics in aquaculture in the near future.

ACKNOWLEGMENTS

Authors would like to acknowlege PID2019-108509RB-I00 project (Ministerio de Ciencia e Innovación), FCJ2018-037566-I grant, PROBISOLE project (Fundación Biodiversidad; PLEAMAR2020-FEMP), ADM-Biopolis and STOLT Sea Farm.

REFERENCES

Abasali H, Mohamad S. 2010. Effect of dietary supplementation with probiotic on reproductive performance of female livebearing ornamental fish. *J Aquac Feed Sci Nutr.* 2(2):11–15.

Acevedo-Rodriguez A, Kauffman AS, Cherrington BD, Borges CS, Roepke TA, Laconi M. 2018. Emerging insights into hypothalamic-pituitary-gonadal axis regulation and interaction with stress signalling. *J Neuroendocrinol.* 30(10):e12590.

Adel M, El-Sayed AFM, Yeganeh S, Dadar M, Giri SS. 2017. Effect of potential probiotic *Lactococcus lactis* subsp. lactis on growth performance, intestinal microbiota, digestive enzyme activities, and disease resistance of *Litopenaeus vannamei. Probiotics Antimicrob Proteins* [Internet]. [accessed 2020 May 10] 9(2):150–156. http://link.springer.com/10.1007/s12602-016-9235-9

Ahmadifar E, Moghadam MS, Dawood MAO, Hoseinifar SH. 2019. *Lactobacillus fermentum* and/or ferulic acid improved the immune responses, antioxidative defence and resistance against *Aeromonas hydrophila* in common carp (*Cyprinus carpio*) fingerlings. *Fish Shellfish Immunol* [Internet]. [accessed 2020 May 8] 94:916–923. https://linkinghub.elsevier.com/retrieve/pii/S1050464819309787

Ai Q, Xu H, Mai K, Xu W, Wang J, Zhang W. 2011. Effects of dietary supplementation of Bacillus subtilis and fructooligosaccharide on growth performance, survival, non-specific immune response and disease resistance of juvenile large yellow croaker, *Larimichthys crocea. Aquaculture* 317(1–4):155–161.

Amir I, Zuberi A, Kamran M, Imran M, Murtaza MH. 2019. Evaluation of commercial application of dietary encapsulated probiotic (*Geotrichum candidum* QAUGC01): Effect on growth and immunological indices of rohu (*Labeo rohita*, Hamilton 1822) in semi-intensive culture system. *Fish Shellfish Immunol.* 95:464–472.

Balcázar JL, de Blas I, Ruiz-Zarzuela I, Vendrell D, Calvo AC, Márquez I, Gironés O, Muzquiz JL. 2007. Changes in intestinal microbiota and humoral immune response following probiotic administration in brown trout (*Salmo trutta*). *Br J Nutr.* 97(3):522–527.

Barb CR, Hausman GJ, Czaja K. 2005. Leptin: A metabolic signal affecting central regulation of reproduction in the pig. In: *Domest Anim Endocrinol.* 2005 Jul;29(1):186–92. doi: 10.1016/j.domaniend.2005.02.024. Epub 2005 Apr 7. PMID: 15927773.

Botić T, Klingberg TD, Weingartl H, Cencič A. 2007. A novel eukaryotic cell culture model to study antiviral activity of potential probiotic bacteria. *Int J Food Microbiol.* 115(2):227–234.

Cao L, Pan L, Gong L, Yang Y, He H, Li Y, Peng Y, Li D, Yan L, Ding X, et al. 2019. Interaction of a novel *Bacillus velezensis* (BvL03) against *Aeromonas hydrophila* in vitro and in vivo in grass carp. *Appl Microbiol Biotechnol* [Internet]. [accessed 2020 May 8] 103(21–22):8987–8999. http://link.springer.com/10.1007/s00253-019-10096-7

Carnevali O, Avella MA, Gioacchini G. 2013. Effects of probiotic administration on zebrafish development and reproduction. *Gen Comp Endocrinol* [Internet]. [accessed 2019 May 1] 188:297–302. https://linkinghub.elsevier.com/retrieve/pii/S0016648013000968

Chabbi A, Ganesh CB. 2015. Evidence for the involvement of dopamine in stress-induced suppression of reproduction in the cichlid fish *Oreochromis mossambicus. J Neuroendocrinol.* 27(5):343–356.

Chabbi A, Ganesh CB. 2016. Neuroanatomical evidence for the involvement of β-endorphin during reproductive stress response in the fish *Oreochromis mossambicus. J Chem Neuroanat.* 77:161–168.

Chauhan A, Singh R. 2019. Probiotics in aquaculture: A promising emerging alternative approach. *Symbiosis* 77(2):99–113.

Corr SC, Hill C, Gahan CGM. 2009. Understanding the mechanisms by which probiotics inhibit gastrointestinal pathogens. *Adv Food Nutr Res.* 56:1–15.

Das S, Mondal K, Haque S. 2017. A review on application of probiotic, prebiotic and synbiotic for sustainable development of aquaculture. *J Entomol Zool Stud.* 5(2):422–429.

Devi G, Harikrishnan R, Paray BA, Al-Sadoon MK, Hoseinifar SH, Balasundaram C. 2019a. Comparative immunostimulatory effect of probiotics and prebiotics in Channa punctatus against *Aphanomyces invadans. Fish Shellfish Immunol.* 86:965–973.

Devi G, Harikrishnan R, Paray BA, Al-Sadoon MK, Hoseinifar SH, Balasundaram C. 2019b. Effect of symbiotic supplemented diet on innate-adaptive immune response, cytokine gene regulation and antioxidant property in Labeo rohita against *Aeromonas hydrophila. Fish Shellfish Immunol.* 89:687–700.

Van Doan H, Hoseinifar SH, Tapingkae W, Seel-audom M, Jaturasitha S, Dawood MAO, Wongmaneeprateep S, Thu TTN, Esteban MÁ. 2019. Boosted growth performance, mucosal and serum immunity, and disease resistance Nile tilapia (*Oreochromis niloticus*) fingerlings using corncob-derived xylooligosaccharide and lactobacillus plantarum CR1T5. *Probiotics Antimicrob Proteins.* 12:400–411.

Dong Y, Yang Y, Liu J, Awan F, Lu C, Liu Y. 2018. Inhibition of *Aeromonas hydrophila*-induced intestinal inflammation and mucosal barrier function damage in crucian carp by oral administration of *Lactococcus lactis. Fish Shellfish Immunol* [Internet]. [accessed 2020 May 8] 83:359–367. https://linkinghub.elsevier.com/retrieve/pii/S1050464818305850

Feng J, Chang X, Zhang Y, Yan X, Zhang J, Nie G. 2019. Effects of *Lactococcus lactis* from *Cyprinus carpio* L. as probiotics on growth performance, innate immune response and disease resistance against *Aeromonas hydrophila. Fish Shellfish Immunol* [Internet]. [accessed 2020 May 8] 93:73–81. https://linkinghub.elsevier.com/retrieve/pii/S1050464819307375

Fukami K, Nishijima T, Ishida Y. 1997. Stimulative and inhibitory effects of bacteria on the growth of microalgae. In: *Hydrobiologia* [Internet]. Vol. 358. Springer [accessed 2020 Jul 27]; pp. 185–191. https://link.springer.com/article/10.1023/A:1003139402315

Fuller R. 1989. Probiotics in man and animals. *J Appl Bacteriol* [Internet]. [accessed 2019 Mar 13] 66(5):365–378. http://www.ncbi.nlm.nih.gov/pubmed/2666378

Geng X, Dong XH, Tan BP, Yang QH, Chi SY, Liu HY, Liu XQ. 2011. Effects of dietary chitosan and *Bacillus subtilis* on the growth performance, non-specific immunity and disease resistance of cobia, *Rachycentron canadum. Fish Shellfish Immunol.* 31(3):400–406.

Ghosh S, Sinha A, Sahu C. 2007. Effect of probiotic on reproductive performance in female livebearing ornamental fish. *Aquac Res* [Internet]. [accessed 2020 May 9] 38(5):518–526. http://doi.wiley.com/10.1111/j.1365-2109.2007.01696.x

Gioacchini G, Giorgini E, Merrifield DL, Hardiman G, Borini A, Vaccari L, Carnevali O. 2012. Probiotics can induce follicle maturational competence: The *Danio rerio* case1. *Biol Reprod.* 86(3):65.

Gioacchini G, Lombardo F, Merrifeld DL, Silvi S, Cresci A, Avella MA, Carnevali O. 2011. Effects of probiotics on zebrafish reproduction. *J Aquac Res Dev.* (SPEC. ISSUE 1). 12:400–411

Gioacchini G, Maradonna F, Lombardo F, Bizzaro D, Olivotto I, Carnevali O. 2010. Increase of fecundity by probiotic administration in zebrafish (*Danio rerio*). *Reproduction.* 140(6):953–959.

Gobi N, Vaseeharan B, Chen JC, Rekha R, Vijayakumar S, Anjugam M, Iswarya A. 2018. Dietary supplementation of probiotic *Bacillus licheniformis* Dahb1 improves growth performance, mucus and serum immune parameters, antioxidant enzyme activity as well as resistance against *Aeromonas hydrophila* in tilapia *Oreochromis mossambicus*. *Fish Shellfish Immunol* [Internet]. [accessed 2020 Jul 27] 74:501–508. https://pubmed.ncbi.nlm.nih.gov/29305993/

Gong L, He H, Li D, Cao L, Khan TA, Li Y, Pan L, Yan L, Ding X, Sun Y, et al. 2019. A new isolate of pediococcus pentosaceus(SL001) with antibacterial activity against fish pathogens and potency in facilitating the immunity and growth performance of grass carps. *Front Microbiol.* 10(June):1384.

Guerra SM, Valcarce DG, Cabrita E, Robles V. 2013. Analysis of transcripts in gilthead seabream sperm and zebrafish testicular cells: mRNA profile as a predictor of gamete quality. *Aquaculture* [Internet]. [accessed 2019 Jan 15] 406–407:28–33. https://www.sciencedirect.com/science/article/pii/S0044848613002226

Guerreiro I, Couto A, Machado M, Castro C, Pousão-Ferreira P, Oliva-Teles A, Enes P. 2016. Prebiotics effect on immune and hepatic oxidative status and gut morphology of white sea bream (*Diplodus sargus*). *Fish Shellfish Immunol.* 50:168–174.

Guluarte C, Reyes-Becerril M, Gonzalez-Silvera D, Cuesta A, Angulo C, Esteban MÁ. 2019. Probiotic properties and fatty acid composition of the yeast *Kluyveromyces lactis* M3. In vivo immunomodulatory activities in gilthead seabream (*Sparus aurata*). *Fish Shellfish Immunol.* 94:389–397.

Gupta A, Gupta P, Dhawan A. 2014. Dietary supplementation of probiotics affects growth, immune response and disease resistance of *Cyprinus carpio* fry. *Fish Shellfish Immunol* [Internet]. [accessed 2020 May 9] 41(2):113–119. https://linkinghub.elsevier.com/retrieve/pii/S1050464814003088

Hai N V. 2015. The use of probiotics in aquaculture. *J Appl Microbiol* [Internet]. [accessed 2020 Feb 10] 119(4):917–935. http://www.ncbi.nlm.nih.gov/pubmed/26119489

Harikrishnan R, Balasundaram C, Heo MS. 2010. Effect of probiotics enriched diet on *Paralichthys olivaceus* infected with lymphocystis disease virus (LCDV). *Fish Shellfish Immunol.* 29(5):868–874.

Hooshyar Y, Abedian Kenari A, Paknejad H, Gandomi H. 2020. Effects of *Lactobacillus rhamnosus* ATCC 7469 on different parameters related to health status of rainbow trout (*Oncorhynchus mykiss*) and the protection against *Yersinia ruckeri*. *Probiotics Antimicrob Proteins.* 12:1370–1384

Hoseinifar SH, Hosseini M, Paknejad H, Safari R, Jafar A, Yousefi M, Van Doan H, Torfi Mozanzadeh M. 2019. Enhanced mucosal immune responses, immune related genes and growth performance in common carp (*Cyprinus carpio*) juveniles fed dietary *Pediococcus acidilactici* MA18/5M and raffinose. *Dev Comp Immunol* [Internet]. [accessed 2020 May 8] 94:59–65. https://linkinghub.elsevier.com/retrieve/pii/S0145305X1830586X

Hoseinifar SH, Sun YZ, Wang A, Zhou Z. 2018. Probiotics as means of diseases control in aquaculture, a review of current knowledge and future perspectives. *Front Microbiol.* 9(OCT):2429.

Jaramillo-Torres A, Rawling MD, Rodiles A, Mikalsen HE, Johansen LH, Tinsley J, Forberg T, Aasum E, Castex M, Merrifield DL. 2019. Influence of dietary supplementation of probiotic pediococcus acidilactici MA18/5M during the transition from freshwater to seawater on intestinal health and microbiota of atlantic salmon (*Salmo salar* L.). *Front Microbiol* [Internet]. [accessed 2020 Jul 19] 10(SEP). https://pubmed.ncbi.nlm.nih.gov/31611864/

Jiang Y, Zhou S, Chu W. 2019. The effects of dietary *Bacillus cereus* QSI-1 on skin mucus proteins profile and immune response in crucian carp (*Carassius auratus gibelio*). *Fish Shellfish Immunol* [Internet]. [accessed 2020 May 8] 89:319–325. https://linkinghub.elsevier.com/retrieve/pii/S1050464819302529

Jinendiran S, Nathan AA, Ramesh D, Vaseeharan B, Sivakumar N. 2019. Modulation of innate immunity, expression of cytokine genes and disease resistance against *Aeromonas hydrophila* infection in goldfish (*Carassius auratus*) by dietary supplementation with *Exiguobacterium acetylicum* S01. *Fish Shellfish Immunol* [Internet]. [accessed 2020 May 8] 84:458–469. https://linkinghub.elsevier.com/retrieve/pii/S1050464818306569

Kazuń B, Małaczewska J, Kazuń K, Żylińska-Urban J, Siwicki AK. 2018. Immune-enhancing activity of potential probiotic strains of *Lactobacillus plantarum* in the common carp (Cyprinus carpio) fingerling. *J Vet Res.* 62(4):485–492.

Kesarcodi-Watson A, Kaspar H, Lategan MJ, Gibson L. 2012. Performance of single and multi-strain probiotics during hatchery production of Greenshell™ mussel larvae, *Perna canaliculus. Aquaculture* 354–355:56–63.

Kim DH, Austin B. 2006. Innate immune responses in rainbow trout (*Oncorhynchus mykiss*, Walbaum) induced by probiotics. *Fish Shellfish Immunol.* 21(5):513–524.

Klakegg Ø, Salonius K, Nilsen A, Fülberth M, Sørum H. 2020. Enhanced growth and decreased mortality in Atlantic salmon (*Salmo salar*) after probiotic bath. *J Appl Microbiol* [Internet]. [accessed 2020 Jul 19] 129(1):146–160. https://onlinelibrary.wiley.com/doi/abs/10.1111/jam.14649

Kong Y, Gao C, Du X, Zhao J, Li M, Shan X, Wang G. 2020. Effects of single or conjoint administration of lactic acid bacteria as potential probiotics on growth, immune response and disease resistance of snakehead fish (*Channa argus*). *Fish Shellfish Immunol* [Internet]. [accessed 2020 Jul 26] 102:412–421. https://pubmed.ncbi.nlm.nih.gov/32387561/

Kuebutornye FKA, Abarike ED, Lu Y, Hlordzi V, Sakyi ME, Afriyie G, Wang Z, Li Y, Xie CX. 2020. Mechanisms and the role of probiotic Bacillus in mitigating fish pathogens in aquaculture. *Fish Physiol Biochem.* 46(3):819–841

Kuebutornye FKA, Wang Z, Lu Y, Abarike ED, Sakyi ME, Li Y, Xie CX, Hlordzi V. 2020. Effects of three host-associated Bacillus species on mucosal immunity and gut health of Nile tilapia, *Oreochromis niloticus* and its resistance against *Aeromonas hydrophila* infection. *Fish Shellfish Immunol* [Internet]. [accessed 2020 May 10] 97:83–95. https://linkinghub.elsevier.com/retrieve/pii/S1050464819311751

Kühlwein H, Merrifield DL, Rawling MD, Foey AD, Davies SJ. 2014. Effects of dietary β-(1,3)(1,6)-D-glucan supplementation on growth performance, intestinal morphology and haemato-immunological profile of mirror carp (*Cyprinus carpio* L.). *J Anim Physiol Anim Nutr (Berl).* 98(2):279–289.

Li H, Zhou Y, Ling H, Luo L, Qi D, Feng L. 2019. The effect of dietary supplementation with *Clostridium butyricum* on the growth performance, immunity, intestinal microbiota and disease resistance of tilapia (*Oreochromis niloticus*). *PLoS One.* 14(12):e0223428.

Li J, Tan B, Mai K. 2009. Dietary probiotic Bacillus OJ and isomaltooligosaccharides influence the intestine microbial populations, immune responses and resistance to white spot syndrome virus in shrimp (*Litopenaeus vannamei*). *Aquaculture* 291(1–2):35–40.

Madreseh S, Ghaisari HR, Hosseinzadeh S. 2019. Effect of lyophilized, encapsulated *Lactobacillus fermentum* and lactulose feeding on growth performance, heavy metals, and trace element residues in rainbow trout (*Oncorhynchus mykiss*) tissues. *Probiotics Antimicrob Proteins.* 11(4):1257–1263.

Makled SO, Hamdan AM, El-Sayed AFM. 2019. Growth promotion and immune stimulation in nile tilapia, *Oreochromis niloticus*, fingerlings following dietary administration of a novel marine probiotic, *Psychrobacter maritimus* S. *Probiotics Antimicrob Proteins* [Internet]. [accessed 2020 May 10]. http://link.springer.com/10.1007/s12602-019-09575-0

Marcon M, Mocelin R, Benvenutti R, Costa T, Herrmann AP, de Oliveira DL, Koakoski G, Barcellos LJG, Piato A. 2018. Environmental enrichment modulates the response to chronic stress in zebrafish. *J Exp Biol.* 221(Pt 4):jeb176735.

Mehdinejad N, Imanpour MR, Jafari V. 2019. Combined or individual effects of dietary probiotic, *Pediococcus acidilactici* and nucleotide on reproductive performance in goldfish (*Carassius auratus*). *Probiotics Antimicrob Proteins* [Internet]. [accessed 2020 May 11] 11(1):233–238. http://www.ncbi.nlm.nih.gov/pubmed/29318466

Miyazaki T. 1998. A simple method to evaluate respiratory burst activity of blood phagocytes from Japanese flounder. *Fish Pathol* [Internet]. [accessed 2020 May 11] 33(3):141–142. http://joi.jlc.jst.go.jp/JST.Journalarchive/jsfp1966/33.141?from=CrossRef

Modanloo M, Soltanian S, Akhlaghi M, Hoseinifar SH. 2017. The effects of single or combined administration of galactooligosaccharide and *Pediococcus acidilactici* on cutaneous mucus immune parameters, humoral immune responses and immune related genes expression in common carp (*Cyprinus carpio*) fingerlings. *Fish Shellfish Immunol* [Internet]. [accessed 2020 May 9] 70:391–397. https://linkinghub.elsevier.com/retrieve/pii/S105046481730551X

Mohammadi G, Rafiee G, Abdelrahman HA. 2020. Effects of dietary *Lactobacillus plantarum* (KC426951) in biofloc and stagnant-renewal culture systems on growth performance, mucosal parameters, and serum innate responses of Nile tilapia *Oreochromis niloticus*. *Fish Physiol Biochem* [Internet]. [accessed 2020 May 10]. http://link.springer.com/10.1007/s10695-020-00777-w

Mohammadian T, Nasirpour M, Tabandeh MR, Heidary AA, Ghanei-Motlagh R, Hosseini SS. 2019. Administrations of autochthonous probiotics altered juvenile rainbow trout *Oncorhynchus mykiss* health status, growth performance and resistance to *Lactococcus garvieae*, an experimental infection. *Fish Shellfish Immunol.* 86:269–279.

Mohapatra S, Chakraborty T, Kumar V, Deboeck G, Mohanta KN. 2013. Aquaculture and stress management: A review of probiotic intervention. *J Anim Physiol Anim Nutr (Berl).* 97(3):405–430.

Naderi Farsani M, Bahrami Gorji S, Hoseinifar SH, Rashidian G, Van Doan H. 2020. Combined and singular effects of dietary primalac® and potassium diformate (KDF) on growth performance and some physiological parameters of rainbow trout (*Oncorhynchus mykiss*). *Probiotics Antimicrob Proteins.* 12(1):236–245.

Nandi A, Banerjee G, Dan SK, Ghosh K, Ray AK. 2018. Evaluation of in vivo probiotic efficiency of bacillus amyloliquefaciens in labeo rohita challenged by pathogenic strain of aeromonas hydrophila MTCC 1739. *Probiotics Antimicrob Proteins.* 10(2):391–398.

Nayak SK. 2010. Probiotics and immunity: A fish perspective. *Fish Shellfish Immunol.* 29(1):2–14.

Noor-Ul H, Haokun L, Junyan J, Xiaoming Z, Dong H, Yunxia Y, Shouqi X. 2020. Dietary supplementation of *Geotrichum candidum* improves growth, gut microbiota, immune-related gene expression and disease resistance in gibel carp CAS III (*Carassius auratus gibelio*). *Fish Shellfish Immunol.* 99:144–153.

Nurhajati J, Atira, Aryantha INP, Indah DK. 2012. The curative action of *Lactobacillus plantarum* FNCC 226 to *Saprolegnia parasitica* A3 on catfish (*Pangasius hypophthalamus* Sauvage). *International Food Research Journal* 19(4):1723–1727.

Palermo FA, Mosconi G, Avella MA, Carnevali O, Verdenelli MC, Cecchini C, Polzonetti-Magni AM. 2011. Modulation of cortisol levels, endocannabinoid receptor 1A, proopiomelanocortin and thyroid hormone receptor alpha mRNA expressions by probiotics during sole (*Solea solea*) larval development. *Gen Comp Endocrinol.* 171(3):293–300.

Panigrahi A, Azad IS. 2007. Microbial intervention for better fish health in aquaculture: The Indian scenario. *Fish Physiol Biochem.* 33(4):429–440.

Panigrahi A, Kiron V, Kobayashi T, Puangkaew J, Satoh S, Sugita H. 2004. Immune responses in rainbow trout *Oncorhynchus mykiss* induced by a potential probiotic bacteria *Lactobacillus rhamnosus* JCM 1136. *Vet Immunol Immunopathol.* 102(4):379–388.

Panigrahi A, Viswanath K, Satoh S. 2011. Real-time quantification of the immune gene expression in rainbow trout fed different forms of probiotic bacteria *Lactobacillus rhamnosus. Aquac Res* [Internet]. [accessed 2020 May 11] 42(7):906–917. http://doi.wiley.com/10.1111/j.1365-2109.2010.02633.x

Ramesh D, Souissi S, Ahamed TS. 2017. Effects of the potential probiotics *Bacillus aerophilus* KADR3 in inducing immunity and disease resistance in *Labeo rohita. Fish Shellfish Immunol.* 70:408–415.

Ramesh D, Vinothkanna A, Rai AK, Vignesh VS. 2015. Isolation of potential probiotic *Bacillus* spp. and assessment of their subcellular components to induce immune responses in Labeo rohita against *Aeromonas hydrophila. Fish Shellfish Immunol.* 45(2):268–276.

Ringø E, Zhou Z, Vecino JLG, Wadsworth S, Romero J, Krogdahl, Olsen RE, Dimitroglou A, Foey A, Davies S, et al. 2016. Effect of dietary components on the gut microbiota of aquatic animals. A never-ending story? *Aquac Nutr* [Internet]. [accessed 2020 Jul 27] 22(2):219–282. https://onlinelibrary.wiley.com/doi/full/10.1111/anu.12346

Rodriguez -Estrada U, Satoh S, Haga Y, Fushimi H, Sweetman J. 2009. Effects of single and combined supplementation of *Enterococcus faecalis*, mannan oligosaccharide and polyhydroxybutyrate acid on growth performance and immune response of rainbow trout *Oncorhynchus mykiss. Aquac Sci.* 57(4):609–617.

Safari R, Adel M, Lazado CC, Caipang CMA, Dadar M. 2016. Host-derived probiotics *Enterococcus casseliflavus* improves resistance against Streptococcus iniae infection in rainbow trout (*Oncorhynchus mykiss*) via immunomodulation. *Fish Shellfish Immunol.* 52:198–205.

Sahandi J, Jafaryan H, Soltani M, Ebrahimi P. 2019. The use of two bifidobacterium strains enhanced growth performance and nutrient utilization of rainbow trout (*Oncorhynchus mykiss*) fry. *Probiotics Antimicrob Proteins* [Internet]. [accessed 2020 Jul 27] 11(3):966–972. https://pubmed.ncbi.nlm.nih.gov/30109493/

Salinas I, Díaz-Rosales P, Cuesta A, Meseguer J, Chabrillón M, Moriñigo MÁ, Esteban MÁ. 2006. Effect of heat-inactivated fish and non-fish derived probiotics on the innate immune parameters of a teleost fish (*Sparus aurata* L.). *Vet Immunol Immunopathol.* 111(3–4):279–286.

Sangma T, Kamilya D. 2015. Dietary bacillus subtilis FPTB13 and chitin, single or combined, modulate systemic and cutaneous mucosal immunity and resistance of catla, *Catla catla* (Hamilton) against edwardsiellosis. *Comp Immunol Microbiol Infect Dis.* 43:8–15.

Saurabh S, Sahoo PK. 2008. Lysozyme: An important defence molecule of fish innate immune system. *Aquac Res* [Internet]. [accessed 2020 May 11] 39(3):223–239. http://doi.wiley.com/10.1111/j.1365-2109.2007.01883.x

De Schrijver R, Ollevier F. 2000. Protein digestion in juvenile turbot (*Scophthalmus maximus*) and effects of dietary administration of *Vibrio proteotyticus. Aquaculture* [Internet]. [accessed 2020 May 10] 186(1–2):107–116. https://linkinghub.elsevier.com/retrieve/pii/S0044848699003725

Schwartz W 1979. R.Y. Stanier, E. A. Adelberg and J. L. Ingraham, The Microbial World (4th Edition). XV, 871 S., 614 Abb., 165 Tab. Englewood Cliffs 1976. Prentice Hall Inc. £ 15.75. *Z Allg Mikrobiol* [Internet]. [accessed 2020 Jul 27] 19(2):151–151. http://doi.wiley.com/10.1002/jobm.19790190223

Servin AL. 2004. Antagonistic activities of lactobacilli and bifidobacteria against microbial pathogens. *FEMS Microbiol Rev.* 28(4):405–440.

Shokryazdan P, Jahromi MF,Navidshad B, Liang JB. 2017. Effects of prebiotics on immune system and cytokine expression. *Med Microbiol Immunol.* 206(1):1–9.

Singh ST, Kamilya D, Kheti B, Bordoloi B, Parhi J. 2017. Paraprobiotic preparation from *Bacillus amyloliquefaciens* FPTB16 modulates immune response and immune relevant gene expression in *Catla catla* (Hamilton, 1822). *Fish Shellfish Immunol.* 66:35–42.

Sudo N. 2006. Stress and gut microbiota: Does postnatal microbial colonization programs the hypothalamic-pituitary-adrenal system for stress response? *Int Congr Ser.* 1287:350–354.

Tan HY, Chen SW, Hu SY. 2019. Improvements in the growth performance, immunity, disease resistance, and gut microbiota by the probiotic *Rummeliibacillus stabekisii* in Nile tilapia (*Oreochromis niloticus*). *Fish Shellfish Immunol.* 92:265–275.

Tang Y, Han L, Chen X, Xie M, Kong W, Wu Z. 2019. Dietary supplementation of probiotic *Bacillus subtilis* affects antioxidant defenses and immune response in grass carp under *Aeromonas hydrophila* challenge. *Probiotics Antimicrob Proteins* [Internet]. [accessed 2020 May 8] 11(2):545–558. http://link.springer.com/10.1007/s12602-018-9409-8

Tang Z, Sun H, Chen TJ, Lin Z, Jiang H, Zhou X, Shi C, Pan H, Chang O, Ren P, et al. 2017. Oral delivery of *Bacillus subtilis* spores expressing cysteine protease of Clonorchis sinensis to grass carp (*Ctenopharyngodon idellus*): Induces immune responses and has no damage on liver and intestine function. *Fish Shellfish Immunol.* 64:287–296.

Taoka Y, Maeda H, Jo JY, Kim SM, Park S Il, Yoshikawa T, Sakata T. 2006. Use of live and dead probiotic cells in tilapia *Oreochromis niloticus. Fish Sci.* 72(4):755–766.

Tinh NTN, Dierckens K, Sorgeloos P, Bossier P. 2008. A review of the functionality of probiotics in the larviculture food chain. *Mar Biotechnol.* 10(1):1–12.

Valcarce DG, Genovés S, Riesco MF, Martorell P, Herráez MP, Ramón D, Robles V. 2017. Probiotic administration improves sperm quality in asthenozoospermic human donors. *Benef Microbes* [Internet]. [accessed 2019 Mar 17] 8(2):193–206. http://www.ncbi.nlm.nih.gov/pubmed/28343402

Valcarce DG, Martínez-Vázquez JM, Riesco MF, Robles V. 2020. Probiotics reduce anxiety-related behavior in zebrafish. *Heliyon* [Internet]. [accessed 2020 Jul 27] 6(5). https://doi.org/10.1016/j.heliyon.2020.e03973

Valcarce DG, Pardo MÁ, Riesco MF, Cruz Z, Robles V. 2015. Effect of diet supplementation with a commercial probiotic containing *Pediococcus acidilactici*(Lindner, 1887) on the expression of five quality markers in zebrafish (*Danio rerio* (Hamilton, 1822)) testis. *J Appl Ichthyol* [Internet]. [accessed 2019 Mar 17] 31:18–21. http://doi.wiley.com/10.1111/jai.12731

Valcarce DG, Riesco MF, Martínez-Vázquez JM, Robles V. 2019a. Diet supplemented with antioxidant and anti-inflammatory probiotics improves sperm quality after only one spermatogenic cycle in zebrafish model. *Nutrients* [Internet]. [accessed 2019 Apr 30] 11(4):843. http://www.ncbi.nlm.nih.gov/pubmed/31013929

Valcarce DG, Riesco MF, Martínez-Vázquez JM, Robles V. 2019b. Long exposure to a diet supplemented with antioxidant and anti-inflammatory probiotics improves sperm quality and progeny survival in the zebrafish model. *Biomolecules* [Internet]. [accessed 2020 Feb 10] 9(8). http://www.ncbi.nlm.nih.gov/pubmed/31382562

Vazirzadeh A, Roosta H, Masoumi H, Farhadi A, Jeffs A. 2020. Long-term effects of three probiotics, singular or combined, on serum innate immune parameters and expressions of cytokine genes in rainbow trout during grow-out. *Fish Shellfish Immunol* [Internet]. [accessed 2020 Jul 27] 98:748–757. https://pubmed.ncbi.nlm.nih.gov/31726098/

Vidal S, Tapia-Paniagua ST, Moriñigo JM, Lobo C, García de la Banda I, Balebona M del C, Moriñigo MÁ. 2016. Effects on intestinal microbiota and immune genes of Solea senegalensis after suspension of the administration of *Shewanella putrefaciens* Pdp11. *Fish Shellfish Immunol.* 58:274–283.

Vílchez MC, Santangeli S, Maradonna F, Gioacchini G, Verdenelli C, Gallego V, Peñaranda DS, Tveiten H, Pérez L, Carnevali O, Asturiano JF. 2015. Effect of the probiotic Lactobacillus rhamnosus on the expression of genes involved in European eel spermatogenesis. *Theriogenology.* 84(8):1321–1331.

Won S, Hamidoghli A, Choi W, Park Y, Jang WJ, Kong I-S, Bai SC. 2020. Effects of *Bacillus subtilis* WB60 and *Lactococcus lactis* on growth, immune responses, histology and gene expression in Nile Tilapia, *Oreochromis niloticus. Microorganisms* [Internet]. [accessed 2020 May 10] 8(1):67. https://www.mdpi.com/2076-2607/8/1/67

Xia Y, Wang M, Gao F, Lu M, Chen G. 2020. Effects of dietary probiotic supplementation on the growth, gut health and disease resistance of juvenile Nile Tilapia (*Oreochromis niloticus*). *Anim Nutr.* 6(1):69–79.

Ye JD, Wang K, Li FD, Sun YZ. 2011. Single or combined effects of fructo- and mannan oligosaccharide supplements and *Bacillus clausii* on the growth, feed utilization, body composition, digestive enzyme activity, innate immune response and lipid metabolism of the Japanese flounder *Paralichthys olivaceus.*

Aquac Nutr. [Internet]. [accessed 2020 Jul 27] 17(4):e902–e911. https://onlinelibrary.wiley.com/doi/full/10.1111/j.1365-2095.2011.00863.x

Yi CC, Liu CH, Chuang KP, Chang YT, Hu SY. 2019. A potential probiotic *Chromobacterium aquaticum* with bacteriocin-like activity enhances the expression of indicator genes associated with nutrient metabolism, growth performance and innate immunity against pathogen infections in zebrafish (*Danio rerio*). *Fish Shellfish Immunol.* 93:124–134.

Yu Y, Wang C, Wang A, Yang W, Lv F, Liu F, Liu B, Sun C. 2018. Effects of various feeding patterns of *Bacillus coagulans* on growth performance, antioxidant response and Nrf2-Keapl signaling pathway in juvenile gibel carp (*Carassius auratus gibelio*). *Fish Shellfish Immunol* [Internet]. [accessed 2020 May 8] 73:75–83. https://linkinghub.elsevier.com/retrieve/pii/S1050464817307271

Zaineldin AI, Hegazi S, Koshio S, Ishikawa M, Bakr A, El-Keredy AMS, Dawood MAO, Dossou S, Wang W, Yukun Z. 2018. Bacillus subtilis as probiotic candidate for red sea bream: Growth performance, oxidative status, and immune response traits. *Fish Shellfish Immunol.* 79:303–312.

Zhang D, Gao Y, Ke X, Yi M, Liu Z, Han X, Shi C, Lu M. 2019. *Bacillus velezensis* LF01: In vitro antimicrobial activity against fish pathogens, growth performance enhancement, and disease resistance against streptococcosis in Nile tilapia (*Oreochromis niloticus*). *Appl Microbiol Biotechnol.* 103(21–22):9023–9035.

Zhong XQ, Liu MY, Xu C, Liu WB, Abasubong KP, Li XF. 2019. Dietary supplementation of Streptococcus faecalis benefits the feed utilization, antioxidant capability, innate immunity, and disease resistance of blunt snout bream (*Megalobrama amblycephala*). *Fish Physiol Biochem.* 45(2):643–656.

Zohar Y, Mylonas CC. 2001. Endocrine manipulations of spawning in cultured fish: From hormones to genes. *Aquaculture* 197(1–4):99–136.

Zorriehzahra MJ, Delshad ST, Adel M, Tiwari R, Karthik K, Dhama K, Lazado CC. 2016. Probiotics as beneficial microbes in aquaculture: An update on their multiple modes of action: A review. *Vet Q.* 36(4):228–241.

Index

B

C

D

E

S

T

V

W

Y

Z